Analysis of Milk and its Products

A Lab Manual

— 2nd Edition —

MILK INDUSTRY FOUNDATION

BIOTECH

2027

BIOTECH BOOKS®

Original Title: Laboratory Manual: Methods of Analysis of Milk and Its Products

2027 Reserved
ISBN 978-81-7622-127-6

Published by
BIOTECH BOOKS
4762-63/23, Ansari Road, Darya Ganj,
New Delhi-110002 (India)
Phone : 011 4300 3222
E-mail: biotechbooks@yahoo.co.in

PRINTED IN INDIA

PREFACE

This is the second edition of a manual that has achieved a distinguished place in the dairy industry and has rendered a service to the industry throughout the world.

Planned originally for members of the Milk Industry Foundation and its predecessor organization, the manual was made available to colleges, and health officers, who in large numbers wanted the privilege of purchasing the book.

The first Laboratory Manual was compiled and published by the association in 1933 to meet a need for a comprehensive volume that would present the laboratory procedures employed in a well-rounded dairy laboratory control program.

In order that such a manual would be truly representative of recognized best practice in the industry, the task of assembly was referred to a group, selected from the membership, who were directly concerned with or actually engaged in laboratory control of the quality of dairy products.

The existence of the need for the manual and the general utility of the first effort has been amply attested, initially by its enthusiastic reception throughout the industry and by the fact that the first edition has gone through four printings up to 1939. The steady demand for copies of the manual since then has exceeded the hopes of the members who were active in fostering its publication.

Within five years of the advent of the first edition of the manual it became apparent that sufficient advancement in dairy laboratory procedure had occurred to warrant consideration of the publication of a revised edition. Accordingly in 1939 it was decided to discontinue further printings of the original manual and the active work of revision was initiated by the Laboratory Manual Committee.

This work progressed satisfactorily until conditions imposed by the outbreak of war made it necessary to curtail activity. The task of revision was actively resumed by the committee in 1945. In many respects it has been fortunate that the completion of this revision was delayed because it has been possible to present many advances in our technology that have been developed in the interim.

The general form of presentation of the text has been retained but the material has been rearranged under a greater number of chapter headings to provide more clarity and to facilitate ease in locating the various topics when using the manual. A consistent effort has been made to cite the best available reference material for the contents of all chapters.

To be highly commended for their persevering efforts in the preparation of the revised edition of the Milk Industry Foundation Laboratory Manual are A. J. Powers, of Borden's Farm Products, New York City, Chairman of the Laboratory Committee; C. J. Austin, Supplee-Wills-Jones Milk Co., Phila., Pa.; A. C. Fay, H. P. Hood & Sons, Boston, Mass.; P. E. LeFevere, Chestnut Farms Dairy, Washington, D. C.; F. M. Scales, Sheffield Farms Co., New York City (until his retirement in 1945) and E. B. Kellogg, Secretary, Milk Industry Foundation, who, though not officially a member of the Laboratory Manual Committee, worked with much zest and diligence at meetings and between meetings to facilitate the task of revision.

Acknowledgement is also made to the American Public Health Association and the Association of Official Agricultural Chemists for their generous permission to use illustrations and material.

The Milk Industry Foundation extends its appreciation to the many members in United States, Canada, England, Australia and Argentina, and other persons in countries throughout the world, who have commended the manual and have made possible its publication and distribution.

CONTENTS

PART I

ORGANIZATION OF A DAIRY LABORATORY

CHAPTER I

THE MILK CONTROL LABORATORY

CHAPTER 2

SUGGESTED SCHEDULE OF ROUTINE LABORATORY PROCEDURE

PART II

MICROBIOLOGICAL CONTROL OF DAIRY PRODUCTS

CHAPTER 3

AGAR PLATE COUNTS

CHAPTER 4

AGAR PLATE COUNTS ON SPECIAL PRODUCTS

CHAPTER 5

DETERMINATION OF SPECIAL TYPES OF ORGANISMS

CHAPTER 6

DETERMINATION OF SANITIZATION OF UTENSILS AND EQUIPMENT

CHAPTER 7

DIRECT MICROSCOPIC EXAMINATION OF DAIRY PRODUCTS

CHAPTER 8

DETECTION OF MASTITIS

CHAPTER 9

REDUCTION TESTS

CHAPTER 10

SPECIAL CULTURE PROPAGATION

CHAPTER 11

DETERMINATION OF YEASTS AND MOLDS

PART III

CHEMICAL CONTROL METHODS FOR DAIRY PRODUCTS

CHAPTER 12

COLLECTION AND CARE OF SAMPLES

CHAPTER 13

Babcock Test for Fat

CHAPTER 14

Roese-Gottlieb Fat Determination

CHAPTER 15

Gerber Test for Fat

CHAPTER 16

MOJONNIER DETERMINATION OF TOTAL SOLIDS

CHAPTER 17

TOTAL SOLIDS DETERMINATION WITHOUT MOJONNIER EQUIPMENT

CHAPTER 18

MOISTURE, SALT, AND FAT DETERMINATION IN BUTTER AND CHEESE

CHAPTER 19

TITRATABLE ACIDITY

CHAPTER 20

HYDROGEN ION DETERMINATION

CHAPTER 21

PHOSPHATASE TEST FOR PASTEURIZATION CONTROL

CHAPTER 22

NEUTRALIZER DETECTION

PART IV

PHYSICAL CONTROL METHODS FOR DAIRY PRODUCTS

CHAPTER 23

SPECIFIC GRAVITY DETERMINATION OF MILK

CHAPTER 24

DETERMINATION OF ADDED WATER

CHAPTER 25

Sediment Tests

CHAPTER 26

Cream Volume Determination

CHAPTER 27

Curd Tension Determination

CHAPTER 28

Viscosity Determination of Dairy Products

CHAPTER 29

Homogenization Efficiency Determination

PART V

MISCELLANEOUS AND SPECIAL TESTS OF DAIRY PRODUCTS

CHAPTER 30

Miscellaneous Tests

PART VI

MICROBIOLOGICAL, CHEMICAL, AND PHYSICAL TESTS FOR NON DAIRY PRODUCTS

CHAPTER 31

Chemical Control Procedures for Washing and Sterilizing Solutions and Brine

CHAPTER 32

Physical Tests Applied to Glass Milk Bottles

CHAPTER 33

Sugar Syrup Tests

CHAPTER 34

Gelatin Examination

CHAPTER 35

VANILLA FLAVOR TESTS

CHAPTER 36

CHOCOLATE AND COCOA TESTING

CHAPTER 37

FRUIT TESTS

CHAPTER 38

TIN DETERMINATIONS

CHAPTER 39

BIOCHEMICAL OXYGEN DEMAND DETERMINATION

PART VII

PREPARATION OF MEDIA AND REAGENTS

CHAPTER 40

CULTURE MEDIA

CHAPTER 41

STAINS

CHAPTER 42

STANDARD SOLUTIONS

CHAPTER 43

INDICATORS AND REAGENTS

APPENDIX

CONVERSION TABLES

ENGINEERING

FIRST AID SUGGESTIONS

ICE CREAM—CALCULATING THE MIX

ICE CREAM—FREEZING THE MIX

ICE CREAM MIX

LEGAL STANDARDS

PROPERTIES OF DAIRY AND RELATED PRODUCTS

INDEX

PART I

ORGANIZATION OF A DAIRY LABORATORY

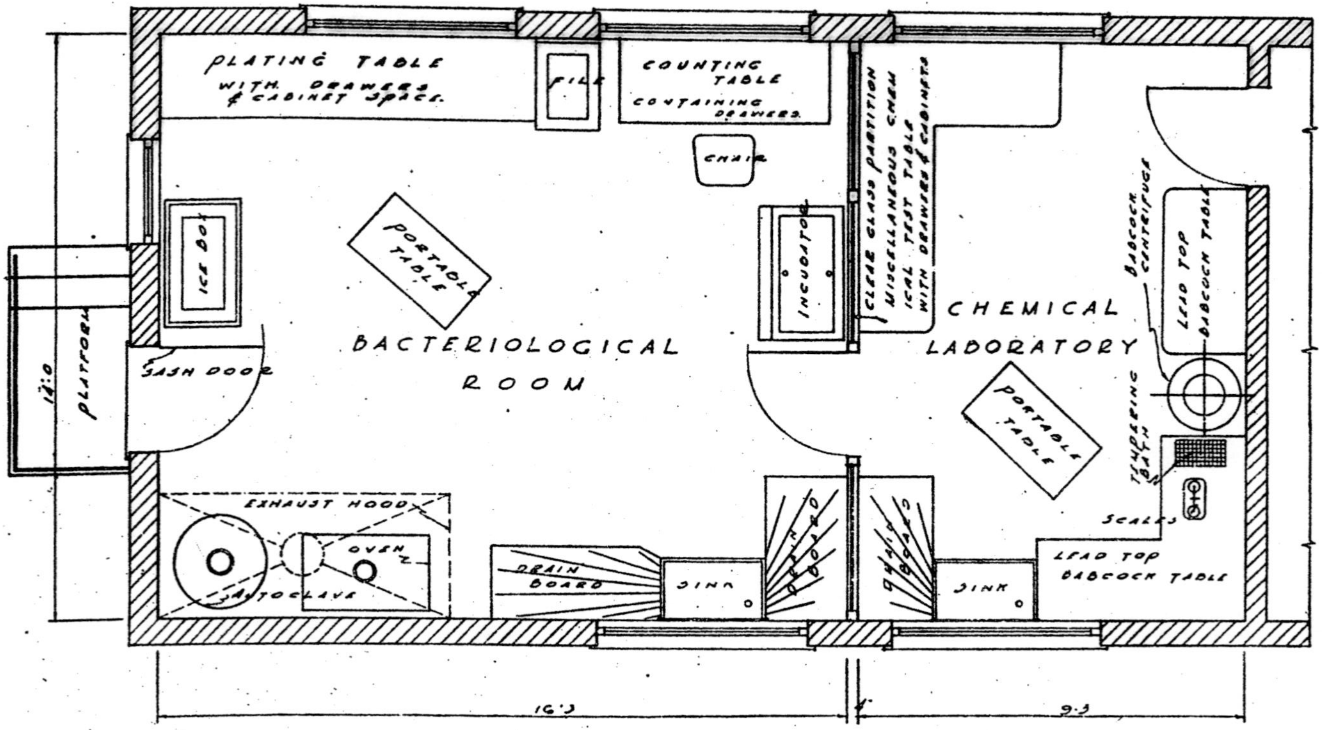

Figure 1
Floor plan covering minimum requirements for milk control laboratory.

CHAPTER 1

THE MILK CONTROL LABORATORY

Laboratory control work is best conducted in light, clean, well ventilated rooms so located as to keep the workers free from interruption and distraction. Laboratory space should not be made a corridor for persons or materials passing from plant to office or to the outside. Better work may be expected of laboratory technicians when undue plant operating noise or excessive noise from outside the building is eliminated.

It is manifestly impossible to recommend a single laboratory layout that would be universally adaptable. The suggested floor plan is illustrative of minimum requirements and provides the facilities to take on additional laboratory control work should the demand be sudden or gradual. Compactness is very desirable for the control laboratory but it is well to bear in mind the eventual expansion that most laboratories experience and to plan accordingly.

Northern exposure is usually to be preferred over southern exposure for natural lighting of the laboratory. Artificial light for the laboratory should be of a general intensity of at least 30 foot-candles, with a minimum of 50 foot-candles of vertical illumination on work tables. Indirect artificial light is recommended, in which case the ceiling should be approximately 12 feet high and have a matte finish for high reflective value. It is recommended that 500 watt lamps be used for indirect lighting or 300 watt lamps for direct lighting. This should provide the desired minimum general light intensity of 30 foot-candles. Help on proper lighting can usually be secured from the local electric company.

For the sake of compactness the laboratory floor plan shows the steam autoclave and the dry heat oven located under an exhaust hood in the bacteriological laboratory. Greater working comfort will be provided and hence more efficient activity by laboratory technicians may be expected when it is possible to install this sterilizing equipment in a separate room or to enclose it in a partitioned compartment which is vented to the outside.

Routine Control Measures

Equipped with the necessary specialized apparatus, the following routine control measures could be carried on in the laboratory illustrated in figure 1 on the opposite page.

1. Standard plate counts.
2. Test for pasteurizability of raw milk and cream supplies.
3. Methylene blue and resazurin tests.
4. Examination for presence and concentration of coliform organisms.
5. Direct microscopic counts.
6. Phosphatase test to indicate adequate pasteurization of milk and cream.
7. Tests for percentage of milk fat and total solids in milk and cream.

Figure 2
Taking bacteriological samples from cans, showing sample shipping case on floor in background.

8. Determination of the distribution of milk fat in homogenized milk (the U. S. P. H. S. index of homogenization).

9. Acidity tests on dairy products.

10. Cream volume determinations.

11. Viscosity and whipping tests on cream.

12. Curd tension tests.

13. Test for sediment in milk.

14. Strength of alkaline and acid cleansing solutions.

15. Tests for available chlorine in sterilizing solutions.

Figure 3
Procedure for taking bacteriological samples from weigh can.

In some of the larger dairy control laboratories where the work is necessarily more diversified there may be occasion to engage in tests for abnormal or adulterated milk, bacteriological control of certified milk products, determination of vitamin C and dissolved oxygen in milk, determination of metallic salts in milk, analysis of cleansing compounds, analysis of boiler feed water, analysis of refrigeration brine, physical tests on milk bottles and milk cans, and many other operations. To this list of routine procedure might be added tests frequently requested by purchasing departments, such as analysis of lubricating oils and greases, paints, oats and other feeds, hand soaps, auto soaps, milk filter cloth, and so on.

Figure 2 illustrates the procedure for taking bacterial samples from cans.

Figure 4
Cooling tray for proper icing of samples during collection.

Figure 3 illustrates the procedure for taking bacterial samples from the weigh can. Milk cans in the foreground contain clean cold water and scalding water for rinsing and sterilizing the stainless steel sampling tube between samples.

Figure 4 illustrates the proper icing of samples during collection. Note drainage outlets for overflow of melted ice to prevent flooding or contaminating bottle caps.

Bacteriological Equipment

Containers

Sampling containers should be water-tight, provide for at least ten ml of sample, and readily permit of effective washing and sterilizing.

Figure 5
Vertical type steam autoclave

Ground glass stopper bottles and screw cap vials have been found useful for this work. Figure 2 illustrates the sampling procedure with the aluminum screw cap glass vial.

Cotton plugged or ordinary cork stoppered containers are not to be used for bacteriological sampling purposes.

Pipettes and Dippers

Both aluminum or stainless steel pipettes and cone-shaped dippers have been used for sampling. A sterile agitator must be provided for properly stirring before sampling with the pipette. (See figure 2) Cone-shape dippers, which are so constructed as to nestle and make a compact package, may be used both as sterile agitator and sample catcher.

Figure 6
Horizontal autoclave showing pressure gage mounted in the top, and the dial type thermometer mounted in the exhaust line below the body of the sterilizer.

Cooling Tray

When taking samples on deck, a cooling tray containing ice and water may be used to advantage as illustrated in Figure 4. Provision should be made for occasional agitation of the ice water to insure rapid cooling. Ice water level is controlled by overflow to prevent flooding of the vials and contamination of the caps.

Shipping Case

When samples are to be in transit only a very short time, an ordinary metal container will suffice, but when much time will be required for transportation of samples it will be found advantageous to use wooden shell metal lined insulated shipping cases that will cut down the loss of ice (figure 2). Much benefit may be derived from the practice of chilling the sample cases before packing the samples.

Description of the necessary specialized apparatus for the listed routine control tests will be found in the sections of the Manual devoted to these tests. Comment that follows will be limited to basic equipment essential to proper operation of the milk control laboratory.

Sterilizing Equipment

A. AUTOCLAVES

(Steam Pressure Sterilizers)

Satisfactory results may be obtained with the large or medium size vertical type autoclave (figure 5) or the horizontal type (figure 6). The source of steam for the autoclave may be direct or regenerated. Direct steam and ground seat door facilitates faster autoclave operation. The vertical type with regenerated steam heating has the advantage of considerably lower initial cost.

A 25-quart aluminum pressure cooker fitted with a steam coil and operated in the same manner as the larger vertical type autoclave has been found to be an economical and effective means of melting and sterilizing small batches of nutrient agar and other media (figure 7). Dependable and correct pressure gages, pressure relief or safety valves, and indicating thermometers are essential for effective and safe operation of autoclaves. (See figure 8.)

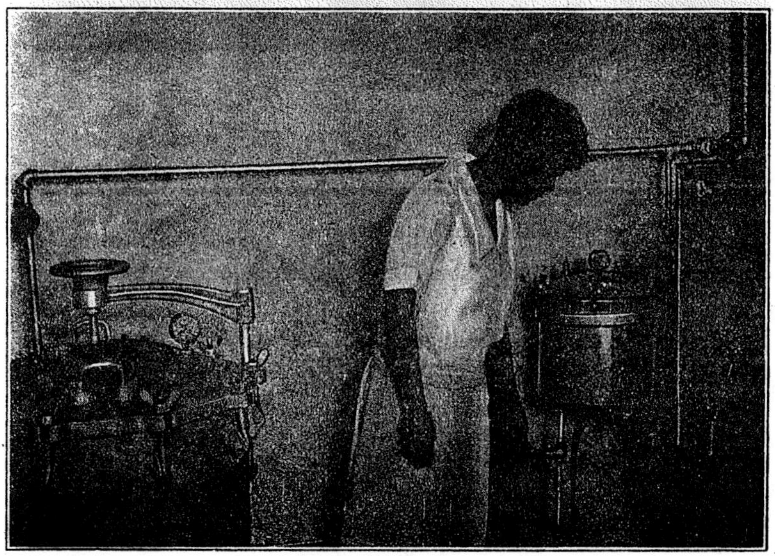

Figure 7
Aluminum pressure cooker or autoclave.

Figure 8
Autoclave showing pressure gage and thermometer mounted on top.

Precaution on Operation of Autoclaves for Effective Sterilization

1. Avoid over crowding the autoclave.

2. Make certain all air has been exhausted before building up pressure.

3. Equip the autoclave with accurate thermometers and pressure gages and watch them while the autoclave is in operation.

4. Do not release the pressure too suddenly lest cotton plugs be blown from the flasks or tubes or the media surge up and contact the plugs or the outside surface of the retaining vessel.

For Safety in Operation

1. Equip the autoclave with a reliable pressure relief valve and be certain that it is in good working order at all times.

2. Provide pressure reducing valves when direct steam operated autoclave is connected to a high pressure line or a line that is subject to significant variation in steam pressure.

3. Make certain that ground seat doors do not permit steam leakage and that gaskets on bolt-on covers are sound.

4. Make certain that gas, electric, or steam operated generators on indirect autoclaves contain a sufficient supply of water before starting the autoclave.

5. Make certain that all pressure is off the autoclave chamber before attempting to open the door.

B. Dry Heat Sterilizing Ovens

Bacteriological glassware may be sterilized in dry-heat ovens operated by gas, electricity, compressed natural gas, kerosene, or gasoline. It is advisable that all glassware be dry before subjecting it to sterilizing temperature in the oven. Observance of this precaution will prevent excessive breakage and promote effective sterilization. It is not advisable to overcrowd the oven with material to be sterilized unless an extended sterilizing period is maintained.

When gas is available the type of gas heated oven shown in figure 9 will be found convenient and efficient. This type of oven may be adjusted to operate on compressed gas. The ordinary household type gas heated oven may be utilized as a dry-heat sterilizer.

Electrically heated ovens will give satisfactory results although the time required to reach sterilizing temperature may be longer than with gas-heated ovens. The feasibility of electric ovens will depend on the power available and the local electric rate. Kerosene stoves and gasoline stoves have been successfully used as a source of heat for sterilizing ovens.

Precautionary Measures to Be Observed in Operation of Gas, Compressed Gas, Kerosene, and Gasoline Heated Ovens

1. The oven should be situated in a well ventilated room so that any escaping unoxidized gas, and the gaseous products of combustion will be carried to the outside. An exhaust hood or a direct flue connection is desirable.

2. Pilot light equipped gas ovens should not be lighted manually if the pilot flame has become extinguished. Always check for leaking gas before attempting to light the pilot. To prevent the extinguishing of pilot lights, direct flue connections to the outside should be equipped with backdraft diverters and the damper should be removed from the

Figure 9
Gas oven dry heat sterilizer installation. Glassware drying cabinet, heated by steam coil, is illustrated at right.

oven flue. For information on automatic safety devices, inquiry should be made to the engineering division of the local gas company.

3. Do not light a gas or gasoline operated oven when the door is closed, and always keep clear of the swinging arc of the door when lighting the oven. Avoid any operating condition that might lead to the confinement of escaping gas or gasoline vapor. (It is good practice to display a sign on or near the sterilizing oven to remind the technician of the necessary operating precautions to be observed.)

4. As a precaution against fire, burners and surrounding parts of kerosene or gasoline stoves should be kept scrupulously clean and free of fuel. The fuel tanks should be carefully examined from time to time for leaks. Fire extinguishers should be near at hand.

5. Check all ovens, especially new installations, to ascertain that the temperature maintained at each shelf level is equal to or greater

than the minimum temperature required for positive sterilization. This applies to electric ovens as well as those operated by gas or oil.

Bacteriological Incubators

Incubators, whether they are standard electric types, water jacketed, or specially constructed cabinets or rooms, should conform to the following specifications:

1. With incubators loaded, the temperature of extreme portions as indicated by reliable thermometers (at least two placed on top and bottom shelf) should be 32 or 35 C (89.6 or 95 F).

2. Proper air circulation should be provided to insure uniformity of temperature without appreciable drying of the nutrient medium. Excessive humidity should be controlled (page 48).

3. Contact points on electric incubators should frequently receive attention to prevent pitting and subsequent arcing, which would lead to excessive temperature variation. Electric relay control in conjunction with contact points has proved effective in eliminating this difficulty.

4. Incubators should be so located that they will not be subjected to extreme variations in room temperature. Proper insulation of incubators will minimize the effect of changes in room temperature.

5. Incubators should be set level, have a reliable source of heat and be protected from excessive vibration of the building.

Bacteriological Dilution Bottles

The type of dilution bottles illustrated in figure 10 on page 14 have been used to advantage in various laboratories. Cotton plugs or cork stoppers are unsatisfactory. Several modifications of rubber stoppers and metal caps have found favor for this purpose.

A satisfactory dilution bottle should be of glass capable of withstanding repeated autoclaving, should have a capacity of at least 120 ml, and should permit of easy washing and rinsing. Dilution bottles of heat resistant glass are higher in initial cost than flint glass but over a long time they will be more economical and will prove more satisfactory.

Figure 10
Satisfactory types of dilution bottles

Bacteriological Dilution Bottle Filler

The simplest forms of dilution bottle fillers are the dipper type and the overflow type. Several types of automatic fillers are available which provide rapid measured filling. Description of an automatic dilution bottle filler follows (See figures 11 and 12).

This filler is made up of a special automatic overflow glass measuring cylinder, a 1¼ to ¼ inch reducing coupling that has been especially machined to accommodate the glass cylinder, a three-way air cock valve, and the necessary pipe, nipples, and fittings. All metal parts are of brass, which may be chromium plated, except the spring, which is made of music wire.

The filler may be attached directly to the tap line with a small brass globe valve to control the rate of flow of the water. The glass cylinder is blown to contain 84 ml of water at room temperature, and the capacity of the filler may be slightly increased by raising the cylinder in the specially machined reducing coupling or further increased by lengthening the nipple between this coupling and the three-way air cock valve. The apparatus may be adjusted to deliver from 99 to 105 ml or even a greater amount of dilution water, depending on the amount of loss by evaporation under local autoclaving conditions.

With proper attention to capacity and the three-way valve adjustment, this apparatus may be depended on to facilitate accurate and rapid filling of dilution bottles.

Figure 11
Automatic filler for dilution bottles

The normal filling time for this apparatus is thirty-two bottles in two and one-half minutes, or approximately thirteen bottles per minute.

Sink for Bacteriological Laboratory

A white enameled sink is attractive and serviceable for the bacteriological laboratory. However, acids and other corrosive chemicals are likely to destroy the enamel finish. More resistant materials for sinks are stainless steel, ceramic, alberene stone and special synthetic materials which are available from laboratory supply houses. A sink of two compartment construction provides washing and rinsing facilities for laboratory glassware. (See figure 13.)

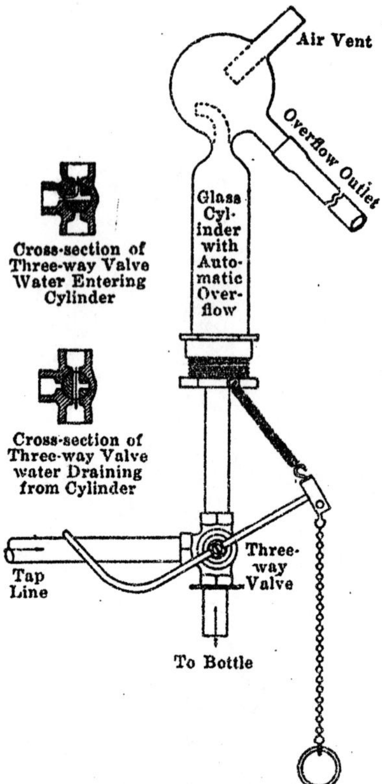

Figure 12
Dilution bottle filler

Bacteriological Glassware Cabinets

A cabinet should be provided for miscellaneous glassware. A wall cabinet, wooden or metal, with sliding glass doors will be found very satisfactory and attractive. The lower portion of a plating table may be utilized as a cabinet for storing glassware.

Plating Table

Several features that should be considered in choosing or constructing a plating table follow:

1. The plating table top should be smooth, level, and relatively impervious to moisture. The surface may be wood, heavy linoleum, phenolic resin finish or any other synthetic material that will not warp and will not tend to scratch Petri dishes. (Stone or glass tops are objectionable for this reason.)

2. The frame may be of wood, riveted or welded metal, or pipe-stem construction. The height should be about 40 inches. A convenient width for wall tables is 24 to 30 inches, but center tables may be from 24 to 40 inches wide. Wooden frames should be protected against warping, but a pipe-stem frame will not warp and may be so constructed as to be readily leveled on any floor. Figure 14 shows the wall type plating table.

3. It is very convenient to have drawers along the working sides of the plating table. Closed cabinets built in under the drawers provide convenient storage room. Toe space (approximately 4 to 6 inches) built into the bottom of the working sides of plating tables will add to the comfort of the technician.

4. Difficulty may be experienced with the setting of nutrient agar in very warm weather and a provision for accelerating this process by cooling a portion of the table surface will be found helpful.

5. Pipette boxes kept off plating table surfaces protect the sterile pipettes from contamination with water and makes them easily accessible. (See figure 13.)

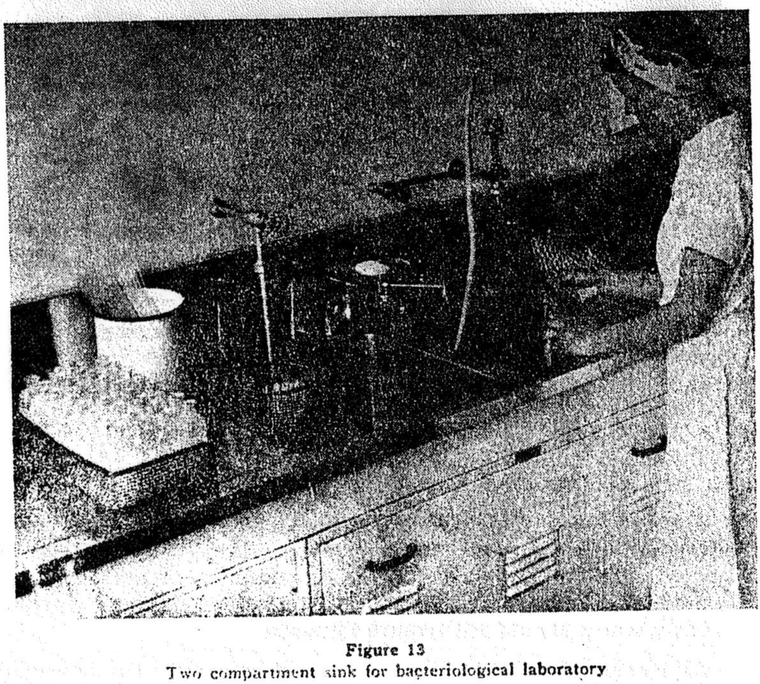

Figure 13
Two compartment sink for bacteriological laboratory

Figure 14
Wall type plating table

6. A Bunsen burner provides the most suitable means of flaming flasks, tubes, and bottles, but if gas facilities are not available, alcohol lamps or "sterno" cans may be used.

7. The plating capacity desired will determine the size of the plating table surface. A wall table will accommodate one plate for every four inches of its length, and a center table will provide room for two plates, for every four inches of its length and width.

Nutrient Agar

The Borden or the Bowman methods (page 505) may be used to advantage when it is necessary to prepare large batches of nutrient agar. In laboratories in which the consumption of nutrient agar is limited, it may be more economical to use the prepared dehydrated product which should not require filtration.

All laboratories should be equipped with apparatus for determining the hydrogen ion concentration of nutrient agar. Colorimetric comparators as well as potentiometers have been found satisfactory for this purpose.

If the conditions in the individual laboratory will permit, nutrient agar may be tubed in approximately 10 ml amounts for pouring into individual plates or a number of plates may be poured from a quantity of agar in a flask. A thermostatically controlled water bath or an Arnold sterilizer may be used to temper the medium to about 113 F. (45 C.), at which temperature it may be held for a limited time before pouring.

Bacterial Plate Counting Apparatus

Agar plates may be counted over Wolffhuegel, Stewart, or Jeffer rulings with ordinary daylight as the source of light, but it is advisable to use some type of uniformly controlled artificial illumination. Satisfactory artificial illumination is provided by the Quebec colony counter (figure 15), in which both direct and reflected light rays are uniformly focused on all portions of the agar plate so that the colonies

Figure 15
Approved type (Quebec) colony counter

stand out well against a background of subdued light. While the Wolffhuegel counting plate appears to be more popular, either the Stewart or Jeffer plate may be used. The lens on the Quebec colony counter magnifies about one and one-half diameters and in its fixed position makes possible the examination of the entire plate.

It is advisable to use a hand tally register for recording the number of colonies as they are counted.

Refrigerator

The size and design of the icebox will depend on the volume of work handled by the laboratory. The desirability of mechanical refrigeration is obvious. When an ice chest refrigerator is to be used it may be advantageous to employ a top icing type, so that cracked ice from incoming sample shipping cases may be utilized.

Babcock Equipment

Automatic Babcock Acid Measuring Devices

The multicharge straight-sided burette which delivers 17.5 ml of "Babcock" sulfuric acid per charge is regarded as a somewhat hazardous apparatus with which to handle concentrated acid because it usually must be filled from a position above eye level. An automatic burette or automatic acid bottle as shown in figure 16 may be safely employed for measuring the acid charge. Automatic acid burettes requiring air pressure for operation should be used with caution.

Cream Scales

Cream scales used for official work should be accurate and sensitive to 0.03 gm. and should be of the style for one bottle with counterpoise weight or the style for one bottle on each pan. For general control work the four-bottle type or the style for two bottles on each pan has been found to be convenient and satisfactory.

Babcock Centrifuges

The centrifuge may be hand operated or powered by steam or electricity. It must be capable of developing and maintaining the officially specified r. p. m. Centrifuge speed should be occasionally checked with a speed indicator with the cover closed to nullify the effect of air resistance. A device for properly heating the interior of the centrifuge is essential.

Babcock Tempering Bath

A water bath (figure 16) should be provided in which the test bottles may be immersed to the level of the top of the fat column and held at about 135-145 F (57-63 C) or at a temperature in accordance with regulations of individual states.

Figure 16

Babcock testing equipment showing, left to right, torsion balance, tilt acid bottle (on back of table), automatic acid burette, lactometer in jar, mixing beaker with thermometer, tempering bath, and mechanical Babcock bottle shaker.

Mechanical Babcock Bottle Shakers

Mechanical Babcock bottle shakers (figure 16) are not absolutely necessary but they promote uniformity of shaking and facilitate speed of testing.

Test Bottle Wash Racks and Washing Machines

Wash racks holding 12 to 24 test bottles speed up the washing operation and protect the bottles from breakage. Mechanical washers are available which force streams of water and detergent solution into the inverted test bottles.

Chemical Laboratory Sink

The chemical laboratory sink should be of alberene stone, ceramicware or lead-lined construction for resistance to acids. It is desirable to use lead traps and drain pipes in connection with chemical sinks

Mojonnier Tester

Some control laboratories use the Roese-Gottlieb method or the Mojonnier modification of this procedure for determining the fat content of milk, cream, and ice cream mix. These tests depend upon ether extraction of the fat and are considered more accurate than the Babcock test. Full description of the equipment necessary for the Roese-Gottlieb test (Mojonnier modification) will be found on page 209. The Mojonnier tester is shown in figure 17.

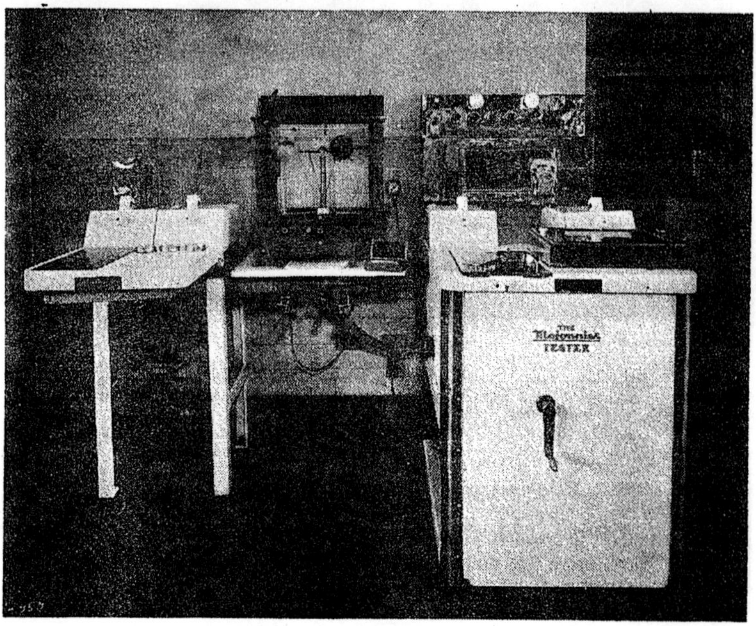

Figure 17
Mojonnier test unit

CHAPTER 2

SUGGESTED SCHEDULE OF ROUTINE LABORATORY PROCEDURE

It is difficult to present blanket specifications for the frequency of laboratory control tests that will fit the operation of every milk plant. The type and frequency of control tests will be determined by the extent of the plant operation, the available laboratory facilities, and other conditions of local character. To secure conditions in which the laboratory control facilities are well balanced with the plant operation, the following schedule of routine control tests is suggested.

Receiving Stations and Milk Processing Plants

A. Babcock Test or Gerber Test

1. On composite samples of producers' milk as often as state law or local practice prescribes.

2. Daily on all incoming shipments of raw milk other than individual producers' supplies.

3. Daily on all plant-separated cream and on all incoming cream shipments before and after standardizing.

4. Daily on raw milk from each storage tank immediately before release for pasteurization.

5. Daily on a representative series of samples of all finished products. (These should include the first bottle of milk, cream, and other products for sale from each filler).

6. Twice weekly on samples of homogenized milk to determine the efficiency of homogenization (Page 356.)

7. On all duplicate or parallel samples of milk, cream, and other products from federal, state, and municipal inspection departments.

B. Lactometer Test

1. On all raw or pasteurized milk having an abnormal fat content.

2. Daily on raw milk from each storage tank immediately before release for pasteurization.

3. Daily on a representative series of all grades of finished product. (This should include the first bottle of milk for sale from each filler).

4. Once a month on each producer's milk.

5. On all duplicate or parallel milk samples from federal, state, or municipal inspection departments.

C. ACIDITY TEST

1. Daily on all tank-car and tank-truck shipments of raw milk before acceptance.

2. Daily on all raw milk from each storage tank immediately before release for pasteurization.

3. On a representative sample of each batch of cream as separated, or shipment as received, and on each lot of cream just before pasteurization.

4. Daily on representative samples of finished product on all grades of milk.

5. Daily on representative samples of packaged cream.

6. Daily on buttermilk and other cultured-milk products.

7. Occasionally on representative samples of "route-return" products.

D. PHOSPHATASE TEST

Daily on representative samples of all pasteurized milk, cream, and butter. (Should the pasteurizing run be prolonged, it is advisable to test a sample taken at the beginning, one at the middle, and another at the end of the run).

E. STANDARD PLATE COUNT

1. At least five times monthly on samples from all producers shipping premium grades of raw milk and at least once monthly on all others.

2. Daily on raw milk from all storage tanks (direct microscopic method may be substituted).

3. On samples representative of all cream shipments as received.

4. Weekly on a representative number of washed, empty milk bottles.

5. Weekly on a representative number of washed, empty milk cans.

6. Daily on representative series of samples of all grades of milk, cream, chocolate drink, and other products offered for sale.

7. Weekly on all grades of milk, cream, chocolate drink, and other products returned from delivery routes.

8. Quarterly on all plant water supplies.

F. Laboratory-Pasteurized or "Pasteurizability" Test of Raw Milk

1. Monthly on each producer's supply. (See page 99).

2. Two or three times a month on incoming mixed supplies of raw milk.

G. Determination of Coliform Organisms

1. At least three times weekly on representative samples of all packaged pasteurized products.

2. Quarterly on all plant water supplies.

H. Direct Microscopic Count
(Breed Method)

1. At least once a month on samples from all producers.

2. Daily on raw milk from all storage tanks unless the standard plate method is preferred.

3. Occasionally on a series of samples taken throughout the pasteurizing operation to disclose the possible development of thermophilic organisms.

4. On raw milk samples when it is desirable to determine such general types of contamination as udder infection (mastitis), dirty utensils, dust and dirt, and faulty cooling .

NOTE: Valid interpretation of the aforementioned types of contamination by direct microscopic examination of mixed raw milk supplies is open to question. Greater accuracy of detection of these conditions may be expected when the sample to be examined represents the milk of one producer or, in case of udder infection, milk of the individual cow. Faulty cooling will magnify any type of bacterial contamination and possibly give rise to a misleading interpretation.

I. Methylene Blue Reduction Test or Resazurin Test

These methods provide a rapid means of checking the quality of raw milk from individual producers or mixed raw milk from storage or shipping tanks and may be used when standard plate and direct microscopic counts are not made.

J. Sediment Test

Monthly on samples of each producer's milk. Repeat on the following two days for all milk showing an unsatisfactory test.

K. TEMPERATURE OBSERVATION

1. Determine the temperature of individual producer's milk as often as the prevailing temperature demands. (In warm weather daily observations may be necessary.

NOTE: The temperature of milk in individual cans should be determined with a quick-acting deck thermometer. Many operators can detect milk on which temperature readings are necessary by touching the outside of the can.

2. Daily on all tank-car and tank-truck shipments leaving the receiving station, and again as received at the processing plant.

3. Daily on milk and cream in storage tanks before release for pasteurization.

4. Daily on representative samples of all grades of milk, cream, and fluid milk by-products in the refrigerator.

5. Twice weekly on representative samples of all fluid products returned from routes.

L. ODOR, FLAVOR, AND APPEARANCE

1. Daily on all cans of milk and cream as received from producers and other sources.

2. On each tank-car or tank-truck shipment before acceptance.

3. Daily on all products at the time of packaging and again after 24 hours storage.

M. CREAM LINE DETERMINATION

1. At least twice weekly on representative filler samples of all grades of milk. The milk bottle gage method (page 340) is recommended.

NOTE: Daily cream-volume readings will aid in detecting cases of overheating, excessive pumping, prolonged agitation in raw or pasteurized milk storage tanks, or accidental mixture of homogenized and unhomogenized milk. Line-run tests by the M.I.F. method (page 338) are effective in determining the point in the processing at which creaming ability of the milk is being impaired.

N. CREAM VISCOSITY DETERMINATION

At least twice a week on packaged cream.

NOTE: Viscosity tests on each batch of cream may be very useful in disclosing processing irregularities.

O. CURD TENSION DETERMINATION

At least once a week on homogenized milk and soft-curd milk.

P. EFFICIENCY OF HOMOGENIZATION (MICROSCOPIC OR U.S.P.H.S. METHODS INDEX) DETERMINATION

At least once a week on the milk run through each homogenizer (see page 356).

Q. STRENGTH OF ALKALI DETERMINATION IN WASHERS

1. Daily on solutions in tanks of bottle washers. One test should be made at the beginning of the run after the alkali charge has been added and properly mixed, and a second sample should be tested at the completion of the bottle washing operation.

2. Daily on solutions from tanks of can washers. Tests should be made at the beginning and end of the can-washing operation and as often in between as is necessary to assure proper strength of solution during the operation.

R. AVAILABLE CHLORINE DETERIMINATION IN STERILIZING SOLUTIONS

1. At least three times during bottle washing: at the beginning, the middle, and at the end of the day's operation.

2. Daily on sterilizing solutions used for flushing or spraying contact surfaces of milk equipment. Test strength of chlorine solution immediately before use, and immediately after it has contacted the milk equipment surfaces.

S. EXAMINATION OF REFRIGERATING BRINE

Monthly for specific gravity, chemical reaction, hydrogen ion concentration, and per cent of corrosion inhibitor (if inorganic chromates, chromic acid, or chrom-glucosates have been added), and for possible presence of free ammonia. (See pages 438-449.)

T. HARDNESS OF FEED WATER AND TOWER WATER DETERMINATION

Quarterly on each water supply used for these purposes. (See page 407).

U. DAILY REFERENCE SAMPLES FOR PASTEURIZING PLANTS

Set aside each day for reference a series of samples of final packages representative of all milk and milk products distributed. Install in the refrigerator a rack of sufficient capacity to accommodate such a series of samples for each day in the week. Discard the week-old samples as those for the current day become available. This practice permits reference to operating performance for the past week (often helpful when checking alleged product defect).

Creameries

A. Babcock, Gerber or Modified Fat Tests

1. On composite samples of producers' milk as often as state law or local practice prescribes.

2. On all shipments of cream as received. (If an individual shipment consists of more than one can, take a representative sample of that shipment. A representative composite should include cream from at least 10 per cent of the number of cans).

3. On each batch of cream ready for churning.

4. Daily on representative samples of the accumulated buttermilk.

5. On single tubs from an appropriate number of representative churnings from each car of butter shipped or from each shipment of butter received.

B. Moisture and Salt Test

1. On each completed churning of butter.

2. On single tubs from an appropriate number of representative churnings from each shipment of butter received.

C. Acidity Test

1. On each batch of cream before and after neutralization.

2. On each batch of cream before churning, to determine the end of the ripening period.

3. On each batch of starter before breaking, to determine the end of the ripening period. (The creatine test may also be used to check ripeness of starters).

D. Bacterial Plate Count

1. Weekly on representative samples of producers' milk used for starter making.

2. On each shipment of sweet cream received.

3. Occasionally on butter churned and butter bought.

4. Weekly on the water supply used for washing butter.

5. Weekly on solutions used for soaking parchments.

6. Weekly on washed cans.

7. Occasionally as a check on contamination of contact surfaces of churns, workers, tables, automatic cutters, wrappers, and walls of the refrigerator. (See page 117).

E. YEAST AND MOLD COUNTS

1. On each churning shipped or received.

2. Occasionally as a check on contamination of contact surfaces of churns, workers, tables, automatic cutters, wrappers, and walls of the refrigerator. (See page 117).

F. MOLD MYCELIA AND SEDIMENT DETERMINATION

1. On each churning shipped or received.
2. On cream shipments received.

G. DETERMINATION OF COLIFORM ORGANISMS

1. Weekly on water supply used for washing butter.
2. Weekly on solutions used for soaking parchments.
3. Occasionally on butter churned and butter bought.
4. Occasionally on pasteurized cream before ripening.
5. Occasionally on starters and pasteurized milk used for propagating starters.
6. Occasionally as a check on contamination of contact surfaces of churns, workers, tables, automatic cutters, wrappers, and condensation from walls of the refrigerator. (See swab method, page 116).
7. Quarterly on all plant water supplies not covered in (1) above.

H. PHOSPHATASE TEST

1. Daily on batches of pasteurized cream intended for churning.
2. On each churning as shipped or received.
3. On each churning after two or three weeks of storage or just prior to sale.

I. METHYLENE BLUE REDUCTASE OR RESAZURIN TESTS

1. Monthly on samples of each producer's milk and repeat on the following two days on milk of producers showing an unsatisfactory test.

NOTE: Direct microscopic examination of producers' supplies may be substituted for methylene blue reductase or resazurin tests.

2. On each shipment of sweet cream if standard plate count is not used.

J. SEDIMENT TEST

1. Monthly on samples of each producer's milk. Repeat on the following two days for all milk showing an unsatisfactory test.
2. Daily on wash water for butter.

K. TEMPERATURE OBSERVATION

1. Determine temperature of sweet cream or milk from individual producers as often as prevailing temperature conditions demand.

L. ODOR, FLAVOR, AND APPEARANCE EXAMINATION

1. On each shipment of cream and milk as received.
2. On each batch of starter milk.
3. On each batch of ripened starter.
4. On each batch of cream before churning.
5. On each churning of butter after chilling.

M. DETERMINATION OF ACCURACY OF WEIGHING EQUIPMENT

Check daily all scales, applying the following tolerances:

Print scales	0.05 per cent
Package scales (tubs)	0.25 per cent
Receiving room scales	0.25 per cent

N. EXAMINATION OF REFRIGERATING BRINE

Monthly for specific gravity, chemical reaction, hydrogen ion concentration, concentration of corrosion inhibitor (if sodium chromate, chromic acid, or chrom-glucosate inhibitor has been used), and for possible presence of free ammonia. (See pages 438 to 449).

Ice Cream Plants

A. BABCOCK, GERBER OR MODIFIED FAT TESTS

1. On composite samples of producers' milk as often as state law or local practice prescribes.

2. On representative samples of cream shipments received each day.

3. On samples representing batches of milk, cream, and condensed milks, as may be required, preparatory to the calculation and the making of the ice cream mix.

4. On storage tank samples of ice cream mix, both before and after standardization, when the Mojonnier test is not available.

5. On finished ice cream frequently enough to provide a satisfactory control on the operation when no Mojonnier equipment is available.

B. MOJONNIER FAT AND SOLID TESTS

1. On composite samples taken from at least 10 per cent of the cans in each shipment (or representative tank samples of tank shipments) of whole, plain, and sweetened condensed milk.

2. Solids determinations on all batches of skim-condensed milk received.

3. On samples of butter representative of individual churnings.

4. On samples representing each batch of mix, before and after standardizing.

5. At least once a week on samples of vanilla, chocolate, and strawberry, of each grade of ice cream produced.

C. STANDARD PLATE COUNT

1. On composite samples taken from at least 10 per cent of the cans of milk and cream of all shipments received (or representative tank samples of tank shipments).
2. Weekly line-run check of the processing operation.
3. Weekly samples representing:
    a. Each flavor of each grade of ice cream made.
    b. Each type of novelty produced.
4. On representative samples of each batch of egg products, nuts and fruits as used.
5. Weekly on a representative number of washed cans and paper containers.
6. Quarterly on all plant water supplies.

D. DETERMINATION OF COLIFORM ORGANISMS

1. Weekly on samples representing:
    a. Each flavor of each grade of ice cream made.
    b. All types of novelties produced.
2. Weekly line-run check of the processing operation.
3. On representative samples of each batch of egg products, nuts, and fruits as used.
4. Quarterly on all plant water supplies.

E. ACIDITY TEST

1. On composite samples taken from at least 10 per cent of the cans of milk, cream, skim milk, whole- and skim-condensed milk in each shipment (or representative tank samples of tank shipments) as received, and prior to use in mix.
2. On each mix after pasteurization, and again before freezing if mix is not frozen the same day.

F. ODOR, FLAVOR, AND APPEARANCE EXAMINATION

1. Daily on each producer's milk shipment.

2. On large shipments, at least 10 per cent of the cans, or on each tank of milk, cream, skim milk, whole- and skim-condensed milk received.

3. On each churning of butter immediately upon receipt and again prior to use.

4. Daily on representative samples of all products to be used in the mix.

5. On each mix on hand each morning, recording also the age of the mix.

6. On representative samples of each batch of flavoring ingredients as used.

G. Methylene Blue Reduction or Resazurin Test

Monthly on samples of each producer's milk.

NOTE: Standard plate counts or direct microscopic counts may be made on individual producer's milk instead of the methylene blue or resazurin tests.

H. Laboratory-Pasteurized or "Pasteurizability" Test of Raw Milk and Cream

Should there be difficulty in obtaining satisfactory standard plate counts in freshly pasteurized ice cream mix it may be advisable to:

1. Laboratory pasteurize and make subsequent plate count of each producer's supply once a month, and apply corrective field inspection to those supplies showing excessive numbers of heat-resistant organisms.

2. Check incoming bulk shipments of raw milk and cream for heat-resistant organisms two or three times a month.

I. Temperature Observation

1. Determine temperature of individual producer's milk as often as the prevailing temperature demands. (In warm weather, daily observations may be necessary).

NOTE: Temperature of milk in individual cans should be determined with a quick-acting deck thermometer. Many operators can detect milk on which temperature readings are necessary by touching the outside of the can.

2. Daily on all tank-car and tank-truck shipments leaving the receiving station, and again as received at the processing plant.

3. Daily on milk and cream in storage tanks before release for pasteurization.

4. Daily on representative samples of all grades of milk, cream, and fluid milk by-products in the refrigerator.

5. Twice weekly on representative samples of all fluid products returned from routes.

J. SEDIMENT TEST

Monthly on samples of each producer's milk. Repeat on the following two days for all milk showing an unsatisfactory test.

K. ALKALI DETERMINATION IN WASHERS

Daily at the beginning and at the end of the operation.

L. AVAILABLE CHLORINE DETERMINATION IN STERILIZING SOLUTIONS

Daily on all solutions before use, and on drainings to determine the residual chlorine after the sterilizing flush or spray.

M. MISCROSCOPIC EXAMINATION OF MIX FOR FAT-GLOBULE SIZE

At least weekly on a representative sample of ice cream mix immediately after homogenization.

N. REFRIGERATING BRINE EXAMINATION

Once a month for specific gravity, chemical reaction, hydrogen ion concentration, concentration of corrosion inhibitor (if sodium chromate, chromic acid, or chrom-glucosate inhibitor has been used), and for possible presence of free ammonia. (See pages 438 to 449).

PART II

MICROBIOLOGICAL CONTROL OF DAIRY PRODUCTS

CHAPTER 3

AGAR PLATE COUNTS

Introduction

The agar plate count is a widely used method of determining the so-called "number of bacteria" in the sample examined. In reality it is only an approximation based on the number of colonies which develop under arbitrary but standard conditions of growth when a known amount of sample is employed. In the interpretation of the results it is highly important that due recognition be given to the limitations imposed by the arbitrary nature of the method of analysis.

Briefly stated, the method consists of:

1. Obtaining and preparing under aseptic precautions a fair representative sample of the material to be examined;

2. Making exact dilutions of the sample in dilution blanks containing a known amount of sterile water, the number and choice of dilutions to be employed depending on the probable magnitude of the bacterial count expected;

3. Transferring with sterile pipettes known amounts of the chosen dilutions to sterile Petri dishes;

4. Pouring into the Petri dish about 10 ml of agar medium which has previously been melted and cooled to about 45 C (113 F), a temperature which is just above the solidifying point of the medium Immediately following the introduction of the fluid medium and before it has had time to solidify, the agar and the diluted sample previously added to the Petri dish are thoroughly mixed. After the medium has solidified, the plates are inverted and placed in the incubator at a prescribed temperature for a prescribed length of time.

5. After suitable incubation, counting the number of visible colonies on the plates under uniform conditions of lighting and magnification, and multiplying by the appropriate dilution to determine the "plate count per milliliter of milk."

Obviously, the type of agar medium used, the temperature and time of incubation, and many other factors might affect materially the number of colonies which could develop, and hence affect the resulting count. It is quite important, therefore, that all the details of preparing the samples, making the dilutions, compounding the medium, technique of plating, conditions of incubation, and facilities for counting the colonies be carefully and accurately standardized. Such standardization is necessary in the interest of uniformity of results from day to day in the same laboratory, and in the comparability of the results with those obtained in different laboratories.

American Public Health Association Standard Methods

The importance of standardization, even of the minute details of the plating procedure, particularly as applied to the analysis of dairy products, was early recognized by the American Public Health Association. As a result there have been published successive editions of *Standard Methods for the Examination of Dairy Products,* in which every step in the procedure is carefully prescribed. A copy of the latest published edition should not only be available but should be carefully studied and frequently used in every laboratory in which bacterial counts are made on dairy products. This is especially important if the results of the laboratory analyses are to have any official status, because most state laws and city milk ordinances recognize the A. P. H. A. methods as the official procedures for making bacterial counts.

There are, of course, special instances in which it may be neither necessary nor even desirable to follow the standard methods in making bacterial counts. Certain products such as tomato or other fruit juices, special types of cheese, acidophilus milk, and many other products may require different types of media or special conditions of incubation in order to foster growth of certain types of prevailing organisms. Likewise if the results are to be used only privately, as for example in some industrial control procedure, many modifications of the standard methods may be justified. In general, however, it is recommended that unless there is some specific reason for deviating from the generally accepted A. P. H. A. standard methods, these procedures should be followed explicity, especially for the determination of bacterial numbers in dairy products.

For making bacterial plate counts on products other than fluid milk or cream, the reader should first refer to the section on pages 56-67 under the heading Agar Plate Counts on Special Products. Such nonfluid products as butter, ice cream, cheese, gelatin, and sugar, require special technic in preparing or measuring the sample, and, as previously indicated, some products may even necessitate a different medium or perhaps a change in the time and temperature of incubation. Except for certain modifications to meet special conditions, much of the actual technic of making agar plate counts is the same for all types of products.

In this Manual the procedure is outlined in detail for making agar plate counts on milk. If it is desired to make agar plate counts on some products other than milk, follow the same procedure with the exceptions noted for the specific product to be analyzed.

Preparation of Materials

A. CLEANING GLASSWARE

Glassware used in bacteriological work should be chemically clean and bacteriologically sterile.

Dilution bottles, flasks, test tubes, and culture dishes may be satisfactorily cleaned by washing in warm soap solution, rinsing thoroughly with warm tap water, and finally with distilled water.

Glassware, particularly pipettes, should occasionally be immersed in a sulfuric acid-dichromate cleaning mixture.

NOTE: The effectiveness of a good cleaning mixture depends on the care with which it is made. Dissolve 40 gm of finely ground potassium dichromate in 150 ml of distilled water and cool. Place the solution in a Kjeldahl flask and add *very slowly* 330 ml of concentrated sulfuric acid. The Kjeldahl flask should be constantly rotated under a stream of cold tap water throughout the period of adding the acid, and the acid should be added only a few milliliters at a time. The mixture should not be allowed to become hot. The maximum yield of brick red crystals will be obtained if the directions are carefully followed.

After removal from the cleaning mixture the glassware should be thoroughly rinsed first in tap water, and then in distilled water.

All glassware should be dried by draining and protected from dust.

In order to facilitate cleaning, glassware should be immersed in or filled with water until the time for washing. This is especially important in the case of pipettes.

B. STEAM STERILIZATION OF MEDIA AND WATER BLANKS

In sterilizing media and water blanks, the autoclave should be operated so as to maintain a temperature of 121 C (249.8 F) for at least 20 minutes. The time of operation should be determined by the amount of material loaded into the autoclave, the volume of the largest container to be sterilized, and the initial temperature of the liquid. To obtain a temperature of 121 C, 15 pounds pressure of saturated steam is required. At no time should the autoclave be overloaded. The pet cock must be left open for a few minutes on starting to allow the steam to replace the air in the autoclave. The sterilizing efficiency of an autoclave is dependent on the fact that pure steam under pressure attains a high temperature. If all the air is not expelled from the autoclave before the pet cock is closed, the resulting mixture of air and steam will not attain the expected temperature. This is the cause of many failures in sterilization. The autoclave should be equipped with an accurate thermometer as well as a pressure gage. See figure 8, page 10.

C. Dry Heat Sterilization of Glassware

Pipettes and culture dishes should be sterilized in a hot air oven. (See figure 9, page 12). To prevent excessive breakage, all glassware should be thoroughly dried before being placed in the hot-air oven. To insure this sterilization, the ordinary hot-air oven, not overloaded, should be operated at a temperature of at least 160 C (320 F) for not less than one hour. Ample time must be allowed for all glassware to reach the sterilizing temperature before beginning the prescribed two-hour period. If pipettes are placed in metal containers for sterilization, the smallest dimension of the metal box should not exceed three inches. If boxes of larger sizes are used, they should not be completely filled with pipettes, thereby reducing the factor of slow heat penetration through the compact mass of glass. (See figure 18).

NOTE: To check the effectiveness of heat penetration, distribute small portions of cotton at various levels in the oven. This cotton will be charred if adequate sterilization temperatures have been maintained.

Figure 18

Round and square type metal containers for sterilizing pipettes. Note that the boxes are filled only approximately to two-thirds capacity to reduce the factor of heat penetration.

D. PREPARATION OF DILUTION BLANKS

The types of bottles which may be used for dilution blanks are illustrated in figure 10 (page 14). Care should be taken to insure the use of non-toxic water in the preparation of water blanks. Tap water is preferable to distilled water if it is not chlorinated and contains no other substances toxic to bacteria. Toxicity of the dilution water employed may be determined by making replicate plates at 5-minute intervals for 30 minutes from a single dilution blank to which some milk has previously been added. A considerable reduction in the counts obtained from successive plates suggests toxicity of the dilution water.

NOTE: If the local tap water is found to be toxic, thereby necessitating the use of distilled water, it is very important to prepare a stock buffered phosphate solution as directed on page 523 to be added to the distilled water used in the dilution blanks at the rate of 1 ml of stock solution to 800 ml of distilled water.

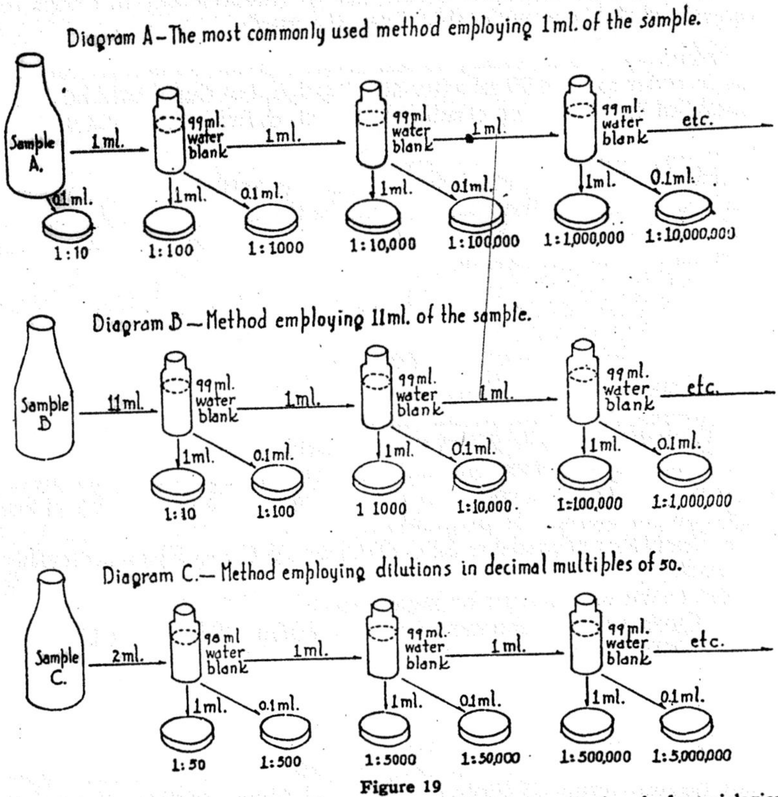

Figure 19

Diagrammatic sketches showing the methods of making various dilutions for bacteriological plating. The choice of the size of the sample, the number and the extent of dilutions to be employed will depend upon the sample to be analyzed.

The amount of water necessary to add to the bottle in order to have 99 ml after sterilization depends on the extent of evaporation from the bottle during sterilization. This in turn depends on the conditions of operating the autoclave and the type of closure of the dilution bottle. Such an amount of water should be added prior to sterilization so that the average contents of ten bottles after sterilization is 99 ml, with a maximum tolerance for individual bottles of plus or minus 2 ml. It is apparent that the conditions of autoclaving with respect to the quality of the steam, the completeness of exhaustion of air, the pressure employed, the rate at which the desired pressure is attained, the time of heating, the rate at which the pressure is lowered, the number of bottles placed in the autoclave, and the looseness of the closures on the bottles should be as uniform from day to day as it is possible to control such factors. The dilution bottles should have a mark etched or scratched at the level indicating 99 ml, and the sterile blanks should be inspected before use to eliminate any in which the water level deviates noticeably from the mark.

Ordinarily it is necessary to add about 102-103 ml before sterilization in order to have 99 ml after sterilization, but this should be determined for the operating conditions of the individual autoclave.

During sterilization the closures of the bottles should be loosened sufficiently to permit expansion within the bottles. After sterilization, the closures of the bottles should be tightened to retard evaporation. Dilution blanks should be protected from dust and used within a few days after sterilization.

The Agar Plate Count

Apparatus

1. Supply of sterile Petri dishes.
2. Supply of sterile 1.1 ml pipettes.
3. Supply of sterile 99 ml water blanks.
4. Standard tryptone-glucose-extract-milk agar which has previously been melted and cooled in a water-bath to 45 C (113 F). (See page 506 for method of preparation).
5. Incubator adjusted to 32 C (89.6 F) 35 C (95 F) or preferably approaching 35 C (depending upon local regulations or customs with respect to the temperature of incubation).
6. Quebec colony counter or its equivalent. (See figure 15.)
7. Tally counter.

Procedure

NOTE: Study diagram A of figure 19, page 41.

1. Dry the outside of the sample bottle or vial and shake vigorously but uniformly 25 times in an up and down direction over a distance of about 1 foot. (The time required for shaking should be

about 7 seconds.) Remove the cap or stopper carefully to avoid contamination.

2. Transfer 1 ml of the sample to a sterile 99 ml water blank by means of a sterile pipette, blowing out the last drop.

NOTE: When samples (for example: butter, cheese, fruit, egg products, sugar, gelatine, etc.) are to be weighed, see page 48.

3. Shake this dilution as in (1).

4. By means of a sterile 1.1 ml pipette transfer 0.1 ml of the dilution to a sterile culture dish. Allow pipette to drain and touch it to a dry spot on the plate to remove the last drop (in this case do not blow out the last drop). This makes a 1:1,000 dilution plate. Transfer the remaining 1 ml to another culture dish. This makes a 1:100 dilution plate. Label the plates accordingly.

5. If further dilutions are required, the same pipette as was used in (4) may be used to transfer 1 ml of the dilution described in (2) to another 99 ml water blank and shake as in (1).

6. Transfer 0.1 ml of the dilution mentioned in (5) to another culture dish by means of a sterile pipette. This makes a 1 : 100,000 dilution plate, and transfer the remaining 1 ml to another culture dish to give a 1 : 10,000 dilution. Similiarly, higher dilutions in multiples of ten may be prepared by continuing the series with a third or fourth dilution blank, and inoculating culture dishes in the same manner. (See figure 19, page 41.)

Figure 20

The technique of inoculating (left) and pouring (right) agar plates. Note that the lids of the Petri dishes are raised just sufficiently to introduce the pipette (left) or the mouth of the agar container (right).

7. After flaming the mouth of the container, pour approximately 10 to 12 ml portions of the agar, which has been previously melted and cooled to 45 C (113 F), from test tubes or flasks directly on the dilutions in the culture dishes, and mix thoroughly with a gentle rotary motion. Melted agar should be cooled in a water bath maintained at 45 C (113 F). See figure 21. Allow the agar to harden, then invert the plates and place in the incubator for 48 hours, ± 3 hours, after which time they are ready to be counted. In case porous-topped culture dishes are used, the plates need not be inverted. An additional 3 to 5 ml of agar should be poured to avoid excessive drying when porous tops are used.

Figure 21

Water bath thermostatically controlled at 45 C (113 F) for cooling the melted agar medium to the proper pouring temperature.

NOTE: Not more than fifteen minutes should elapse between the time the milk is diluted and the time the last plate is poured. See results in tables on page 46.

8. All colonies should be counted that are recognizable with a reading glass four to five inches in diameter and magnifying 1.5

diameters. The use of a colony counter and a tally counter as described on page 19 is recommended. (See figure 15.) Only plates containing between 30 and 300 colonies are to be used in calculating counts. If such plates are not available the fact should be noted in recording the results.

NOTE: The dilutions explained above are for the purpose of illustration It may be found that other dilutions are more suitable for the work at hand. Figure 19, page 41, shows diagrammatically the method of making various combinations of dilutions.

Experience with any supply of milk will indicate what dilutions are best suited, and it may be possible to use fewer dilutions. For any counting other than routine, duplicate plates should be made from each dilution. In routine plating of milk from individual dairy farms, especially those in which a fairly accurate knowledge of the general quality is based upon frequently repeated tests, it is common practice to employ only a single plate of a single dilution.

Reporting Results of Agar Plate Counts

Since agar plate counts are at best an approximation and represent only the number of colonies which develop under arbitrary conditions of growth, the results should not imply greater accuracy than the method of analysis justifies, and, further, they should not be quoted as representing the number of bacteria per milliliter of the sample. Only two significant figures should be employed in the reported count and this should be reported as "Standard Plate Count at 35 C (or 32 C)........." For example, if 259 colonies are counted on a 1:100 dilution plate at 35 C the count should be reported as "Standard Plate Count (35 C) 26,000 per milliliter" and not as 25,900 bacteria per milliliter." The latter method implies greater accuracy than the method of analysis justifies, and also implies that all the bacteria in the sample have been enumerated.

Precautions in Plating

Many of the difficulties encountered in making agar plate counts may be avoided if serious attention is given to standardizing in great detail every step in the procedure. The following list of precautions is by no means complete but may call attention to a few of the more important points:

1. Keep samples below 40 F (4.4 C) until plated.

2. Any samples received at the laboratory that are warm or are not surrounded by ice or iced water, or any other unavoidable abnormal circumstances surrounding the collection or plating of the samples, should have these facts reported in conjunction with the results.

3. Shake all samples thoroughly and uniformly.

4. In using a pipette, insert it into the liquid only about one-half inch, and on withdrawing scrape the tip on the inside of the container to eliminate the unmeasured drop of liquid which may be hanging to the end of the pipette. This is particularly important when a 0.1 ml quantity is transferred.

5. Once the analysis is begun, it should be completed as soon as is consistent with careful work. Do not allow the sample container to stand at room temperature on the laboratory table for a prolonged period before withdrawing the sample; do not allow the diluted sample to remain in the dilution blank long before inoculating the Petri dishes; do not allow the inoculated Petri dishes to remain long before pouring the agar. The agar should be solidified within a few minutes after it is poured into the Petri dish. In warm weather it may be necessary to hasten solidification of agar plates by transferring them to a cold place for a few minutes. Not more than 15 minutes should elapse from the time the samples are placed on the plating table until the time the last agar plate has solidified. This obviously limits the number of samples which can be handled at one time. The importance of care in keeping the samples cold and of plating the samples promptly after the dilutions are made is emphasized by the data in tables 1 and 2.

TABLE 1[2]

THE EFFECT OF TIME, TEMPERATURE AND GROWTH ON THE RESULTS OBTAINED BY THE PLATE METHOD ON MILK OF LOW BACTERIAL CONTENT

	Plated immediately	Agar poured after 15 min.	Held in dilution bottle 15 min.
Original count	5,300	....	....
	5,900	....	....
Sample held at room temperature for 2 hours	13,200	16,600	18,000
	13,500	16,000	19,200

TABLE 2[2]

THE EFFECT OF TIME, TEMPERATURE AND GROWTH ON THE RESULTS OBTAINED BY THE PLATE METHOD ON MILK OF HIGH BACTERIAL CONTENT

	Plated immediately	Agar poured after 15 min.	Held in dilution bottle 15 min.
Original count	6,000,000	6,200,000	7,700,000
	5,400,000	7,500,000	9,000,000
Sample held at room temperature for two hours	28,000,000	35,700,000	35,800,000
	28,100,000	34,000,000	35,600,000

6. Control plates should be prepared to check the sterility of the Petri dishes, the agar, and the dilution water. This involves two plates, one containing agar only, and one containing 1 ml of dilution water from a sterile blank, plus the agar. If either of these plates shows appreciable growth, the value of the results obtained with that batch of Petri dishes, agar, or dilution blanks should be questioned.

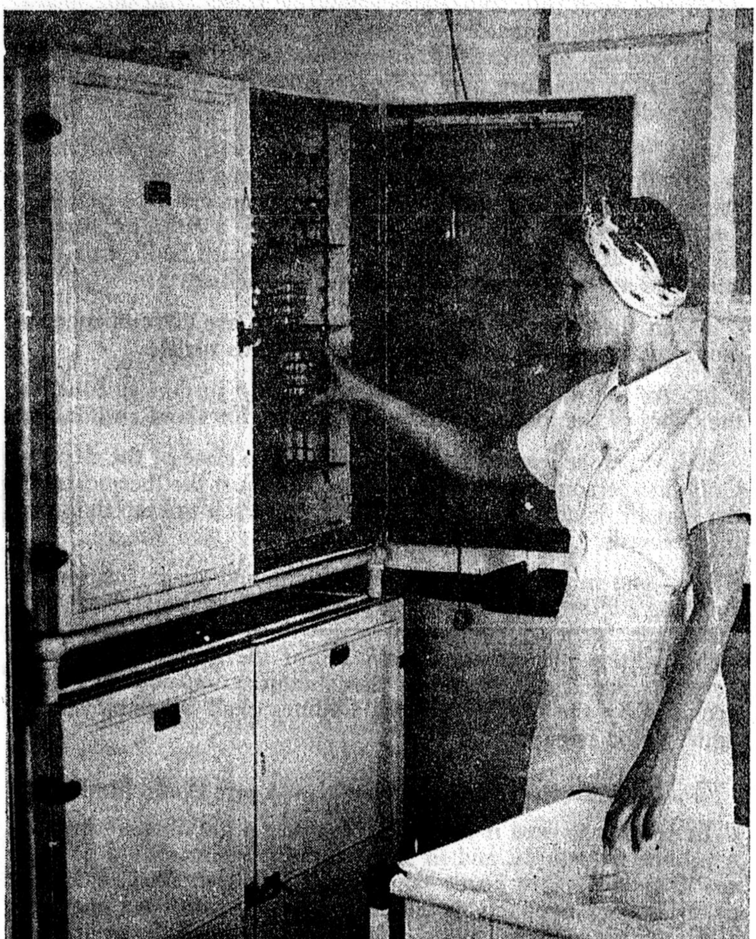

Figure 22
Bacteriological incubator.

7. Control the humidity of the incubator to avoid the drying of the agar in the Petri dishes. It is usually necessary to keep a pan of water in the incubator, but care should be taken that the humidity is not too high as indicated by condensation on the walls of the incubator.

8. Do not overload the incubator. Arrange the piles of plates as shown in figure 22, so that there is a free space for circulation of air around each pile. No plates should be closer than one inch from the walls of the incubator or from other piles of plates. Plates should not be moved or disturbed during incubation because droplets of condensation moisture may fall on the medium and promote growth of spreaders.

9. Uniformly well-regulated temperature of incubation should be provided. At least two thermometers should be used in each incubator: one in the top and one near the bottom stack of plates. The thermometer bulb should be submerged in a tube or small flask of water. It is recommended that a daily record be kept of the temperatures at three regular readings during each day:

a. In the morning when the incubators are first opened, as an index of the temperature during the previous night;

b. At the middle of the day, preferably just after lunch when the doors of the incubator have been closed for about an hour, and

c. Just before leaving the laboratory at the close of the day's work. Only one person should be designated to adjust the thermostatic controls, and it is also advantageous for this person to keep the temperature records.

10. The importance of uniform light and magnification in the counting of the colonies on the plates justifies the recommendation that some form of equipment be provided for this purpose. The eighth edition of *Standard Methods for the Examination of Dairy Products* specifically requires that "illumination equivalent to that provided by the Quebec Colony Counter shall be employed." See figure 15 on page 19.

Gravimetric Samples for the Agar Plate Method

In some instances it is necessary to use a gravimetric rather than a volumetric sample, and in such cases the count should be reported as "Standard plate count per gram." This method is frequently used for such products as butter, ice cream, condensed milk, cheese, fruits, egg products, sugar, gelatin, powdered milk, and nuts. In weighing a sample it is usually preferable in the interest of accuracy to employ an 11 gram sample in a 99 gram (ml) water blank rather than a 1 gram sample.

A. PROCEDURE FOR FLUID SAMPLES

1. Carefully counterpoise a 99 gram (ml) water blank on balances that are sensitive to 30 mg. It is also advisable to place on the balance pan with the water blank a piece of sterile paper upon which the stopper of the dilution bottle may be placed when it is removed. See figure 23.

2. Advance the weights on the sliding scale of the balance 11 gm or place an equivalent weight on the opposite balance pan. Remove the stopper from the dilution blank and place it on the balance pan so that its inner surfaces are not contaminated.

3. With a sterile pipette of adequate capacity remove slightly more than 11 gm of a well mixed representative sample, and rapidly introduce approximately 10 gm into the open dilution blank. The final portion should be slowly added drop by drop until an accurate balance is reached. Replace the stopper. This gives an initial dilution of 1 : 10.

4. Shake the bottle thoroughly and proceed to make appropriate dilutions as described in the diagrammatic sketch figure 19 on page 41. In all other details the plating procedure is the same as described on pages 42 to 48.

B. PROCEDURE FOR SAMPLES IN GRANULAR OR POWDER FORM

1. Place a large piece of sterile paper on a balance that is sensitive to 30 mg. If necessary counterpoise the weight of the paper and then advance the weight on the sliding scale 11 gm or place an equivalent weight on the opposite pan of the balance.

2. With a sterile spatula rapidly transfer approximately 10 gm of a well mixed sample to the center of the sterile paper. The final portion should be added slowly until an accurate balance is reached.

3. Carefully remove the paper and the weighed sample from the pan and transfer to a 99 gm (ml) dilution blank. This gives an initial dilution of 1 : 10.

4. Shake thoroughly and proceed to make appropriate dilutions as described in the diagrammatic sketch in figure 19 on page 41. In all other details the plating procedure is the same as described on pages 42 to 48.

Simplified Procedure for Making Bacteria Counts

A. BURRI TECHNIQUE[9]

Apparatus

1. Quantity of test tubes, about 6" x 5/8".

Figure 23.

Weighing 11 grams of a fluid sample directly into a dilution blank containing 99 grams of sterile water, particularly useful for samples of heavy cream, condensed milk, etc., with high viscosity or for melted ice cream containing considerable incorporated air. (See diagram B, figure 19, page 41)

2. Welded platinum-iridium inoculating loop, welded platinum-iridium wire 0.32 mm diameter (B. & S. gauge 28), loop 1 mm inside diameter, standardized to deliver 0.001 ml of milk.

3. Bacteriological incubator.

Reagents

Burri medium (Standard APHA medium except containing 2 percent agar, see below.)

Procedure

The quantitative smear culture method described by Burri consists essentially in using a loop for measuring the milk and smearing the contents of the loop on dry agar slants. This method merits widespread use and gives excellent results with a minimum of time and material. It is especially suited to the examination of milk with a low bacterial count within the range in which the other rapid methods (Breed smears, methylene blue tests) are not satisfactory.

It is necessary to use 2 per cent agar in order to obtain a gel that is firm and free from excess moisture. Into each tube are added six to eight milliliters of this agar. After sterilization for 20 minutes at 15 pounds pressure, the tubes are placed at a nearly horizontal position so as to give a slant 3 inches in length. The tubes should not be used until the surfaces of the slants are thoroughly dry (about a week). Wet slants may cause spreading of the colony growth.

In securing the inoculum the loop is immersed not deeper than from 2 to 3 mm into the milk. The inoculation of the slant is made by touching the loop to the middle of the slant, about 1 cm from the bottom, again in the center of the slant and finally in the middle about 1 cm from the top. The inoculum is then spread by zigzagging with the loop across the slant, beginning at the bottom and continuing to the top.

After incubation for 48 hours at the temperature desired (usually 32 or 35 C (89.6 or 95 F), the colonies on the slants are counted. The number of colonies found on the slant should be multiplied by 1,000 to give the total count per ml of milk examined.

NOTE: The advantage of the method is the ease with which a large number of tests can be made in the plant or in the field with relatively little equipment (a supply of agar slants, an inoculating loop, and an alcohol lamp).

B. MEYERS AND PENCE MODIFICATION OF AGAR PLATE METHOD

Where large numbers of samples are to be plated for purposes of routine control, it is frequently desirable to reduce the amount of glassware, dilution bottles, and medium employed. Meyers and Pence (7) have described a simplified procedure which substitutes standardized platinum loops (0.01 ml and 0.001 ml) for the pipettes and dilution blanks, and substitutes a special oval test tube for a Petri dish. Less than half the amount of agar is used in the tube as is required for a Petri dish and approximately twice as many samples can be analyzed per man hour as is possible with the standard technique.

Apparatus

1. Platinum-iridium loop standardized to deliver 0.01 ml with 3 inch shank attached to suitable handle.

2. Platinum-iridium loop standardized to deliver 0.001 ml with 3 inch shank attached to suitable handle.

3. Oval test tubes, Pyrex 152 mm in length and the two diameters of 23 mm and 11.5 mm.

4. Test tube racks.

5. Incubator.

6. Quebec colony counter.

Reagents

Ingredients for the preparation of tryptone-glucose-extract-milk agar. See page 506.

Procedure

1. Prepare standard tryptone-glucose-extract-milk agar as described on page 506. Place approximately 4 ml quantities of agar in the oval tubes fitted with cotton stoppers. Sterilize at 15 pounds pressure for 15 minutes.

2. Melt the sterile agar in oval tubes and transfer to a water bath maintained at not less than 45 C (113 F). It is important to leave the tubes in the water bath until the temperature of the agar has reached an equilibrium with the temperature of the water bath.

3. Shake the milk samples vigorously 50 times. (If samples are in tube vials, several samples may be shaken simultaneously.)

4. Select a loop (0.01 or 0.001 ml) appropriate for the expected bacteria count of the sample under analysis. Sterilize the loop in the flame and allow to cool in the air.

5. Insert the loop not more than 2 to 3 mm into the milk sample and transfer the loopful of milk to the oval tube of agar using aseptic precautions. Shake the loop several times in the agar to dislodge the milk. Replace the cotton plug and gently shake the agar in the tube to mix the milk and agar.

6. Slant the agar in the tube by placing it on the table with the mouth end of the tube on a strip of metal or wood 1/8 inch in thickness so that the agar flows to a point 2.5 to 3.0 inches from the bottom of the test tube.

7. When the agar has solidified, the tubes are placed in a wire rack in a horizontal position so that the slanted surfaces are in an inverted position. The racks are carefully placed in the incubator to keep the slanted surfaces facing downward, in a position comparable to an inverted Petri dish, thereby reducing the contamination from condensate running over the surface of the agar. Care should be taken to avoid moving the racks during the incubation period.

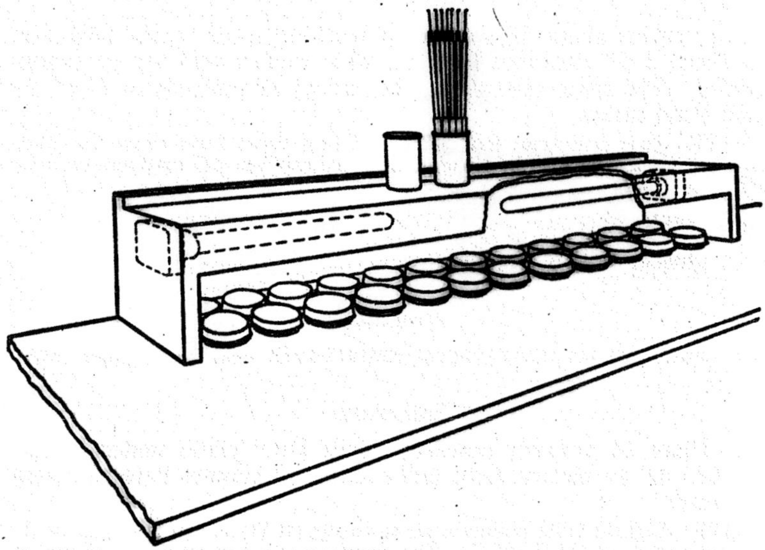

Fig. 24
Heating Canopy for Modified Plate Method (Borden)

8. Incubate the tubes 48 hours at 32 or 35 C, whichever is the local custom.

9. Colony counts are made in the usual manner by placing the tube over a well lighted colony counter.

C. Modified Plate Method (Borden)

This method has been used by Borden Farm Products Division laboratories since 1938 and has been found very helpful in handling large numbers of bacterial plates made from raw and "laboratory pasteurized" milk samples.

Apparatus

1. Platinum-ruthenium or platinum-iridium wire (0.3 mm diameter) loops made to deliver 0.001 ml of milk. The loops are formed in a true circle around a No. 55 carbon steel standard twist drill, have an inside diameter of 1.32 mm, have shanks from one to two in. long and are fixed to a suitable holder.

2. Pyrex Petri dishes, 15 x 60 mm.

3. Heating canopy (see figure 24 and 35) with heat source from two, 12 in., 40 watt electric show case bulbs as illustrated. Canopy is aluminum foil lined, approximately 34" long x 7" wide x 5½" high,

overall, covers about 32½" x 6" of working space on the table, and has about 3½" clearance between table surface and top of canopy opening. The space covered by the canopy accommodates 24 of the small Petri dishes.

NOTE: It is important that source of heat come from above the Petri plates otherwise moisture from the molten agar medium will condense on inner surface of covers.

4. Incubator.

5. Quebec colony counter.

Reagents

1. Standard tryptone-glucose-extract-milk agar (see page 506).

Procedure

1. Place 24 properly marked, sterile Petri plates under canopy and turn on the electric light bulbs about 10 minutes before plating is to start.

NOTE: Lighted bulbs maintain air in canopy at 115 to 120 F (46 to 49 C) and plates at about 113 F (45 C). The canopy works best on wood, plastic, or linoleum top tables—stone or glass surfaced tables may take too long to warm up and are not as practical to use.

2. After Petri plates have reached proper temperature pour about 4 ml of medium at 45 C (113 F) into each plate and replace plate under canopy which will maintain the medium in molten condition.

3. Sterilize wire loops in a gas flame (about eight is a convenient number), place in one receptacle and allow them to cool.

4. Agitate milk sample, remove closure, and with the sample held at a 45° angle insert the sterile loop into the container with the loop horizontal to the surface of the milk and remove a loopful of milk.

NOTE: The loop is always touched to the surface of the milk in the horizontal plane and never in the vertical plane.

5. Remove cover from a Petri dish and transfer loopful of milk to the dish by drawing the loop back and forth in the molten medium, immediately replace cover and swirl the contents of the plate to properly distribute the 0.001 ml of milk.

6. The mixture of milk and medium is permitted to solidify outside of the canopy.

7. Place used loop in other receptacle, select another sterile loop and proceed as before.

8. After completion of the eighth plating remove the used loops, sterilize them as a group in the gas flame, return them to the other receptacle and proceed as before.

9. Incubate the plate for 48 hours at 32 C or 35 C, whichever is the local custom.

10. Count plates with the aid of the Quebec colony counter, multiplying number of colonies by 1000 to arrive at the modified plate count.

NOTE: The area of the 15 x 60 mm petri plate is 20 sq. cm. and, if necessary, estimates on high count samples may be made on this basis.

NOTE: These modified techniques are particularly advantageous for plating milk samples that have been pasteurized in the laboratory.

Another method involving the culturing of milk samples in roll tubes has been described by Prouty and Bendixen (8). The inoculated tubes of melted agar are rapidly whirled in a horizontal position by means of a special machine. The agar film thus solidifies around the inner surface of the tube and thereby facilitates counting and examining the colonies.

REFERENCES

1. Corbin, C. I.: Methods of Identifying the Possible Reasons for High Count Through Characteristic Colonies. *Proc. 20th Ann. Conv. Internat. Assoc. Milk Dealers,* Laboratory Section, (1927) pp. 56-66.

2. Albus, W. R.: Some Observations on the Plate-Count Method of Enumerating Bacteria in Milk. *Proc. 21st Annual Convention, I. A. M. D.* (1928) pp. 80-87.

3. Schacht, F. L.: Faulty Technic as Practiced by Laboratory Workers. *Proc. 23rd Ann. Conv. Ibid.* (1930).

4. Powers, A. J.: Standard Plate Method. *Ibid.*

5. Schacht, F. L. and Robertson, A. H.: Accuracy of Counting on Agar Plate. *Amer. J. Pub. Health 22:* 80 (1932).

6. *Standard Methods for the Examination of Dairy Products,* 8th edition, 1941, American Public Health Association, New York.

7. Meyers, R. P., and Pence, J. A.: A Simplified Procedure for the Laboratory Examination of Raw Milk Supplies. *J. Milk Technol. 4*:18-25 (1941).

8. Prouty, C. C. and Bendixen, H. A.: A Comparison of the Roll-Tube and Standard Plate Methods of Making Bacterial Counts of Milk. *Ibid. 7*:5-16 (1944).

9. Dorner, W.: Bacterial Flora of Aseptically Drawn Milk. N. Y. Agr. Expr. Sta. Technical Bulletin 165, 1930.

CHAPTER 4.

AGAR PLATE COUNTS ON SPECIAL PRODUCTS.

The plating procedure described on pages 42-48, is specifically applicable to the preparation of bacterial plates for fluid samples which may be easily handled with a pipette. Other products such as cheese, butter, and milk powder, present a slightly different problem, particularly from the standpoint of preparing the sample for the first dilution blank. For certain products it is preferable or perhaps necessary to use a gravimetric sample as described on page 48. Unless it is desirable for special reasons to use another medium or perhaps another incubation temperature to favor the development of certain kinds of microorganisms, the technique of making agar plate counts after the first dilution would be identical with that described for milk on pages 42-48.

To avoid duplication, only the special techniques involved in preparing the first dilution for each of these special products will be discussed. The reader is referred to the diagrams in figure 19 on page 41 and the directions on pages 42-48 for completing the procedure of making agar plate counts on these products.

Butter

Although agar plate counts are sometimes made on butter, interpretation of the analysis is complicated by the fact that bacteria are commonly added to the product in the form of a butter culture. Yeast and mold counts are more commonly employed as a sanitary index. (See page 157). In those instances in which bacteria counts are made, however, certain special precautions should be observed in procuring the sample and in preparing the dilutions.

Apparatus

1. The usual plating equipment.
2. Sterile butter or cheese triers.
3. Dilution blanks warmed to 45 C (113 F).
4. Wide-mouth sterile sample jars.
5. Sterile 11 ml pipettes.

Procedure

1. Method of Sampling

A. BUTTER IN CHURN

After the butter has been worked and is ready to be removed from the churn, collect samples for microbiological analysis by means of a sterile trier, taking samples from 8 to 10 different locations in the churn, so that the total volume secured will not be less than 4 ounces. Place the butter samples in sterile jars with screw or glass top closures.

B. BUTTER IN TUBS OR PACKAGES

Remove a total of at least 4 ounces of butter by boring diagonally from two different parts of the tub or package with a sterile trier, the plugs of butter being not less than 2 inches in length and including the surface portion. Transfer the butter from the trier to the sterile sample jar with a sterile spatula or spoon.

C. PRINT BUTTER

Because of difficulty in obtaining the same amount of exposed surface of butter of one, one-half, and one-quarter pound prints, print butter should be sampled by means of a trier. By using a small trier and plugging the end of the print, a plug of 3 to 4 inches may be obtained. Such a plug should weigh at least one-half ounce. Transfer the butter from the trier to a sterile sample jar with a sterile spatula or spoon.

The American Butter Institute recommends that the print or roll be cut with a sterile spatula or knife into one-quarter pounds, one of which is then placed into a sterile jar.

2. Care of Samples.

Immediately place all samples in cracked ice, or in a refrigerator where the temperature does not exceed 4 C (39.2 F). Plate as soon as possible.

Refrigerate samples which are procured from stations remote from the laboratory at 4 C (39.2 F) or below until shipped. At no time prior to analysis should samples of butter containing less than 10 per cent salt in the serum be permitted to become warmer than 4 C (39.2 F), nor samples of butter containing over 10 per cent to become warmer than 15.6 C (60 F).

Cool and pack the samples for shipping so as to prevent breakage, and maintain a low temperature during transit. The samples upon arrival at the laboratory should be plated as soon as possible. Record the period of time elapsed between sampling and plating.

3. Preparation for Plating.

Place the sample jar containing the butter to be tested in a water bath maintained at 40 to 45 C (104 to 113 F) and shake until the butter is of thin, creamy consistency. Do not allow the time required for the sample to reach a creamy consistency to exceed 15 minutes.

4. Preparation of Dilutions.

Warm the 11 ml pipettes to be used, by drawing 11 ml of warm (40-45 C) sterile dilution water into the pipette; discharge; then draw 11 ml of thoroughly mixed melted butter into the pipette, and discharge this into 99 ml of sterile water. Care must be taken to have as little butter as possible adhere to the outside of the pipette.

Proceed with making dilutions as shown in diagram B, figure 19 on page 41. Pour tryptone-glucose-extract-milk agar in the Petri dishes and incubate plates as directed for milk on page 48. The interval between the time the butter is melted and the time the agar is poured should not exceed 15 minutes.

REFERENCE

Laboratory Manual—Methods of Analysis. Published by the American Butter Institute.

Cheese

Agar plate counts on cheese are frequently unsatisfactory because of the wide variability of duplicate tests. This is due in part to the difficulty of preparing the sample, and in part to the tendency of bacteria to colonize in the cheese.

Apparatus

1. The usual plating equipment (page 42).
2. Sterile mortar and pestle.
3. Sterile medium fine sand.
4. Sterile cheese trier, knife, and paper.
5. Balances sensitive to 0.1 gm.

Procedure

1. With a sterile trier remove a 2 inch plug from the cheese.
2. With a sterile knife, cut away about ½ inch of that portion of the plug near the rind.

3. Transfer the plug to a balance pan which has been covered with sterile paper and weigh 11 grams of the cheese. (It is advisable to use portions of three or more plugs taken from different parts of the cheese).

4. With a sterile knife cut the weighed sample into several small pieces.

5. Transfer about one third of the sample to a sterile mortar.

6. Add about one-half teaspoonful of sterile sand to the mortar, and grind the cheese thoroughly for 2 to 3 minutes.

7. Continue with a little more cheese and a little more sand until all the sample has been thoroughly ground.

8. Transfer the cheese and sand to a sterile 99 gm (ml) dilution blank.

9. Shake vigorously for 2-3 minutes.

In making subsequent dilutions, it is advisable to select a pipette with a large tip in order to avoid difficulty with the sand. In order to prevent the rapidly settling sand from carrying the sample to the bottom of the bottle while the pipetteful is being withdrawn, it is helpful to keep the sand in suspension by a rotary motion of the dilution bottle while pipetting. The wide variation in the agar plate counts on cheese necessitates the use of a wide range of dilutions in plating.

Since the organisms of interest in various cheeses are of different types, special media and special incubation temperatures more appropriate to the growth requirements are commonly employed. Casein agar is frequently used and plates are incubated four days at room temperature 21 C (69.8 F), (see page 522).

Cheese Spreads, Materials of Pasty Consistency, and Fruits

Materials having a pasty consistency such as cheese spreads, cottage cheese, creamed cheese, fruits, and other semisoft materials cannot be satisfactorily sampled with a cheese trier and cannot be handled with a pipette.

Apparatus

1. The usual plating material (see page 42).
2. Sterile parchment papers (4 and 6 inches square).
3. Sterile forceps.
4. Balance sensitive to 0.1 gm.

Procedure

1. Place on the pan of a suitable balance (sensitive to 0.1 gm) a piece of sterile parchment paper 6 inches square. Using sterile forceps, place a smaller piece of sterile parchment paper (4 inches square) on top of the first piece of sterile paper.

2. With a sterile spatula, transfer enough of the pasty product to the center part of the smaller piece of sterile paper until the weight has been increased 11 gm.

3. With another sterile spatula, transfer as much of the weighed sample as can easily be removed from the paper to a sterile wide-mouth 99 gm (ml) dilution blank. Rinse off the cheese (or other material) adhering to the spatula in the dilution water. With sterile forceps and the aid of a sterile glass rod, transfer the piece of paper on which the cheese was weighed to the dilution blank. Thorough disintegration of the pasty sample must necessarily depend upon vigorous agitation. Proceed to make dilutions as shown in diagram B, figure 19 on page 41. Unless special types of bacteria requiring special media are to be studied, use standard tryptone-glucose-extract-milk agar in the plates and incubate 48 hours at 32 or 35 C (89.6 or 95 F) as desired.

Condensed Milk

As in the case of heavy cream, the high viscosity of condensed milk renders a volumetric sample unsatisfactory. Weigh an 11 gram sample as directed on page 48, make the dilutions according to diagram B, figure 19 on page 41, and proceed with the plating technique as directed for milk. See pages 42-48.

Cream

For cream with 40 per cent milk fat or less it is customary to use a volumetric sample, and in such a case the procedure is the same as that described for milk. For creams with fat contents in excess of 40 per cent, the high viscosity results in unsatisfactory and inaccurate volumetric samples because of the adherence of cream to the pipette. Many workers prefer to use a gravimetric sample for heavy viscous cream.

For heavy viscous cream, prepare a 1 : 10 dilution by weighing 11 gm of thoroughly mixed sample into a 99 gm (ml) water blank as described on page 48. Subsequent dilutions should be prepared according to the diagram B, figure 19 on page 41. It is advantageous to use dilution blanks which have previously been warmed to 30-32 C (86-89.6 F). When warm dilution blanks are employed, it is important to complete the plating quickly to avoid growth in the dilution blanks. The elapsed time from diluting to plating should not exceed five minutes. The remainder of the technique is the same as for milk.

Evaporated Milk

For evaporated milk use the volumetric procedure described on page 42 and shown diagrammatically in figure 19, page 41.

Granulated Materials

See Powdered Materials (page 63).

Ice Cream

Apparatus

1. Sterile spatulas or cheese triers.
2. Sterile large spoons (tablespoons).
3. Sterile glass rods.
4. Sterile sample jars with mouth wide enough to admit the spatulas or spoons employed.
5. Water bath maintained at 45 C (113F).
6. Sterilize in a test tube a butter boat of sufficient size to hold a 1 gram sample of melted ice cream (for California method).

A. BULK ICE CREAM

Procedure

1. With a sterile spoon remove and discard about 1 inch of the surface ice cream in the package.

2. Insert a sterile spatula or cheese trier deep into the ice cream at some point on the newly exposed surface, and transfer a suitable amount (not less than 50 ml.) to the sterile wide-mouth usually 2 ounce sample jar. A sterile glass rod should be used for scraping the plug of ice cream off the spatula or trier into the bottle. Refrigerate the sample until ready for use.

3. When ready for plating, place the sample jar in a water bath maintained at 45 C (113 F) to melt the ice cream. Rotate the sample jar occasionally to hasten the melting and also to facilitate the release of the air incorporated in the ice cream. The melting process should be completed as rapidly as possible, and in no case should exceed 15 minutes. Only a few samples should be melted at a time to avoid delay in plating after the ice cream is melted, otherwise the melted samples should be transferred to ice water.

Choice of Volumetric or Granimetric Sample for Ice Cream

On melting, some samples of ice cream persistently retain the incorporated air, and the melted product has a distinct foamy structure. For such ice creams, a volumetric sample is unsatisfactory, and it is essential that a gravimetric sample be employed. Most ice creams melt readily to a sufficiently fluid consistency that a volumetric sample may be satisfactorily measured. Most laboratory workers prefer to use a volumetric sample if possible in order to save time. Parallel counts made with gravimetric and volumetric samples of ice creams which melt readily without retaining appreciable amounts of air, show relatively little difference in results, owing no doubt to the fact that the inherent error of the plate method overshadows any slight difference in the two methods of measuring samples. Marked differences are frequently observed in the results of parallel gravimetric and volumetric determinations on samples which melt to foamy consistency. In order to avoid the inconsistency of making part of the counts on a volumetric basis and part on a gravimetric basis, some laboratories use the gravimetric method on all ice creams plated.

A certain amount of air is inevitably retained in all melted ice cream, even though it may not be evident to the eye. In general it is preferable to use the larger amount of sample (11 gm or 11 ml) rather than a smaller sample (1 gm or 1 ml).

If the volumetric or the gravimetric method is chosen, transfer 11 ml or 11 gm of melted ice cream to a 99 ml dilution blank and proceed with making the dilutions as illustrated in diagram B, figure 19, page 41, and with the plating procedure as described on pages 42-48. The same medium is used for ice cream as for plating milk.

B. CALIFORNIA METHOD

A method of weighing samples of ice cream recommended by the California State Department of Agriculture involves the use of an approximate 1 gram sample placed in a sterile weighing boat.

1. Using chemical balances, determine the weight of the test tube containing the sterile butter boat. See figure 25.

2. Under aseptic conditions remove the stopper from the test tube, and by carefully tilting the tube, cause the end of the butter boat to protrude sufficiently from the tube to permit the addition of the melted ice cream sample to the sterile butter boat.

3. By means of a sterile pipette, add as near 1 gm of a suitably prepared sample of melted ice cream as can be estimated.

4. Replace the cotton stopper in the test tube and reweigh.

5. The difference between the two weighings should be recorded as the exact amount of ice cream employed in the sample. (Since this weight will not necessarily be exactly 1 gm, it is necessary to adjust accordingly the ultimate plate count obtained.)

6. After removing the cotton stopper and flaming the mouth of the test tube, tilt the tube so that the butter boat containing the weighed sample can slide directly into a 99 gm sterile dilution blank which has been momentarily opened for the purpose.

7. The dilution should be tentatively regarded as a 1 : 100 dilution, and subsequent dilutions and plates prepared as described on pages 42-48 and shown diagrammatically in part A of figure 19, page 41.

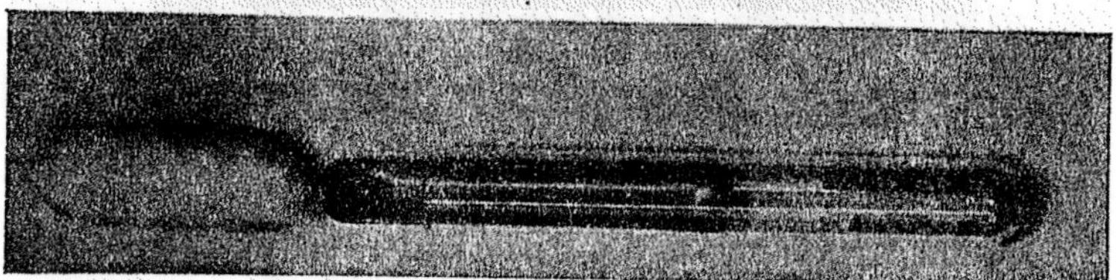

Courtesy of the American Public Health Association

Figure 25.

Culture tube with butter boat as used by the California State Department of Agriculture in their official ice cream control work.

C. SAMPLING SINGLE FLAVOR BRICK OR PACKAGED ICE CREAM

1. In procuring a sample from brick or packaged ice cream of one flavor, cut the brick in two using a sterile knife.

2. With a sterile spoon or cheese trier remove a suitable portion from the center of a freshly exposed surface, avoiding the removal of any of the ice cream that is in contact with the package. See figure 26.

3. Transfer the ice cream to a sterile wide-mouth bottle with the aid of a sterile glass rod.

4. The procedure for melting the sample and preparation of dilutions is the same as described for bulk ice cream on page 61.

D. SAMPLING BRICK ICE CREAM WITH TWO OR MORE FLAVORS

If it is desired to obtain individual samples of each of the flavors represented, the procedure is the same as for sampling single flavored brick ice cream except that separate samples are taken of each flavor.

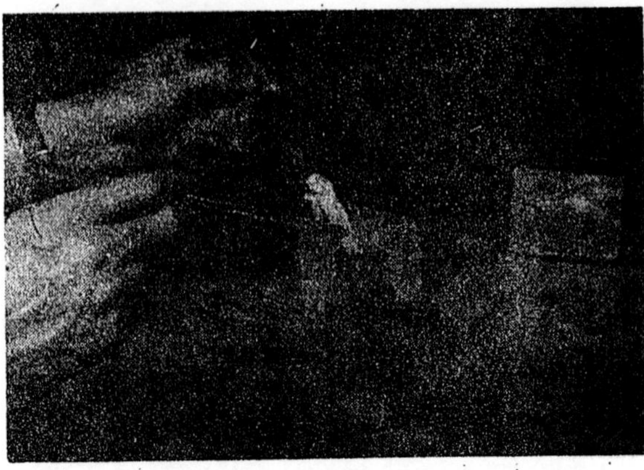

Figure 26.
Sampling brick ice cream from surface freshly exposed by cutting the brick in two with a sterile knife.

1. Cut the brick in two with a sterile knife, and use a sterile spoon or cheese trier to remove a suitable sample of one of the flavors from the freshly exposed surface, avoiding the removal of any of the ice cream which was exposed to the package, or any of the other flavors of ice cream.

2. Transfer the sample to a sterile wide-mouth jar with the aid of a sterile glass rod. The same procedure is followed for each of the other flavors, using a separate sterile spoon or cheese trier, glass rod, and sample bottle.

The procedure for melting, making dilutions, and plating are the same as described for bulk ice cream (page 61).

If it is desired to obtain a mixed sample of all of the flavors represented in a multiflavored brick, proceed as follows:

1. Cut a slice about three-fourths inch thick from the center of the brick, using a sterile knife.

2. With the same knife, remove the slice from the brick and place it horizontally into a sterile dish or on a sterile piece of paper.

3. With another sterile knife, cut an equal number of cubes of the same size from each of the flavors represented.

4. With the aid of a sterile glass rod transfer all of them to a wide-mouth dilution bottle. In cutting the cubes avoid taking any of the ice cream which has been in contact with the package.

5. Another method is to cut the brick in two with a sterile knife, then by means of a sterile cheese trier remove plugs of equal length from each of the flavors.

6. The procedure for melting the mixture and for making the dilutions is the same as described for bulk ice cream (page 61).

E. Sampling Novelty Ice Cream

1. Place an individual mold, chocolate covered ice cream, or Eskimo pie on a sterile dish or paper.

2. With a sterile knife cut the novelty into pieces small enough to be admitted to the sterile wide-mouth sample bottle. In the case of chocolate coated novelties, if it is desired to test only the ice cream, the chocolate coating may be removed with a sterile knife. Another sterile knife should then be used for cutting the ice cream into small pieces.

3. The procedure for melting the product and preparing the dilutions is the same as described for bulk ice cream (page 61).

F. Sampling Sherbets and Ices

The procedure for sampling sherbets and water ices as well as the method of melting and making dilutions is the same as described for ice cream (page 61).

Powdered Materials

(Such as Gelatin, Milk Powder, Chocolate, Cocoa, Sugar, Nuts)

Apparatus

1. The usual plating material (see page 42).
2. Sterile paper.
3. Sterile spatula.
4. Balance sensitive to 30 mg.
5. Sterile water blanks.

Procedure

1. Place a piece of sterile paper on a balance sensitive to 30 mg.

2. With a sterile spatula transfer to the paper 11 gm of a well mixed representative sample of the material to be tested. See figure 27.

3. Carefully remove the paper from the balance, avoiding contact with the sample, and slowly pour the material directly into a 99 gm (ml) dilution blank.

Figure 27.
Weighing 11 grams of a sample in powder form for bacteriological analysis.

4. Dilution and plating techniques are the same as for milk. (See page 41.)

Special Precautions for Gelatin

In the case of gelatin, the dilution blank should previously be warmed to 40-45 C (104-113 F) in order to facilitate solution of the gelatin. After vigorous shaking, place the dilution blanks in a water bath at 40-45 C (104-113 F) and shake again every few minutes until the gelatin has dissolved. The process should require not longer than 15 minutes.

Special Precautions for Dry Milk Products

Following the addition of 11 gm of the dry milk sample to a 99 ml dilution blank, shake the bottle vigorously 25 times, up and down, through an excursion of about one foot during a period of about seven seconds. One teaspoonful of sterile solid, resistant glass beads may be used in each bottle to aid in breaking up the milk particles.

Dilutions should be made according to the following table:

1:10—prepared by transfer of 11 gm dry milk into a 99 ml sterile dilution blank.

1:100—transfer 11 ml of the preceding (1:10) dilution into a 99 ml sterile dilution blank.

1:1000—transfer 11 ml of the preceding (1:100) dilution into a 99 ml sterile dilution blank.

Higher dilutions are handled similarly.

Plating and incubation are the same as for milk (see p. 41).

Avoid counting undissolved milk particles as pin-point colonies. When this difficulty occurs, check plates under a low power microscope. Convert actual colony count to bacterial estimate *per gram* by application of the proper dilution factor.

REFERENCE

The Grading of Nonfat Dry Milk Solids and Sanitary and Quality Standards. Bull. 911 (revised). American Dry Milk Institute, Inc., Chicago, Illinois. 1948.

CHAPTER 5

DETERMINATION OF SPECIAL TYPES OF ORGANISMS

Acidophilus

Bacterial counts should be made frequently on acidophilus milk as an index to the potency of the product.

Apparatus

1. Usual plating materials (see page 42).
2. Carbon dioxide generating and measuring equipment.

Reagents

Tomato juice agar or trypsin digest agar (see pages 519 & 520).

Procedure

1. Prepare dilutions of 1 to 100,000, 1 to 1,000,000 and 1 to 10,000,000 from a volumetric sample as shown in figure 19, page 41).

2. Pour plates with tomato juice agar or with trypsin digest agar.

3. After the agar has solidified, place the plates in an air-tight container, from which at least 10 per cent of the air is to be replaced with carbon dioxide (see figure 28 on page 69). A pressure cooker or a desiccator jar, the lid of which is equipped with glass-stoppered inlet and an outlet tube, serves well for this purpose. Determine the approximate cubical capacity of the container, and replace from 10 to 50 per cent of the volume with carbon dioxide from a suitable generator or carbon dioxide pressure tank (technique described on page 71).

4. Close the stop cocks on the inlet and outlet tubes of the container to entrap the gas, and place the container with its Petri dishes in the incubator at 37 C (98.6 F) for three to five days.

5. Count the filamentous colonies under favorable light conditions.

NOTE: To determine contamination of acidophilus milk, make appropriate dilutions (usually 1 : 10, 1 : 100, and 1 : 1,000) and use plain beef-extract agar instead of the special medium recommended above. Incubate the plates without exposure to a CO_2 atmosphere at 37 C (98.6 F) for 48 hours. The colonies which develop on these plates are usually contaminants since the *Lactobacillus acidophilus* does not grow well, if at all, on plain beef-extract agar.

Brucella

A. DESCRIPTION OF THE SPECIES

Although *Brucella abortus*[1] is commonly associated with infectious abortion in cattle, *Brucella suis* with abortion in swine, and *Brucella melitensis* with abortion in sheep, all three types of organisms have been isolated from cows' milk, and all three are infectious for man. Laboratory workers should handle cultures of these organisms with extreme care. As pointed out by Evans,[2] the disease in man may last only a few days or may persist for several years and may vary from a mild unrecognized form to a serious illness. Owing to the variations in symptoms, correct diagnosis is difficult and confusion frequently results. Mild attacks may be diagnosed as influenza, and more severe attacks as typhoid, influenza, malaria, tuberculosis, or rheumatism. The organisms may infect human beings through the skin as a result of handling infected animals or their excretions, or through the digestive tract from consuming raw dairy products containing the infectious agent. Pasteurization readily destroys all three types of the Brucella group.

The incidence of undulant fever in human beings is substantially higher among such persons as veterinarians, farmers, and packing-

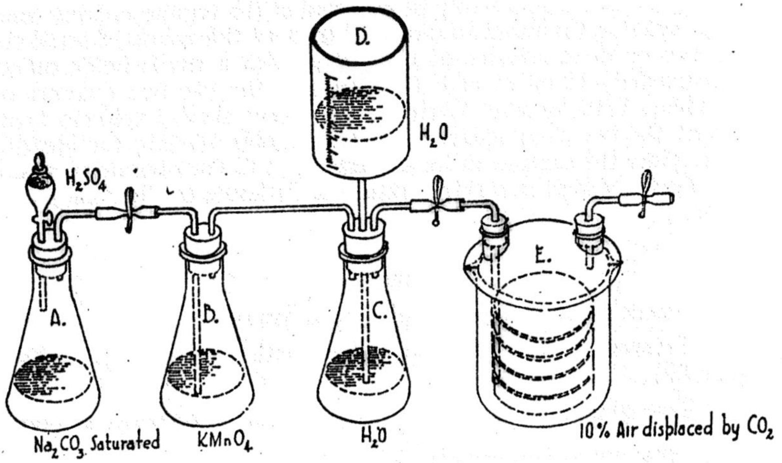

Figure 28.

Diagram of assembly for making carbon dioxide and displacing a known percentage of the air in the closed container (*E*) in which plates are to be incubated. Note that carbon dioxide developed in flask *A* by introducing sulfuric acid into saturated sodium carbonate is washed in potassium permanganate. The pressure created by the accumulated gas in flask *C* forces a measured volume of water into cylinder *D*. By releasing the two pinchcocks on the right, an amount of water may be re-admitted to flask *C* from cylinder *D* equivalent to 10 per cent of the volume of container *E*. The gas is entrapped by closing the pinchcocks, and container *E* is disconnected from the assembly and placed in the incubator. (A cylinder of CO_2 gas may be substituted for flasks *A* and *B*.) (Also see fig. 29, page 71, for an alternative method of providing CO_2 atmosphere for incubating plates.)

house workers (whose occupations bring them in close contact with infected animals) than is the incidence among persons who consume raw milk containing the organisms. The prevalence of infectious abortion among dairy cattle, the frequency with which the organisms may be isolated from raw market milk, and the large number of people who consume raw milk daily are not in harmony with the relatively small number of recognized cases of undulant fever in man. The evidence seems to indicate that the digestive tract is not the optimum portal of entry of these organisms and that in most instances the contraction of the disease through the drinking of infected raw milk must presume either a low resistance or the ingestion of a large number of organisms.

The Brucella organisms are characteristically rather difficult to isolate, particularly on the primary isolation from the body fluids such as milk, blood, urine, and feces. The difficulty of isolation is due in part to the special requirements of the organism and in part to the fact that they may constitute only a small percentage of the original bacterial flora of the sample.

B. Isolation of Brucella from Milk

Collection of the Milk Sample

Samples should preferably be collected at the regular milking time. After washing the flanks, udder, and teats of the animal, discard the first two or three streams of milk, then with a sterile bottle collect approximately 10 ml of milk from each of the two rear quarters of the udder. With another sterile bottle collect similar volumes from each of the two front quarters. After suitable labeling for identification, place the samples in ice and transport to the laboratory, where they should be kept in a refrigerator for 24 hours to allow the cream to rise.

Apparatus

1. Sterile glass rods with right angle bends.
2. Pressure cooker (small size) fitted with one way stopcock (see figure 29).
3. Two graduated cylinders, 500 ml, fitted with rubber stoppers.
4. Ring stand and clamps.
5. Glass tubing.
6. Rubber tubing and pinchcocks.
7. Cylinder of carbon dioxide (CO_2) gas.
8. Sterile Petri dishes.
9. Sterile pipettes, 1 ml, graduated in 0.1 ml.

Reagents

1. Tryptose agar or liver agar (Difco dehydrated), made according to directions of manufacturer.
2. Crystal violet solution (0.0143 per cent aqueous stock solution).
3. Tryptose agar slants.

Procedure

1. To the desired quantity of sterile tryptose agar or liver agar, which has been melted, add a suitable quantity of the stock (0.0143 per cent) solution of crystal violet to give a final concentration of 1 : 700,000 (1 ml of stock solution for each 100 ml of medium). Pour agar plates and allow the medium to solidify.
2. With a sterile pipette, transfer to each of two agar plates 0.1-0.2 ml of the cream layer from the sample of milk which has previously been refrigerated for 24 hours.
3. Spread the cream evenly over the agar with a sterile glass rod bent at right angles.

Courtesy of the American Public Health Association

Figure 29.

Apparatus for obtaining carbon dioxide and culture container for the isolation of *Brucella abortus*. Apparatus shown consists of an ordinary pressure cooker with the pressure gage removed, a CO_2 tank, and two graduated cylinders one of which is inverted over the other. A one-way metal stopcock is inserted in the pressure gage opening for removing air by means of a vacuum pump and receiving the CO_2 collected and measured over a column of water in the graduated cylinder. The two cylinders are connected together by means of glass tubing. The gas coming from the tank is received in the lower cylinder, forcing the water upward into the inverted cylinder. When the required amount of gas is collected in the lower cylinder, the flow of gas from the tank is closed. The cock in the pressure cooker is opened allowing the collected gas to flow into the cooker. There are three sizes of the cooker on the market. The average laboratory will need only the smaller one. (Also see figure 28, page 69, an alternative device to produce CO_2 atmosphere for incubation of plates.)

4. Place the plates in the container (pressure cooker or desiccator jar) and displace 10 per cent of the air of the pressure cooker with CO_2 as directed in the legend under figure 29. The apparatus shown in figure 28 may be used if a cylinder of CO_2 is not available.

5. Incubate the plates at 35 to 37 C (95 to 98.6 F) for 5 days.

6. The colonies have a clear light blue color and are from 2-7 mm in diameter. Transfer such colonies to tryptose agar slants or to another plate containing crystal violet liver agar.

7. Identify the cultures by means of the agglutination test, using specific agglutinating sera which may be obtained from biological supply houses.

NOTE: The isolation of *Brucella* species from market milk is more difficult than isolation from samples drawn aseptically from the udder. This is due to the fact that the bacteriostatic dye (crystal violet) in the concentration employed does not inhibit the common milk-souring streptococci. These streptococci interfere with the isolation because they rapidly lower the pH to a point which discourages the growth of *Brucella* species. The isolation of *Br. melitensis* and *Br. suis* from the milk is accomplished by the same procedure as outlined for isolating *Br. abortus*.

C. THE AGGLUTINATION METHOD USING MILK SERUM (RAPID SLIDE OR PLATE METHOD)

Apparatus

1. Darkfield illumination box fitted with a ruled glass plate.
2. Pipettes, 0.2 ml, graduated to 0.01 ml.
3. Standardized dropper pipette.
4. Sterile test tubes.
5. Sterile pipettes.

Reagents

1. Standardized antigen (obtainable from biological supply houses).

2. Rennet solution, 1 per cent.

Procedure

1. Preparation of samples.

(a) Collect the samples under aseptic conditions, preferably from individual quarters, filling the test tubes or vials only to about one-half their capacity. (Do not use strippings.)

(b) Place the samples in the refrigerator 6 to 8 hours to allow the cream to rise.

(c) With a sterile pipette withdraw the cream and transfer to another sterile tube, then add about 2 drops of 1 per cent rennet solution for each 5 ml of cream.

(d) Mix thoroughly and place the tubes in the incubator at 37 C in a slanting position for about 2 hours until the serum has separated.

(e) Remove the clear serum, carefully avoiding the inclusion of particles of casein, and place the samples of serum in the refrigerator for 6-8 hours to allow them to become clear. The serum from sour milk, colostrum milk, or milk in which proteolysis of the curd has occurred cannot be used satisfactorily in the test.

2. Make dilutions of the serum with a 0.2 ml pipette graduated in 0.01 ml. Transfer to one of the etched squares on the glass plate 0.08 ml of milk serum. To the successive adjacent squares in the row transfer 0.04, 0.02, 0.01, and 0.005 ml of serum respectively.

3. Thoroughly shake the bottle of antigen and remove a dropperful. With the dropper in a vertical position, add one drop of antigen to each amount of milk serum on the plate. Quickly mix the serum and the antigen using a clean toothpick. Begin the mixing operation with the mixture containing the smallest amount (0.005 ml) of serum, and proceeding to the next higher mixture in the series. During the mixing process, spread the serum-antigen mixture over about three fourths of the area within the etched square. After mixing all the dilutions of serum and antigen, carefully tilt the plate backward and forward for about five minutes. Place the plate on the illuminating box and observe the mixtures for evidence of agglutination or clumping of the cells in the antigen.

In performing the test care should be taken to use only perfectly clean glassware and to rinse all pipettes immediately after use.

Interpretation

For samples taken from individual quarters, the presence of infection is indicated by complete flocculence or agglutination of the antigen with 0.04 ml of the milk serum. If the milk was taken from a composite sample from several cows, infection is indicated by agglutination of the antigen with 0.08 ml of milk serum.

REFERENCES

1. *Standard Methods for the Examination of Dairy Products*, American Public Health Association, 1790 Broadway, New York, N. Y.

2. Evans, Alice C.: The Chronic Brucellosis Patient, *Amer. J. Nursing 39*: 1-4 (1939).

3. Anderson, Gaylord, W.: Undulant Fever in Man and Its Relation to Bang's Disease in Live Stock, *J. Milk Tech.* 1:26-29 (1937)

4. Kerby, Grace P. and Calder, Royall M.: Culture Media for Brucella, *J. Bact. 40* : 637 (1937).

5. Meyer, K. F., and Eddie, B.: Laboratory Infection Due to Brucella, *J. Infect. Dis., 68* : 24-32 (1941).

Coliform Group

A. SIGNIFICANCE

The so-called coliform group includes all the non-spore-forming gram-negative rods capable of fermenting lactose with the formation of gas. In the interpretation of the significance of their presence in dairy products, recognition should be given to the following generalizations:

1. Coliform types are almost always found in fresh raw milk as a common contaminant.

2. These organisms are usually incapable of surviving pasteurization, although occasional strains are encountered which do survive.

3. The absence of these organisms in milk as it leaves the pasteurizer and their presence in the same bottled pasteurized milk warrant an investigation of the effectiveness of the sterilization of equipment employed following pasteurization and the possibilities of contamination following pasteurization, particularly in the process of bottling. The presence of coliform types in the bottled pasteurized product should always be checked with samples taken of the same milk leaving the pasteurizer before conclusions are drawn that contamination has occurred subsequently to pasteurization.

4. Owing to the ease with which coliform types grow in milk, samples should not be over three to four hours old at the time of analysis or there should be convincing evidence that the product has been kept under excellent refrigeration, certainly below 40 F (4.4 C). If samples are not fresh and if there is no satisfactory evidence of effective refrigeration, the value of numerical determinations of coliform types is greatly diminished. Robinton and Genung (1) question the value of the coliform test in cream because of the rapid growth of the organisms in cream.

5. As pointed out by Weber (2) and Tiedman and Smith (3), non-heat-resisting coliform types may gain entrance to pasteurized milk from various sources, such as: a. ineffective sterilization of coolers, pipelines, pumps, fillers, and bottles; b. cracks and crevises in equipment which harbor and protect the organisms from the cleaning and sterilizing operations; c. hand contamination of milk-contact surfaces of equipment subsequent to sterilization; d. rinsing of equipment subsequent to sterilization with water containing coliform types; e. washing equipment used for pasteurized milk (pipe lines, filler parts, etc.) in the same wash tank after having washed

equipment used for raw milk; and f. indirect hand- and air-borne contamination from droplets of condensate which drain over the contaminated outside surfaces of the cold pipelines and the filler bowl and then drop into the open bottle. Rubber moats or metal drip deflectors attached to each of the filler spindles have been employed to reduce contamination from this source (see figures 30, 31, 32).

Although in water analysis for coliform types it is customary to make a confirmatory test following a positive presumptive test, *Standard Methods for the Examination of Dairy Products* stipulates that the standard test for the coliform group in routine pasteurization control be limited to the presumptive test. According to standard methods, positive presumptive tests on pasteurized milk should be followed by samples taken directly from the pasteurizing vat at the end of the holding period and at such other subsequent points as may reveal the portal of entry of these organisms. Samples of hot milk taken from pasteurizing vats should be cooled at a rate comparable to that provided by the cooling equipment employed in the plant. Samples which show a positive presumptive test when taken directly from the pasteurizing vat after pasteurization should also be tested by the phosphatase test (see page 265) as an index to the effectiveness of pasteurization. It is frequently helpful to take samples of the first milk going through the equipment after sterilization, after a shutdown of significant duration, and also at the end of the day's run. Fouts and Freeman (4) have shown diagrammatically the procedure to follow in making the routine presumptive and confirmatory test of dairy products.

B. PRESUMPTIVE TEST

(Liquid Medium)

Apparatus

1. Sterile pipettes, 10 ml and 1 ml, graduated in 0.1 ml.
2. Sterile dilution blanks, 99 ml.

Reagents

Either one of the following media may be used.

1. Brilliant green lactose peptone bile (2 per cent) prepared in 10-ml quantities in Durham fermentation tubes (see page 514).

For all tubes into which quantities of 1 ml or less of milk are to be inoculated, prepare the medium as directed on page 515.

For all tubes to be inoculated with 10 ml of milk, prepare a special batch of the medium containing one and one-half times the proportions of the various ingredients shown in the appendix, and

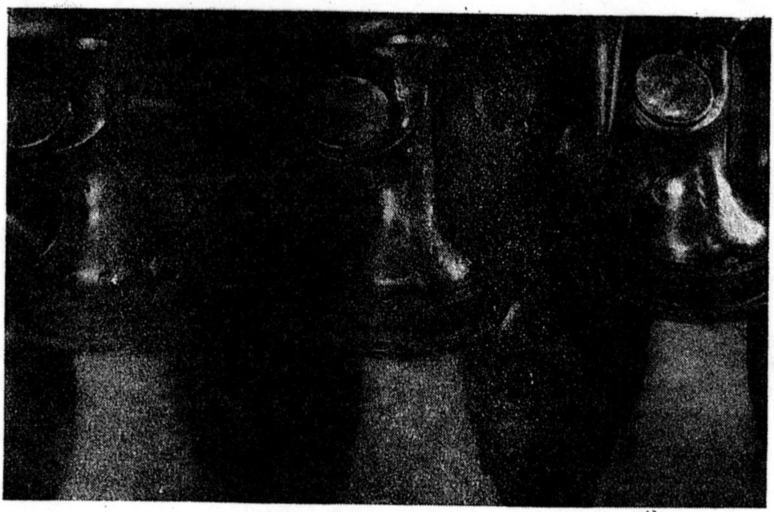

Figure 30.
Rubber moats attached to filler spindles to deflect water of condensation away from bottles entering or leaving the filler.

place in Durham fermentation tubes containing 20 ml of the medium. The medium may be prepared from the dehydrated form.

2. Formate ricinoleate lactose peptone broth prepared in 10-ml quantities in Durham fermentation tubes (see page 517).

For all tubes into which quantities of 1 ml or less of milk are to be inoculated, prepare the medium as directed in the appendix.

For all tubes to be inoculated with 10 ml of milk, prepare a special batch of the medium containing one and one-half times the proportions of ingredients shown in the appendix and place 20 ml in the Durham fermentation tubes. The medium may be prepared from the dehydrated form.

Procedure for Determining the Most Probable Number

The decimal multiples or dilutions of 1 ml of milk to be employed will depend on the nature and history of the sample and also on the technician's familiarity with samples from the same source. According to standard methods, it is usually desirable to plant three or more dilutions or decimal multiples of 1 ml in order to have at least one negative and one positive tube in the series.

1. Using either one of the two media (brilliant green lactose peptone bile or formate ricinoleate broth):

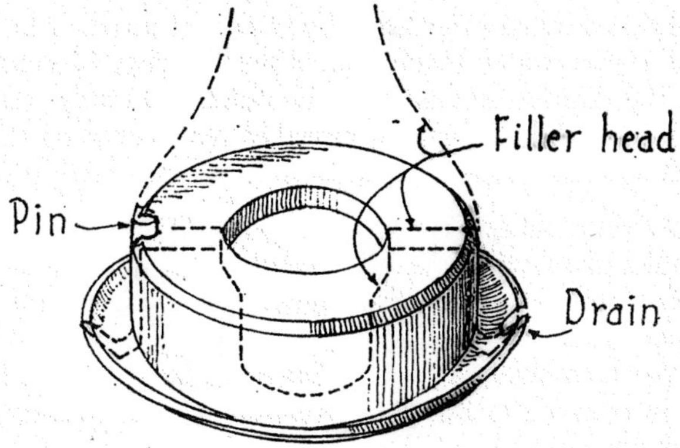

Figure 31.
Diagrammatic sketch of the rubber moats shown in Figure 30.

a. Inoculate 10 ml of milk into each of five tubes containing 20 ml of the concentrated liquid medium.

b. Inoculate 1 ml of milk into each of five tubes containing the regular concentration of broth.

c. Similarly, inoculate 0.1 ml of milk into another series of five tubes.

NOTE: Some technicians prefer to inoculate a single fermentation tube with 1 ml of milk and run a larger number of samples representing a broader picture of the day's output of the plant.

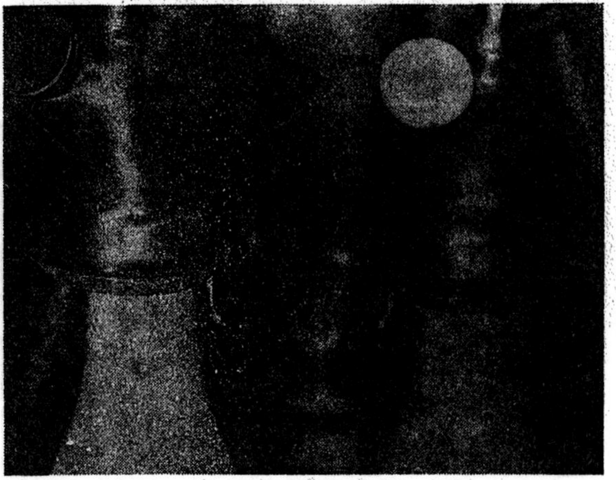

Figure 32.
Metal moats brazed onto filler spindles to deflect condensation away from the bottles entering or leaving the filler.

d. If higher dilutions are deemed advisable, transfer 1 ml of milk to a 99-ml sterile water blank and, after shaking, inoculate 1 ml of this 1 : 100 dilution into each of five tubes. Similarly, tubes containing 0.001 ml of milk may be prepared by inoculating 0.1 ml of this 1 : 100 dilution. Higher dilutions in series are rarely necessary.

2. Incubate the tubes at 35 to 37 C (95 to 98.6 F) for 48 hours. Those samples showing gas in the inverted vial after 48 hours are regarded as positive presumptive tests. For a description of the mathematical method of determining the most probable number of coliform types from the results when five tubes are used in each series, the reader is referred to *Standard Methods for the Examination of Dairy Products*.

TABLE 3

TABLE OF MOST PROBABLE NUMBERS USING FIVE TUBES WITH EACH DILUTION

More complete tables will be found in the *Standard Methods of Water Analysis*, 9th Ed., 1946, pp. 205.

Positives with 10 ml	Probable No. per 100 ml	Positives with 10 ml	1 ml	Probable No. per 100 ml	Positives with 10 ml	1 ml	0.1 ml	Probable No. per 100 ml	Positives with 10 ml	1 ml	0.1 ml	Probable No. per 100 ml
0	0	0	0	0	0	0	0	0	1	0	0	2
1	2.2	0	1	1.8	0	0	1	1.8	1	0	1	4
2	5.1	0	2	3.7	0	0	2	3.6	1	0	2	6
3	9.2	0	3	5.6	0	0	3	5.4	1	0	3	8
4	16	0	4	7.5	0	0	4	7.2	1	0	4	10
5	18+	0	5	9.6	0	0	5	9	1	0	5	12
		1	0	2.0	0	1	0	1.8	1	1	0	4
		1	1	4.1	0	1	1	3.6	1	1	1	6.1
		1	2	6.2	0	1	2	5.5	1	1	2	8.1
		1	3	8.4	0	1	3	7.3	1	1	3	10
		1	4	11	0	1	4	9.1	1	1	4	12
		1	5	13	0	1	5	11	1	1	5	14
		2	0	4.5	0	2	0	3.7	1	2	0	6.1
		2	1	6.9	0	2	1	5.5	1	2	1	8.2
		2	2	9.4	0	2	2	7.4	1	2	2	10
		2	3	12	0	2	3	9.2	1	2	3	12
		2	4	15	0	2	4	11	1	2	4	15
		2	5	17	0	2	5	13	1	2	5	17
		3	0	7.9	0	3	0	5.6	1	3	0	8.3
		3	1	11	0	3	1	7.4	1	3	1	10
		3	2	14	0	3	2	9.3	1	3	2	13
		3	3	18	0	3	3	11	1	3	3	15
		3	4	21	0	3	4	13	1	3	4	17
		3	5	25	0	3	5	15	1	3	5	19
		4	0	13	0	4	0	7.5	1	4	0	11
		4	1	17	0	4	1	9.4	1	4	1	13
		4	2	22	0	4	2	11	1	4	2	15
		4	3	28	0	4	3	13	1	4	3	17
		4	4	35	0	4	4	15	1	4	4	19
		4	5	43	0	4	5	17	1	4	5	22
		5	0	24	0	5	0	9.4	1	5	0	13
		5	1	35	0	5	1	11	1	5	1	15
		5	2	54	0	5	2	13	1	5	2	17
		5	3	92	0	5	3	15	1	5	3	19
		5	4	160	0	5	4	17	1	5	4	22
		5	5	180+	0	5	5	19	1	5	5	24

TABLE 3 (CONTINUED)

TABLE OF MOST PROBABLE NUMBERS USING FIVE TUBES WITH EACH DILUTION

More complete tables will be found in the *Standard Methods of Water Analysis*, 9th Ed., 1946, pp. 205.

Positives with 10 1 0.1 ml ml ml	Probable No. per 100 ml	Positives with 10 1 0.1 ml ml ml	Probable No. per 100 ml	Positives with 10 1 0.1 ml ml ml	Probable No. per 100 ml	Positives with 10 1 0.1 ml ml ml	Probable No. per 100 ml
2 0 0	4.5	3 0 0	7.8	4 0 0	13	5 0 0	23
2 0 1	6.8	3 0 1	11	4 0 1	17	5 0 1	31
2 0 2	9.1	3 0 2	13	4 0 2	21	5 0 2	43
2 0 3	12	3 0 3	16	4 0 3	25	5 0 3	58
2 0 4	14	3 0 4	20	4 0 4	30	5 0 4	76
2 0 5	16	3 0 5	23	4 0 5	36	5 0 5	95
2 1 0	6.8	3 1 0	11	4 1 0	17	5 1 0	33
2 1 1	9.2	3 1 1	14	4 1 1	21	5 1 1	46
2 1 2	12	3 1 2	17	4 1 2	26	5 1 2	64
2 1 3	14	3 1 3	20	4 1 3	31	5 1 3	84
2 1 4	17	3 1 4	23	4 1 4	36	5 1 4	110
2 1 5	19	3 1 5	27	4 1 5	42	5 1 5	130
2 2 0	9.3	3 2 0	14	4 2 0	22	5 2 0	49
2 2 1	12	3 2 1	17	4 2 1	26	5 2 1	70
2 2 2	14	3 2 2	20	4 2 2	32	5 2 2	95
2 2 3	17	3 2 3	24	4 2 3	38	5 2 3	120
2 2 4	19	3 2 4	27	4 2 4	44	5 2 4	150
2 2 5	22	3 2 5	31	4 2 5	50	5 2 5	180
2 3 0	12	3 3 0	17	4 3 0	27	5 3 0	79
2 3 1	14	3 3 1	21	4 3 1	33	5 3 1	110
2 3 2	17	3 3 2	24	4 3 2	39	5 3 2	140
2 3 3	20	3 3 3	28	4 3 3	45	5 3 3	180
2 3 4	22	3 3 4	31	4 3 4	52	5 3 4	210
2 3 5	25	3 3 5	35	4 3 5	59	5 3 5	250
2 4 0	15	3 4 0	21	4 4 0	34	5 4 0	130
2 4 1	17	3 4 1	24	4 4 1	40	5 4 1	170
2 4 2	20	3 4 2	28	4 4 2	47	5 4 2	220
2 4 3	23	3 4 3	32	4 4 3	54	5 4 3	280
2 4 4	25	3 4 4	36	4 4 4	62	5 4 4	350
2 4 5	28	3 4 5	40	4 4 5	69	5 4 5	430
2 5 0	17	3 5 0	25	4 5 0	41	5 5 0	240
2 5 1	20	3 5 1	29	4 5 1	48	5 5 1	350
2 5 2	23	3 5 2	32	4 5 2	56	5 5 2	540
2 5 3	26	3 5 3	37	4 5 3	64	5 5 3	920
2 5 4	29	3 5 4	41	4 5 4	72	5 5 4	1600
2 5 5	32	3 5 5	45	4 5 5	81	5 5 5	1800+

NOTE: In the interest of uniformity this table, consisting of portions (rearranged and amplified) of the table presented in *Standard Methods of Water Analysis*, Eighth Edition, 1936, has been substituted for that of a previous (Seventh) edition of the *Standard Methods for the Examination of Dairy Products.*

C. PRESUMPTIVE TEST

(Solid Medium)

Apparatus

1. Sterile Petri dishes.
2. Sterile pipettes, 1.1 ml.
3. Dilution blanks, 99 ml.

Reagents

Either one of the following solid culture media may be used for the presumptive test instead of the liquid media.

Although these media may be prepared from the ingredients, it is recommended in *Standard Methods for the Examination of Dairy Products* in the interest of uniformity that they be used in the dehydrated form.

1. Violet red bile agar (see page 518).
2. Desoxycholate agar (see page 515).

Procedure for Presumptive Test (Quantitative)

1. Inoculate Petri dishes with 1 ml and 0.1 ml quantities of milk. Other Petri dishes may be inoculated with 1:100 or higher dilutions if deemed desirable.

2. Pour approximately 10 ml of the melted agar medium, cooled to 45 C (113 F) into each inoculated Petri dish and agitate sufficiently to mix the sample and the medium.

3. After the medium has solidified, pour an additional 3 or 4 ml of agar over the surface, thereby forming a film to discourage the development of surface colonies.

4. Incubate the plates in an inverted position at 35 to 37 C (95 to 98.6 F) for 20 to 24 hours.

5. After incubation examine the plates for dark-red subsurface colonies of at least 0.5 mm in diameter. Only typical colonies should be counted and the number multiplied by the dilution to determine the number of coliform types per milliliter of milk.

NOTE: The choice of the use of the plates or the fermentation tubes for making the presumptive test should depend on the probable number of coliform types expected in the sample. In general the plate method is more accurate when counts in excess of 5 coliform types per milliliter are expected, but is unreliable when fewer coliform types are present. Skinner, Marwin, and Honaas (5) found more false positives in formate ricinoleate broth (dilution extinction method) than with brilliant green bile broth, due primarily to species of Proteus.

D. COMPLETED TEST

Since a postive presumptive test is not certain evidence of the presence of coliform types, further confirmation of the presumption may be desirable.

Apparatus

1. Platinum or nichrome inoculating needle.
2. Microscope with oil-immersion objective.
3. Glass slides.

Reagents

1. Poured sterile plates of either eosin-methylene blue agar (see page 516) or Endo agar (see page 516).
2. Gram stain (see page 525).
3. Nutrient agar slants (see page 506).
4. Lactose broth fermentation tubes (see page 518).

A. Procedure to Be Used When Liquid Medium Was Employed in the Presumptive Test

1. Using a sterile inoculating loop, streak some of the liquid from the Durham fermentation tubes showing gas in the presumptive test on either an eosin-methylene blue agar plate or on an Endo agar plate. Care should be taken in making the streak culture to insure a suitable distribution of individual colonies at some point along the streak.

2. Incubate the plate at 35 to 37 C (95 to 98.6 F) for 18 to 24 hours.

3. From a satisfactory plate select well isolated, small, flat to concave colonies rather closely adherent to the agar, each with a dark center and surrounded by a distinct metallic sheen. Using a sterile inoculating needle, transfer one or more of these to (a) lactose broth fermentation tubes, or (b) nutrient agar slants. Incubate the lactose broth fermentation tubes at 35 to 37 C (95 to 98.6 F) for 48 hours and the nutrient agar slants for 24 hours.

4. After incubation make a Gram stain from the growth on the agar slant, and examine under the oil-immersion objective for short, non-spore-forming, gram-negative rods.

5. After incubation examine the lactose broth fermentation tube for the presence of gas.

The presence of gas in the lactose fermentation tube and the presence of gram-negative, non-spore-forming rods, with the absence of any spore-forming types in the growth from the agar slant, constitute a completed test. If these conditions are not met, the completed test is regarded as negative.

B. Procedure to Be Used When Solid Medium Was Employed in the Presumptive Test

If violet red bile agar or desoxycholate agar plates were used for the presumptive test instead of the liquid medium, the procedure for the first step in the completed test is as follows:

1. Select a typical, deep red, subsurface colony from the plate used in the presumptive test, and transfer to a lactose broth fermentation tube.

In the interest of reliability several colonies should be selected, and a similar number of lactose fermentation tubes inoculated.

2. If none of the lactose broth tubes shows gas after 48 hours incubation at 35 to 37 C (95 to 98.6 F), the completed test may be regarded as negative.

3. As soon as possible after the lactose fermentation tubes begin to show gas, proceed to run the completed test according to the directions given when the liquid media (brilliant green lactose peptone bile or formate ricinoleate broth) were used in the presumptive test.

E. DETERMINATION OF COLIFORMS IN SPECIAL PRODUCTS

Colon determinations are frequently made on various granular or powdered ingredients of ice cream such as gelatin and milk powder.

Gelatin

Under aseptic conditions weigh 1 or 11 gm (see figure 23, page 50) of gelatin and place in a 99-ml sterile dilution blank. Shake thoroughly to wash the gelatin granules but do not attempt to dissolve the gelatin. While rotating the dilution blank to keep the gelatin in suspension, remove five 1-ml portions, and place 1 ml into each of five fermentation tubes containing brilliant green lactose peptone bile (2%) or ricinoleate broth and proceed the same as for milk.

Egg Powder, Nuts, and Miscellaneous Products

Proceed the same as for gelatin.

Milk Powder

Under aseptic conditions, reconstitute the milk powder to the original composition of milk (see page 66) and proceed the same as for fluid milk.

REFERENCES

1. Robinton, E. D., and Genung, E. F.: The Effect of Temperature on Coliform Organisms in Milk and Cream, *J. Milk Techn1., 8:* 97-100 (1945).

2. Weber, C. W.: Coliform Organisms in Pasteurized Milk, *Ibid. 8:* 217-222 (1945).

3. Tiedeman, W. D., and Smith, S. E.: Practical Application of Several Coliform Tests to Pasteurized Milk, *Ibid. 8:* 323-330 (1945).

4. Fouts, E. L., and Freeman, T. R.: Coliform Organisms in Dairy Products and Their Control, *Ibid. 8:* 89-94 (1945).

5. Skinner, C. E., Marwin, R., and Honaas O.: A Comparison of Two Standard Media for the Detection and Enumeration of Escherichia and Aerobacter in Milk, *Ibid. 7:* 200-205 (1944).

6. Fournelle, H. J., and Macy, H.: A Study of the Coliform Group in Ice Cream, *J. Dairy Sci. 25:* 475-487 (1942).

7. McCrady, M. H.: The Numerical Interpretation of Fermentation Tube Results, *J. Infect. Dis., 17;* 183 (1915).

8. McCrady, M. H.: Tables for Rapid Interpretation of Fermentation Tube Results, *Canad. Pub. Health J., 9;* 201 (1918).

9. Hoskins, J. K.: The Most Probable Number of *B. coli* in Water Analysis, *J. Am. W. W. Assn., 25;* 867 (1933).

10. Hoskins, J. K.: Most Probable Numbers for Evaluation of Coli Aerogenes Tests by Fermentation Tube Method, *Pub. Health Rep., 49;* 393 (1934).

11. Sherman, J. M., and Wing, Helen U.: The Significance of Colon Bacteria in Milk with Special Reference to Standards, *J. Dairy Sci.,* 16:165 (1933).

12. Yale, M. W., and Eglinton, Richard: The Value of the Colon Test as a Means of Detecting Unsanitary Conditions on the Farm, *Proc. Internat. Assoc. Dairy & Milk Inspectors, for 1935,* 1936 pp. 116-125.

13. Hoskins, J. K.: The Most Probable Number of B. coli in Water Analysis, *J. Amer. Water Works Assn., 25:* 867 (1933).

Pathogenic Streptococci*

A. DISCUSSION OF HEMOLYSIS AND PATHOGENICITY

The term streptococcus is used as a descriptive term of morphology to indicate that the organism in question is essentially spherical in shape and grows characteristically in chains, roughly similar to a string of beads. The term is also used generically to include in the systematic classification of bacteria substantially all of those organisms which commonly grow as chains of cocci or spheres. Some of the species belonging to the genus Streptococcus are pathogenic for man, some pathogenic for certain animals, some pathogenic for both, and others are definitely nonpathogenic. The streptococci may be roughly classified according to their ability to produce hemolysis of blood when colonies develop on a blood agar plate. Considerable confusion has resulted from early attempts to use hemolysis as a criterion of pathogenicity. More recent work has shown that other criteria must be employed because both pathogenic and nonpathogenic strains may be found in the hemolytic and nonhemolytic types of streptococci.

*For the detection of mastitis in milk, see Chapter 8, page 137.

In dairy work primary interest is centered around two particular species of streptococci. One of these is the hemolytic type *Streptococcus epidemicus,* commonly responsible for septic sore throats in human beings, and the other is a nonhemolytic type, *Streptococcus agalactiae,* which is responsible for a substantial percentage of the chronic mastitis of dairy cattle. Milk handlers who are victims of septic sore throat or carriers of the organism may contaminate the milk, more specifically, milkers may contaminate the bovine udder, set up a mammary infection, and thereby render dangerous the milk subsequently produced by the infected animal. Pasteurization destroys the organism responsible for septic sore throat.

Although the incidence of mastitis due to the nonpathogenic types of streptococci (principally *S. agalactiae*) is rather high, the occurrence of mastitis due to the pathogenic types of streptococci is fortunately relatively rare. The importance of these few cases is magnified however by the serious explosive outbreaks which they do cause, particularly in areas where raw milk predominates.

In case of a milk-borne epidemic of septic sore throat, it is important to proceed rapidly to identify the herd, the individual cow or cows, and the person or persons responsible for initiating the infection. After identification of the suspected focus of the infection, it is important to establish the identity of the cultures from the suspected source and those isolated from the victims of the epidemic. It is not sufficient merely to locate one or more cows in a herd having mastitis caused by a hemolytic streptococcus. Under the pressure of quick localization of the cause of an epidemic, the milk from herds containing such animals should of course be excluded, but until the evidence is more complete, the focus of the infection should not be regarded as identified.

Lancefield[1] in a comprehensive study has subdivided the large group of hemolytic streptococci into nine serological groups designated A, B, C, D, E, F, G, H, and K. According to Brown,[2] it may be stated very generally, that most of the human pathogens belong to Group A, that many of the members of groups B and C are of importance in animal diseases, that human beings may harbor hemolytic streptococci of any of the serologic groups with or without symptoms of the disease, and that cow's milk frequently contains hemolytic streptococci of groups B, C, D, and E, of which those of group D are resistant of pasteurization.

Lancefield's group A certainly includes the hemolytic streptococci responsible for scarlet fever and septic sore throat. Although sporadic cases of human infection may occasionally be caused by hemolytic streptococci belonging to groups other than group A, these are regarded as opportunist pathogens of little or no epidemiologic significance. Groups A, B, and C are of most interest because of their association with human and animal diseases, and each of these groups contains numerous species. It should be clear from this discussion that the mere identification of a hemolytic streptococcus should not be regarded as final and conclusive evidence that the focus initiating an epidemic has been found.

The reader is referred to the excellent monograph by Frost and Englebrecht,[3] in which the streptococci of interest in the handling of milk are classified and all the tests, media, and procedures which are useful in case of an epidemic traceable to the milk supply are outlined. Although the following material is quoted with special permission from their monograph, it is not possible to present all the details essential to a complete understanding of the principles involved in the various tests.

B. COLLECTION OF SAMPLES

If feasible, samples should be taken from individual quarters of the udder. If this is not feasible because of the number of animals to be examined, pooled samples aseptically drawn from as many as ten animals may be collected. The milk from individual animals in the group should be taken subsequently if the preliminary results from a pooled sample warrant it. Sterile wide-mouth bottles of about two ounce capacity and fitted with a suitable stopper are used in collecting the samples. The udder, teats, and flanks of the animal are cleaned with a wet cloth, and the first stream of milk from each teat is discarded. If pooled samples are taken, one stream from each quarter of each of ten cows is taken for the sample. If individual quarter samples are taken, a larger portion should be procured from each quarter. In all cases care should be exercised to preclude contamination from outside sources.

Samples should be packed in ice for transportation to the laboratory. If samples cannot be examined immediately they may be preserved by the addition of approximately 30 per cent of glycerol or by adding 0.1 ml of a 1 per cent solution of brilliant green per 10 ml sample.

C. DETECTION

(Procedure of Frost and Englebrecht[3])

(For directions in preparing the various media and making certain tests, see pages 87 to 92).

1st day. Pour blood agar plates (dilution of the milk sample 1:20).

2nd day. (a) Fish representative or suspicious colonies into serum broth.

(b) Examine suspicious colonies for encapsulated streptococci (see page 91).

3rd day. (a) Repeat examination of the original blood agar plates for new or other suspicious colonies.

(b) Make Gram stain (see page 525) from the serum broth culture (2a). If desirable to make serological classification according to the Lancefield grouping, a drop of the broth culture should be tested against the appropriate antiserum.

(c) Remove 0.5 ml of the broth culture for the hemolysis test (see page 92). If this test is negative, discard the culture. If the test is strongly or only moderately positive (++), inoculate from the broth culture to

(d) Blood agar slants (see page 510), and to

(e) Various test substances, including dextrose broth and other fermentable carbohydrate media such as trehalose and sorbitol for fermentation (see page 512); also include

(f) Sodium hippurate medium (see page 513).

4th or 5th day. Examine agar slants for capsules (see page 91).

8th day. (a) Determine pH of dextrose broth.

(b) Determine fermentation of other test substances such as the fermentable carbohydrate media.

(c) Determine the hydrolysis of sodium hippurate.

(For the non-hemolytic streptococci in milk, the same procedure is used except that the tests for hemolytic titer and capsules are omitted.)

D. IDENTIFICATION REACTIONS

SUMMARY OF BIOCHEMICAL CHARACTERISTICS OF HEMOLYTIC STREPTOCOCCI OF THE VARIOUS (LANCEFIELD) SEROLOGIC GROUPS

Precipitation Group	pH in 1 per cent dextrose broth	Hydrolysis of sodium hippurate	Fermentation of Trehalose	Fermentation of Sorbitol
A	4.6 - 5.4	−	+	−
B	4.2 - 4.6	+	+	−
C	4.6 - 5.0	−	−	+
D	4.2 - 4.6	+ or −	+	+
E	4.6 - 5.8	−	+	−
F	4.6 - 5.0	−	+	+
G	4.6 - 5.0	−	+	−
H	4.4 - 4.8	−	+	−
K	4.6 - 4.8	−	−	−

Antisera are available on the market for serological groups A, B, and C.

The identification of species of the streptococci depends upon their action in various media, the presence or absence of capsules, the hemolysis of blood agar, hydrolysis of sodium hippurate, optimum growth temperature, and various other characteristics.

The following is the classification of the hemolytic streptococci as given by Frost and Englebrecht.[8]

Detailed Descriptions of Hemolytic Streptococci (Beta Type*)

A Key to the Hemolytic Species of Genus Streptococcus

I. The hemolytic group—produce beta type of hemolysis on blood agar plates

a. Acid in lactose
- b. Acid in mannite
  - c. Acid in salicin...infrequens
  - cc. No acid in salicin..hemolyticus i
- bb. No acid in mannite
  - c. Acid in salicin
    - d. Hydrolyzes sodium hippurate
      - e. High acid in dextrose (pH 4.7-4.4).....mastitis
    - dd. No hydrolysis in sodium hippurate
      - ee. Low acid in dextrose (pH 5.6-4.8)
        - f. Forms capsules and giant colonies
          - g. Acid in sorbitol but not in trehalose zooepidemicus
          - gg. No acid in sorbitol but does form acid in trehalose............epidemicus
        - ff. Does not form capsules..................pyogenes erysipelatis scarlatinae
  - cc. No acid in salicin
    - e. Hydrolyzes sodium hippurate..............asalignus
    - ee. No hydrolysis of sodium hippurate..anginosus

aa. No acid in lactose
- b. Acid in mannite
  - c. Acid in salicin...alactosus
  - cc. No acid in salicin.......................................hemolyticus iii
- bb. No acid in mannite
  - c. Acid in salicin
    - d. Forms capsules ....equi
    - dd. Does not form capsules..........................equisimilis
  - cc. No acid in salicin..subacidus

a. Hemolytic Streptococci

Of particular interest in the hemolytic group are *Streptococcus epidemicus* and *Streptococcus pyogenes.* For complete details of the characteristics of these and other species, the reader is referred to Frost and Englebrecht,[8] but the following principal characteristics may be helpful.

Streptococcus epidemicus

Morphology.—This species is indistinguishable under the microscope from other streptococci; ordinarily not in long chains, fre-

*See definitions of various types of hemolysis on page 89.

quently in twos and fours; capsules may frequently be demonstrated by the wet India ink method. On blood agar the organism is strongly hemolytic (beta type), producing characteristic colonies in 16-18 hours at 35 to 37 C (95 to 98.6 F), and reaching their maximum size and distinctive appearance in 48 hours. Colonies are about 1 mm in diameter, are dense, and are surrounded by a clear zone of hemolysis varying from 4.3 to 4.8 mm in diameter.

Fermentation reactions.—Forms acid in lactose, salicin, trehalose, dextrose, and saccharose. Does not form acid in mannite, raffinose, inulin, or sorbite.

Dextrose broth.—Final pH about 5.3.
Litmus milk turned slightly acid.
Sodium hippurate.—Not hydrolyzed.
Fibrin of clotted human blood.—Dissolved promptly.
Thermal death point.—125.6 F (52 C) in 10 minutes exposure.
Not soluble in bile.

Pathogenicity.—This organism is the recognized cause of septic sore throat, a disease that is usually milk borne, although there is some indication that the disease may be spread by contact. The organism has frequently been found in the throats of human carriers. *S. epidemicus* may easily be confused with *S. zooepidemicus*, which is not so clearly associated with septic sore throat but in certain instances may definitely be a cause of mastitis in cattle.

Streptococcus pyogenes

Morphology.—Length of chains is variable, formation of capsules is doubtful but commonly absent, blood agar, strongly hemolytic (beta type), colonies after 48 hours at 35 to 37 C (95 to 98.6 F) are 0.3 to 0.8 mm in diameter with a hemolyzed zone averaging about 3 mm in diameter.

Fermentation reactions.—Forms acid in lactose, salicin, trehalose, dextrose, galactose, levulose, mannose, maltose, and saccharose. It does not form acid in mannite, raffinose, inulin, sorbite, arabinose, or xylose.

Dextrose broth, final pH 4.8 to 5.4.
Litmus milk, usually marked acidity with no coagulation.
Sodium hippurate is not hydrolyzed.
Fibrin of clotted human blood quickly dissolved by many strains.

Pathogenicity.—*S. pyogenes* is the common cause of pyogenic diseases, wound infections, and abscesses, commonly associated with erysipelas, middle ear infection, and mastoid and sinus infections. So far as is known the only lower animal infected is the cow. The organism was found twenty-nine times in milk by Frost and Englebrecht,[3] and in one case a cow was found to be carrying this organism in considerable numbers.

Streptococcus mastitidis

Another hemolytic streptococcus of the beta type frequently encountered in milk is *S. mastitidis.* It is apparently a strong competitor of the nonhemolytic *S. agalactiae* as a common cause for chronic mastitis in dairy cattle. This type resembles *S. pyogenes,* a well known human form, except that *S. mastitidis* is capable of hydrolizing sodium hippurate and produces more acid in dextrose broth. Frost and Englebrecht[3] do not believe the presence of this beta hemolytic type, *S. mastitidis,* in milk is of any public health significance.

Various types of hemolysis

Beta—Streptococci are said to produce typical beta hemolysis when the hemolysis of the red blood cells around the colony is so complete that they cannot be detected under the microscope. Other streptococci produce various intergrades of hemolysis designated by such terms as:

Alpha—Partial hemolysis of corpuscles with a greenish discoloration surrounding the colony, forming an indefinitely bounded zone 1 to 2 mm in diameter, and surrounded by a second narrow, clearer, not discolored, partly hemolyzed zone.

Alpha prime—Intermediate between the alpha and beta type so far as surface colonies are concerned. It has hemolyzed zones and no apparent discoloration. The subsurface colonies show a hazy hemolyzed zone, which, on microscopic examination, show a few blood corpuscles, especially near the colony.

Delta—Characterized by methemoglobinization of the red cells without any hemolysis. Such hemoglobin as has leaked into the medium is rendered colorless. Deep colonies (48 hrs.) appear olive green in color and are surrounded by a pale green zone slightly narrower than the diameter of the colony.

Gamma—No hemolysis of blood or discoloration of the agar.

b. Non-hemolytic Streptococci

The following classification of the non-beta hemolytic types of streptococci is taken from Frost and Englebrecht's monograph.[3]

Detailed Descriptions of Non-hemolytic Streptococci
(Alpha, Gamma, and Other Types)

A Key to Non-hemolytic Species of the Genus Streptococcus

II. The non-beta hemolytic group; i.e., those that produce alpha, alpha prime, etc., type of colony or fail to produce any change in blood agar plates; i.e., form gamma colonies

a. Acid in lactose
  b. Acid in mannite
    c. Acid in salicin
      d. Acid in raffinose

e. Acid in inulin..inulinaceus, (Orla-Jensen)
ee. No acid in inulin ....fecalis, (Andrewes & Horder)
dd. No acid in raffinose
e. Acid in inulin..distendens, (Hammer)
ee. No acid in inulin
f. Acid in saccharose ..liquefaciens, (Orla-Jensen)
ff. No acid in saccharose ..lactis (Lister), (Lohnis)
cc. No acid in salicin....non-hemolyticus i, (Holman)
bb. No acid in mannite
c. Acid in salicin
d. Acid in raffinose
e. Acid in inulin
f. Alpha colony ......bovis, (Orla-Jensen)
ff. Gamma colony ......cardioarthritidis, (Small)
ee. No acid in inulin
f. Alpha colony ......bovinus, (Frost & Engelbrecht)
ff. Gamma colony ......anhemolyticus, (Zangemeister)
dd. No acid in raffinose
e. Acid in inulin..pneumosimilis, (Frost & Engelbrecht
ee. No acid in inulin
f. Hydrolyzes sodium hippurate ..Agalactiae (Guillebeau), Migula
ff. No hydrolysis of sodium hippurate
g. From human source ....mitis, (Andrewes & Horder)
gg. From dairy products ..cremoris, (Orla-Jensen)
cc. No acid in salicin
d. Common in human mouth ....salivarius, (Andrewes & Horder)
dd. Produces a soluble toxin specific to measles....morbilli, (Ferry & Fisher)
ddd. Grows at high temperature..thermophilus, (Orla-Jensen)
aa. No acid in lactose
b. Acid in mannite
c. Acid in salicin..........non-hemolyticus ii, (Holman)
cc. No acid in salicin....non-hemolyticus iii, (Holman)
bb. No acid in mannite

c. Acid in salicin
  dd. No acid in raffinosé
    e. Acid in inulin..equarius, (Frost & Engelbrecht)
    ee. No acid in
      inulin ....equinus, (Andrewes & Horder)
cc. No acid in salicin
  d. Acid in raffinose
    e. Acid in inulin..ignavus, (Holman)

Streptococcus agalactiae

Of particular interest in the dairy laboratory is the nonhemolytic *Streptococcus agalactiae,* which is perhaps the most common cause of chronic mastitis in dairy cattle. The following characteristics are of value in its identification.

Morphology.—Has no distinguishable morphological characteristics which differentiates it from other streptococci.

Blood agar—Alpha hemolysis, showing a greenish discoloration after 18-24 hours incubation.

Fermentation reactions.—Acid is formed in lactose, salicin, trehalose, dextrose, galactose, levulose, mannose, maltose, dextrin, glycerin, and saccharose.

Acid is not formed in mannite, raffinose, inulin, sorbite; arabinose, rhamnose, or xylose.

Dextrose broth, final pH 4.4-4.8.

Methylene blue reduction is variable, some strains fail to reduce it.

Sodium hippurate is hydrolyzed.

Pathogenicity.—Not known to be pathogenic for man, but very prevalent in the udders of dairy cattle as the most common cause of mastitis.

E. TESTS FOR STUDY OF STREPTOCOCCI

a. Hydrolysis of Sodium Hippurate

Inoculate a tube of sodium hippurate medium (see page 513) with and incubate 5 days at 35 to 37 C (95 to 98.6 F). The hydrolysis is detected by adding 0.5 ml of a 12 per cent ferric chloride solution in 2 per cent HCl to 2 ml of the culture (which shows definite growth) in a clean test tube. Shake well and allow to stand at room temperature for 30 minutes. The appearance of a heavy brown precipitate which does not clear on standing indicates that hydrolysis has occurred. Positive reactions are obtained with typical bovine strains of streptococci.

b. Capsule Test

Prepare a streak culture on a fresh, blood agar slope (see page 510 from a 24-hour broth culture of the organism. Incubate this over night in a moist chamber, which is made by using a wide-mouth bottle with a moist piece of cotton in the bottom, and is closed with a glass or rubber stopper. Sometimes capsules can be demonstrated in the broth cultures or direct from the blood agar plate. In urgent cases, this is tried to save time, but it is of diagnostic significance only when the results are positive. The addition of ascites fluid or of more serum to the blood agar favors the production of capsules.

The India ink method, as recommended by Brown,[2] is used for capsule detection. An India ink made in Germany, which is designated as "Pelikan-Tusche Nr. 541," is used. This ink is prepared for bacteriological work and is free from bacteria, whereas the ordinary India ink often contains capsulated organisms.

The culture should not be over 16 to 18 hours old when used, and preferably from 6 to 8 hours. Place a loopful (3 mm) of sterile broth on a clean glass slide, and in this suspend a small portion of the streptococci removed from the blood agar slant with a sterile loop. The portion should come from near the serum at the base of the slant, as the capsules, if present, are more pronounced in the growth at this point. Add a similar loopful of the India ink to the preparation on the slide, before it becomes dry, and place a clean cover slip over the preparation. The weight of the cover slip spreads the drop evenly on the slide, and the drop should be just large enough to form a thin moist film under the cover slip. Study the preparation immediately under the oil immersion lens. If capsules are present, definite light areas should show around the organisms.

e. Hemolytic Test

Make up a 5 per cent suspension of rabbit blood corpuscles as follows: To 0.5 ml of the corpuscles add 9.5 ml of sterile physiological sodium chloride solution (8.5 gm per liter of distilled water). To determine the hemolytic titer, mix 0.5 ml of a 5 per cent suspension of the washed rabbit blood corpuscles with 0.5 ml of a 24-hour serum broth culture in a three-inch test tube (75 x 10 mm) and allow to stand at 35 to 37 C (95 to 98.6 F) for two hours. At the end of this time record the results. A negative test is recorded with a zero sign and a positive test with a plus sign. Four plus signs are used for complete hemolysis, and one, two, or three plus signs for partial hemolysis, the number depending on the degree. Discard all those cultures which show negative or one plus hemolysis. Otherwise, subsequent tests are made. If the so-called pseudohemolytic streptococci are wanted for further study, the cultures showing negative or one plus hemolysis are kept.

NOTE: For description of the detection of mastitis in milk, see Chapter 8, page 137.

Protein Digesting Bacteria

In the analysis of various dairy products, particularly cheese and butter, it is frequently desirable to determine not only the number but the percentage of protein digesting types of bacteria present.

Apparatus

Usual plating materials (see page 42).

Reagents

1. Casein agar (see page 522).
2. One per cent HCL solution.
3. Litmus milk (see page 523).

Procedure

Appropriate dilutions are prepared in the manner described for the specific product to be analyzed. In pouring the plates, use casein agar as described on page 522. Incubate the plates at 21 C (69.8 F), approximately room temperature, for five days.

After making a count of the total number of all colonies on the plate, flood the plate with a dilute acid solution (1 per cent hydrochloric acid) until the casein suspended in the medium becomes cloudy and opaque. Pour off the excess acid and count the number of colonies that are surrounded by a clear halo or zone in which the casein has been digested.

Another method frequently employed to determine the percentage of protein-digesting bacteria is as follows:

1. Make plates from appropriate dilutions of the sample, using any suitable medium such as tryptone-glucose-extract-milk agar or casein agar.

2. After suitable incubation, select an area on the plate containing 15-25 well separated colonies which are representative of the various kinds of colonies present. Inoculate tubes of litmus milk from each of the colonies in this selected area.

3. After incubating the inoculated tubes of litmus milk for five days at room temperature, determine the percentage of the total tubes which show evidence of proteolysis. Such tubes are usually alkaline in reaction as shown by the indicator, and the milk is changed to a semiclear watery liquid in which the casein has been converted to soluble compounds.

NOTE: The percentage of acid producers may also be determined simultaneously from the number of litmus milk tubes showing a typical acid fermentation.

REFERENCES AND SUGGESTED READING

1. Lancefield, Rebecca C.: A Serological Differentiation of Human and Other Groups of Hemolytic Streptococci, *J. Exp. Med. 57:* 571-595 (1933).

2. Brown, J. Howard: A Simplified Method of Grouping Hemolytic Streptococci by the Precipitin Reaction, *J. Amer. Med. Assoc. 111:* 310-311 (1938).

3. Frost, W. D., and Englebrecht, Mildred A.: *The Streptococci: Their Descriptions, Classification, and Distribution, with Special Reference to Those in Milk.* 172 pages. Published by Willdof Book Co., Madison, Wis., 1940.

4. Morgan, M. E.; Anderson, E. O., and Plastridge, W. N.: Mastitis and the Plate Count of Milk: I. A Quantitative Study of the Growth of *Streptococcus Agalactiae* in Various Plating Media, *J. Milk Technol. 4:* 245-252 (1941).

5. Morgan, M. E.; Anderson, E. O., and Plastridge, W. N.: Mastitis and the Plate Count of Milk. II. The Influence of *S. Agalactiae* Mastitis Upon the Standard Plate Count of Milk, *Ibid. 5:* 67-76 (1942).

6. Bryan, C. S.; Darby, C. W.; Mallman, W. L., and Corbett, A. C.: The Role of Disinfectants in the Control of Mastitis, *Ibid. 5:* 77-85 (1942).

7. Mattick, A. T. R.; Shattock, P. M. F., and Jacob, M. Moreira: The Relationship of Methods of Bacteriological Examination to the Eradication and Control of Mastitis *(Streptococcus Agalactiae):* I. Use of Enrichment Technique in Revealing Streptococcal Infections of the Cow's Udder; II. *Streptococcus Agalactiae* Infections in Heifers, *J. Dairy Research 12:* 139-154 (1941).

8. Foley, Edward J.: Concise Methods for the Detection of Streptococci of Lancefield's Group B or C in Milk Samples, *J. Milk Technol. 5:* 94-98 (1942).

9. Frayer, J. M.: Mastitis—Laboratory Tests and Their Interpretation *Ibid. 7:* 89-97 (1944).

10. Whiteside, W. H.: Observations on a New Test for the Presence of Mastitis in Milk, *Canad. Pub. Health J. 30:* 44 (1939).

11. Murphy, J. M. and Hanson, J. A.: A Modified Whiteside Test for the Detection of Chronic Bovine Mastitis, *Cornell Vet.: 31:* 47-55 (1941).

Ropy Milk Organisms

A. TESTING PROCEDURE

Sampling. The sample of milk, carefully collected in a sterile container, should have an air space of about one-half inch between the surface of the milk and the cover of the container. Although a one-ounce vial may be used, experience has shown that a larger sample, such as a half pint, is more likely to reveal the presence of organisms producing the typical ropy type of fermentation.

Incubation. Incubate the milk at 60-65 F (15.6-18.3 C) and examine at intervals of 12, 24, and, if necessary, 48 hours. If there is difficulty in maintaining this temperature, it is suggested that room temperature of about 70 F (21.1 C) be employed. The latter temperature permits the free growth of the acid-forming organisms which have a tendency to inhibit the development of the ropy milk types. However, if there is an appreciable number of ropy milk organisms, it is not likely that the acid formers will mask the presence of the ropy milk fermentation.

Testing. After the incubation period, the ropy condition of the milk may be detected with a platinum needle or toothpick. Dip the point of the needle or toothpick into the surface of the cream in spots near the side of the container, and gently raise the point to see if there is evidence of ropiness as indicated by the stretching of the cream into threads at least one-half inch long. The natural viscosity of the cream will cause it to adhere as the toothpick or needle is raised from the cream, but the column between the surface and the toothpick will break before it can be drawn out into the thin threads which are so characteristic of ropiness. Not only should the cream be tested around the circumference of the surface, but also at several spots in the center, because the ropiness may be localized in areas where colonization of the organisms occurs.

Care should be exercised at the 12-hour examination to use a sterile platinum needle or a fresh toothpick for each individual sample in order to avoid possible contamination of subsequent samples with ropy types of organisms, which would lead to erroneous conclusions at 24- and 48-hour observations. The degree of ropiness in the individual patron's samples should be recorded by the marks +, ++, +++, and ++++.

B. DETECTING SOURCES OF ROPY MILK CONTAMINATION

If water samples are to be tested, 1 or 2 ml of the water may be added to sterile whole or skim milk. Milk which has been boiled for ten minutes may be used. Incubate and test as has been described.

Water from the cooling vat, from mud holes in the barnyard or pasture, or condensation water from equipment may be tested in this way.

If there is no condensation water present, sterile water may be poured over the equipment and this inoculated into sterile milk.

Samples of milk may be taken also at different stages of processing, the same as for line-run test.

C. EXAMINATION OF PRODUCERS' MILK FOR ROPINESS

To determine the extent of a ropy-milk outbreak among the farms delivering to a particular plant, half-pint samples of the first milk dumped from a can should be collected in sterile containers. These should be incubated and examined as previously described.

Care should be exercised to prevent carrying contamination from one sample to another. It is best to take samples in the final container; or, if a dipper is used, it should be well scalded between uses.

Under the conditions of an extensive outbreak, ropiness may be found in a considerable percentage of farmers' milk. Elimination of the trouble over so large a field is difficult. Ordinarily, thorough pasteurization of the milk and sterilization of the equipment is dependable as a means of avoiding trouble from ropiness. Care must be taken to wash all pipe lines, pumps, filler parts, etc., used for handling pasteurized milk in a separate vat from that used for washing equipment handling raw milk. In this case great care must be taken to avoid recontaminating the milk after pasteurization. However, occasionally the ropy milk organisms become so numerous that even with the utmost precaution the finished product becomes more or less ropy. When this condition obtains, it may be necessary to eliminate some of the most heavily contaminated sources of supply until the trouble is brought under control.

NOTE: The Burri method also has been found very useful in detecting sources of ropiness in milk. For further reference see page 49.

D. CORRECTIVE PROCEDURE

A general improvement of sanitary conditions is the most important factor in establishing control of this trouble and therefore the following corrective procedures on the farm may be helpful:

1. The barn should be thoroughly cleaned by removing all old bedding, manure, dirt, and, where possible, dust from beams and other structures. Lime should be freely sprinkled on the floor, and either a fresh coat of whitewash should be applied to the interior or all surfaces should be treated with a strong germicidal solution.

2. The cows should be curried and cleaned to remove the possibility of particles of dirt dropping into the milk pail.

3. After cleaning the cows' flanks, the udders should be wiped with a chlorine solution (30 p.p.m.).

4. At milking time the air should be free from dust, and the feeding of dry feeds at this time should be avoided.

5. The milk pails, strainers, and cans should be thoroughly washed inside and out, then sterilized with steam, boiling water, or a chlorine rinse of at least 100 p.p.m. not more than one hour before filling with milk. The cans and utensils should be immediately drained and the chlorine solution discarded. This preliminary rinsing of cans on the farm is recommended as an extra precaution in addition to the cleaning and sterilizing of the cans at the plant. Sterilization by the dipping of equipment in a vat should not be employed.

6. In order to determine whether individual cows are responsible for the contamination, a sample of the fore-milk should be collected directly from the udder of each cow in a small sterile container suitable for incubation. This is not to imply that ropy milk organisms are normal inhabitants of the interior of the udder but that they may be introduced into the milk from exterior surfaces of the udder. Great care should be exercised so that no contamination is introduced in taking the sample. The samples should be incubated and tested as already described.

In case only a few cows seem responsible for the ropy milk, it is good practice to place them side by side in the barn, milk them last, use extra precautions with them, and keep their milk separate for disposal.

7. If milk is cooled in a water tank, the vat should be drained, thoroughly cleaned, and then chlorinated with a solution containing at least 250 p.p.m. In times of drought the cooling vat, owing to neglect in changing the water, is very likely to be a source of contamination. The water in the vat should be tested as previously described, and if found to contain ropy milk types of organisms, the vat should be emptied daily and chlorinated with each refilling (10 p.p.m.). The chlorine should be added at least 30 minutes before use in order to sterilize the water. If there is a continuous stream containing these organisms running into the vat, another source of water should be sought, or if the temperature is low enough, the vat may be filled and chlorinated as indicated and the water then by-passed to the drain.

8. The barnyard and pastures should be carefully inspected and put in good sanitary condition. All mud holes and wet spots should be filled in and, if necessary, a grade established for drainage. If any wet area cannot be dried owing to springs, it should be fenced so that the cattle will not have access to it. In the case of a stream, its banks should not retard the flow of the water,thereby creating a stagnant condition. If this cannot be corrected, the objectionable areas should be fenced.

If cows are placed in a clean barn, and the precautions taken regarding feeding, watering, sterilizing utensils, cooling the milk, and providing a satisfactory barnyard and pasture, the trouble should decrease and disappear in a short time. If these precautions do not correct the condition, the offending cows should be identified and completely isolated. In very persistent cases immediate results may not be expected, but two weeks should show a definite improvement.

Sporogenes Test

Apparatus

1. Sterile test tubes.
2. Pipettes, 10 ml.
3. Water bath at 80 C (176 F).

Reagent

Sterilized paraffin.

Procedure

1. Place 5 ml, 10 ml, and 15 ml respectively of milk into each of three sterile test tubes.

2. Heat the three tubes of milk in the water bath at 80 C (176 F) for 10 minutes to destroy all vegetative cells and drive off the incorporated air.

3. Just before the completion of the heating process, carefully superimpose a layer of melted sterile paraffin directly on the surface of each tube of milk. The depth of the paraffin layer should be equivalent to about one and one-half times the diameter of the test tube.

4. Solidify the paraffin quickly by placing the tubes in iced water.

5. Incubate the tubes 48 hours at 37 C (98.6 F) and observe for "stormy fermentation," in which the paraffin plug has been forced upward by gaseous fermentation.

Interpretations

If a stormy fermentation is observed in the tube containing 5 ml of milk, excessive manurial pollution is indicated. If the 5 ml tube shows no change, but stormy fermentation is observed in the tube containing 10 ml of milk, it is assumed that the care used in production was fair. If stormy fermentation obtains only in the tube containing 15 ml of milk, it is assumed that manurial pollution of the milk was reduced to a reasonable minimum. North[2] has modified this method for the use of Smith fermentation tubes and eliminated the use of paraffin.

REFERENCES

1. Weinzirl, J. and Veldee, M. V.: A Bacteriological Method for Determing the Manurial Pollution of Milk, *Amer. J. Public Health 5;* 862-866 (1915).
2. North, C. E.: Has Sanitation of Certified Milk any Practical Value? *Certified Milk 9;* Jan. 3-5, (1934), and *9* Feb. 6-8, (1934).

Thermoduric and Thermophilic Bacteria

Thermoduric bacteria are those which are capable of enduring a heat exposure usually lethal to vegetative cells. The term as commonly applied in dairy bacteriology refers to organisms surviving the thermal exposures employed in pasteurization. Thermoduric organisms and thermophilic organisms should not be confused. Thermophilic organisms are those capable not only of withstanding high temperatures but actually growing at high temperatures. Milk containing thermoduric types may show little or no reduction in bacterial count after pasteurization, whereas milk containing thermophilic organisms may actually show a higher bacteria count after pasteurization by the holding method. Thermophilic bacteria have not been found to develop appreciably at the thermal exposures employed in the short-time-high-temperature pasteurization, however they may develop in certain sections of a regenerator after prolonged operation.

High counts in the finished bottle of pasteurized milk are frequently traceable to milk from one or more sources in which the percentage of organisms capable of surviving pasteurization is high. These may be either thermoduric or thermophilic. If the bacteria count of the finished pasteurized milk is as high or almost as high as that of the raw milk, the trouble is likely due to thermoduric types of bacteria. If, on the other hand, the count on the finished product is significantly greater than that of the raw milk, there are likely to be found thermophilic bacteria in one or more of the sources of supply or perhaps in the pasteurizing equipment. Thermoduric organisms are quite common in milk supplies whereas the true thermophilic types occur less frequently, although they are by no means rare.

Inefficient cleaning and sterilizing on some dairy farms create a favorable chain of circumstances which daily tend to kill the weaker organisms and favor the survival of the more heat resistant types. These may be lodged in the roughly soldered joints of equipment or in the cracks of old rubber parts of a milking machine. Unless the temperature of the sterilizing water, the volume of water employed, and the time of exposure are adequate to destroy practically all the vegetative cells present in the equipment, there will inevitably be a selective breeding of the heat-resistant types of bacteria in that equipment.

Occasionally high heat-resistant counts may be traced to the milk from an individual cow in the herd. Certain strains of the micrococci and other types which become chronic inhabitants of the milk cisterns, particularly of the old cows, have been found to be thermoduric but not thermophilic. The common mastitis organisms, however, are incapable of surviving pasteurization.

In any case, a high thermoduric count reflects conditions which deserve attention, identification, and correction.

A. THERMODURIC (LABORATORY PASTEURIZED) COUNT

Since the dealer in pasteurized milk must meet the legal bacterial standards set up for the finished product as delivered to the customer, he should be vitally interested in locating the source or sources of raw milk containing organisms resistant to pasteurizing temperatures. With increasing emphasis on milk with a low bacteria count, more and more attention has been paid to the thermoduric count sometimes called the "laboratory pasteurized count" (L. P. count) or the "heat resistant count" (H. R. count).

Apparatus

1. Usual plating materials (page 42).

2. Water bath maintained at 143.5-144 F (62.0-62.2 C) (see figure 33).

3. Screw cap sample vials.

Procedure

1. Place test tubes with tight stoppers or screw-cap vials, containing approximately uniform amounts of the raw milk samples to be tested, in a suitable tube rack. One extra tube of milk is equipped with a thermometer (pilot tube).

2. Pre-heat samples in a water bath so that the milk will reach uniformly a temperature of 143 F (61.6 C) in not less than 3 nor more than 5 minutes.

3. Transfer the rack of tubes immediately to an insulated, thermostatically controlled water bath operated at 143.5-144 F (61.9-62.2 C) and hold for 33 to 37 minutes depending on local custom or requirement. (See figure 33).

4. Hold the temperature of the milk in the pilot tube at 143 F throughout the holding period. The temperature of the water bath and of the milk may be varied to meet local pasteurizing practices.

5. Transfer the samples immediately to ice water for cooling to 50 F (10 C).

6. Plate, using a dilution of 1 : 10, 1 : 100, or 1 : 1,000, depending on the expected count.

The procedure for making the dilutions and pouring the plates is the same as shown diagrammatically in figure 19 on page 41 and described on pages 42 to 48.

In some laboratories where large numbers of samples are analyzed daily for thermoduric count, a method employing a platinum loop standardized to deliver 0.01 ml or 0.001 ml of the laboratory pasteurized milk is used to inoculate melted agar in the Petri dishes instead of employing the usual dilution technique. (See page 53, also figure 34, page 101 and figure 35, page 102). Such loops may be purchased from scientific supply houses. In this connection it has been found economical to employ small Petri dishes (50 to 75 mm) as a means of saving agar medium, or to use the Meyers and Pence method described on page 51.

Samples of milk which show excessive thermoduric counts may be regarded as coming from a farm which needs field inspection. There is a high correlation between excessively high thermoduric

Figure 33.

Equipment for laboratory pasteurization of raw milk samples. Note: (1) open pre-heating bath on the right, thermostatically controlled at 150 F (65.5 C); (2) water jacketed holding tank thermostatically controlled at 143.5 to 144 F (62.0 to 62.2 C); heavily insulated lid and holes to accommodate two thermometers, one for the water temperature and one for the milk temperature in the pilot tube.

Courtesy of the American Public Health Association

Figure 34.

Calibrated loop for use in measuring milk.

Figure 35.

Showing the direct inoculation of small size Petri dishes of agar, using a standardized (0.01 or 0.001 ml) platinum loop, thereby eliminating the use of dilution bottles.

counts and the findings of an experienced field man on critical inspection of the conditions on the producer's farm. Elimination of the milk from such a farm or the correction of the conditions responsible for the high thermoduric counts frequently yields surprisingly favorable effects on the bacterial count of the finished pasteurized milk.

B. THERMOPHILIC COUNT

The sources of the heat-loving or thermophilic bacteria are not so clearly understood as are the sources of the more common thermoduric types. In identifying the cause of trouble with thermophils, it is advisable not only to test the milk from individuals farms but also to take line-run samples through the various steps of the plant processing. Particular attention should be given to samples from any point in the equipment where high temperatures persist for prolonged periods, such as the foam in the top cooler pan or the foam in a spray vat.

Apparatus

1. Usual plating material (see page 42).
2. Incubator maintained at 55 C (131 F).

Procedure

1. Make appropriate dilutions of the sample as shown diagrammatically in figure 19 on page 41, and pour plates using standard tryptone-glucose-extract-milk agar as described on page 506. The amount of agar introduced into each plate should be increased to about 20 ml to offset the evaporation which will result from the higher temperature of incubation.

2. Incubate the plates 48 hours at 55 C (131 F). Some workers prefer to operate the incubators at 60 C (140 F) for thermophilic counts. A pan of water should be kept in the incubator to retard excessive drying of the agar.

3. In counting the colonies it is frequently necessary to employ higher magnifications than are commonly used for routine work. In reporting the results, the temperature of incubation should be indicated.

C. IDENTIFYING THE SOURCE OF THERMOPHILS

The plant operator is usually more interested in quick and effective identification and elimination of the source or sources of thermophils than in the actual comparative numbers of these organisms present in a given series of samples. The following procedure enables reasonably quick identification of the most prolific sources of thermophils.

1. Samples of milk in test tubes or screw-capped vials are placed in an insulated water bath thermostatically controlled at 143.5-144 F (61.9-62.2 C) and held for not less than four hours. Samples which are not available until late in the day are held in the water bath at 143.5 F (61.4 C) over night. If thermophilic organisms are present in any of the samples, the prolonged holding of the milk at high temperatures provides opportunity for them to develop in great numbers. The actual period of holding (4 to 16 hours) is not important, because exact counts are not to be made. (The same samples used for routine thermoduric counts may be returned to the water bath and the incubation continued for subsequent determination of the presence of thermophilic types of bacteria.)

2. After prolonged incubation of the milk at pasteurizing temperature, the sample is examined for thermophils either by direct microscopic examination of a stained smear or by the plating procedure.

For direct microscopic examination, a loopful of the incubated milk is spread over about 1 sq. cm of a clean glass slide, dried, stained, and examined according to directions given on page 119. No effort is made, however, to enumerate the organisms, but only to determine if they are present in large numbers in the smear. Long rods which stain poorly are frequently seen in large numbers in such smears.

For the plating procedure, thoroughly mix a loopful of the incubated milk in a freshly poured plate of melted agar. The plate may be incubated at temperatures of 55 C (131 F) or higher, or it may be incubated at the temperature ordinarily employed for routine plating. Usually 24 hours' incubation is enough to identify the samples containing the thermophils, although all plates showing no growth after 24 hours should be returned to the incubator for an additional 24 hours.

No effort is made to count the colonies, but only to report the presence or absence of thermophils. Unless the plates are quite heavily seeded the sample should be regarded as negative for thermophils.

Figure 36 shows a typical plate made from milk containing thermophilic types of bacteria. Note the small colonies. Figures 37-47 inclusive show the various microscopic fields observed under different operating conditions.

REFERENCES

1. Breed, R. S.: Thermophils in Pasteurized Milk and Their Relation to Certain Types of Equipment. *Proc. 22nd Ann. Conv. I. A. M. D. 1929, Lab. Sec.* pp. 5-33.

2. Breed, R. S.: Heat-Resistant and Heat-Loving Bacteria in Pasteurized Milk. *Proc. 21st Ann. Conv. I. A. M. D. 1928, Lab. Sec.* pp. 16-30.

3. Eisman, L. P.: Thermoduric and Thermophilic Organisms in Milk. *J. Milk Technol. 5:* 102-105 (1942).

4. Boyce, E. B., Lythgoe, H. C., Ruggles, E. K., and Lane, R.: The Use of Laboratory Pasteurization in Solving Milk Problems. *Ibid. 5:* 146-150 (1942).

5. Hileman, J. L., Moss, C., and Stead, B.: Thermoduric Bacteria in Milk. III. The Effect of Changing Agar and Temperature of Incubation for Plate Counts on the Problem of Thermoduric Bacteria in Milk. *J. Dairy Sci. 24:* 799-806 (1941).

6. Macy, H., and Erekson, J. A.: Seasonal Variations in Thermoduric Organisms. *Proc. 34th Ann. Conv. I. A. M. D. 1941, Lab. Sec.* 127-135.

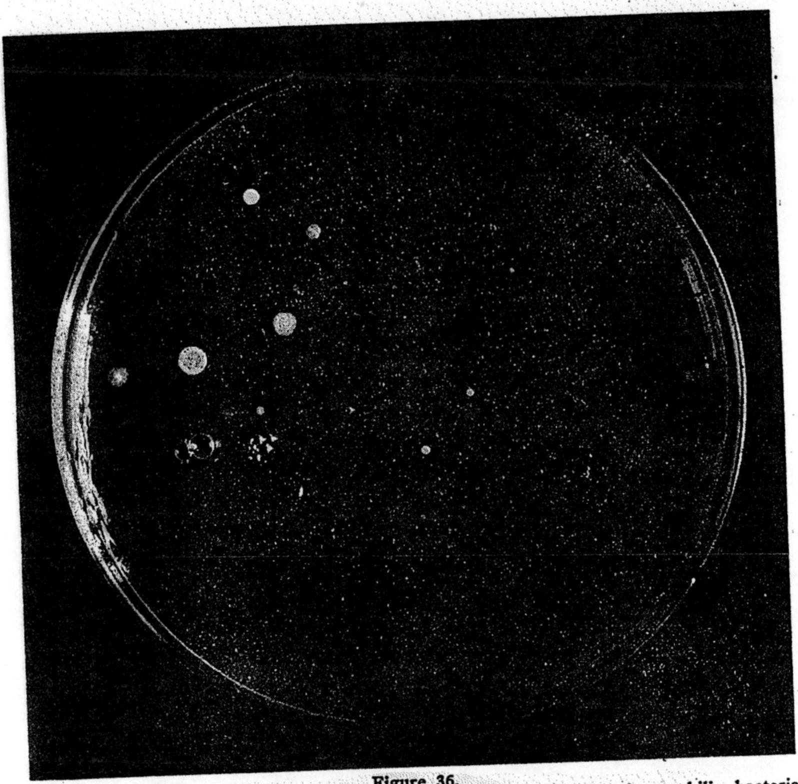

Figure 36.

Agar plate prepared from pasteurized milk known to contain thermophilic bacteria. Numerous pinpoint colonies have developed on the plate, not all of which show in the photograph. Figures 36 to 47, courtesy of the New York State Agricultural Experiment Station.

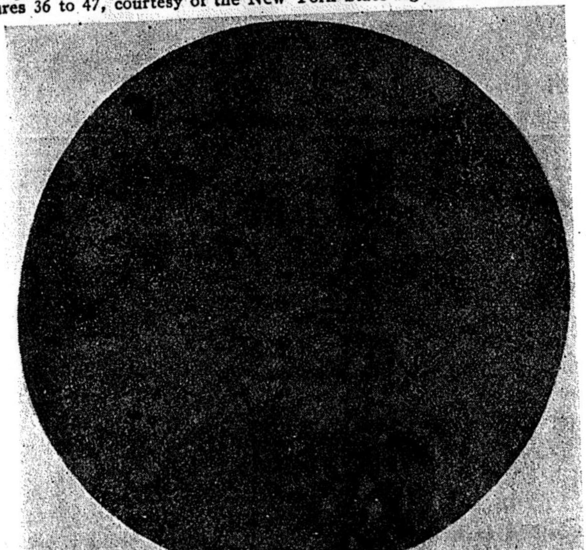

Figure 37.

Microscopic appearance of slide prepared from a pint bottle of pasteurized milk. It shows the largest number of presumably heat-resistant and heat-loving bacteria found in any sample. x600.

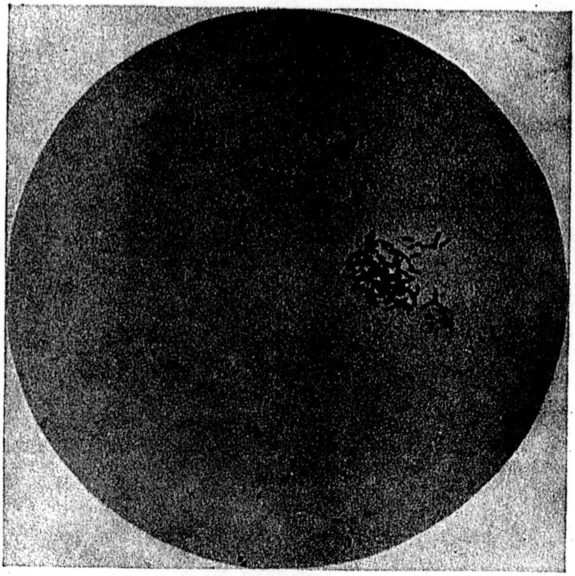

Figure 38.
Microscopic appearance of pasteurized milk taken at end of run where the filter sediment showed as in Figure 41. x600.

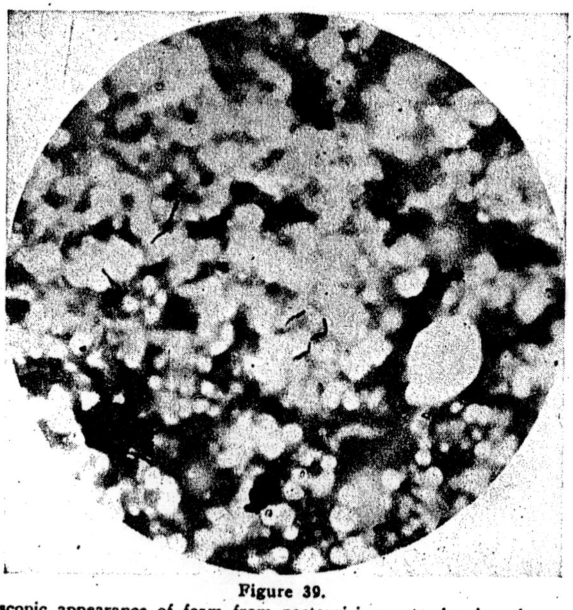

Figure 39.
Microscopic appearance of foam from pasteurizing vat, showing thermophilic bacilli. x600.

Figure 40.

Microscopic appearance of milk stone scraped from pasteurizing vats. x600. Numerous bacilli and spores are evident.

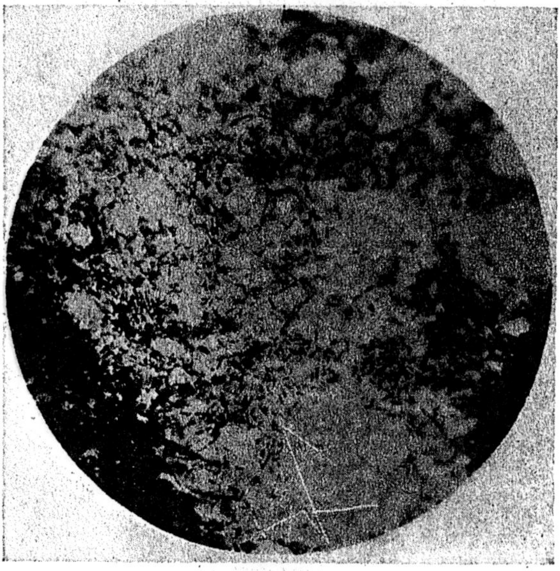

Figure 41.

Microscopic appearance of sediment from filter cloth after use in filtering pasteurized milk at 143° F. for seven hours. x600.

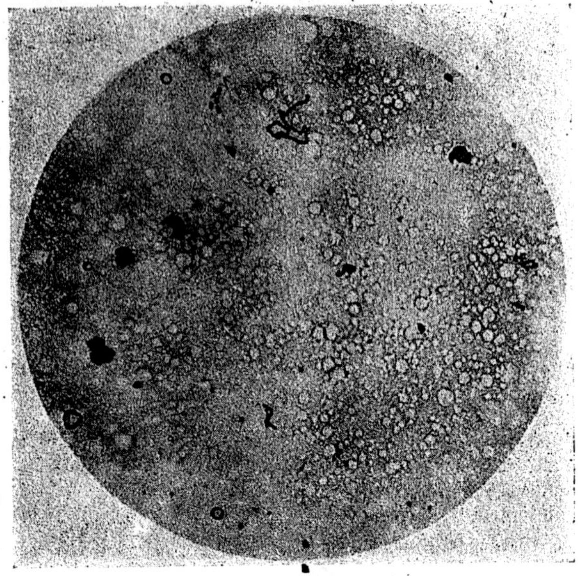

Figure 42.
Microscopic appearance of drainings from filter operated at from 100 to 120° F. x600.

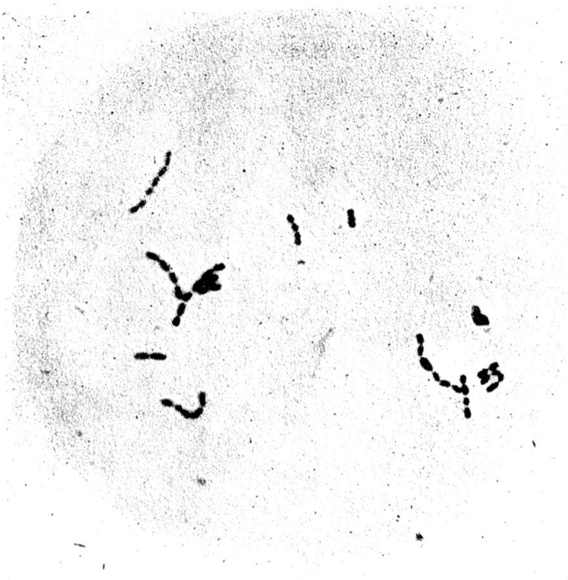

Figure 43.
Microscopic appearance of *Streptococcus thermophilus* in a bouillon culture. x1000.

Figure 44.

Microscopic appearance of bottled pasteurized milk during a brief period when it ontained numerous spore-bearing bacteria that did not show on the routine control plates. :600.

Figure 45.

Microscopic appearance of sample of pasteurized milk taken at end of a six-hour run in a pasteurizing plant. Routine plates made from the milk from this plant showed many pinpoint colonies. x600.

Figure 46.

Microscopic appearance of a sample of pasteurized milk taken at the end of a five-hour run. Two fields of the microscope are combined so as to show the variety of types of bacteria present. Numerous pinpoint colonies were developing from the milk from this plant on routine plates at this time. x600.

Figure 47.

Duplicate of figure 46 except that it shows the condition of the milk at the end of the run of a second unit in the same plant. x600.

CHAPTER 6

DETERMINATION of SANITIZATION of UTENSILS and EQUIPMENT

Bacterial Counts of Containers

A. Glass and Paper Milk Containers

Apparatus

1. Usual plating material (See Page 42).

Note: 1. In certain instances special dilution water may be necessary if the bottles or containers contain residual alkali or chlorine. 2. If the dilution blanks are to be used in rinsing bottles containing chlorine, add 4 ml of a 0.1 N sodium thiosulfate solution to the 1 ml of $K_4H_2PO_4$ before diluting to 800 ml. Dilution blanks prepared from such buffered solutions should be sterilized in the usual manner. 3. It is frequently necessary to buffer the water used in preparation of the dilution blanks in order to offset the alkalies that may not be thoroughly rinsed out of the bottles in soaker-type washers. Water may be buffered by adding 34 gm of dihydrogen potassium phosphate (KH_2PO_4) to 500 ml of water, then add 175 ml of N NaOH and dilute to 1 liter with distilled water. Adjust this solution to pH 7.3. Dilute 1 ml of this stock solution to 800 ml of boiled and cooled distilled water. 4. Note optional broth rinse method, see *Standard Methods for the Examination of Dairy Products*, 9th ed.

Procedure

1. Flame the mouth of a dilution bottle and pour 100 ml of sterile water into each milk bottle to be tested.

2. Stopper the milk bottle carefully with a sterile paper or metal cap and shake vigorously twenty-five times, making sure to wet the entire inner surface of the bottle.

3. By means of a sterile pipette, transfer 1 ml of the rinse water to a culture dish, and pour agar plates in the usual manner.

4. After 48 hours incubation, all recognizable colonies on the plate should be counted with the aid of an approved colony counter.

Calculations and Standards

Assuming that 1 ml of the 100 ml of rinse water was plated, the number of colonies multiplied by 100 is considered the plate count per bottle. Bottles containing not more than one organism per miliiliter of capacity are considered satisfactory. The usual standards are not in excess of 1,000 for a quart, 500 for a pint, and 250 for a half-pint bottle.

Note: Volumes other than 100 ml are frequently used in different laboratories as a matter of local convenience. If low counts are expected in either glass or paper containers it is advisable to use a lesser volume of rinse water (30 ml) and plate 10 ml by distributing approximately equally among three Petri dishes (a slightly larger volume of agar should be poured into each plate to offset the added dilution). The three plates should be counted as a unit and the total number of colonies multiplied by three since one-third of the dilution water was employed.

B. DETECTING CONTAMINATION ON THE POURING LIP OF BOTTLES

Apparatus

1. The usual plating equipment (See Page 42).

Procedure

1. Pour a suitable number of agar plates and place the plates in a refrigerator until the agar has become very firm.

2. Invert a freshly washed but dry milk bottle over the agar so that the lip of the bottle is firmly imprinted on the surface of the agar without breaking the agar gel. Rotate the bottle slightly to dislodge as many of the organisms as possible. Repeat the procedure for as many bottles as desired.

NOTE: Care should be taken to select bottles that are well drained and do not contain free water. After incubation of the plates at 35 C (95 F) for 48 hours, study the number and character of the colony development. Special media may be used for the identification of various organisms of special interest.

NOTE: These tests are frequently made with agar plates containing 10 per cent of sterile blood, and the subsequent colony development observed for hemolysis. Special attention should be given to the identification of colon types of organisms on such plates as an index to contamination of the bottles after sterilization. For this purpose eosin methylene blue agar or Endo's agar plates may be used (see page 516).

C. MILK CANS

Apparatus

1. Usual plating materials (see page 42).

Reagents

100 ml quantities of sterile water.

Procedure

1. Carefully introduce directly from a bottle 100 ml of sterile water into the can to be tested.

2. Replace the can cover and shake vigorously twenty-five times, making sure to wet the entire inner surface of the can and cover. Remove the cover and pour the rinse water back into a sterile con-

tainer, avoiding as far as practicable contamination from the lip of the can.

3. By means of a sterile pipette, transfer 1 ml of the rinse water to a culture dish and pour agar plates in the usual manner. If high counts are expected, it may be necessary to dilute this rinse water before plating (See figure 19, page 41).

4. After 48 hours incubation, all recognizable colonies on the plate should be counted with the aid of an approved colony counter.

Calculations and Standards

Assuming that 1 ml of the 100 ml of rinse water was plated, the number of colonies on the plate multiplied by 100 is considered as the plate count per can and cover. Plate counts of not more than 40,000 per 10-gallon can, 20,000 per 5-gallon can, or 8,000 per 2-gallon can are considered satisfactory (approximately 1 per ml capacity of the container).

In the interpretation of tests on washed containers it is important to recognize that a milk bottle or can which exceeds the arbitrary standard has not been cleaned and sterilized as thoroughly as experience would indicate to be reasonably possible. A quart bottle containing 5,000 bacteria or a 10-gallon can containing 200,000 would theoretically increase the bacteria count of the milk only about 5 per ml, which is too small an increase to be measured. This increase is not the important point; it is important, however, that such a bottle or can has evidently been subjected to an inefficient cleaning and sterilizing procedure, and that there is need for serious attention to the defects revealed.

NOTE: Volumes of sterile rinse water such as 500 ml or 1000 ml are frequently employed in some laboratories as a matter of local convenience. Suitable adjustment of the calculations would be necessary to determine the total count, i.e. multiply the number of colonies on the plate by 500 or 1000.

D. DISINTEGRATION OF PAPER BOARD FOR BACTERIAL ANALYSIS

Apparatus

1. Sterile scissors, sterile forceps.
2. Balances, sensitive to 30 mg.
3. Special 10 ml pipette with large opening (3-4 mm) in tip.
4. High-speed food mixer (Stevens. See figure 48).
5. Cup and agitator assembly of the mixer (sterilized).
6. Supply of sterile water.
7. Sterile Petri dishes, pipettes, dilution blanks and agar for plating.

Procedure

1. With sterile scissors and forceps, cut from well selected areas of the paper container or paper stock approximately 10 gm of sample. Weigh the sample in a counterpoised sterile Petri dish and record the exact weight.

Figure 48.

High speed disintegrator for paper samples which are to be analyzed bacteriologically. Note: The metal cup containing the high speed agitator is demountable from the motor assembly. The cup and the dilution water are sterilized separately in the autoclave, and the sterile water and paper placed in the assembled machine just prior to disintegration.

2. Transfer the sample to the sterilized cup assembly of the mixer. Add 500 ml of sterile water.

3. Connect the cup assembly to the mixer and disintegrate the paper for 5 to 10 minutes at high speed.

4. Using a sterile pipette with a specially large opening at the tip, transfer 10 ml of disintegrated paper pulp, distributing it approximately equally among three sterile Petri dishes. Pour agar plates. Incubate the plates at 32 or 35 C (89.6 to 95 F) for 48 hours and count the total number of colonies on all three plates as a unit.

Calculation and Standard

Multiply the total number of colonies on all three plates by five, since 10 ml of the disintegrated pulp suspension contained ⅕-gm. Thus if the total number of colonies on all three plates is 100, the paper stock should be reported as having a plate count of 500 per gram. Paper stock having a plate count of 500 per gram or less may be regarded as satisfactory. With the improvement in sanitary control methods in paper mills, most of the paper stock used in making paper containers for milk will consistently show plate counts of less than 250 per gram. Sanborn (1) and Ball (2) have discussed the public health aspects of sanitary control of paper containers and have described in detail the various procedures and devices employed to prepare the samples for analysis.

Tests for Sanitary Condition of Equipment

A. Miscellaneous Equipment

Larger pieces of equipment such as vats, coolers, and fillers may be checked for sterility by one of four procedures.

Procedure 1

Take a series of samples of the first milk which comes in contact with the surfaces of the equipment and observe the results for evident increases in bacterial counts. Samples taken of the first milk as it starts over the top of the cooler and the same milk as it leaves the cooler will frequently show pronounced increases in bacterial count if the cooler has been grossly neglected in the matter of cleaning and sterilizing. Similarly, a comparison of the counts of samples of the first milk entering a filler bowl with that of milk in the first bottle will usually reflect any undesirable cleaning and sterilizing methods on this equipment.

For large vats it is helpful to compare bacterial counts of the first milk collected at the inlet of the vat with the count on a sample taken from the vat as soon as the bottom is well covered. The smaller the amount of milk allowed to enter the vat before the second sample is taken, the more satisfactory will be the comparison of the counts. Although by this method the milk does not come in contact with all of the interior surfaces of the vat, such a comparison will usually reveal an unsatisfactory job of cleaning and sterilizing.

Procedure 2

Sterility tests on some pieces of equipment such as sections of pipe lines, pumps, small coolers, ground valves in the bottom of vats or in pipe lines, may be made by pouring a known quantity of sterile water over the surfaces or through the equipment. The plate counts on this water after exposure reveal the effectiveness of the sterilizing methods.

Procedure 3

Sterility tests on a connected series of pieces of equipment may be made by running through a relatively large volume of water which has been rendered sterile by the following method:

a. Add sufficient chlorine disinfectant to a suitable volume of clean water so as to give approximately 5 parts per million of available chlorine. Permit the chlorine to act for about 15 minutes to allow time for destruction of organisms present in the water.

b. Add a calculated amount of sodium thiosulphate to neutralize the chlorine present. (Prepare a 13 per cent solution of $Na_2S_2O_3 \cdot 5H_2O$ and add 1 ml of this solution to each gallon of water containing 5 p.p.m. of chlorine.)

Make bacterial counts of the sterilized water before and after it has been passed through the assembled equipment. Large increases in counts indicate ineffective sterilization.

Procedure 4

Prepare a series of cotton swabs on wood applicators as shown in figure 49. Place several of these in a test tube fitted with a cotton stopper or fruit jar and sterilize in the dry air oven or in the autoclave. Also prepare and sterilize a number of test tubes containing 5 ml of sterile phsysiologically normal saline solution (8.5 gm NaCl per liter). Some technicians prefer to use Ringer's solution, made by dissolving in 200 ml of distilled water, NaCl 1.8 gm; $CaCl_2$ 0.048 gm; KCl 0.084 gm; $NaHCO_3$ 0.06 gm.

a. Remove a swab and insert into a tube of sterile saline or Ringer's solution to moisten the swab.

b. Swab carefully and thoroughly a 2-inch square area (4 square inches) of the metal surface of the equipment to be tested. Occasionally remoisten the swab by dipping into the tube.

In testing drinking glasses, swab five times around the inner surface and five times around the outer surface of the lip of the glass. (See figure 49.)

c. Place the swab into the tube and shake vigorously to dislodge the bacteria removed from the surface of the equipment.

d. Place 2 ml of the solution into the culture dish and make agar plates in the usual manner. Count the colonies which develop after 48 hours incubation.

NOTE: The same swabs may subsequently be used to inoculate special media as a rough qualitative test for the presence of colon types of organisms on equipment. In this case the swab is used to inoculate fermentation tubes containing selective medium for the presumptive coliform test (see page 80).

Interpretation

None of the methods of making sterility tests on larger pieces of equipment lends itself to the establishment of fixed standards. The interpretation of the results is largely a matter of judgment fortified by repeated experience with tests on the individual piece

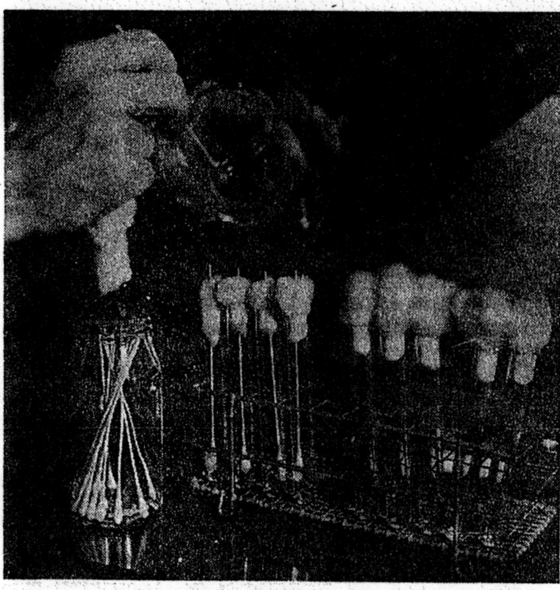

Figure 49.
The swab method of making sterility tests (described in Procedure 4, page 116).

of equipment involved. The important objective of such tests is to detect poor sterilizing methods and any of these methods serves effectively to that end.

B. CHURNS

Apparatus

1. Sterile cotton swabs.
2. Sterile spatulas.
3. Sterile Petri dishes.

Reagents

1. Sterile dilution blanks.
2. Special plain nutrient agar medium containing 2.5 per cent of air-dried agar instead of the usual 1.5 per cent in regular medium.

Procedure

1. About 10 ml of nutrient agar (containing 2.5 per cent air-dried agar) is melted and cooled to 41 to 43 C (105.8 to 109.4 F).

2. A small area on the inside of the churn is moistened with sterile water applied with a sterile cotton swab.

3. The agar is applied directly to the moistened surface and allowed to solidify.

4. The solidified agar is carefully removed with a sterile spatula and placed in a Petri dish with that portion upward which was in contact with the churn.

5. The agar disc is incubated at 21 C (69.8 F) for four days. Plates should be counted in less than four days if it becomes evident that the colonies are so numerous as to be in danger of growing together.

6. The results are expressed in terms of colonies per square centimeter and are determined from a measured area of the disc.

By using an acidified medium comparable to that employed in the determination of yeasts and molds (but containing 2.5 per cent air-dried agar), the method outlined may be used for determining yeast and mold contamination in the churn.

It is possible to apply this method of examination to almost any type of equipment, such as pails, cans, weigh vats, cooler troughs, and ice cream freezers.

NOTE: This procedure was described by B. W. Hammer and H. C. Olson in *Iowa State College Research Bulletin* No. 141, before the introduction of tryptone-glucose-extract-milk agar. The use of the new medium (containing 2.5 per cent of air-dried agar) in this procedure increases the counts obtained per square centimeter, and thereby serves as a better index to the effectiveness of the sterilization of the churn.

REFERENCES

1. Sanborn, J. R.: Public Health Compliance by Manufacturers of Paper for Packaging Perishable Foods, *J. Milk Technol. 5:* 88-93, (1942).

2. Ball, C. O.: Aspects of Public Health Legal Control of Paper Containers, *Ibid. 5:* 6-17 (1942).

CHAPTER 7

DIRECT MICROSCOPIC EXAMINATION OF DAIRY PRODUCTS

(Breed Technique)

Market Milk

A. General Description

Briefly stated, the Breed method consists of preparing a stained film of known volume, spread over a known area, and examining it with a standardized, microscope, illustrated in figures 50 and 51.

Apparatus

1. Microscope with the field of the oil immersion objective standardized to a known diameter. (See page 121).
2. Stage micrometer (needed only for the original standardization of the diameter of the microscopic field).
3. Staining jar.
4. Slide-holding forceps.
5. Supply of clean glass slides, preferably with etched area for labels.
6. Guide plate, ruled in square centimeters.
7. Capillary pipettes (0.01 ml) or platinum loop, standardized to deliver 0.01 ml.
8. Platinum or nichrome wire with handle.

Reagents

1. Methylene blue stain (see page 526).
2. Xylol.
3. Alcohol, 90 per cent.
4. Water.

NOTE: If Newman-Lampert stain (see page 526) is used, no other reagents are necessary.

B. Standardization of the Microscope

When purchasing a new microscope, the manufacturer on request will usually standardize without extra charge the diameter of the oil immersion field to a value suitable for use in the Breed method. The object in standardizing the microscope is to provide a field the area of which will provide a convenient factor in the subsequent calculations of the number of bacteria in the samples to be examined. For example, if the area of the field were 1/3299 sq cm, it would be neces-

Courtesy of the American Public Health Association

Figure 50.

Samples of milk (at left) from which preparations are to be made for direct microscopic examination. Guide plate and necessary equipment for making slides in center and at the right.

Courtesy of the American Public Health Association

Figure 51.

Basket containing prepared slides (at left) ready for staining by use of staining solution in adjacent round jar. Microscope with slide ready for counting the bacteria in center with record sheet and tally counter at right of microscope.

sary to multiply the cumbersome figure of 3299 in all calculations. It greatly facilitates calculations if the area of the field is adjusted to represent 1/3000 sq cm or some other similarly convenient value for multiplication.

The following table shows the diameters of microscopic fields corresponding to areas in convenient fractional values of a square centimeter.

Diameter in mm.	Area in fractions of a sq. cm.	Factor for the average number of bacteria per field
0.206	1/3,000	300,000
0.178	1/4,000	400,000
0.160	1/5,000	500,000
0.146	1/6,000	600,000

The procedure for standardizing the diameter of the microscopic field to one of the desired diameters in the preceding table is as follows:

1. Locate and center the lines of a stage micrometer under the low power objective of the microscope. Relocate and center the lines using the high-dry objective; then finally locate and center the lines under the oil-immersion objective. The lines on the stage micrometer are so arranged that the wide spaces represent 0.1 mm, and the narrow spaces 0.01 mm. Adjust the stage micrometer so that approximately one-half of the field under the oil immersion objective is marked by the small (0.01 mm) spaces and the other half of the field marked by one of the large (0.1 mm) spaces. Carefully shift the stage micrometer so that the line marking the large space just barely touches the side of the microscopic field, then count the number of small lines visible in the other half of the field. If one can see five small spaces (0.05 mm) and one full large space (0.1 mm) the width of the field is 0.15 mm.

2. Measure the diameter of the field first with the draw tube of the instrument as far down as it will go, and then measure it again with the draw tube pulled upward as far as it will go. Within the two limits thus established for a maximum and minimum diameter of the microscopic field, select one of the diameters shown in the table. By a process of trial and error, adjust the length of the draw tube to a point which will give exactly one of the diameters shown in the table. Record the tube length and the magnification of the eye-piece employed: Once this standardization has been done it is not necessary to do it again unless the eye-piece or the objective is changed on the microscope. If the optical system has not been changed, the desired area of the field can be obtained at any subsequent time by setting the tube length at the indicated value.

The optical system of most microscopes functions optimally at a given tube length (usually 160 mm). It is preferable therefore to use an eye-piece which will give a desired diameter of field when the tube length is not too far from the optimum tube length for which the instrument is adjusted. A special ocular disc having cross lines within a circle may be inserted in the eye-piece, and the diameter of the circle standardized. The use of such a disc eliminates the edges of the microscopic field, which are usually out of focus.

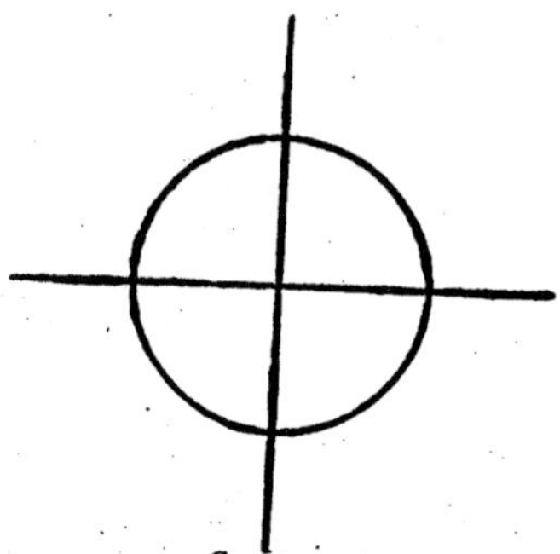

Courtesy of the American Public Health Association

Figure 52.

Form of guide lines used on special ocular micrometer.

C. PREPARATION OF THE SMEAR OF FILM

1. Transfer 0.01 ml of milk to a clean glass slide by means of a capillary pipette graduated to deliver 0.01 ml (figure 53) or by means of a standardized platinum loop. (See figure 34, page 101).

2. Place the slide over a guide plate (figure 54) marked off in square centimeter areas. With a sterile needle or straight wire, spread the milk evenly over exactly one square centimeter on the glass slide. Allow to dry in a level position in a warm place protected from dust and flies. The drying may be slightly hastened by the gentle application of heat, but this must not be overdone or the smear is likely to come off the slide. Drying should be completed within 5 minutes. After the smear is thoroughly dried it is ready to stain.

D. STAINING OF SMEAR

1. Dip the dried smear in xylol long enough to dissolve the fat (usually one to two minutes).

2. Drain and thoroughly dry.

3. Dip the slide into 90 per cent ethyl alcohol for one to two minutes to fix the smear.

Courtesy of the American Public Health Association

Figure 53.

Capillary pipette delivering 0.01 ml.

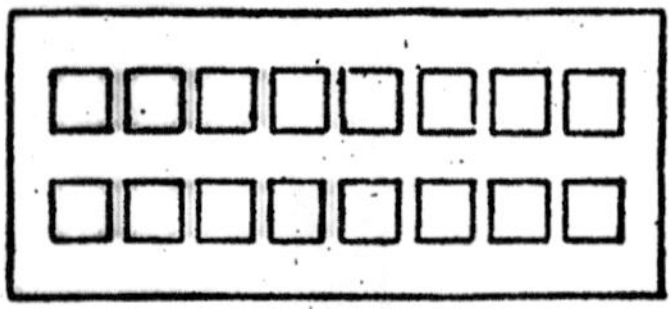

Courtesy of the American Public Health Association

Figure 54.

Special (2 by 4½ in.) guide plate.

4. Drain the excess alcohol.

5. Dip the slide into the staining solution long enough to stain properly (usually about one minute). Avoid overstaining. Decolorize with alcohol if overstained.

6. Rinse in water and thoroughly dry the slide before examining.

NOTE: (1) If the Newman-Lampert technic is used in staining, dip the slide momentarily into the Newman-Lampert solution, remove, and dry thoroughly before dipping into water to remove the excess of stain. This technic substitutes the one operation for the first five steps previously listed in the staining process.

NOTE: (2) Another method which involves the use of Newman's solvents has been found helpful. This method consists of preparing the Newman-Lampert solution as originally described by the authors but eliminating the methylene blue stain from the preparation. Mix 54 ml of alcohol and 40 ml of tetrachlorethane (technical). Warm the mixture in a water bath for ten minutes at 140 F (60 C), then cool by placing the flask in tap water for ten minutes, and add slowly (drop by drop) 6 ml of glacial acetic acid with constant agitation.

Use this preparation, known as Newman's solvents, as a single fixing agent for the smears. The staining procedure in this case is as follows:

a After the smear has been thoroughly dried, immerse in Newman's solvents for thirty seconds to one minute.

b Remove and dry the smear (drying is important).

c Immerse the dried smear in methylene blue stain for thirty seconds to one minute.

d Immerse the slide in water to remove excess stain, blot dry and examine. The advantage of this procedure is that smears rarely if ever come off the slide during the staining process.

E. Examination of Smear

1. Focus the oil immersion objective on the smear and adjust the light to the desired intensity.

2. Count the individual bacteria cells (or the clumps of cells) in the field and record the value. Repeat the count on representative fields well distributed in various parts of the smear. Record a value for each field even though no organism may have been seen. The number of fields to be counted depends on the quality of the milk and

the standardization of the microscope. As a usual routine, count 60 fields if the area of the field is standardized to 1/6,000 square centimeter; 50 fields if 1/5,000; 40 fields if 1/4,000; and 30 fields if 1/3,000. If each field contains a large number of organisms it would not be feasible to attempt to count so many fields. On the other hand, if the milk is of good quality, most, if not all, of the fields will contain no bacteria, and in such cases even a larger number of fields may be examined with ease. In the examination of each field it is important to change the focus with the fine adjustment in order to insure visibility of all bacteria present in each microscopic field.

In some instances it is more convenient to count the clumps of bacteria as a unit instead of enumerating the individual cells in the clump. If counts of clumps are made, the report should indicate that fact.

F. CALCULATIONS

1. From the total number of individual bacteria (or clumps) counted and the number of fields observed, determine the average number of bacteria (or clumps) per field.

2. Multiply the average number of individual bacteria (or clumps) per field by the appropriate factor, 600,000, 500,000, 400,000 or 300,000, depending on the diameter to which the field of the microscope has been standardized. See table of factors on page 121.

Thus if the microscopic field has been standardized to a diameter of 0.206 mm, the area of the field is 1/3,000 sq cm, and in one microscopic field there is observed 1/3,000 of the milk sample spread over the square centimeter of area. Since there is only 0.01 ml of milk in the entire smear, the observer sees 1/300,000 of an entire milliliter of milk. Hence the factor of 300,000 for the average number of bacteria in one field gives the number for a whole milliliter of milk. Similarly the factors of 600,000, 500,000, and 400,000 might be explained.

For example, if the total number of bacteria observed in 30 fields were 150, the average per field would be 150/30 or 5 bacteria per field. If the factor for the microscope was 300,000, the count would be 5 × 300,000 or 1,500,000 per milliliter. If clumps instead of individual cells are counted, the calculations are the same except that the report of the number should indicate that it represents a "clump count."

In reporting bacterial counts by the direct microscopic method, care should be taken to avoid giving a fictitious idea of the accuracy of the counts. The values should be reported only in round numbers.

NOTE 1: The arithmetical calculations may be somewhat shortened by simply adding four ciphers to the total number of bacteria observed in 30 fields if the microscope is standardized to 1/3,000 sq cm. This is possible because with such a microscope, the factor for one field is 300,000, hence for 30 fields it would be 300,000 divided by 30 or 10,000. Similarly, add four ciphers to the total number of organisms: in 40 fields if the factor is 400,000; in 50 fields if the factor is 500,000; or in 60 fields if the factor is 600,000.

NOTE 2: Attempts have been made to simplify the procedure of counting the bacteria in Breed smears. Notable among these is the suggestion of Levowitz (3) whereby the smear is prepared in the usual way, but the operator counts the number of bacteria (or clumps) observed in one or more strips across the entire smear. In this way it is not necessary to keep count of the number of fields examined. The diameter of the field is standardized to 0.200 mm, hence the strip across the smear represents one-fiftieth of the entire area of the smear. Since there is 0.01 ml of milk on the smear, the quantity of milk actually observed is 1/5000 of a ml. The total number of organisms (or clumps) observed in one strip would therefore be multiplied by 5000.

Buck (4) has described a mechanical device for the mechanical stage of the microscope which facilitates spacing the fields equally while searching the slide.

Mallmann and Bryan (5) have recommended that a standardized platinum loop delivering 0.01 ml of milk be recognized as a standard procedure to replace the capillary pipettes. To improve the lighting conditions, these authors also recommend that a Wratten M filter No. E-22 be placed between the light source and the object to sharpen the contrast of between the organisms and the background.

Special stains have been recommended by Mallman and Bryan (5) (see page 524 and also by Gray (6).

Other Dairy Products

The general sanitary conditions surrounding the production and processing of various dairy products in certain instances may be more clearly revealed by direct microscopic examination than by making the usual plate counts. Excessively high bacterial counts by the direct microscopic method accompanied by acceptably low plate counts either suggest the masking of ingredients of poor bacteriological

REFERENCES

1. Breed, R. S.: The Determination of the Number of Bacteria in Milk by Direct Microscopic Examination, *Centralbl. F II Abt. 30*:337-340 (1911).
2. *Standard Methods for the Examination of Dairy Products*, 9th edition, 1948.
3. Levowitz, David: Reproductible Data by Microscopic Method, *International Association of Milk Dealers, Association Bulletin 10*:139-150. (1944).
4. Buck, T. C., Jr.: A Microscopic Field Spacer, *Jour. Milk Tech., 7:* 341-342 (1944).
5. Mallmann, W. L. and Bryan, C. S.: A Study of the Methods for the Microscopic Examination of Raw Milk with Suggested Improvements, *Ibid. 7*: 315-321 (1944).
6. Gray, P. H. H.: Two Stain Method for Direct Bacteria Count, *Ibid. 6*: 76 (1943).

quality by high processing temperatures, or reflect the presence of organisms incapable of growth under the conditions of plating. In the latter case the direct microscopic method is particularly useful in the detection of thermophils, psychrophlis, or other particular types of organisms which require special conditions for growth. The value of the microscope in the examination of various dairy products other than milk has not been fully appreciated or explored.

A. BUTTER

a. Hammer and Nelson Method[1]

Apparatus and Reagents

1. The equipment required for Breed direct microscopic method for examining milk. (See page 119).
2. Separatory funnels with accessory equipment as shown in figures 55 to 61 inclusive.
3. Conical beaker, 250 ml.

Sampling

Methods of sampling and care of samples of butter as outlined on page 56.

Procedure

1. Melt and thoroughly mix in a 250-ml beaker the entire sample of butter, or enough of it to yield 10 ml of the melted product.

Melt by placing the beaker in a water bath at a temperature of 120-130 F (48.8-54.4 C) and slowly agitate with a thermometer used as a stirring rod, until the temperature reaches 110 F (43.3 C) (see figure 55).

2. Transfer 10 ml of the melted fat at a temperature of 110 F to a 30-ml separatory funnel (see figures 56 and 57), using a pipette that has been warmed by passing it through a flame.

NOTE: The size of the portion centrifuged is of no great importance but a standard volume seems advisable and one of 10 ml has some advantages over smaller amounts. These advantages are:

(a) Enough serum collects to permit the removal of the serum without including any considerable percentage of fat; and (b) the volume of serum secured provides ample material for plating in case the microscopic picture makes this advisable.

3. Place the separatory funnel in the special support shown in figure 56. A rubber stopper around the upper part of the funnel rests against the wall of the support and the funnel is carried in this way. The assembled funnels, funnel supports, and special centrifuge head clamped to the top of a ring stand are shown in figure 58. Very little breakage of funnels has been encountered and most of this has

Figure 55.
Preparing butter sample for analysis. Figures 55 to 61 courtesy Agricultural Experiment Station, Iowa State College of Agriculture and Mechanic Arts.

involved the tip of the funnel which projects into the lower portion of the support.

The centrifuging should throw out the serum completely without causing a separation in the serum itself. When centrifuging has been done properly the fat layer will be clear and the serum layer uniform throughout (see figure 59). With a centrifuge in which the stopcocks of the funnels are 14½ inches apart with the machine in operation, centrifuge one minute at a speed of 1,000 revolutions per minute. Samples of butter of different characters behave differently, and some types are easily overcentrifuged so that there is a separation in the serum. When this occurs a solid pack of material at the bottom of the funnel may make it difficult to start a flow through the stopcock. It is then necessary to remove the

Figure 56.
Separatory funnels, funnel supports, and centrifuge head.

serum completely and thoroughly mix it, if a representative portion is to be secured for spreading on the slides. In general, the centrifuging of butter made from poor quality cream must be carefully done or there will be a separation in the serum; presumably, this is due to the content of particles of casein or some casein derivative that readily centrifuges out. With butter made from good quality sweet cream, the centrifuging can be greatly in excess (in either time or speed) of that suggested without any undesirable effect.

4. Immediately after centrifuging, remove the funnel and draw off the serum (see figure 60). Unless this is done before the fat begins to congeal, much difficulty is encountered. The serum should be removed very slowly so that the plane between the fat and serum is kept as flat as possible. When the serum is drawn off rapidly the portion along the wall of the funnel tends to lag and the fat is pulled down to the stopcock when there is still a considerable percentage of the serum in the funnel. Remove the serum as completely as possible without including objectionable amounts of fat.

A short test tube (see figures 60 and 61) makes a satisfactory container in which to collect the serum, since portions of the serum can be easily removed from it with a Breed pipette.

5. Thoroughly agitate the collected serum and transfer 0.01-ml portions to microscope slides with a Breed pipette. Ordinarily the serum is spread over 1, 4, or 9 sq cm. If the area is larger than 1 sq cm sterile skim milk or water, free from stainable bacteria cells, is added with a loop to facilitate even distribution. The spreading is conveniently done with a transfer needle bent at an angle of 90 degrees (see figure 61). When skim milk or water is to be used it should be placed near the 0.01 ml of serum and thoroughly mixed with the serum before an attempt is made to cover the desired area

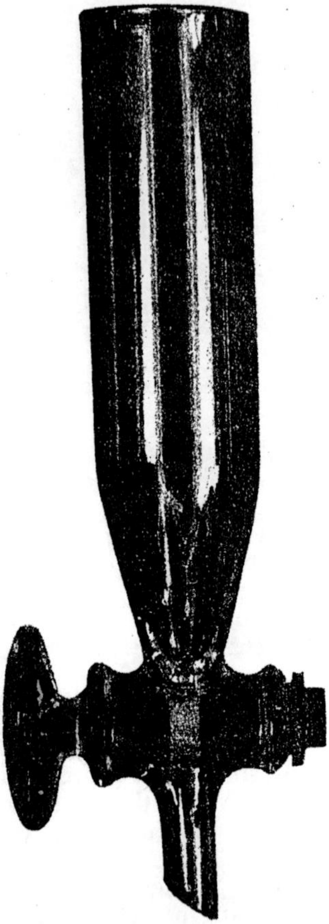

Figure 57.
Special separatory funnel for centrifuging melted butter.

The area over which the 0.01 ml of serum should be spread depends on the bacterial content of the butter and the information desired. If a study of the morphological types of organisms present is the main consideration it is an advantage to have the organisms rather numerous in each field, whereas if an estimation of the number is desired the organisms should not be so numerous that counting is difficult. With butter having a high bacterial content, counting may be facilitated by spreading the 0.01 ml of serum over an area as large as 9 sq cm.

A card with the desired area laid out in black and surrounded by a broad red border aids in the making of smears that are reasonably

Figure 58.
Assembled funnels, funnel supports, and centrifuge head clamped to the top of a ring stand.

exact in area and sufficiently uniform in thickness. If a portion of the background is white the desired results are not so easily secured.

6. Dry the slides rapidly on a level surface well protected from flies and dust.

7. Fix and stain the smear as outlined on page 122.

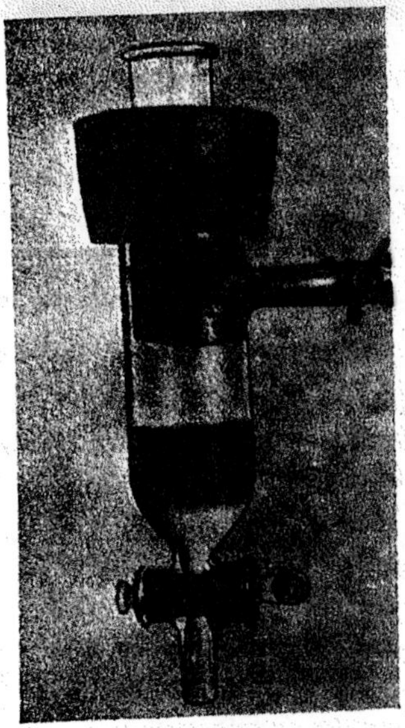

Figure 59.
Centrifugal butter showing clear fat layer and uniform serum layer.

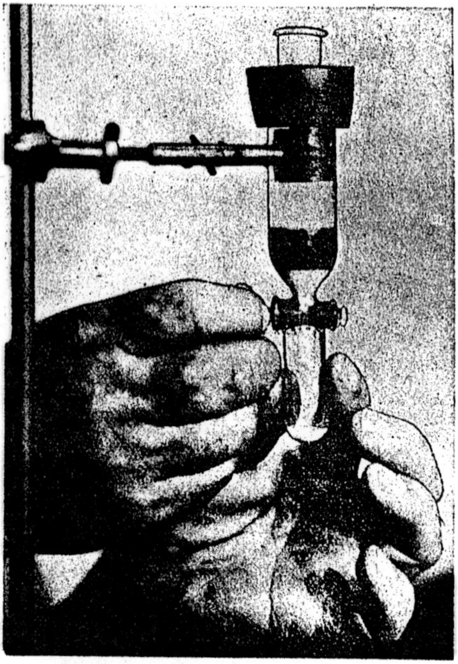

Figure 60.
Removing the serum from the separatory funnel.

8. Examine the stained smears under the oil immersion lens for the general morphological types and the approximate number of organisms present.

9. Estimate the number of organisms per milliliter of butter by

(a) determining the average number per field,

(b) calculating the number per milliliter of serum from the area over which 0.01 ml of serum was spread and the diameter of the microscopic field, and

(c) dividing by 0.9 to change from a basis of 1 ml of serum to that of 1 ml of butter. Steps (b) and (c) may be combined for routine work.

Example: Multiply average number of bacteria per field by the factor for the microscopic field (see page 121), then by the number of square centimeters over which the smear has been spread (1, 4, or 9), then divide by the factor 0.9 to correct for the specific gravity of butter. The result is the number of bacteria per gram of butter.

Figure 61.
Equipment used in making preparations for the serum.

b. Fay Method[2]

Apparatus

1. Microscope with oil immersion field standardized to a diameter of 0.157 mm.
2. Standard glass slide (3 x 1 inch).
3. Flat (whisky type) bottle containing water at 176 F (80 C).

Reagents

1. Methylene blue stain (see page 526).
2. Xylol.
3. Ethyl alcohol, 70 per cent.
4. Mayer's egg albumin. (To prepare, mix equal parts of glycerin and egg albumin, beat thoroughly, then filter through cotton. Add 1 per cent of sodium salicylate and keep refrigerated.)

Procedure

1. Place 0.1 ml of melted butter on a thoroughly cleaned glass slide.
2. Add one drop of xylol and one drop of Mayer's egg albumin, and mix the three ingredients thoroughly using a sterilized platinum or nichrome wire. Spread the mixture evenly over the entire area of the 3 x 1 inch glass slide.

3. Fix the smear by placing the slide for 10 to 15 minutes on a flat (whisky type) bottle containing water at 176 F.

4. Immerse the slide in 70 per cent alcohol for 10 to 20 seconds.

5. Immerse in xylol 1 minute, then allow the slide to dry thoroughly before placing in the stain.

6. Immerse in methylene blue stain 1 minute.

7. Dip the slide twice (quickly) in water to remove excess stain and allow to dry.

8. Examine with an oil immersion lens, the diameter of the field of which has been standardized to 0.157 mm (see page 121).

9. To obtain the total count, multiply the average number of bacteria in a representative number of fields by 1,000,000.

NOTE: For samples that are low in bacterial count, a strip across the short (1 inch) dimension of the slide may be examined, and the total number of organisms observed in the entire strip multiplied by 5,000 will give the number per ml of melted butter. If 0.2 ml of sample is employed instead of 0.1 ml, the factor would be 2,500. The actual factor (for 0.1 ml sample) is 4,886, and in order to have the factor exactly 5,000, the field should have a diameter of 0.152 mm instead of 0.157 mm. Owing to the difficulty of detecting organisms at the extreme edges of the field, however, it is not necessary to readjust the diameter to 0.152 mm. When counting the organisms in a strip it is not necessary to keep account of the number of fields examined.

c. Hussong's Qualitative Method[3]

Apparatus

1. Microscope, with oil immersion lens.
2. Clean glass slide.

Reagents

1. Acetone.
2. Methylene blue stain (see page 526).

Procedure

1. Place a small piece of butter on a clean glass slide and gently warm over a low flame or hot plate.

2. Add a few drops of acetone, and mix thoroughly either by gently moving the slide or by means of a sterile needle. When the mixture is complete, gently pour off the acetone-fat-water mixture, leaving the casein and other solids on the slide.

3. A second extraction is made in the same manner with acetone.

4. Allow the smear to dry.

5. Stain with methylene blue and examine with the oil immersion lens of a microscope.

NOTE: This method is intended for the gross microscopic examination of prevailing types of bacteria and not for enumerating the bacteria in butter. For making the yeast and mold count, see Chapter 11, page 157.

B. ICE CREAM

Direct microscopic examination is made of ice cream by using the same technique as that described on page 119 for milk, with the following precautions.

a. More difficulty is encountered with the small capillary pipettes in measuring 0.01 ml of the viscous ice cream mix. For this reason some workers prefer to use a standard loop (see figure 34, page 101).

b. Owing to the higher fat content, extra care must be exercised in treating the dried smear with xylol and alcohol to insure the removal of the fat. It is important to dry the slide after treatment with xylol and alcohol, and if fat is still evident on the smear, repeat the xylol and alcohol treatment, making sure that the drying process is also repeated.

NOTE: (1) In order to overcome the difficulty with the heavy smears of ice cream some technicians prefer to spread the 0.01 ml sample over 2 sq cm (1 × 2 cm) instead of 1 sq cm. In this case the factor per field should be appropriately adjusted. For example, if the factor for the microscope is 300,000, the adjusted factor should be 600,000. With samples of heavy viscous ice cream, care must be used when a loop is employed in measuring the sample.

NOTE: (2) If difficulties are encountered in measuring the 0.01 ml sample and with the removal of the fat from the smear, the method described by Fay[4] may be helpful. This method involves the use of 0.1 ml instead of 0.01 ml of sample. Place the 0.1 ml of melted ice cream on a thoroughly cleaned (3 × 1 inch) glass slide and add two drops of sterile water. Thoroughly mix with a sterile needle and spread the mixture over the entire area of the slide. After the smear has dried, fix and stain in the same manner as described for milk on page 122. Standardize the field of the oil immersion lens to a diameter of 0.157 mm. Multiply the average number of bacteria per field by 1,000,000 to determine the number per milliliter. With samples having a low bacteria count, the total number of organisms observed in a strip across the short (1 inch) dimension of the slide may be multiplied by the factor 5,000.

C. CREAM

Direct microscopic examination of cream is made in the same manner and with the same precautions as those described for ice cream. Special care must be given to the removal of fat from smears prepared from samples having a high fat content.

D. EVAPORATED, CONDENSED, AND POWDERED MILKS

Direct microscopic examinations for evaporated, condensed, or powdered milks may be made by reconstituting the product to the original composition of milk with sterile water and using aseptic precautions.

After the product has been reconstituted, follow the same procedure in making the smear, staining, and examining as described on page 119 for fluid milk.

REFERENCES

1. Hammer, B. W., and Nelson, J. A.: Bacteriology of Butter: II A Method for the Microscopic Examination of Butter. *Iowa State College Research Bul.* 137 (1940), pp. 106-120.

2. Fay, A. C.: A Method for the Microscopic Examination of Butter, *J. Dairy Sci., 18* (9): (1935) 603-604.

3. Hussong, R. V.: A Method for the Microscopic Examination of Butter: *B. W. Hammer Panegyric,* Collegiate Press, Inc., Ames, Iowa, 1937, pp. 73-74.

4. Fay, A. C.: A Modification of the Method for the Direct Microscopic Examination of Ice Cream and Other Dairy Products, *J. Dairy Sci. 16*: 311 (1933).

CHAPTER 8

DETECTION OF MASTITIS

General Discussion

A summary of the available information in regard to the detection of mastitis indicates that there is no one test which is entirely satisfactory when used alone. Physical examination of the cows is useful. The bromthymol blue test is a valuable aid, as is the strip cup with a black cloth over the screen. Direct microscopic examination will show the presence of leucocytes and occasionally the organisms causing the trouble. Large numbers of leucocytes may or may not indicate mastitis. A combination of streptococci and leucocytes is probably the most conclusive evidence to be found with a microscope. A pronounced reaction with bromthymol blue is usually accompanied by a large number of leucocytes or leucocytes and streptococci. In some cases a bright yellow color (indicating an acid condition) has been noted with bromthymol blue when the microscope showed an enormous number of leucocytes and long chain streptococci.

It is safe to say that milk which shows lumps or other abnormalities should not be used. If the microscopic test shows large numbers of leucocytes or leucocytes and streptococci, or if the bromthymol blue test is strongly positive, the milk from such cows should be excluded from the supply.

The following are a number of tests which have been found useful in the detection of animals infected with mastitis. Usually it is advisable to employ a combination of several of these tests and to regard the results as preliminary or presumptive, to be confirmed by more exact bacteriological procedures, such as isolation of the causative organism on blood agar and identification as outlined on pages 83 to 92.

Black Cloth or Strip Cup Test

Apparatus

Strip Cup. A piece of finely woven black cloth stretched tightly over a cup and held in place by a rubber band may serve satisfactorily as a strip cup. Strip cups are also available from dairy supply houses, equipped with a finely woven metal screen in one-half of the open area of the cup.

Procedure

Milk the first two or three streams of milk onto the cloth (or wire mesh) and observe carefully for fibrinous, stringy, or granular specks.

Interpretation

Positive findings are quite conclusive evidence of mastitis and the results of more precise methods usually confirm such positive findings. Milk from such animals should be excluded, and the animals segregated completely from the herd.

Bromthymol Blue Test (Thybromol Test)

This test is based on the observation that pathologically abnormal cells in the udder secrete milk that is more alkaline than milk secreted by normal cells. If the infection in the udder is sufficiently extensive and involves a sufficiently large percentage of the secreting tissue, the resulting reaction of the milk will be dominated by the more alkaline portion secreted.

Apparatus

Test tubes

Reagent

Bromthymol blue—solution of 0.2 per cent in 47.5 per cent alcohol. (Neutralize the indicator solution to a dark grass green color using N NaOH.)

Procedure

To 3 ml of fore-milk (preferably collected directly from one quarter of the udder) add 0.5 ml of the bromthymol blue solution.

Interpretation.

A dark blue-green color suggests mammary infection. Positive findings with the bromthymol blue test are reasonably reliable and are usually confirmed by more precise tests. Negative findings however are inconclusive.

Catalase Test

Apparatus

1. Test tubes graduated with milliliter markings (not less than 20 ml capacity, see note under interpretation.)

2. Rubber stoppers, one hole, each fitted with a short piece of glass tubing which has been drawn to a capillary opening. The tube should be inserted so that the capillary end protrudes from the large end of the stopper.

Reagent

Hydrogen peroxide (H_2O_2) solution, one per cent.

Procedure

1. Place 15 ml of milk in a graduated test tube, then fill the tube completely with 1 per cent H_2O_2 solution.
2. Insert the rubber stopper containing the capillary tipped tube.
3. Invert the test tube and place in a cup or beaker.
4. Incubate at 98.6 F (37 C) for 3 hours.

The liberation of oxygen from the H_2O_2 by the enzyme catalase may be used as a rough quantitative index of abnormal secretion in the udder.

Interpretation

More than 2.5 ml of gas collected in the top of the tube is presumed to indicate an abnormal condition.

NOTE: In the absence of graduated test tubes, the volume of gas in the inverted tube after incubation may be measured by making a mark on the tube with a wax pencil, and after emptying the tube measuring the amount of water required to fill the tube to the mark.

Field Test for Chlorides

This test is based on the assumption that the selective permeability of the cells of the udder is impaired by the presence of infection. As a result, the interchange of materials from the blood is abnormal, and one of the manifestations of this abnormality is the high percentage of chlorides in the milk. One of the fundamental assumptions upon which this test is based, namely, that the normal chloride content of milk is 0.14 per cent, has been subject to question. It is perhaps erroneous to assume that the percentage of any constituent of milk is constant.

Apparatus

1. Test tubes.
2. Pipettes, 1 ml.
3. Pipettes, 5 ml.

Reagents

1. Silver nitrate ($AgNO_3$) solution, aqueous, 0.1341 per cent.
2. Potassium chromate (K_2CrO_4) solution, aqueous, 10 per cent.

Procedure

1. To exactly 5 ml of a 0.1341 per cent solution of $AgNO_3$, add two drops of a 10 per cent solution of K_2CrO_4. This gives a red color.
2. To this mixture add exactly 1 ml of milk to be tested.

Interpretation

If the milk contains an abnormally high percentage of chlorides (more than 0.14 per cent), a yellow color will appear within a few minutes. Persistence of the red color indicates normal chloride content.

Quantitative Test for Chlorides

Apparatus

1. Pipettes, 10 ml.
2. Burette.

Reagents

1. Potassium chromate (K_2CrO_4) solution, aqueous, 10 per cent.
2. Silver nitrate ($AgNO_3$), 0.1 N.

Procedure

1. Dilute exactly 10 ml of milk with 30 to 40 ml of distilled water.
2. Add 8 to 10 drops of a 10 per cent solution of K_2CrO_4.
3. Titrate with 0.1 N $AgNO_3$ solution to the first faint shade of reddish color, that is, to the first "off-color" from the original yellow.
4. Multiply the number of milliliters of 0.1 N $AgNO_3$ solution used by 0.0355 to determine the percentage of chlorides.

Interpretation

Values in excess of the arbitrary limit of 0.14 per cent chlorides are presumed to be high and to indicate mastitis. As previously mentioned, this arbitrary limit is subject to some question and should be interpreted with caution. Some workers prefer to employ arbitrary standards higher than 0.14 per cent as indicative of abnormal secretion.

Direct Microscopic Test

INCUBATED MILK

Although samples of milk taken directly from individual quarters of the udder may be examined under the microscope for the presence of long chained streptococci, the chances of finding the organisms will be greatly increased if the milk is incubated before the examination is made. The period of incubation however should be carefully regulated so as to give the streptococci a chance to grow, and to prevent other types from overgrowing and obscuring the presence of the organism sought.

Apparatus

1. Sterile test tubes.
2. Microscope, slides, staining materials, and other necessary equipment for making direct microscopic examination (see page 119).

Collection of Samples

After washing and wiping the flanks, udder, and teats of the animal, collect the first few streams of milk from one quarter of the udder. Label for identification. Collect similar samples from the other quarters. Since these samples should be incubated for a definite length of time (12 to 16 hours), place them in the incubator so that the terminating hour of incubation will be convenient for examination. For example, samples placed in the 37 C (98.6 F) incubator at 5 p.m. will have completed 16 hours of incubation at 9 a.m. the following day. Samples collected at earlier times during the day may be placed in the refrigerator immediately after being drawn from the animal and held until the appropriate time to begin the incubation period. Similarly the samples may be placed in the refrigerator for a few hours following the incubation period if it is not convenient immediately to make the microscopic examinations. These periods of refrigeration before or after the incubation should not exceed 8 hours. The actual period of incubation should be carefully regulated. Incubation for periods less than 12 hours are not likely to give the udder streptococci an opportunity to grow sufficiently to facilitate detection, and incubation for more than 16 hours frequently enables other types of bacteria to grow into such numbers as to dominate the microscopic picture.

Procedure

1. After incubation of the samples from the individual quarters for 12 to 16 hours, prepare Breed smears in the manner described on pages 122-124.

2. Examine the smears for long chains of streptococci, but no effort should be made to enumerate the bacteria since the sample has been incubated.

3. Count the number of leucocytes (white blood cells) in a representative number of fields and multiply by the factor for the microscope (see page 121).

Interpretation

As the samples were collected under aseptic conditions and in a sterile container (test tube) having a small opening to exclude contamination from the air or the flanks of the cow, it is reasonable to assume that organisms observed in the smear actually came from the interior of the udder. The fact that long chain forms of streptococci are associated in a high degree with mastitis justifies a strong suspicion that infection may exist in the quarter of the udder represented by the sample. Such a conclusion is not without question, however, and should be confirmed by repetition of the examination, and by further isolation and identification of the causative organism as described on pages 90-92. Of the various preliminary or presumptive tests for mastitis, the presence in aseptically drawn milk of long chain streptococci, associated with large numbers of the leucocytes, is perhaps the most conclusive evidence of mastitis that can be obtained without the actual isolation and identification of the organisms.

The results of microscopic examination of samples taken from the weigh vat or from cans of milk are not to be relied on. The fact that the milk has been subjected to extraneous contamination discourages the conclusion that the streptococci observed necessarily came from the udder. The presence of long chain streptococci accompanied by excessive leucocyte counts, even in samples taken under these questionable conditions, at least warrants further investigation, but not a definite conclusion that mastitis exists. The high leucocyte count may be from a "stripper cow," and the streptococci may have come from contamination of the equipment.

LEUCOCYTE COUNT

Leucocytes are normally present in milk, but when found in excessive numbers they may be the result of an infection or a traumatic injury, or the milk may be from cows early or late in their lactation period (strippers). Milk from an aged cow in which the epithelial structure of the udder is sloughing off may show large numbers of epithelial cells that are sometimes confused with leucocytes. Epithelial cells are usually less densely stained than the leucocyte; the cell has the appearance of lacking body and substance, and the edges are usually more frayed and irregular, and transmit more light. With experience the technician learns to differentiate between epithelial cells, leucocytes, and phagocytes. A phagocyte may be defined as a leucocyte which has functioned in its capacity to engulf and digest an invading organism.

Various arbritrary standards for numbers of leucocytes have been employed by different workers as indicating an infectious condition. A reasonably conservative standard is 500,000 leucocytes per milliliter of milk from individual quarters as an index of udder infection. High leucocyte counts in themselves are not conclusive, but when accompaned by long chain streptococci, and especially when leucocytes (phagocytes) are found which have engulfed the streptococci, the evidence of infection is clearer.

Leucocyte counts may be made by the Breed technique either on milk freshly drawn from the udder or on the incubated specimens (see page 141). The average number of leucocytes per microscopic field is determined and multiplied by the factor appropriate for the microscope (see page 121).

Hotis Test

The Hotis test is based on the observation that *Streptococcus agalactiae,* when growing in milk, produces a characteristic colony-like mass of cells adhering to the sides of the test tube. By the introduction of an acid indicator (bromcresol purple), the identification of these colonies or "buttons" is facilitated by their characteristic yellow color.

Apparatus

1. Sterile test tubes.
2. Pipettes, 10 ml graduated to 0.1 ml.
3. Pipettes, 1 ml graduated in 0.1 ml.
4. Incubator, 37 C (98.6 F).

Reagent

1. Bromcresol purple, 0.5 per cent aqueous solution.

Procedure

1. Place in a sterile test tube 9.5 ml of milk which has been drawn aseptically from an individual quarter of the udder.
2. Add 0.5 per cent solution of bromcresol purple, and mix the milk and the indicator.
3. Incubate the samples 24 hours at 37 C (98.6 F).
4. Observe for the presence of small yellow masess of growth adhering to the sides of the tube. These are usually composed of threads of *Streptococcus agalactiae,* and indicate udder infection.

Hotis Test

(Foley Modification)

Apparatus

1. Sterile test tubes, 150 x 15 mm, plugged with cotton.
2. Chemical balance.
3. Incubator, 37 C (98.6 F).
4. Pipettes, 1 ml.

Reagents

1. Sodium azide.
2. Bromcresol purple.
3. Crystal violet.
4. Distilled water.

Preparation of Solution

1. Dissolve in approximately 100 ml of water the following: sodium azide 1.0 gm, bromcresol purple 5.0 gms, crystal violet 0.1 gm.
2. Make up the final volume to one liter.
3. Place 1 ml quantities of the solution in test tubes, plug the tubes with cotton stoppers, and sterilize either by heating in the autoclave at

15 pounds pressure for 10 minutes or by placing in boiling water for 30 minutes.

Procedure

1. Add 20 ml of milk to be tested to a tube containing the dye solution, mix by gentle shaking, and place in the incubator at 37 C (98.6 F).

2. After 16 hours incubation examine tubes for typical yellow centers of growth characteristic of *Streptococcus agalactiae*. Tubes showing slight acid changes after 16 hours incubation should be returned to the incubator for an additional 8 hour incubation and re-examined.

NOTE: Crystal violet serves to inhibit micrococci, and sodium azide serves to inhibit coliform types which frequently interfere with the Hotis test as originally described.

Whiteside Test

(Murphy and Hanson Modification)

Apparatus

1. Eye droppers (selected to give drops of uniform size).
2. Window pane glass (size to be determined by number of samples to be run; usually 6 inches square is a satisfactory size).
3. Green colored blotter approximately the size of the glass plates used.
4. Glass stirring rods, approximately four inches long.
5. Test tubes or screw cap vials for collecting samples.

Reagent

1. Sodium hydroxide solution, N NaOH.

Procedure

1. Collect and label samples of fore-milk from each of the individual quarters of the animal. Complete the test and reading on a single sample before beginning the test on the next quarter sample.

2. Using an eye dropper, place exactly 5 drops of the milk on the glass plate (placed over the green blotter as a background).

3. Immediately add to the milk exactly 1 drop of the N NaOH solution (using another eye dropper which has been selected to yield drops of the same size as the one employed for the milk sample).

4. Stir the milk-alkali solution with a glass rod for about 20 seconds. The reaction occurs during the stirring process and takes the form of a precipitate or a viscid mass. In general, the opacity of the mixture in which the precipitate is finally suspended is inversely proportional to the amount of the precipitate.

Interpretation of Results

In order to be classified as negative, a sample should be entirely free of precipitate. If there is some degree of precipitation, the results are expressed as 1, 2, 3, and 4. A reading of 4 is characterized by a thick viscid mass that does not readily break up. A 3 reaction at first forms a viscid mass and then breaks up into a slightly opaque fluid and a precipitate composed of large particles or a solid mass of a scum-like nature. In a 2 reading the precipitate is composed of particles smaller than those in a 3 reaction. The precipitate tends to clump but does not form a mass or scum as in a 3 reaction. In a 1 reaction the precipitate is composed of small particles which are usually dispersed through the mixture but which may show a tendency to clump.

The test parallels rather closely the leucocyte count in its ability to detect the presence of udder infection and lends itself better to field use than does the microscopic examination for leucocytes.

CHAPTER 9

REDUCTION TESTS

The length of time required for certain reducible dyes to change in color when added to milk affords a simple, quick, and economical method of differentiating good and poor milk. The principal dyes used for this purpose are methylene blue and resazurin. In general it may be stated that if one of these dyes is added to milk with a high bacterial population, the dye will be reduced in a much shorter time than will be required if the bacterial content of the milk is low. It is assumed that the time required to effect reduction of the dye will be commensurate with the relative number of bacteria in the milk samples tested.

This assumption is not entirely correct because there are factors other than the number of bacteria present in the samples which may influence the reduction time. Chief among these factors are the kind of organisms present, the number of leucocytes in the milk, the period of exposure to light, the amount of dissolved oxygen, and the tendency of the milk fat to sweep the organisms to the surface as the cream layer forms in the test tube. Certain organisms, such as *Streptococcus lactis,* are active in their reducing capacities, whereas others, such as *S. agalactiae* and *Bacillus subtilis,* have poor reducing capacities. It is therefore obvious that the speed of reduction depends not only on the number of organisms present in the milk but likewise on the kind of organisms. Large numbers of leucocytes and exposure of the milk to sunlight, or even to an electric light, tend to shorten the reduction time. Agitation, resulting in increasing the amount of incorporated oxygen, and the sweeping action of the rising fat tend to retard the reduction time.

An understanding of the many factors affecting the results of reduction tests discourages attempts to correlate certain reduction times with specific bacterial numbers. Although these limitations must not be overlooked, the reduction tests do serve as excellent screening tests to identify the sources of poor milk. When used in combination with other tests, such as the direct microscopic examination or the plating method, the reduction tests may constitute an important part of an effective quality control program.

Methylene Blue Test

Apparatus

1. Water bath or incubator suitably controlled at 37 C (98.6 F).

2. Sterile test tubes fitted with rubber stoppers (size 12 x 150 mm).
3. Sterile pipettes or dippers, 10 ml.
4. Sterile pipettes, 1 ml.
5. Test tube racks.
6. Ice water bath for refrigerating samples after adding dye.
7. Tempering bath for preheating tubes to 37 C (98.6 F).

Reagents

Methylene blue solution prepared by dissolving one methylene blue thiocyanate tablet in 200 ml of sterile (or freshly boiled) water in a sterile amber colored bottle. It is preferable to allow the mixture to stand several hours to insure complete solution. Prepare a fresh solution weekly or oftener.

NOTE: The methylene blue thiocyanate tablets should be certified by the Biological Stain Commission.

Procedure

1. Arrange the sterile test tubes with rubber stoppers in a test tube rack, and add 1 ml of the methylene blue dye solution to each tube.

2. With a sterile 10-ml pipette or 10-ml dipper, add 10 ml of the milk to be tested to one of the tubes, and label appropriately for identification of the sample. Replace the stopper.

3. Place the tube into a test tube rack in ice water until all the samples have been measured. (Because of the sweeping action of fat rising in the milk, the tubes should not be held in the ice water any longer than is necessary to accumulate a convenient number of samples, and in no case longer than two hours.)

4. Transfer the rack of tubes to a tempering water bath (see figures 62) of sufficient capacity and operated at about 40 C (104 F) so as to raise the temperature of the tubes to 37 C (98.6 F in five minutes). (Uniform conditions for tempering within the prescribed time limit are important.) A thermometer placed in a pilot tube should be used as a guide to the efficiency of tempering conditions.

5. As soon as the milk has been properly tempered, invert the tubes a few times to insure complete mixing of the dye and uniform creaming, then place the rack of tubes immediately into the incubator or water bath operated at a temperature of 37 ± 0.5 C (see figure 62). Do not agitate the samples after the incubation period has begun. Record the time when incubation begins.

NOTE: Do not use an incubator or water bath in which the heating element consists of electric lights unless these lights are shielded so as to prevent the light from reaching the samples. Similarly in preparing the samples for incubation, avoid direct sunlight or other strong light by shielding the tubes.

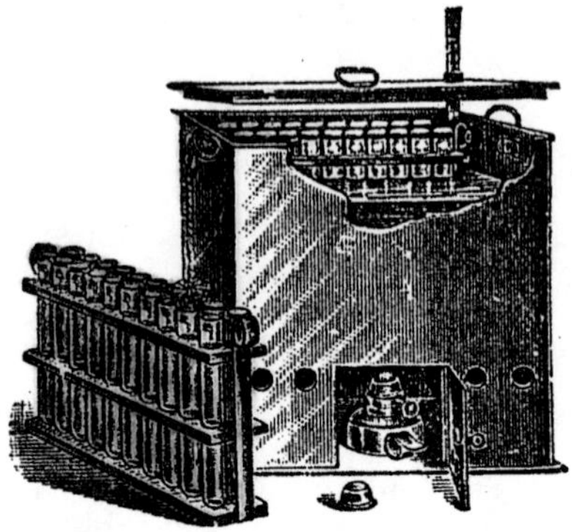

Figure 62.
Water bath for reduction test. Courtesy Dairy Industries Supply Co

6. Observe the tubes either at sufficiently frequent intervals to determine the actual time that each sample reduces or make observations only at predetermined intervals such as after one, two, six, and eight hours. In the latter case record the identities of tubes which have reduced since the previous observation.

NOTE: Sometimes the end-point of reduction is difficult to determine because of uneven disappearance of color throughout the tube. In such cases it is recommended that when four fifths of the visual portion of the contents of the tube has turned white the sample be regarded as reduced.

Modifications of the Methylene Blue Technique

Owing to the fact that many bacteria are swept to the surface of the test tube as the fat rises in the milk, inconsistencies are frequently observed between the reduction time and the bacterial counts determined by other methods. In order to offset this factor of the sweeping action of the fat, the so-called "English Modification" proposes that the tubes be gently inverted each half hour during the incubation period.

Another modification has been proposed by Johns[13] whereby the tubes are examined after two, four and six hours, at which times all samples that have not reduced are gently inverted. By this method an improvement was observed in the correlation between the reduction times and the bacterial numbers as determined by direct microscopic examination.[10, 11]

Precautions

1. When the bacterial population of a sample of milk is sufficiently low as to require considerably in excess of eight hours to

decolorize, the kind of organisms may overshadow the number of organisms as a determining factor in reduction time. For this reason it is not desirable to attempt fine differentiation of samples having reduction times beyond an arbitrary limit of eight to ten hours. Abele[11] found that a substantial percentage of the samples that were passed by a reduction time of six hours or longer were subsequently found to have plate counts in excess of the 200,000 per ml permitted.

2. Under conditions of field use it may not always be practical to use sterilized test tubes and pipettes. A 10-ml dipper is frequently substituted for the sterile pipettes and the dipper is subjected to "practical sterilization" by keeping it submerged in hot water (near boiling) between uses (see figure 63). The test tubes and stoppers are likewise subjected to practical sterilization by placing them in actively boiling water for ten minutes prior to use. Do not use chlorine solutions for sterilization of apparatus for this test, because chlorine may influence the test.

NOTE: The sampling dipper, illustrated in figure 63, can be used for collection of samples for both the methylene blue and the reszaurin tests. The handle extends 23.5 inches above the top of the dipper.

Figure 63.
Stainless steel, round bottom dipper with overflow and pouring lip. Courtesy of Sediment Testing Supply Co.

3. Metal stoppers may be used in place of rubber stoppers provided they are leak proof.

4. Test tubes without lips are preferred for this test and should be of uniform bore to facilitate free interchange of stoppers.

Resazurin Test

The use of resazurin as an oxidation-reduction indicator, originally developed in Germany,[1, 2] was further studied by Ramsdell and his associates.[3] This dye is more electropositive and more sensitive than methylene blue to slight changes in the oxidation-reduction potential induced by bacteria, leucoytes, and light. Comparisons of reduction times with resazurin and with methylene blue indicate that in general resazurin is reduced in about three-fourths the time required to reduce methylene blue. This ratio however is not fixed and varies with different samples. Golding and Jorgensen (12) have reported on the correlation of the resazurin grade with the standard plate count.

Resazurin reduction occurs in at least two stages, the first an irreversible change from the blue resazurin to the pink resorufin, and the second a reversible change from the pink resorufin to the colorless hydroresorufin.

Apparatus

1. Water bath at 37 C (98.6 F) equipped with suitable cover to protect the samples from light.
2. Pipettes, 1 ml, graduated in 0.1 ml divisions.
3. Sterile test tubes.
4. Sterile pipettes or dippers, 10 ml.

Reagents

Prepare a 0.05 per cent aqueous stock solution of resazurin. The period of utility of the solution may be prolonged if sterile water is used, and aseptic precautions are employed in its preparation. The stock solution should be kept in a tightly stoppered amber colored bottle under refrigeration. Only sterile pipettes should be used in transfering the solution from the bottle.

Procedure

1. Place 10 ml of milk in a sterile test tube. Add 0.1 ml of the 0.05 percent aqueous solution of resazurin dye.

NOTE: Some workers prefer to prepare a 0.005 per cent solution of the dye and use 1 ml, because of greater ease of measuring the larger amount under the conditions at a receiving platform where speed is important. Others prepare sterile vials containing 0.05 mg. of dry resazurin.[14]

2. After gentle mixing, place the tube in the water bath at 37 C (98.6 F) and replace the cover on the water bath to exclude light from the tube (see figure 64).

3. Record the time that incubation is begun.

4. Observe at suitable intervals to determine the rate or time of reduction to the pink color (resorufin) or to the colorless (white) hydroresorufin.

Reading the Results

Several procedures have been proposed for reading the end-point of resazurin tests. Ramsdell[3] prepared a set of ten color standards representing the progressive color changes from the blue to the full pink stage. These were prepared from various mixtures of undyed china clay and china clay dyed with bromcresol green, crystal violet, and alkaline phenolphthalein. Specific directions for making the mixtures were not given by Ramsdell. The tubes were given odd numbers from 1 to 19, with No. 1 indicating the blue color and No. 19 the full pink color. Comparison of the standards with the milk tubes containing the dye enabled the designation of twenty intergrades of color changes.

Figure 64.

Insulated water bath thermostatically controlled at 37 C (98.6 F) for incubating resazurin tubes. Note that the insulated lid helps control the temperature and excludes light from the tubes during incubation.

Another method commonly employed to differentiate merely acceptable and unacceptable milks is to identify those samples which have reduced to a full pink, or to the colorless stage, after an arbitrary period (for example, one hour) of incubation. Although this method may be advantageous from the standpoint of practicability, it substitutes convenience for precision. According to Frayer[4] and Johns[5, 6] when readings are made on this basis the results correlate poorly with the methylene blue reduction test and hardly at all with the plate count.

A third method is to determine the length of time required for the milk to change from the blue to the full pink color. Readings made at 15-minute intervals with removal of the tubes which show the full pink color (or white), give results which closely parallel those obtained with the methylene blue reduction test except that the time required is materially less. Various workers have correlated the results of resazurin tests with other methods of determining quality, but the establishment of standards must await the accumulation of more experience than the existing data afford.

REFERENCES

1. Waldbauer, Olga: The Resazurin Reductase Test, *Kiserlet, Kozlem. 34*: 128-132 (1931) (abstracted in *Chem. Abs. 26*:2797 (1932).

2. Mundinger, E., and Wolf, E.: Die Azurufinprobe, eine verbesserte Reduktase-probe. *Molk. Ztg.* (Hildesheim) *47*:171-173 (1932).

3. Ramsdell, G. A., Johnson, W. T. Jr., and Evans, F. R.: Investigations of Resazurin as an Indicator of the Sanitary Condition of Milk, *J. Dairy Sci. 18*:705-717 (1935).

4. Frayer, J. M.: The Resazurin Test—A Preliminary Study, *Vermont Agri. Exp. Sta.* Bul. 435 (1938).

5. Johns, C. K.: Place of the Methylene Blue and Resazurin Reduction Tests in a Milk Control Program, *Amer. J. Pub. Health 29*:239-247 (1939).

6. Johns, C. K.: The Methylene Blue and Resazurin Tests for Milk, *Canadian Dairy and Ice Cream J. 18*:27 (1939).

7. Schacht, F. L., and Nickols, R. E.: Studies of the Resazurin-Rennet Test, *J. Milk Tech. No. 1. 4*:281-284 (1941).

8. Keenan, J. A., Barrett, W. D., and Rutan, H.: The Resazurin Test—Preliminary Studies on Its Practicalities and Possibilities. *Ibid. 1*:22-25 (1937).

9. Collins, M. A.: Application of the Resazurin Test in Determining the Quality of Raw Milk and Cream, *Ibid. 2*:236-244 (1939).

10. Johns, C. K.: Observations Concerning the Methylene-blue Reduction Test, *Ibid. 8*:80-84 (1945).

11. Abele, C. A.: The Methylene-blue Reduction Test as a Means of Estimating the Bacterial Content of Milk to Determine Its Suitability for Pasteurization or as a Basis of Grading, *Ibid. 8*:67-79 (1945).

12. Golding, N. S., and Jorgensen, Irene: A Correlation of the Resazurin Grade with the Standard Plate Count of Raw Milk, *Ibid. 8*:189-195 (1945).

13. Johns, C. K.: Measuring the Bacteriological Quality of Milk. *J. Milk Tech. No. 2*: 175-180 (1939).

14. Golding, N. S.: The Resazurin Reductase Test Using Prepared Sterile Dry Vials, *J. Dairy Sci. 30*: 521 (1947).

CHAPTER 10

SPECIAL CULTURE PROPAGATION

Propagation of Butter Cultures in the Bacteriological Laboratory

Apparatus

1. Erlenmeyer flasks (200 ml capacity) with cotton plugs.
2. Pipettes (1.1 ml capacity).
3. Culture cabinets.

NOTE: The temperature of incubation of starters is most important for consistent and satisfactory results. No culture cabinet should be used in which proper incubation temperature within a range of ±2 F (1.1 C) cannot be maintained.

4. Funnel for filling flasks with skimmilk.

Reagent

1. Lactic culture (powdered or liquid).

Procedure

1. Sterilize three flasks (200 ml capacity) containing 100 ml of skimmilk in the autoclave at 15 pounds for 15 minutes.

NOTE: Use a funnel in filling to avoid wetting the mouths of the flasks. Excessive wetting of cotton plugs may be avoided by covering them with parchment paper or aluminum foil.

2. Cool the flasks to 80 F (26.6 C) for powdered culture and to 74 F (23.3 C) for liquid culture.

3. Inoculate each flask with the amount of the culture recommended by the laboratory where the culture was prepared.

4. Incubate the three flasks at the inoculating temperatures stated in (2) until the curd is formed. Succeeding propagations should be inoculated and incubated at 70-72 F (21.1-22.2 C).

5. Remove from the incubator and hold at 40 F (4.4 C) or lower until the next transfer is made.

6. Transfers should be numbered and each number carried as a separate culture. Depending on the condition of the different cultures, the best of the group is chosen for inoculating the mother culture. The condition of the cultures governs the amount of inoculation. Two per cent of inoculum is generally necessary for the first propagations, and after four or five propagations 1 per cent will usually suffice.

Starter Making

Apparatus

1. Arnold sterilizer (large enough to hold six 2-liter culture flasks).
2. Culture cabinet (see note page 153).
3. 2-liter culture flasks.
4. Supply of sterilized cotton plugs with parchment or foil hoods.
5. Enamelware dipper.
6. 60-ml pipettes.

Procedure

First Propagation

1. Thoroughly wash and rinse all flasks, pipettes, dipper, and thermometer. Fill the culture flasks about two-thirds full of milk (preferably whole milk). Place the partially filled flasks sealed with cotton plugs with all the other utensils in the Arnold sterilizer, and apply flowing steam for not less than one hour. As an alternate sterilization method, heat the milk in flasks in a water bath to 190 F (87.7 C) using a pilot flask to determine the temperature. Hold the milk at this temperature 30 minutes.

2. Cool the milk to 68-70 F (20-21.1 C).

3. Flame the mouth of the culture bottle or the tube containing the laboratory culture. Remove the plug and transfer with a sterilized pipette 20 or 30 ml of the laboratory culture into one flask containing the sterilized milk at inoculating temperature, mix thoroughly by rotating, and replace the sterilized plug to seal the flask.

4. Place the flask in the culture cabinet and let it incubate at 68-70 F (20-21.1 C), undisturbed until a soft curd has formed. This usually requires from 18 to 24 hours for the first propagation.

5. Hold below 40 F (4.4 C) until ready to make the second propagation.

Second Propagation

1. Prepare and sterilize the utensils and the milk as outlined above in "First Propagation." Have at least six flasks of sterilized milk for this propagation.

2. Cool the milk to 68-70 F as before.

3. Shake the flask containing the first propagation until the curd is reduced to a uniform liquid. With the sterile pipette, transfer one-half of a pipetteful (30 ml) to the first bottle, one pipetteful (60 ml)

to the second, 90 ml to the third, 120 ml to the fourth, 150 ml to the fifth, and 180 ml to the sixth flask of sterilized milk. Mix thoroughly, flame the mouth of the bottle, and seal.

4. Place the inoculated bottles in the culture cabinet (or in the water bath of the culture cabinet) and incubate at 68-70 F, undisturbed until a soft curd has formed.

Third Propagation—Preparation of Mother Starter

1. Prepare a sufficient number of flasks of sterilized milk to provide an adequate quantity of mother starter to seed the batch starter and also retain one flask of mother starter for daily transfers.

2. Thoroughly shake the six flasks of the second propagation. Test each bottle of the second propagation for flavor and aroma by pouring a small amount into a cup, and tasting and smelling, replacing the plugs in each flask as soon as sampled. (Use aseptic precautions in removing the sample for tasting.)

3. Select the flask having the most desirable flavor, aroma, and body, and use this flask for inoculating the third propagation.

4. Be sure that the sterilized milk has been cooled to 68-70 F before making the transfers. Mix the inoculated flasks thoroughly and incubate them at 68-70 F without disturbing, until a soft curd has formed. This will usually require approximately 18 hours. The mother starter is now ready for inoculating the batch starter.

NOTE: The temperature of incubation of starters is most important for consistent and satisfactory results. No culture cabinet should be used in which the proper incubation temperature cannot be maintained within a range of ± 2 F (1.1 C).

Fourth Propagation—Preparation of Batch Starter

1. Thoroughly wash and rinse the starter can or starter vat. Add a sufficient quantity of milk (preferably whole milk) to provide inoculation of the batch starters and pasteurize at 190 F (88 C) for one hour. Cool to 68-70 F. Allow the cans to stand at this temperature for at least ten minutes to insure uniform temperature before inoculation.

2. Under aseptic conditions, transfer approximately 2 per cent of inoculum to the sterilized milk and thoroughly mix. If 40-quart cans are used, place them in a water bath properly insulated and containing sufficient volume of water to maintain a temperature within the limits of 68-70 F for the period of incubation. Have sufficient volume of water in the tank to immerse the cans up to the throat.

3. To insure full development of flavor, make the incubation period long enough to allow at least two or three hours time after the curd has formed.

4. After incubation, transfer the cans to cold water and cool to 40 F (4.4 C) until used.

5. If a water jacketed vat is used for the preparation of the batch starter, have it sufficiently well insulated and provided with a tight cover to enable accurate control of the temperature of incubation and as a protection against contamination during the period of incubation. Do not open the vat during incubation. After incubation, cool the culture to 40 F with as little agitation as possible.

CHAPTER 11

DETERMINATION OF YEASTS AND MOLDS

Determination in Butter

Total bacterial counts cannot logically be used in determining the general conditions surrounding the manufacture and handling of butter because cultures of specific organisms are frequently added to the cream and occasionally directly to the butter itself, with the result that the bacterial content of the finished butter is thus influenced. Yeast and mold counts of butter have, accordingly, been suggested because these microorganisms should be present in very small numbers, if at all, in butter that is efficiently made from pasteurized cream and properly handled in thoroughly cleansed equipment.[1]

Apparatus

1. Sterile screw-cap or glass-stopper glass jars of suitable size (4 oz. size is convenient).
2. Petri dishes and covers (100 x 15 mm).
3. Pipettes (1.1 ml and 10 ml in 1/10ths).
4. Water blanks 90 ml.
5. Water bath 45 C (113 F).
6. Incubator 21-25 C (69.8-77 F).
7. Autoclave.
8. Potentiometer for electrometric pH determinations, or spot plate and bromphenol blue indicator solution for colorimetric determinations.

Reagents

1. Sterile potato dextrose agar (Bacto or equivalent) (see page 518.
2. Sterile U. S. P. tartaric acid solution (10 per cent).

Preparation of Dilutions

1. Warm the sample of butter contained in the sterile jar, as well as a sterile water blank, to about 40 C (104 F), in a water bath of 43-45 C (109.4-113 F). The time required for melting the butter should not exceed 15 minutes.
2. Agitate thoroughly so as to obtain uniform mixing of the serum, water, and fat.

3. With a previously warmed sterile 10 ml pipette, transfer 10 ml of butter to a 90 ml sterile water blank, the water in which is at 37-40 C (98.6-104 F) (11 ml of butter may be added to 99 ml of water to give the same proportionate 1 to 10 dilution).

4. Shake this dilution 25 times in the usual manner just before inoculating Petri dishes with the different dilutions in duplicate.

5. The suggested dilutions are: 1 : 2 (5 ml of the 1 : 10 dilution); 1 : 10 (1 ml of the 1 : 10 dilution); and 1 : 100 (0.1 ml of the 1 : 10 dilution.

Incubation and Colony Counting

The Petri dishes containing the different dilutions are flooded with the melted adjusted potato dextrose agar (see page 519). Not more than 30 minutes should elapse from the time of preparing dilutions to the pouring of the potato dextrose agar on the plates. After solidification of the agar the plates are incubated for 5 days at 21 or 25 C (69.8 or 77 F).

At the end of the incubation period, count the colonies in the same manner as counting bacterial colonies in the plate count for milk, if interested only in the total yeast and mold count. Generally, it is desirable to differentiate between molds and yeasts.

Make a separate count of the yeast colonies, which usually will be characterized as smooth, moist, elevated, surface colonies.

After counting the typical yeast colonies, count the mold colonies. Mold colonies are easily recognized by their profuse growth of hyphae.

Although the acidity of the medium is supposed to inhibit the growth of bacterial colonies, some may develop in spite of the acid. Usually these can be distinguished from the yeast colonies because they are smaller. If in doubt as to yeast colonies or bacterial colonies, the colony should be confirmed by microscopic observation of stained smears. Yeasts are much larger than bacteria under the microscope, and show buds.

Reporting Results

The number of yeast and mold colonies per milliliter of butter should be reported as the total yeast and mold count, although in control work the separate yeast and mold counts are sometimes informative. The colony counts obtained from the 1 : 2 dilution (5 ml of the 1 : 10 dilution) will be multiplied by the factor 2, those from the 1 : 10 dilution (1 ml of the 1 : 10 dilution) by the factor 10, and those from the 1 : 100 dilution (0.1 ml of the 1 : 10 dilution) by the factor 100, to give the actual colony counts per milliliter of butter.

Standards

Different standards for acceptable total yeast and mold counts per milliliter of butter have been suggested by various laboratories. The following are tentative standards, based upon the results of actual creamery control work:

Total yeast and mold count per ml of butter	Sanitary index
1 to 20	Good
21 to 50	Fair
51 to 100	Poor
Over 100	Very poor

NOTE: The foregoing method is recommended by the American Butter Institute, and is taken from their *Laboratory Manual* with permission.

Parson's Method for Visual Demonstration of Mold in Cream

The age of cream to be used in butter making may be in part revealed by the number and size of mold mycelia. The following procedure is designed to concentrate the stained mold mycelia on a small organdy filter disc for visual inspection. See figure 65.

Apparatus

1. Pipette 9 ml.
2. Cream sample jars (1¾" outside diameter) 12 2-ounce.
3. Enamel cup 1 qt.
4. Filtering device.

NOTE: An effective filtering device can be made from a "Superior" or new-type "Illinois" sediment tester by cutting through the tube about 2½" from the bottom. One of these testers is so constructed that the bottom portion holding the filtering disc can be removed, and it then works very satisfactorily as a filtering device.

5. 1-qt. lard pail or its equivalent for collecting filtrates from the mold determinations.
6. 6-in. diameter tin or glass funnel.
7. 1-gallon bottle.

Reagents

1. 120 gm. dry borax (approx. 4½ oz.).
2. 225 gm. crystalline disodium phosphate (approx. 8 oz.).
3. 30 gm. sodium hexametaphosphate (approx. 1 oz.).
4. 0.4 gm. medicinal methylene blue powder.

NOTE: Packages of complete reagent mixed in dry powder form sufficient to make 3½ quarts of reagent solution may be purchased from the Sediment Testing Supply Co., 20 E. Jackson Blvd., Chicago; Schaar & Co., 754 West Lexington St., Chicago; Moseley Laboratories, Indianapolis, Ind.; Iowa Dairy Laboratories, 2412 Maryland Pike, Des Moines, Iowa.

5. 3½ quarts clean, soft water. (If tap water is hard, use distilled water or condensed steam free from traces of oil and boiler compound.)

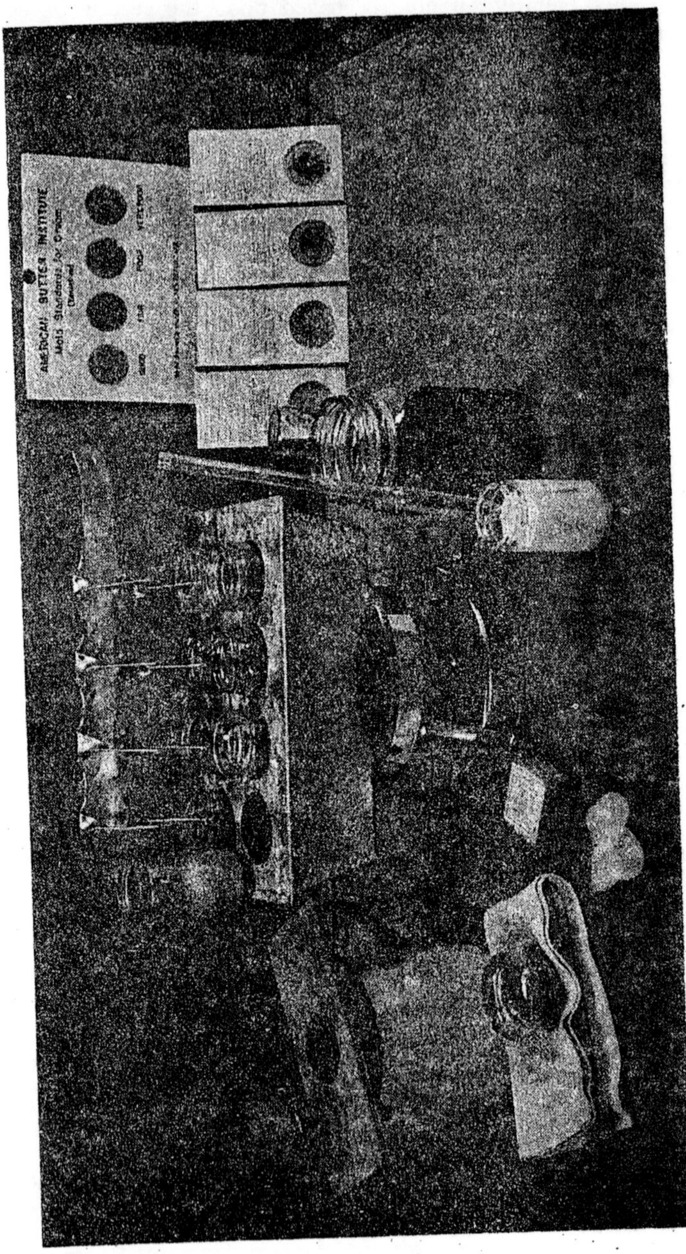

Figure 65.

The equipment used in the Parsons method of detecting molds in cream. Courtesy of Dr. C. H. Parsons and of the American Butter Institute.

NOTE: The ingredients given above should first be mixed together in dry form. It will be found advantageous to transfer the dry powder into a quart cup nearly filled with warm water, stir for 10 or 15 minutes, and pour the solution into a gallon bottle through a funnel containing a tightly woven cloth. The cup is again filled with warm water and the operation repeated until the powder is completely dissolved and the amount of solution is no more than 3½ quarts. It is important that all the powder be dissolved.

Procedure

Cream should be thoroughly stirred before taking the sample. The amount of cream required for the mold test can be drawn from the sample taken for the fat test. Vat samples from the creamery should always be taken before neutralization and after heating to 100 to 110 F (37.7-43.3 C).

Nine ml of the well mixed sample are pipetted into a 2-ounce cream sample jar. Add 17.6 ml of the reagent solution (a 17.6 ml acid dipper or pipette can be used). The jar containing the mixture of cream and reagent is then transferred to a water bath having a temperature of not less than 160 F (71.1 C) and not more than 180 F (82.2 C). The mixture in the jar is then stirred with a gentle back and forth motion for a period of 3 minutes, using a metal strip ¾" wide or a special stirring device with blades ¾" wide so constructed as to allow several samples to be stirred at the same time.

The jar is allowed to remain in the water bath for at least another ½ minute, after which it is removed and rotated four or five revolutions to bring the mold to its center. The contents of the jar are next poured through a filtering device in which an organdy disc has been properly mounted so as to insure an even distribution of mold filaments over the filtering surface (1 inch diameter). No washing need be given the filter disc.

Press the bottom of the filter screen on a pad of dry cloth to absorb the residual solution or fat on the filter disc. Remove the disc from the filtering device and compare it with photographic standards (obtainable from American Butter Institute, 110 North Franklin St., Chicago 6, Ill.). Report the results as Good, Fair, Poor, Very Poor. See figure 66.

NOTE: The visual method, modified in certain respects, has been applied experimentally to butter, but requires further study before its general use can be recommended.

Summary of Steps in the Foregoing Method

1. Mix sample thoroughly.
2. Transfer 9 ml to a cream sample jar.
3. Add 17.6 ml of mold reagent solution.
4. Place cream sample jar containing the mixture in a water bath having a temperature of 160-180 F (71.1-82.2 C).

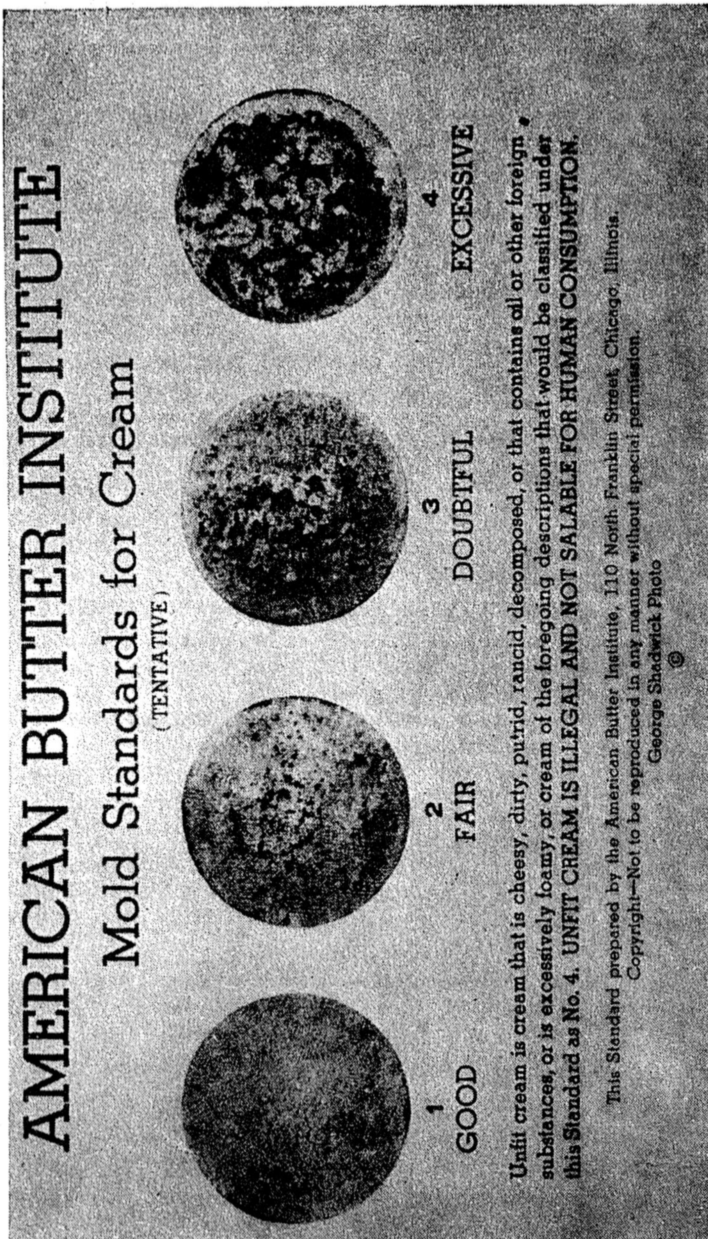

Figure 66.
Tentative standard for the mold in cream as developed by the American Butter Institute. Reproduced with the permission of the American Butter Institute.

5. Stir for 3 minutes with a gentle back and forth motion.

6. Allow the jar to remain in the water bath undisturbed for at least ½ minute.

7. Remove the jar from the water bath and gently rotate four or five revolutions.

8. Transfer the contents to the filtering device.

9. After the solution has passed through the filter, press the bottom of the filter screen on the pad of dry cloth.

10. Remove the filtering disc from the filtering device.

11. Compare with standards and classify.

NOTE: It is very important that the temperature in the water bath be *not less than 160° F (71.1° C) and not more than 180° F (82.2° C)*, and that the reagent containing the cream at the end of the test be at least 140° F (60° C) and not over 150° F (65.5° C). If the temperature is lower than 140° F (60° C), the mold may be contaminated with curd and, in addition, filtration through the organdy disc may be found difficult.

Wildman Method of Detecting Mold in Butter

Apparatus

1. Compound microscope, equipped with good objectives and oculars, giving magnifications of approximately 90, 180, and 500 diameters. For convenience of use the lenses should be adjusted so as to be parfocal. A mechanical stage is highly desirable. It is essential that the combination giving the low magnification be capable of adjustment to give an area of the field of view of 1.5 sq mm (a circle whose diameter is 1.382 mm). With the higher powers, the working distances must be ample to allow the free use of the blood-counting cell.

2. Ocular drop-in disc cross-ruled in sixths of the ocular diaphragm opening, for estimating lengths of mold filaments.

3. Howard mold-counting cell constructed like a blood-counting cell but with unruled central disc about 19 mm in diameter.

Procedure

1. Clean the special Howard cell so that Newton's rings are produced between the slide and the cover glass. Remove the cover and place a small drop of the well mixed sample upon the central disc. Using a knife blade or scalpel, spread the drop evenly over the disc, and cover with the glass so as to give an even spread to the material.

It is of the utmost importance that the drop be taken from a thoroughly mixed sample and spread evenly over the slide disc. Otherwise, when the cover glass is put in place, the insoluble material and consequently the molds may be more abundant at the center of the mount. Avoid using a drop that is much greater than is sufficient to fill the space between the center disc and cover glass. Discard any mount showing uneven distribution, absence of Newton's rings, or liquid that has been drawn across the moat and under the cover glass.

2. Place the slide under the microscope and examine with such adjustment that each field of view covers 1.5 sq mm. This area, which is of vital importance, may frequently be obtained by adjusting the draw tube in such a way that the diameter of the field becomes 1.382 mm (see page 121). In case such adjustment by means of the draw tube is not possible, it is sometimes necessary to make an accessory drop-in ocular diaphragm with the aperture accurately cut to the necessary size. The diameter of the area of the field of view can be determined by use of a stage micrometer, or by employing the rulings on the blood counting cell. In order to use the latter method it is necessary to bear in mind that a square whose diagonal is 1.382 mm has sides of approximately 0.977 mm. Hence the millimeter scale on the blood counting cell can be used by such adjustment that the circle of the field of view cuts off the necessary amount from each corner of the millimeter ruled square. When the instrument is properly adjusted, the quantity of liquid examined per field is 0.15 cu mm (0.00015 ml).

3. Examine at least twenty-field fields from each of two or more mounts taken in such manner as to be representative of all sections of the mount. Observe each field, noting the presence or absence of mold filaments and recording the results as positive or negative. Calculate and report the percentage of fields containing mold filaments.

Mold Mycelia in Butter

(Tentative Method)

Reagent

1. Gum solution.—Make up 1 liter of a 0.75 per cent solution of carob bean gum with 2 per cent of formaldehyde added as a preservative.

NOTE: The dry gum solution may be conveniently prepared by first mixing it in 10-15 ml of 95 per cent alcohol and stirring this mixture rapidly into the water.

2. Gently heat the solution to boiling to drive off alcohol and air, and sustain the heating for 25-30 minutes.

3. Add formaldehyde on cooling.

4. Use the clear supernatant solution, free from cells, left when the cellular elements in the gum gradually settle out. A similar solution made with gum tragacanth may also be used for the purpose.

Procedure

1. Make a careful examination of the surface of the butter to insure freedom from surface mold growth and note any mold growth visible.

2. In order to remove the possibility of contamination of any surface mold not visible, scrape off and discard ⅛ inch of the surface, after which take a sample from the exposed surface.

3. Weigh 1 gm of butter into a ¼-teaspoon measure.

4. Measure 7 ml of the hot gum solution, and with the spoon bottom side up over a 50-ml beaker, pour 2 or 3 ml of the hot solution over the spoon. (This quantity is usually sufficient to loosen the butter and cause it to slide into the beaker.)

5. Use the remainder of the 7 ml of solution to rinse the rest of the fat from the spoon.

6. Stir the mixture until the solution is well mixed and fat globules are 0.1 - 0.2 mm in diameter. (The stirring necessary to obtain a uniform sample must be determined by experience.)

7. Mount a portion of the mixture on the mold-counting slide and estimate the mold as directed above.

8. Report no field positive unless the combined length of the two longest filaments exceeds ⅙ of the diameter of the field.

Practical Determination of the Keeping Quality of Butter

The following practical method of determining the keeping quality of butter, as a useful general index, consists of scoring samples of butter at a temperature of about 50 F (10 C); holding these samples in small jars or in prints at a temperature of 70 F (21.1 C) for a period of eight days, and scoring these samples again at the end of this period after the temperature of the butter has been again adjusted to about 50 F (10 C). The difference between the first and second scores indicates the degree of deterioration which the butter has undergone during the test period.

Experience has shown that this test indicates, in a period of eight days, the degree of deterioration that butter is likely to suffer while in trade channels, the test approximating the variable and detrimental conditions which the butter may encounter before it finally reaches the consumer.

Apparatus

1. Sterile screw-top glass jars, 2-ounce capacity.
2. Sterile table knife or spatula, for taking samples.

3. Incubator capable of maintaining a uniform temperature of 70 F (21.1 C).

4. Cold room or refrigerator at a temperature of about 50 F (10 C).

Procedure

A 2-ounce sterilized screw-top glass jar is solidly filled within about one-half inch of the top with butter from the churn by means of a sterile table knife or spatula. (Quarter pound or one pound prints may be used for samples as an alternative to the jars.) The sample is transferred to an incubator carrying a uniform temperature of 70 F, where it is held for a period of eight days. At the end of this time the sample is transferred from the incubator to a refrigerator at a temperature of about 50 F, where it is held over night. On the following morning the sample is again scored. In addition to recording the difference between the score at the beginning and at the end of the test period, outstanding flavor defects, such as old cream, storage, metallic, tallowy, rancid, oily, fishy, cheesy, limburger, and weedy, should be noted.

NOTE: The temperature in the incubator should not rise more than 2 or 3 degrees above 70 F. (21.1° C.). If the butter oils off, the test will have no value.

It is recommended that the sample be scored at a temperature of about 50 F. (10° C.), both before and after the keeping test, as experience has shown that more consistent results are obtained by so doing.

NOTE: The procedure outlined is recommended by the American Butter Institute and has been taken from their Laboratory Manual with permission.

Determination of Yeasts and Molds in Soft Cheeses

Apparatus

1. Wide-mouth dilution bottles of approximately 150-200 ml capacity.
2. Sterile spoon or spatula.
3. Torsion or other sensitive balance.

Reagent

Potato-dextrose agar acidified to a pH of 3.5 with a sterile 10 per cent solution of tartaric or lactic acid.

NOTE: The acidity of the agar should not be adjusted to pH 3.5 until after the sterilized medium has been cooled to approximately 45 C (113 F) (see page 519).

Procedure

1. Mix the sample thoroughly with a sterile spatula.

2. Weigh 11 gm of the cheese directly into a 99 milliliter water blank.

3. Shake the mixture vigorously.

4. Make various dilutions (see figure 19, page 41).

5. Pour plates with the potato-dextrose agar adjusted to a pH of 3.5 (see page 519).

6. Incubate the plates five days at 21-25 C (70-77 F).

7. Count as molds all colonies with long filaments or aerial mycelia. Count as yeasts colonies which are definitely not mold colonies. Yeast colonies are generally smooth and moist. A low power microscope may be used in identifying doubtful colonies.

Microbial Control of Parchment Wrappers and Liners

Treatment of Liners for Salted Butter

Make a saturated solution of brine (in excess of 36 per cent) by dissolving salt in hot (200 F) (93.3 C) water. The saturation of the solution is indicated when salt remains undissolved in the bottom of the container after the solution has been thoroughly stirred. The parchment should remain in this brine solution for at least 30 minutes.

Treatment of Liners for Unsalted Butter

Submerge the liners in boiling water and allow to soak for at least one hour.

REFERENCES

1. Laboratory Manual—Methods of Analysis, American Butter Institute, Chicago, Ill.

PART III

CHEMICAL CONTROL METHODS FOR DAIRY PRODUCTS

CHAPTER 12

COLLECTION AND CARE OF SAMPLES

Milk and Cream

The sample collected for chemical analysis must be truly representative of the entire batch of milk or cream that is being sampled. Since milk fat is of lower specific gravity than the other constituents of milk or cream, it tends to rise to the surface, hence thorough mixing of milk and cream is essential to obtain a sample that can be regarded as representative of the entire batch. Whenever practical, this mixing may be accomplished by pouring the entire batch from one container to another three or four times. Larger batches may be thoroughly agitated by hand or by mechanical means.

A. Sample from Individual Cow. Milk the cow dry, because the last strippings are the richest in milk fat. Pour the entire contents of the milking from one container to another three or four times, and sample about 200 ml in a clean, dry bottle provided with a tight fitting rubber stopper or other closure impervious to fat.

B. Sample from Individual Can. Pour the milk or cream from one can to another three or four times, or, when this is not practical, mix thoroughly with a hand agitator. Be certain that any milk fat adhering to the neck and under the shoulder of the can is well mixed with the remainder of the milk or cream. Collect about 200 ml in a clean, dry, tightly stoppered bottle. (Cork, cotton, or other absorbent materials should not be used for sample bottle closures.)

C. Sample from Several Cans. When practical, dump the contents of the cans into a vat, mix thoroughly, and sample from the vat. When such procedure is not practical, a composite sample may be taken from the cans after the milk or cream has been properly agitated. To be truly representative, the composite sample should be made up of proportionate amounts, and may be collected by inserting a pipette or milk thief and removing a column of milk or cream representing the full depth of the product in each can. Do not sample with a pipette or milk thief unless the milk or cream has been thoroughly agitated.

When a shipment consists of a large number of cans, it is common practice to make up a composite sample from at least 10 per cent of the cans, well distributed throughout the shipment. Such a sample would be representative only if the cans had been filled with well-agitated milk from a tank in which the entire batch had been collected.

D. Sample from Tank Car, Tank Truck, or Storage Truck. Practically all milk storage tanks, tanks on railroad cars, and some tank trucks are equipped with built-in, motor-driven agitators with which the milk or cream may be thoroughly mixed. Hand agitation of milk or cream in tanks with top ports is possible but not as reliable as agitation by a portable, motor-driven mixer. Milk in a tank may be properly agitated by the application of a stream of filtered air. After proper mixing of the milk or cream the sample may be dipped from the tank, removed through the stop cock in the tank door or from a valve in the discharge line from the tank as the tank is being emptied.

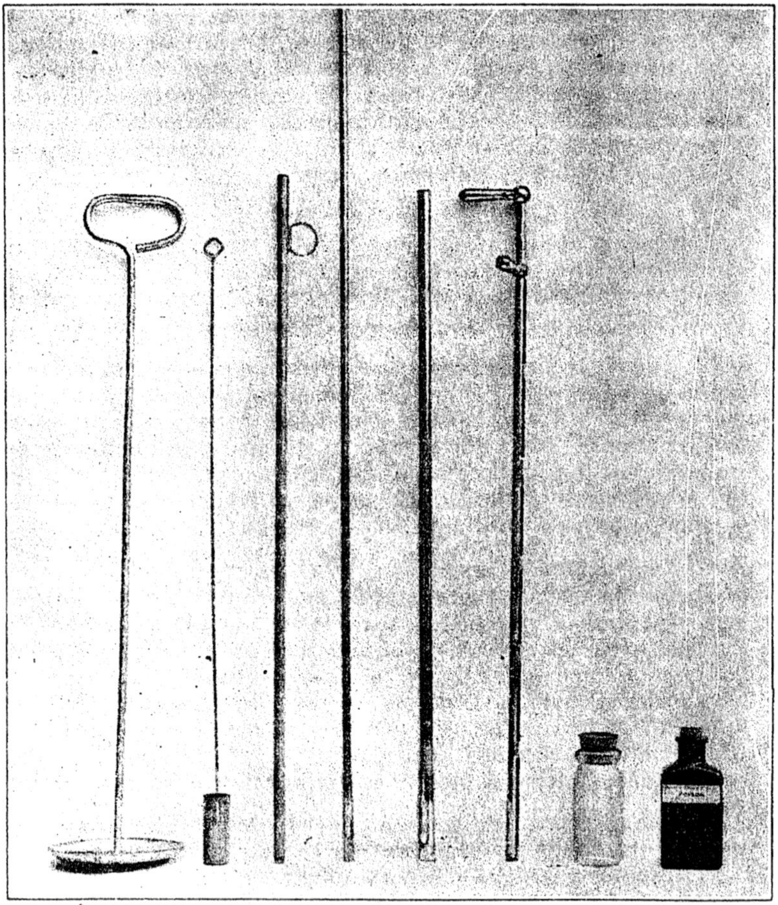

Figure 67.
Equipment used in taking samples. Courtesy of the Pennsylvania State College.

E. Drip Sample from Pipe Line. A composite sample representative of the milk received each day at the plant may be collected by a drip sampler installed at a position where adequate mixing of the milk can be expected in the pipe line leading from the weigh tank. The composite drip sample is satisfactory only if the sampler delivers an amount of milk proportionate to the total weight of milk received. The drip sampler should be at a high point in the line, so installed that it will deliver milk only during periods of forward flow from the weigh tank. Usually a ten-quart milk can will provide ample capacity for the composite drip sample. Icing of composite drip samples during accumulation is desirable, especially in warm weather. The composite drip sample should be thoroughly mixed before samples are removed for testing.

F. Sampling of Milk and Cream Under Abnormal Conditions. In general, avoid the sampling of milk or cream in such abnormal conditions that uniform distribution of the milk fat is not attainable. This applies to milk or cream partially churned, frozen, partially frozen, or curdled. Should it be necessary to sample abnormal milk or cream, proceed as indicated:

G. Partially Churned Milk. Heat the milk gently to 110 F (43.3 C), agitate thoroughly, and remove the sample while the milk is under constant agitation. A truly representative sample cannot be guaranteed under these circumstances. It will not be practical to warm up a batch of milk of any considerable volume, hence the attempted sampling of larger batches than one or two quarts of partially churned milk is to be discouraged.

H. Frozen Milk and Cream. Defrost slowly with agitation in a water bath maintained at a temperature not over 130 F (54.4 C). Rapid defrosting at higher temperatures will induce an "oiling off" of the milk fat and further complicate the problem of securing a representative sample. Do not attempt to include frozen milk or cream from the weigh vat as a part of the representative portion of the daily composite sample.

I. Sour or Curdled Milk. If the curd is not too hard, pour several times from one container to another through a fine wire sieve to break up the curd particles and thus promote uniform mixing. Sample immediately while under constant agitation. It would not be practical to sample a volume greater than one or two quarts of sour or curdled milk by this method. Sour cream is not likely to have curd lumps because of the low casein content of cream, hence it can usually be sampled nearly as readily as sweet cream.

Composite Milk Samples

Dairymen are usually paid for milk on the basis of fat test. It is not practical to sample and make a daily test of the milk as delivered to the receiving station by each patron, hence the practice of accumulating a composite sample over a ten- or fifteen-day period has become customary. The volume of the individual composite sample should not be less than 140 ml and it should be collected during the ten- or fifteen-day period by placing into the patron's composite sample bottle approximately proportionate amounts of the patron's daily milk delivery.

Apparatus

1. Sample dipper with long handle (about 30 inches) of capacity not less than 10 ml if the sampling period is 15 days and not less than 15 ml if the period is 10 days.

2. Sample bottles, wide mouth, round, colorless glass, of at least eight-ounce capacity, provided with tight-fitting rubber stopper and an identification device such as wired-on metal number tag or numerals legible and permanently printed on the bottle or the stopper, if the latter is permanently attached to the bottle. A sufficient number of bottles should be provided to collect samples for the current composite period and for holding the remnants of the tested samples from the preceding period or periods as may be required by state law. See figure 68.

Figure 68.
Composite sample bottle. Courtesy of Mojonnier Brothers Company.

3. Bottle cabinet, so constructed that the composite samples are held in a fixed vertical position, protected from strong light, maintained at a temperature not exceeding 65 F (18.3 C) or below 35 F (1.6 C), and provided with a means for locking the samples when they are not under the supervision of the plant operator. See figure 69.

Figure 69.
Composite sample bottle rack used by Bowman Dairy Company.

4. Tempering bath for composite sample bottles.
5. Mixing beaker 8 to 12 fluid ounce size.
6. Accurate Fahrenheit thermometer, range 20 to 220 F.

Reagents

1. Preservative: Corrosive sublimate tablets containing not less than 0.3 gm (4.6 grains) of mercuric chloride per tablet, or other equally effective preservative.

2. Disinfectant: Four parts of 40 per cent solution of formaldehyde to 96 parts of water for the purpose of soaking rubber stoppers for at least ten minutes after washing and before re-use.

Procedure

1. Place one or not more than two corrosive sublimate tablets or a sufficient amount of other preservative in each clean, dry sample bottle.

2. Each day during the composite sampling period transfer a dipperful of each patron's well-mixed milk from the weigh vat to the sample bottle bearing the patron's number.

PRECAUTION: It is very important that the milk in the weigh vat be thoroughly mixed to insure uniform distribution of the fat before sampling; pouring milk from cans through a strainer into the weigh vat does not always insure proper mixing. The practice of dumping cooled night milk before uncooled morning milk will facilitate mixing of the fat if the weigh vat is so constructed as properly to mix the milk. If proper mixing is difficult it is suggested that baffle plates or grids be placed over the strainer or that a power driven agitator be installed in the weigh vat.

3. Replace the stopper securely to prevent evaporation, and briefly mix the contents of the sample bottle with a rotary motion. Proper mixing is important to incorporate preservative in the freshly added (daily) milk portion and to prevent the fat from forming a tough, leathery layer.

4. Store composite samples under lock and key, in the absence of strong light, and at a temperature range between 65 and 35 F (18.3 and 1.6 C).

5. Prior to testing, composite milk samples should be gently tempered to 95 to 100 F (35 to 37.8 C) in a water bath in which the temperature of the water never exceeds 120 F (48.9 C). Upon reaching proper temperature the composite milk samples should be thoroughly mixed by pouring back and forth, at least three times, between the sample bottle and a clean, dry vessel, avoiding, as much as possible, the incorporation of air.

PRECAUTION: Heating the milk sample above 100 F (37.8 C) tends to cause "oiling-off" of the fat, which makes accurate sampling impossible. Hard fat (high melting point) may be encountered during certain seasons of the year and in certain localities, depending upon the type of ration fed to the cows. In such cases, it will be advisable and feasible, without danger of causing "oiling-off," to temper these composite samples to 105 to 110 F (40.5 to 43.3 C) and to measure the test portion at this temperature. All fat adhering to the sides of the composite bottle must be mixed with the milk sample; this may be facilitated by the use of a notched rubber disc attached to a glass or metal rod. The rubber disc is used as a squeegee to wipe the fat from the interior sides of the sample bottle so that it may be incorporated with the milk.

6. When the contents of the composite sample bottle are thoroughly mixed, immediately and without cooling the milk measure the 17.6 ml portion into the Babcock milk-test bottle. Cool the contents of the test bottle to 60 F (15.6 C) and proceed as described under Babcock fat tests for milk (page 179), or Gerber test (page 221).

Composite samples should be tested as often as the state law or city ordinance prescribes. Generally this is either every ten days or twice a month. In the case of a month of thirty-one days, the last composite sample period is extended to eleven or sixteen days, depending on the required frequency of testing. Some states require duplicate testing of composite samples with the provision that payment to dairymen be based on the average of the two tests.

Ice Cream Mix and Ice Cream

A. Ice Cream Mix. Thoroughly agitate the ice cream mix and proceed as described for milk and cream.

B. Bulk Ice Cream. For regular routine test, transfer about 100-gram sample to a sample bottle by means of a spoon or spatula. For a more representative sample it is recommended that several trierfuls of ice cream be removed from different parts of the can and transferred to the sample bottle.

C. Package Ice Cream. Whenever feasible, transfer the entire contents of the package to the sample bottle. If this is not practical, cut cross section slices from different parts of the package and transfer them to the sample bottle.

Butter

A. Size of Samples. For factory control: 4-8 oz when not shipped. For Outside laboratory control: not less than 4 oz except 12 oz for combined chemical and sediment analysis. Samples that duplicate those taken by official inspectors should be of the same size and taken in the same manner as by the inspectors. To prevent evaporation or other changes of composition while in transit, all sample containers should be completely filled.

B. Butter from Churn. Cut off the top layer of the butter to avoid incorporation of surface moisture, and with the aid of a clean spatula or spoon collect 10 to 12 small portions from locations at the center and the ends of the churn. Or, insert a long trier near each end and in the center of the churning and with a knife, metal spatula, or spoon, remove four portions of the butter about 1 inch long from each trierful. Place the portions of butter in the sample bottle and replace the cap immediately.

C. Butter from Tubs. Remove a full core of butter by boring with a clean, dry trier from top to bottom at a point midway between the center and the edge of the tub. Give the trier one complete turn and withdraw a full core. Transfer to sample bottle with the aid of a spatula and replace the stopper. Moisture adhering to the outside of the trier should not find its way into the sample bottle. For extreme accuracy of sampling, proceed as indicated but remove three trierfuls from similar locations in the tub and make a composite sample of the three cores. The trier may be warmed for boring very hard butter and care should be exercised to have the trier dry before each use.

D. Butter from Individual Prints or Rolls. With the aid of a table knife or metal spatula, cut the print or roll transversely into halves and remove a slab not less than one inch thick from the freshly cut surface. Transfer the butter directly to the sample bottle and immediately seal.

Cheese

A. Hard Cheese. With cheese trier bore one or more plugs from the side to the center of the cheese. As an alternate method the cheese may be bored in three places, (a) from the top center, (b) one inch from the edge, and (c) at a point midway between these two places. Borings may extend halfway or entirely through the cheese.

Plugs of hard varieties of cheese may be cut in strips or grated in order to facilitate transfer to sample bottles.

B. Soft Cheese. Soft or semi-soft cheeses may be satisfactorily sampled by collecting two or three plugs with a cheese trier from different parts of the batch. The sample bottle should be immediately sealed to prevent evaporation.

Dry Milk

Because dry milk solids absorb moisture very readily, it is important that all utensils and containers used for sampling be ***clean and dry*** and that sampling be done as speedily as possible. Avoid sampling if the hands and clothing are not clean and dry, if the atmosphere is humid, if the cold-storage room is damp, or if the utensils or containers have just been brought from a cold room into a warm room.

Dry milk should always be sampled at points beneath the exposed surfaces. A representative sample of a given quantity of dry milk solids includes at least three generous equal portions of the batch, or if made up from individual package units, a combination of the first, last, and intermediate packages taken during the filling operation. Containers used for mixing the composite sample should have a capacity of at least twice the volume of the sample.

With a dry spoon of suitable size, turn over the contents by lifting from the bottom of the container at the rate of one spoonful per second for a period of one minute, at the same time slowly rotating the container at an angle of 45 degrees. It is important to include in the mixing that product in the lowest edge of the container. Precautions should be taken to minimize the possibility of moisture absorption during mixing. Immediately after mixing, fill a one pound sample container, using only single or double friction top cans. Tap the closure if samples are to be shipped or held before analysis.

Evaporated Milk

Evaporated milk usually may be sampled by the same procedure as milk and cream, except that thorough mixing should be assured by pouring the entire sample back and forth approximately thirty times from one container to another.

Condensed Milk

Sample condensed milk at a temperature not below 75 F (24 C), after thorough mixing by vigorous hand or mechanical agitation. Condensed milk in containers that has been stored for a long time may show some sugar crystallization and separation by settling and in such cases it is necessary to place the container in water at 120 F (49 C), and stir the contents, taking care not to incorporate excessive air, until a homogeneous distribution of milk fat and milk sugar has been obtained.

REFERENCES

1. *Laboratory Manual*—Methods of Analysis, American Butter Institute, Chicago, Ill., 1947.

2. *The Grading of Dry Milk Solids*—Including Methods of Analysis, American Dry Milk Institute, Chicago, Ill.

CHAPTER 13

BABCOCK TEST FOR FAT

Babcock Test for Fat in Milk

Apparatus

1. Babcock milk test bottles, 6.5 inch, 18 gm, 8 per cent, graduated in 0.1 per cent divisions, and conforming with local state requirements; or 6 inch, 18 gm, 10 per cent, graduated in 0.2 per cent divisions and conforming with Wisconsin state requirements. (6.5 inches is the average length of the Babcock test bottle.)

2. Babcock pipette (17.6 ml), conforming with local state requirements.

3. Babcock acid measuring device (17.5 ml charge).

4. Babcock centrifuge, electric or steam driven (equipped with heating element if electrically driven).

5. Dividers or knife edge calipers, sharply pointed, with sufficient tension to retain test setting.

6. Tempering bath which may be maintained at 131-140 F (55-60 C), fitted with rack to hold bottles in fixed vertical position and with a properly attached Fahrenheit thermometer.

7. Mechanical shaker for Babcock bottles (not essential but desirable to promote uniformity of mixing of milk and acid and to facilitate speed of testing).

8. Accurate speed indicator, separate or permanently attached to the centrifuge.

9. Sulfuric acid hydrometer for use at 68 F (20 C).

10. Fahrenheit thermometer, accurate to 1 degree, maximum graduation 2 degrees. Graduations not more than 40 degrees per inch, and range 20 to 220 degrees.

11. Cup for mixing milk and cream samples of at least 1200 ml.

Reagents

1. Babcock sulfuric acid (H_2SO_4), sp. gr. 1.82-1.825 at 68 F (20 C), stored in tightly stoppered containers.

2. Clean distilled water or clean low-carbonate soft water, for addition to test bottles.

Procedure

1. Mark the test bottles plainly with pencil or other marking that will not be easily smudged or removed. Test bottles must be clean, and graduation lines must be clearly visible.

2. Temper the milk sample to 60-70 F (15.6-21.1° C) and mix thoroughly by pouring at least three times from one vessel to another.

PRECAUTION: Preserved composite samples or others in which the milk fat may be in an abnormal physical condition should be prepared and measured for testing as explained on pages 172-175.

3. Measure 17.6 ml of the milk into the Babcock pipette by adjusting the top of the milk meniscus to the level of the graduation line, and transfer the contents to the test bottle by allowing the milk to drain freely from the pipette, and then blowing the last drop from the pipette see figure 83).

NOTE: When the practice is to drain the pipette as illustrated in figure 70, a rubber band fastened around the pipette just below the bulb will protect the bulb and facilitate the escape of air from the test bottle. The length of the delivery tube should be sufficient to pass the lower opening of the test bottle neck. Before inserting the pipette nozzle into the bottle neck the nozzle should be wiped free of milk, care being taken not to remove any milk from the bore. Use of more than one pipette when measuring a series of samples saves time.

4. Add 17.5 ml of Babcock H_2SO_4 at 60-70 F (15.6-21.1 C) by tipping the test bottle and allowing the acid to flow gently down the neck of the bottle as it is being slowly rotated.

PRECAUTION: Babcock acid and milk should be at approximately the same temperature when mixed. If acid is added too fast, air bubbles will be entrapped in the test bottle neck; if acid is allowed to drop directly into the milk there may be a violent reaction which will project curdled milk into the bottle neck and may even cause some of the contents to spatter out. The 17.6 ml charge will be found satisfactory when the acid is adjusted to the prescribed specific gravity, but a variation in the amount of acid may be necessary for good results should the acid be slightly stronger or weaker than prescribed.

5. Mix the acid and the milk with a gentle rotary movement of the bottle, by hand or by means of a mechanical shaker, until the milk solids other than fat are thoroughly digested. When curd particles are no longer visible and the bottle contents have assumed a dark chocolate color, the mixing may be regarded as complete. (See figure 71.)

PRECAUTION: When mixing, the operator should hold the neck of the bottle away from himself and others, and should avoid violent shaking to prevent spattering of bottle contents which would render the test inaccurate and possibly cause injury to personnel.

6. Immediately place the test bottles in the centrifuge, taking care to balance the load and, maintaining a temperature of 135-150 F (57.2-65.6 C) in the centrifuge, whirl them for 5 minutes at a speed consistent with the following recommendations:

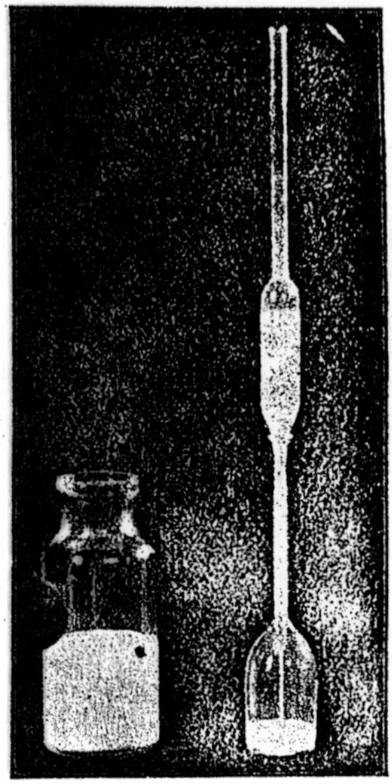

Figure 70.

Method of transferring 17.5 ml of milk to test bottle.
Courtesy *Milk and Milk Products* by Combs, Eckles, and Macy.

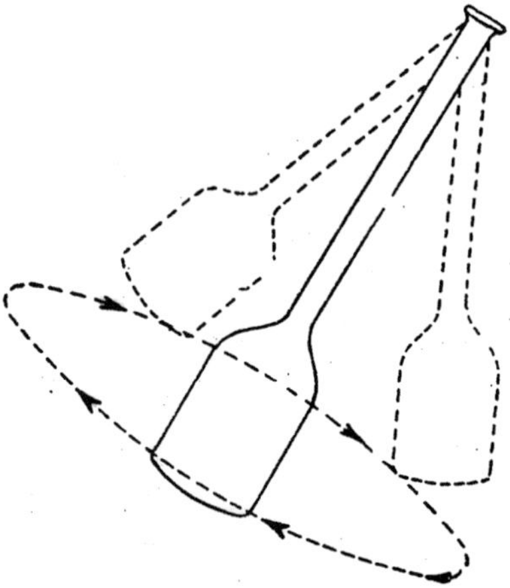

Figure 71.

Rotary motion for most effectively fixing the milk and acid. Figures 71 and 72, courtesy Penn. State College.

Diameter of centrifugal head in inches	R.P.M.
10	1074
12	980
14	909
16	848
18	800
20	759
22	724
24	693

NOTE: The diameter of the centrifugal head is measured from the inside bottoms of opposite cups when they are extended in the horizontal whirling position. The r.p.m. of the centrifuge, with the cover closed, should be checked daily with speed indicator. Obviously a centrifugal head of less than 14 inches in diameter would not accommodate the regular Babcock milk test bottle which has an average length of 6.5 inches. Based on Herrington's[1] calculations the speeds should be: 10", 1171; 12", 1033; 14", 934; 16", 859; 18", 800; 20", 751; 22", 711; 24", 677.

After the 5-minute whirling, add soft water at about 140 F (60 C) to bring the level of the bottle contents to a point just below the lower end of the bottle neck. Centrifuge again for 2 minutes, and add water to bring the level to near the top of the graduated portion of the neck.

NOTE: If necessary to use hard water, add two or three drops of H_2SO_4 per pint of the hot water to prevent foam from forming on top of the fat column when the water is introduced into the test bottle.

7. Centrifuge for 1 minute, and then immerse the bottle in a water bath at 131-140 F (55-60 C) so that the water level of the bath comes slightly above the upper meniscus of the fat column. Allow the bottles to remain until the fat columns are in equilibrium and the lower meniscus has assumed final form (at least 3 minutes—about 3 to 5 minutes).

NOTE: Practical experience indicates that accurate results may be obtained by reading the test bottle directly from the heated centrifuge, provided a minimum temperature of 140 F (60 C) is maintained in the centrifuge at all times and the centrifuge is closed after the removal of each bottle.

8. Remove the bottles one at a time, wipe, and with dividers or calipers read the test directly by simultaneously measuring the length of the fat column from its lowest point to the highest extremity of its upper meniscus. (See figure 72). Duplicate tests should agree within 0.10 per cent. When there is any doubt regarding the accuracy of any test the sample should be retested in duplicate.

NOTE: In *Standard Methods for the Examination of Dairy Products*[2] it is stated "The fat column, at time of measurement should be translucent, of golden-yellow or amber color, and free from visible suspended particles. Reject all tests in which fat column is milky or shows the presence of curd or charred matter, or in which the reading is indistinct or uncertain."

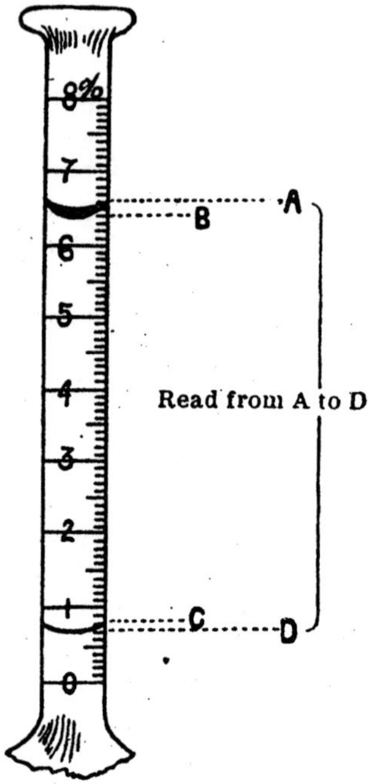

Figure 72.
In reading the percentage of fat in the milk test, the meniscus should be included.

Babcock Test for Fat in Homogenized Milk

It has long been recognized that the standard Babcock test procedure for fat in non-homogenized milk (pages 180 to 183) does not yield comparable results when applied to homogenized milk. General experience has indicated that somewhat lower fat tests (usually not more than 0.10 per cent lower) may be expected on homogenized milk as compared to non-homogenized milk when the standard Babcock test is applied. In addition to this, the application of the standard

REFERENCES

1. Herrington, B. L.: A Note Regarding the Speeds of Babcock Centrifuges. *J. Dairy Sci., 27,* 857-859 (1944).
2. *Standard Methods for the Examination of Dairy Products,* 9th ed., 1948.
3. Babcock, S. M.: A New Method for the Estimation of Fat in Milk, Especially Adapted to Creameries and Cheese Factories, U. of Wis. Bul. 24, 1890.

Babcock procedure to homogenized milk very often produces a charred or curd-like formation immediately beneath the fat column in the neck of the test bottle. This would cast doubt on the accuracy of the reading and constitute sufficient cause for the rejection of the test in question (see note page 182). This situation led to the development of a number of modifications of the Babcock procedure designed to eliminate the formation of charred material in the neck of the Babcock test bottle and to promote closer agreement between the fat test of the same milk before and after homogenization. Lucas and Trout[1] compared the various modifications recommended and suggested a Babcock procedure for testing homogenized milk in which they believe is embodied the best features of these modified Babcock methods. They state, "Employing the suggested technique, the arithmetic-average fat test of 36 samples of homogenized milk was 0.018 per cent lower than that of the same test applied to similar milk not homogenized. The arithmatic-average variations of the 36 tests from the Mojonnier tests were +0.058 and +0.045 per cent for non-homogenized and homogenized milk, respectively."

Modified Babcock Test for Fat in Homogenized Milk

(Lucas and Trout)

Apparatus

Same as listed for Babcock Test for Fat in Milk on page 179, except:

1. Sulfuric acid measuring device to permit addition of 17.5 ml charge of acid in measured portions of 8, 5, and 4.5 ml respectively.
2. Mechanical shaker for Babcock test bottles.

Reagents

1. Babcock sulfuric acid (H_2SO_4), sp. gr. 1.83-1.835 at 68 F (20 C), stored in tightly stoppered containers.
2. Clean distilled water or clean low carbonate, soft water for addition to test bottles.

Procedure

Same as for Babcock Test for Fat in Milk on pages 181 to 183 except:

1. Temper the H_2SO_4 and the homogenized milk to 70 F (21.1 C).
2. Add the H_2SO_4 to the milk in the test bottles in three portions 8, 5, and 4.5 ml respectively, mixing the acid and the milk by rotary motion for at least 15 seconds after each addition of the acid.
3. Place the test bottles in the mechanical shaker and shake for at least 2 minutes before centrifuging.

Trout and Lucas[2] compared the Babcock, Gerber, Minnesota, Pennsylvania, and Mojonnier methods for determining the percentage of fat in homogenized milk (see figure 73) and concluded that homogenization of milk does not affect the Mojonnier fat test. The modified Babcock method (Lucas and Trout) may be used with much assurance of accuracy in testing homogenized milk—24 tests in duplicate averaged within 0.012 per cent of those of non-homogenized milk made by same method and within +0.043 per cent of the Mojonnier average.

Homogenization does not affect the Gerber test. The average Gerber tests of milk before and after homogenization were found to be 0.09 per cent higher than those secured by the Mojonnier method. In making the Gerber test on homogenized milk the following factors were striking:

(a) the clarity of the fat column and supporting liquid.
(b) the identical reading of the duplicate tests.
(c) the consistent check with tests in the non-homogenized milk.
(d) the complete freedom from any char formation.

The Minnesota method yielded average tests of homogenized milk within +0.027 per cent of those of non-homogenized milk but the tests varied from those of the Mojonnier method on the average by -0.433 per cent; hence, it would seem the test could not be recommended for testing homogenized milk. The Pennsylvania method, yielding tests on homogenized milk in these studies markedly lower than the Mojonnier method, cannot be recommended for testing homogenized milk.

REFERENCES

1. Lucas, P. S. and Trout, G. M. A Suggested Modified Babcock Procedure for Testing Homogenized Milk. *J. Dairy Sci., 30:* 95-102 (1947).
2. Trout, G. M. and Lucas, P. S. A Comparison of the Babcock, Gerber, Minnesota, Pennsylvania, and Mojonnier Methods for Determining the Percentage of Fat in Homogenized Milk. *J. Dairy Sci., 30:* 145-159 (1947).

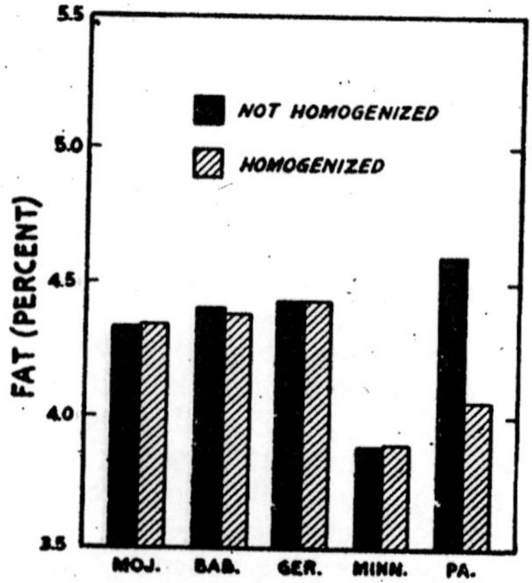

Figure 73.
Fat percentage in milk as determined by the Mojonnier, modified Babcock, Gerber, Minnesota, and Pennsylvania methods (average 24 trials).

Babcock Test for Fat in Cream

Apparatus

1. Babcock cream test bottles (6 inch, 9 gm, 50 per cent), graduated to 0.5 per cent and conforming with local state specifications.

NOTE: For special check work or standardizing purposes the standard 9 inch, 18 gm, 30 per cent test bottles graduated to 0.2 per cent will be found useful.

2. Cream pipette with large tip (9 ml capacity is a convenient size or 17.6 ml when the 18 gm, 30 per cent bottle is used).

3. Cream test scales sensitive to 30 mg.

4. Two 9 gm weights.

5. Other Babcock apparatus listed as items (3) to (10) under Milk on page 180.

Reagents

1. Babcock sulfuric acid (H_2SO_4), sp. gr. 1.82-1.825 at 68 F (20 C), stored in tightly stoppered containers.

2. Clean distilled water, or clean low-carbonate soft water for addition to test bottles.

3. Glymol, a colored mineral oil, to flatten the meniscus, sp. gr. 0.85 at 68 F (20 C). (See page 539).

Procedure

1. Mark the test bottle plainly with a pencil or other marking that will not easily be removed or smudged.

2. Adjust the sample to about 70 F (21.1C) and mix thoroughly by pouring at least four times from one vessel to another.

PRECAUTION: Cream in which fat has become hardened may be brought into proper condition for mixing by slowly and thoroughly warming to a temperature not exceeding 100 F (37.8 C). Should the fat be separated or churned, heat the cream slowly to 100 F. Mix thoroughly and immediately weigh out the test portions. In such cases check samples are imperative.

3. With the aid of the cream pipette, weigh exactly 9 gm of the sample into the test bottle which has been balanced on the cream scale, exercising care to prevent the cream from spilling on the exterior of test bottle or scale pan.

PRECAUTION: The cream scales must be set level and protected from vibration and air currents. Weights should be kept scrupulously clean, calibrated once a week with a check weight, and these check weighings should be recorded.

4. Add 9 ml of soft water to the weighed portion and mix thoroughly by the customary rotary motion. (See note on hard water, page 183).

5. Add slowly 17.5 ml of Babcock H_2SO_4 at 60–70 F (15.6–21.1 C) in such a manner as to wash all cream from the neck into the body of the bottle. Mix thoroughly with a gentle rotary motion or by means of the mechanical shaker until visible curd has been dissolved and the contents assume a coffee color.

PRECAUTION: Never add acid above 70 F (21.1 C), and avoid violent shaking when mixing the contents, thus preventing error in the test and possible injury to the operator.

NOTE: If the procedure required by the state law should differ from the foregoing, the state method should be followed.

6. Place the test bottles in the centrifuge, taking care to balance the load, and maintaining a centrifuge temperature of 150 F (65.6 C), whirl for 5 minutes at a speed consistent with the recommendation given on page 183. Add sufficient soft water at 135-150 F (57.2-65.6 C) to bring the level of the contents to within one-half inch of the lower graduation mark on the neck of the test bottle. Centrifuge for an additional 2 minutes and then bring the level of the contents to a point approaching the top of the graduated column. Centrifuge for 1 minute and then immerse the test bottles in the 131-140 F (55-60 C) bath so that the fat columns are below the surface of the tempering water. Allow 3 to 5 minutes for the fat column to attain equilibrium and for the lower meniscus to assume final form, and then carefully add glymol by allowing a few drops to run down the neck of the bottle from a pipette or dropper and to flow gently over the top of the fat column.

7. Read the test directly by measuring, with sharply pointed dividers or knife-edge calipers, the length of the fat column from the lowest extremity to the point of junction with the glymol. (See figure 74)

NOTE: It is preferable to employ a tempering bath for reading cream tests. However, if the number of tests to be read does not exceed four, satisfactory results may be obtained by reading the bottles directly from the heated centrifuge as has been described under milk testing on page 183.

Duplicate tests should agree within 0.5 per cent when the 6 inch, 9 gm, 50 per cent test bottle is used.

NOTE: The following is quoted from *Standard Methods for the Examination of Dairy Products*, 9th ed., 1948.

"The fat column at time of measurement should be translucent, of golden-yellow or amber color, and free from visible suspended particles. Reject all tests in which fat column is milky or shows the presence of curd or of charred matter, or in which the reading is indistinct or uncertain."

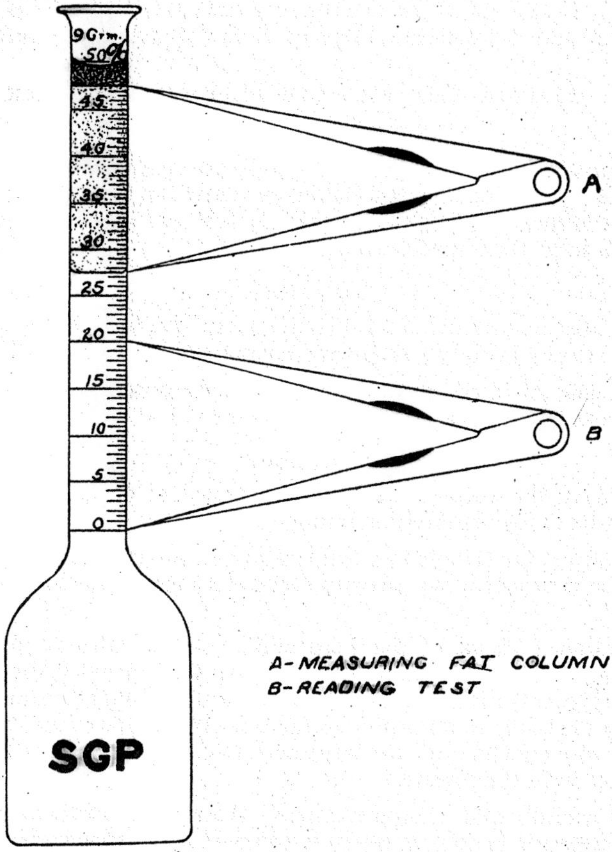

Figure 74.
Correct method of reading the cream test. Courtesy Pennsylvania State College.

Tests for Fat in Skim Milk or Buttermilk

The regular Babcock test for skim milk and buttermilk does not yield the total amount of milk fat in the product because, under the conditions of the test, it is impossible to centrifuge out all of the fat. The Babcock test for skim milk will customarily read from 0.07-0.10 per cent low, and in the case of buttermilk the reading may be 0.30 per cent or more too low. This situation led to the development of a test by the American Association of Creamery Butter Manufacturers, which is reported to show very close agreement with the Roese-Gottlieb method for analysis of milk fat in skim milk and buttermilk. This test is known as the American Association test. Description of both tests follow but of the two the American Association test is preferable for milk fat determinations on skim milk and buttermilk.

A. BABCOCK TEST FOR SKIM MILK AND BUTTERMILK

Apparatus

The usual Babcock equipment for milk (see page 179) with double-necked 18 gm, skim milk test bottles graduated in 0.01 per cent divisions, permitting readings up to 0.25 or 0.50 per cent, and conforming with local state specifications.

Reagents

1. Babcock sulfuric acid (H_2SO_4), sp. gr. 1.82-1.825 at 68 F (20 C), stored in tightly stoppered containers.

2. Clean distilled water, or clean low-carbonate soft water for addition to test bottles.

Procedure

1. Mark the bottles plainly with a pencil or other marking that will not be easily removed or smudged.

2. Adjust the sample to a temperature of 60-70 F (15.6 to 21.1 C) and mix thoroughly by pouring several times from one vessel to another.

3. Draw 17.6 ml of the skim milk into the Babcock pipette by adjusting the top of the skim milk meniscus to the level of the graduation, and transfer it to the skim milk test bottle through the large neck, slanting the bottle in a position to facilitate the escape of air, thus preventing any tendency for the large neck to clog or overflow. Blow the last drop from the pipette.

4. Carefully add, in approximately three equal portions, 18 to 20 ml. of Babcock H_2SO_4 at 60-70 F (15.6-21.1 C), thoroughly mixing each portion with the skim milk.

PRECAUTION: Exercise extreme care to prevent particles of coagulated casein from entering the small or graduated neck, as pressure may be built up that would cause some of the contents to be ejected through the large neck.

5. Balance the test bottles in the centrifuge, taking care to place the large neck toward the center of the machine, to cause the fat to be forced into the calibrated neck as a result of centrifuging. Maintain a temperature of 135-150 F (57.2-65.6 C) in the centrifuge and whirl the test bottles for 10 minutes at a speed consistent with the recommendations on page 182.

After the 10-minute whirling, add clean soft water through the large neck at not less than 150 F (65.6 C) to bring the level of the contents to the lower end of the bottle neck, and whirl for an additional 2 minutes. Carefully add sufficient hot water to bring the fat up into the graduated portion of the small neck, and whirl again for 1 minute.

6. The test bottles may be read directly from the heated centrifuge or after tempering in a water bath at 131-140 F (55-60 C) for 3-5 minutes. With the aid of a pair of sharply pointed dividers read the fat column from its lower extremity to the top of its upper meniscus.

NOTE: Should no fat column appear in the graduated neck, the indications are that insufficient acid was added, or the centrifuge was not up to speed, or that proper temperature was not maintained during the test.

B. AMERICAN ASSOCIATION TEST FOR FAT IN SKIM MILK AND BUTTERMILK

Apparatus

1. The usual Babcock equipment for skim milk, with 18 gm. skim milk test bottle, the graduated neck reading up to 0.50 per cent.

2. Graduated cylinder, 10 ml.

Reagents

1. Babcock sulfuric acid (H_2SO_4), sp. gr. 1.82-1.825 at 68 F (20 C).

2. Normal butyl alcohol.

3. Clean distilled water or clean, low-carbonate soft water for addition to test bottles.

Procedure

1. Measure 2 ml of normal butyl alcohol into the 18-gm skim milk test bottle. Add 9 ml of skim milk or buttermilk and 7 to 9 ml of Babcock H_2SO_4. It may be necessary to vary the amount of acid in order to obtain a fat column of golden yellow to light amber color.

2. Mix the contents of the bottle thoroughly and centrifuge for 6 minutes at the proper speed and temperature (see page 183).

3. Add sufficient clean, soft water (140 F or 60 C) to bring the contents to the base of the neck and whirl for 2 minutes.

4. Add enough of the hot water to bring the fat column up into the graduated portion of the small neck of the bottle, and whirl again for 2 minutes.

5. Read directly from heated centrifuge or after tempering in water bath (see note page 183). Double the reading to obtain the per centage of butterfat, since 9 ml of sample was used in an 18-gm test bottle.

NOTES: (a) The fat column obtained in this test may contain small amounts of lecithin and other substances soluble in normal butyl alcohol. Check tests indicate, however, that results by the use of this method correspond more closely to the Röese-Gottlieb analysis for milk fat than do those obtained by the Babcock method on skim milk or buttermilk.

(b) Should there be any difficulty in cleaning the test bottles, introduce a small amount of lukewarm water, then carefully add a little H_2SO_4, rinse well with this mixture, and then rinse out the bottle with clean, hot water.

Pennsylvania Test for Fat in Chocolate Milk or Drink[1]

Apparatus

1. Usual Babcock equipment for milk and cream. (See pages 180 and 187).

2. Burette with ground-glass stopper and capable of delivering 2-ml portions of ammonium hydroxide.

Reagents

1. Ammonium hydroxide (NH_4OH) (28 to 29 per cent NH_3).

NOTE: It is important to use NH_4OH of correct strength. (Keep in tightly stoppered bottle.)

2. Normal butyl alcohol, b. p. 242.6 F (117 C).

3. Dilute commercial sulfuric acid (H_2SO_4), sp. gr. approximately 1.72-1.74 at 68 F (20 C).

NOTE: The acid is diluted by adding 3.5 parts by volume of commercial H_2SO_4 (sp. gr. 1.82-1.83) to one part of water, in a heat-resistant container.

4. Glymol, a colored mineral oil, to flatten the meniscus, sp. gr. 0.85 at 68 F (20 C).

Procedure

1. Carefully mix the chocolate milk sample, warming if necessary, to redisperse the fat, using care not to churn the product and to bring all the cocoa fiber into suspension.

3. Quickly weigh 18 gm of the well-mixed sample into an 8 per cent Babcock milk-test bottle, previously balanced on a cream scale. Take precautions to prevent chocolate from settling in the well-mixed sample.

3. Add 2 ml of the NH_4OH to the test bottle, using the burette, and mix thoroughly.

4. Add 3 ml of normal butyl alcohol and mix for at least 1 minute.

5. Introduce 17.5 ml of the diluted H_2SO_4 into the test bottle and thoroughly mix with a rotary motion until the contents are completely digested.

NOTE: Be careful to wash all of the sample from the neck of the test bottle while adding the reagents. A small amount of sample left adhering to the inside of the neck might collect at the base of the fat column and make it difficult to obtain a correct reading.

6. Centrifuge for 5 minutes at the speed and temperature indicated for milk (page 183).

7. Add hot water 130-140 F (54.4-60.0 C) to bring the contents to within one-fourth inch of the base of the neck. Do not mix the contents of the bottle at this stage as it will drastically lower the fat reading.

8. Whirl for 2 minutes.

9. Add enough of the hot water to keep the fat well within the graduated portion of the neck, and whirl for 1 minute.

10. Place the test bottles in a water bath at 130 F (54.4 C) and temper for not less than 3 minutes.

11. Allow a few drops of glymol to run down the inside of the neck of the bottle just before reading. Measure the length of the fat column from the bottom of the lower meniscus to the sharp line of demarcation between the glymol and the fat.

NOTE: If glymol is not available, measure from the bottom of the top meniscus to the bottom of the lower meniscus.

CORRECTION TABLE FOR VARYING AMOUNTS OF SAMPLE AND DIFFERENT TYPES OF TEST BOTTLES USED IN THE PENNSYLVANIA METHOD

1. Using 18-gm milk-test bottle:
   - a. 18-gm sample, read direct.
   - b. 9-gm sample, multiply reading obtained by 2.
   - c. 4.5-gm sample, multiply reading obtained by 4.
2. Using 9-gm cream-test bottle or 9-gm ice cream-test bottle.
   - a. 9-gm sample, read direct.
   - b. 18-gm sample, divide reading obtained by 2.
   - c. 6-gm sample, multiply reading obtained by 1.5.
   - d. 4.5-gm sample, multiply reading obtained by 2.

GENERAL USE OF THE PENNSYLVANIA METHOD FOR DETERMINING THE FAT CONTENT OF DAIRY PRODUCTS

This method is especially adapted to the testing of dairy products to which sugar or chocolate has been added, and can be used in testing other dairy products for fat.

Apparatus required, reagents, and general procedure are substantially the same as have been given in detail for chocolate milk, with the exceptions noted under the various products.

Milk. Measure 17.5 ml into an 8 per cent milk-test bottle. (Test not recommended for testing homogenized milk. (See figure 73, page 186).

Cream. Weigh 9 gm into a 9 gm, 50 per cent cream-test bottle.

Cream Containing Added Sugar or Salt. Weigh 9 gm into a 9 gm, 50 per cent cream-test bottle.

Skim Milk, Buttermilk and Whey. Measure 17.5 ml into a skim milk or milk-test bottle.

Evaporated Milk. Weigh either 4.5 or 6.0 gm of sample into an 8 per cent milk-test bottle or into a 9 gm, 20 per cent ice cream-test bottle.

Sweetened Condensed Milk. 1. To facilitate weighing into the test bottle, the sample should be mixed with an equal weight of water as follows: Balance two beakers on opposite pans of cream scales. Add at least 1 ounce of sweetened condensed milk to one of the beakers. Add enough water to the other beaker to balance the scales. Thoroughly mix the sweetened condensed milk and water.

2. Weigh 9 gm of this mixture into a 9-gm, 20 per cent ice cream-test bottle.

3. Multiply the results by 2 because of the dilution.

Dried Whole Milk. 1. Weigh 20 gm into a flask and add 140 gm of water at 125 F (51.6 C). (This makes a dilution of which one eighth is dried milk.)

2. Shake vigorously until all lumps disappear and a homogeneous mixture is obtained.

3. Cool to approximately 70 F (21.1 C).

4. Weigh 18 gm into an 8 per cent milk-test bottle.

5. Multiply the results by 8 because of the dilution.

Dried Cream. 1. Weigh 40 gm into a flask and add 40 gm of water at 125 F (51.6 C). (This makes a dilution of which one half is dried cream.)

2. Shake carefully to prevent "oiling-off" until all lumps disappear and a homogeneous mixture is obtained.

3. Weigh, while still warm, 9 gm into a 9-gm, 50 per cent cream-test bottle. (Weighing while still warm aids in preventing partial churning and subsequent "oiling-off" of the fat.)

4. Multiply the results by 2 because of the dilution.

Ice Cream or Ice Cream Mix.

NOTE: Especially for testing ice cream or mix containing chocolate.

1. Melt ice cream at room temperature and, if necessary, heat to eliminate the foam.

2. Warm ice cream mix to approximately 70 F (21.1 C).

3. Reduce large particles in fruit and nut ice cream to a finely divided state, with the aid of a mortar, and force through a fine mesh screen.

4. Weigh 9 gm of the representative sample into a 9-gm, 20 per cent ice cream-test bottle or an 8 per cent milk-test bottle.

PRECAUTION: Keep the fruit or nut ice cream thoroughly mixed in order to get a representative sample.

Butter. 1. Warm in a closed container a representative sample until it has the consistency of thick cream.

2. Weigh 4.5 gm of a representative sample into a 9 gm., 50 per cent cream-test bottle.

3. Heat the test bottle to bring the butter from the inside of the neck down into the bottle.

Cheese. 1. Finely dice or cut into thin strips a representative sample of cheese and thoroughly mix.

2. Weigh 4.5 gm into a 9 gm, 50 per cent cream-test bottle.

3. Add 8 to 10 ml of water and heat in a water bath at 150-180 F (66-82 C) with intermittent shaking, to facilitate dispersion of the sample.

REFERENCES

1. Swope, W. D.: The Pennsylvania Method for Determining the Percentage of Fat in Dairy Products, Pennsylvania State College, Bulletin 412, 1941.

2. Doan, F. J.: A Modified Babcock Test for Estimating the Fat Content of Chocolate Milk, 24th Annual Convention, Laboratory Section I.A.M.D., 1931.

Modified Babcock Tests for Milk Fat in Ice Cream and Ice Cream Mix

NOTE: These modified Babcock tests for ice cream and ice cream mix have been selected as the best available at this time.

A. Sulfuric Glacial Acetic Acid Method

NOTE: This method should be used only for ice cream mix since its use on melted samples of ice cream yields low tests. This test is not to be used on ice cream mix containing chocolate.

Apparatus

1. Usual Babcock equipment for milk and cream. (See pages 180 and 187).

Reagents

1. Commercial sulfuric acid (H_2SO_4), sp. gr. 1.82-1.83 at 68 F (20 C).

2. Glymol, a colored mineral oil, to flatten the meniscus, sp. gr. 0.85 at 68 F (20 C).

3. Glacial acetic acid.

Procedure

1. Mark the bottles for identification on the etched portion of the base.

2. Mix the sample thoroughly by pouring from one container to another at least four times.

3. With the cream scales level, balance the test bottle or bottles on the scales

4. Weigh into an 18 gm. 8 per cent milk-test bottle at a temperature of about 70 F (21.1 C) 9 gm of the product, using the pipette and being careful that none of the products gets on the outside of the bottle or on the scale pan.

5. Add 13 ml of glacial acetic acid to the bottle and mix thoroughly.

6. Add 9 ml commercial H_2SO_4 and mix thoroughly.

7. Place the bottle in a water bath at about 170 F (76.7 C) and leave for 5 minutes.

8. Remove, shake thoroughly, and place the bottle in a heated centrifuge. See that the centrifuge is balanced. Whirl 5 minutes at the prescribed rate of speed for centrifuging. See page 182.

9. Add soft, hot water at 180 F (82.2 C) until the bottle is filled to the lower end of the neck.

10. Whirl again for 2 minutes.

11. Add soft, hot water at 180 F again until the bottle is filled nearly to the top of the graduated portion.

12. Whirl again for 1 minute.

13. Place the bottle in a hot water bath at 135-140 F (57.2-60.0 C). The bottle should remain in the water bath for at least 3 minutes before being read or until the column is in equilibrium and the lower fat surface has assumed final form. The surface of the water in the water bath should be as high as the top of the fat column in the test bottle. Add a few drops of glymol just before reading the test. Let it run down the side of the bottle neck onto the top of the fat column.

14. Read the test by means of sharply pointed dividers. Read from the lower extremity to the top of the milk fat column. For the purpose of measurement, the surface separating the glymol and the milk fat is regarded as representing the upper limit (top of the milk fat column).

15. Retest all samples if they
    1. contain specks;
    2. contain charred particles;
    3. have a milky fat column;
    4. give indistinct or uncertain readings;
    5. fail to agree within 0.1 per cent in duplicate samples.

Calculations

The percentage of fat is obtained by multiplying the bottle reading by 2, because of the use of a 9 gm sample in an 18 gm, 8 per cent milk-test bottle.

REFERENCE

Crowe, L. K., and Davis, H. P.: Babcock Testing Principles and Uses, *University of Nebraska,* Circular, 35 (1927).

B. Butyl Alcohol (Nebraska) Test

Apparatus

1. Usual Babcock equipment for milk and cream. (See pages 179 and 186).

Reagents

1. Reagent A: Nine parts by volume of normal butyl alcohol and one part of ammonium hydroxide (NH_4OH), c. p.

NOTE: 90 ml of the former and 10 ml of the latter reagent makes a convenient quantity. It is stable when kept in a tightly stoppered bottle.

2. Reagent B: Equal parts of sulfuric acid, sp. gr. 1.82–1.83, and ethyl alcohol, 95 per cent. Pour the acid slowly into the alcohol in a glass beaker or other glass container that will withstand high temperature. The mixture will become extremely hot and will boil, but no trouble need be experienced in preparing it if the acid is poured carefully down the side of the beaker so that it will run under the alcohol. After the acid has been added, or during its addition, the mixture should be stirred with a glass rod and finally cooled to room temperature before being used. Such a mixture, even after being kept as long as eight weeks in a tightly stoppered bottle, has produced satisfactory results.

3. Glymol, a colored mineral oil, to flatten the meniscus, sp. gr. 0.85 at 68 F (20 C).

Procedure

1. With the ordinary torsion balance, weigh 9 gm of ice cream mix or melted ice cream, into an 18 gm, 8 per cent milk-test bottle.

2. Add 5 ml of reagent A. Mix the contents of the bottle thoroughly by shaking. Add 30 ml of reagent B (cooled to room temperature before being used). A small variation in the amount of this reagent apparently does not affect the test. Consequently, if the capacity of the base of the bottle is small, 28-29 ml may be used to prevent the mixture from rising into the neck of the bottle while heating, which makes it difficult to mix the contents of the bottle. If possible, bottles that will hold the regulation 30 ml should be selected and retained for this test. After adding reagent B, again mix the contents of the bottle thoroughly by shaking, and continue mixing until the curd is all dissolved. The contents of the bottle should become clear and free from curd but will not become dark until heated in the water bath.

3. Place the test bottles in a water bath at a temperature between 175 and 180 F (79.4 and 82.2 C) and heat for 15 minutes. Shake the contents of the bottles at least three times during this heating period.

4. Remove from the water bath and centrifuge at the regular speed (see page 182) for 5 minutes. Shake the contents of the bottle thoroughly. If the fat column is not up to the neck of the bottle, add water, at a temperature of not less than 180 F (82.2 C), until the fat column is raised to within about one-quarter inch of the base of the neck. Centrifuge for 3 minutes. If any curd is apparent in the mixture at this time, shake the contents of the bottle thoroughly. Add water at 180 F (82.2 C) until the fat column is raised so that the extremities will be between the 0 and 8 per cent graduations on the neck of the bottle. Centrifuge for 1 minute.

5. Remove from the centrifuge and place the bottles in a water bath for 5 minutes at a temperature of 135-140 F (57.2-60.0 C). Add glymol and read the percentage of fat with dividers as in the Babcock cream test. Measure from the bottom of the lower meniscus to the line between the glymol and the fat. Multiply the reading by 2, to obtain the true percentage of fat in the ice cream sample, since only a 9-gm sample was used in an 18-gm bottle. Duplicate tests should always be made and the reading of these duplicates should check to within 0.1 per cent of each other, making a maximum variation of 0.2 per cent when the reading is doubled. The fat column in the finished tests should be clear and golden yellow in color. A small amount of fluffy brown precipitate may appear below the fat column when testing samples of fruit ice cream. This precipitate does not mix with the fat sufficiently to hinder reading.

REFERENCE

Crowe, L. K.: Testing Ice Cream for Butterfat, *University of Nebraska*, Bul. 246 (1930).

C. Illinois Nonacid Babcock Method

Apparatus

1. Usual Babcock equipment for cream. (See page 187).
2. Shallow water bath provided with means for heating.

NOTE: The heat may be supplied by a gas burner or a jet of steam.

3. Burette graduated to 0.5 ml or 2.5 ml pipette.

Reagents

1. Reagent A: 75 ml of ammonium hydroxide (NH_4OH), c. p., 35 ml of normal butyl alcohol, 15 ml of 95 per cent ethyl alcohol or denatured grain alcohol (S. D. 30).

NOTE: Formula S.D. 30 is a mixture of 30 per cent methyl and 70 per cent ethyl alcohol.

2. Reagent B: 200 gm trisodium phosphate, 150 gm sodium acetate, 1 liter of water.

NOTE: Trisodium phosphate and sodium acetate of commercial grade may be used. Clean tap water may be used instead of distilled water. The reagent remains in solution at ordinary temperatures but will partially crystallize on becoming cold. If crystallization takes place, the crystals should be dissolved by warming before the reagent is used. The reagent does not seem to deteriorate on long standing.

3. Glymol, a colored mineral oil, to flatten the meniscus, sp. gr. 0.85 at 68 F (20 C).

Procedure

1. Weigh 9 gm of the well-mixed sample of ice cream or ice cream mix into an 18 gm Babcock milk-test bottle (8 or 10 per cent scale) or a 9 gm 20 per cent ice cream-test bottle.

2. Add exactly 2.5 ml of reagent A from a burette or pipette. Mix thoroughly.

3. Add from 9 to 10 ml of reagent B and again mix thoroughly. The reagent may be measured in an ordinary 9-ml acid measure or in a 10-ml pipette.

4. Place the test bottle in a shallow water bath and heat the bath to boiling, continuing the heating for several minutes. Shake the contents of the bottle two or three times while heating.

5. Usually after from 15 to 30 minutes, the fat will separate and form a clear yellow layer on top of the liquid. The heating must not cease until the fat layer has definitely separated from the dark portion of the liquid and has become clear.

6. After all the fat has separated, place the test bottle in the centrifuge and whirl 5, 2, and 1 minutes, adding hot water between centrifugings as in the regular Babcock milk test except that the water must not be treated with acid.

7. Place the bottle in a hot water bath at 130-140 F (54.4-60.0 C) for five minutes.

8. Add glymol and read the test, measuring from the bottom of the fat column to the line between the glymol and the fat.

9. If an 18 gm milk-test bottle is used, multiply the reading by 2. If a 9 gm, 20 per cent ice cream-test bottle is used, read the test directly.

REFERENCE

Overman, O. R., and Garrett, O. F.: Nonacid Babcock Method for Determining Fat in Ice Cream, *University of Illinois*, Bul. 360

D. Minnesota Test (Nonacid Method)

NOTE: Especially for testing ice cream or mix not containing chocolate.

Apparatus

1. Usual Babcock equipment for cream. (See page 187).

2. Supply of 9 gm, 20 per cent ice cream bottles.

3. Water bath in which bottles may be held at or near the boiling point.

Reagents

1. Minnesota test reagents: 645 gm sodium salicylate, 355 gm potassium carbonate, 160 gm sodium hydroxide. Completely dissolved in 3 quarters of water.

NOTE: The mixture of dry materials may be purchased in sealed containers which prevent absorption of moisture and caking.

2. One quart of isopropyl alcohol ("Petrohol") is added to the 3 quarts of chemical solution and the reagent is ready for use.

NOTE: The reagent must be stored in glass bottles, tightly closed to prevent evaporation and absorption of carbon dioxide.

3. Glymol, a colored mineral oil, to flatten the meniscus, sp. gr. 0.85 at 68 F (20 C).

Procedure

1. Weigh 9 gm of ice cream or ice cream mix into a 9 gm, 20 per cent ice cream-test bottle.

2. Add 15 ml of the Minnesota reagent.

3. Place test bottle in a water bath at or near the boiling point and digest for 12 to 15 minutes, having the bottles in a rack held at least 2½ inches above the bottom of the bath. Shake the mixture in the test bottle vigorously at the time when at least half the contents of the bottle have turned dark brown (usually after about 2½ minutes). Shake vigorously again about 1 minute later.

NOTE: Some care may be necessary when starting to shake the bottles the second time as the isopropyl alcohol in the reagent may boil off through the neck of the bottle, taking with it some of the mixture.

4. When the fat appears in a clear layer on top or when the solution becomes clear, centrifuge for ½ minute at the prescribed speed (see page 182).

5. Add hot water at 130-150 F (54 to 66 C) to bring the fat into the graduated scale of the test bottle and centrifuge for one-half minute.

6. Temper the ice cream test bottle in a water bath at 130 to 140 F (54.4-60 C) for 5 minutes. Add glymol and read the percentage of fat directly.

NOTE: If an 18 gm, 8 per cent milk-test bottle is used, the reading must be multiplied by 2.

E. Kniaseff Test[1]

(With modifications by H. H. and T. A. Selby[2])

Apparatus

1. Usual Babcock equipment for testing cream. (See page 186).
2. Nine inch, 9 gm, 50 per cent Babcock cream-test bottles of proved accuracy by mercury calibration.

NOTE: Bottle should be accurate within 0.1 per cent in the ranges 50 to 40 per cent, 50 to 30 per cent, and 50 to 20 per cent.

Reagents

1. Reagent 1: 1,000 ml distilled water; 132 gm sodium hydroxide, A. C. S.; 42.8 gm ammonium sulfate, A. C. S.
2. Reagent 2: 391 ml ethyl alcohol, U. S. P.; 24 ml N-butyl alcohol, b. p. 241-243 F (116 to 118 C); 24 ml ammonium hydroxide, A. C. S.; 94 ml ethyl ether, A. C. S.; 94 ml petroleum ether, b. p. 79-88 F (26.1-31.1 C).

NOTE: Reagent 2 undergoes some change in composition due possibly to amine formation. It is recommended, therefore, that this solution age one week before being used. Reagent 2 should be stored under screw caps with liners of metal foil instead of ground-glass stoppers since the ordinary ground-glass closure permits the low boiling fraction to evaporate.

3. Glymol, a colored mineral oil, to flatten the meniscus, sp. gr. 0.85 at 68 F (20 C).

Procedure

1. Weigh 9 ± 0.02 gm duplicate samples into 9 in, 9 gm, 50 per cent Babcock cream-test bottles.
2. Add 8 ± 0.1 ml of reagent 1. Shake 5 seconds with rotary movement.
3. Add 5 ± 0.1 ml of reagent 2.
4. Digest 5 minutes at 183.2 ± 1.8 degrees F (84 ± 1 degree C), shaking 5 seconds after 1.5, 3.5, and 5 minutes. If foam rises above 40 per cent mark, remove bottle for an instant, then reimmerse.
5. Centrifuge 30 ± 5 seconds relative centrifugal force of 165 × gravity at base of bottle at 149 ± 9 F (65 ± 5 C).
6. Add tap water at 158 ± 9 F (70 ± 5 C) to base of neck. Do not agitate.
7. Centrifuge for 30 ± 5 seconds as before.
8. Add water to 50 per cent mark.
9. Centrifuge for 60 ± 5 seconds.
10. Place in water bath at 140 ± 3.6 F (60 ± 2 C), so adjusted that the temperature falls to 134.6 F (57 C) in 10 to 20 minutes.
11. When temperature reaches 134.6 ± 0.9 F (57 ± 0.5 C), add glymol and read.[1]

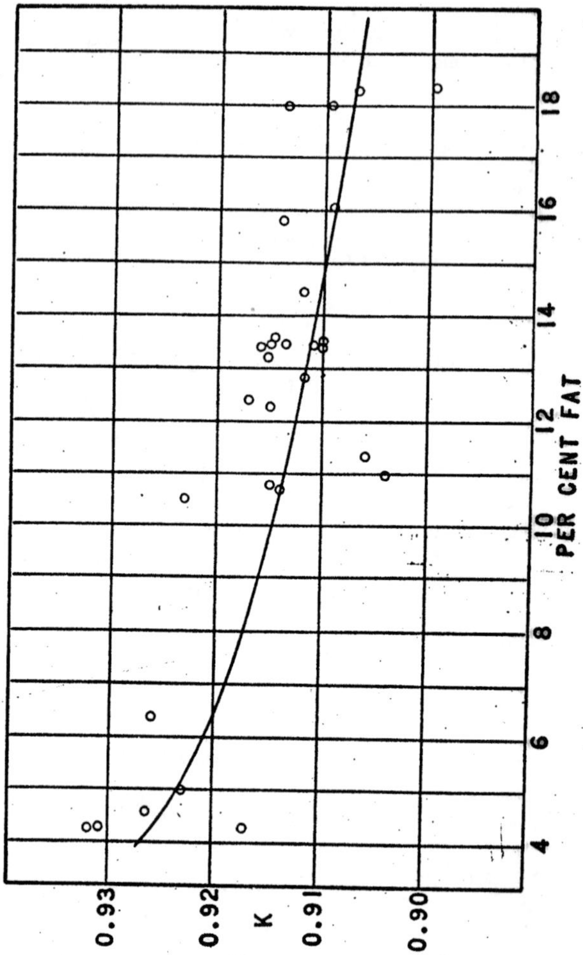

Figure 75.
Chart for determining per cent fat by the Kniaseff test. (See paragraph 12 of procedure.)

12. Multiply reading by appropriate factor, K, obtained from figure 75.

Selby and Selby[2] suggest that any Kniaseff fat test, run as outlined, will probably be within 0.4 per cent of the true value, or that mean of duplicates will be within 0.25 per cent.

REFERENCES

1. Kniaseff, V.: A New Modified Babcock Method to Determine Fat in Ice Cream, *Ice Cream Trade J.*, *30*:29 (1934).
2. Selby, H. H., and Selby, T. A.: Kniaseff Fat Test, *Ind. & Eng. Chem., Anal. Ed.* *11*:393-394 (1939).

Modified Pennsylvania Test for Fat in Ice Cream and Ice Cream Mix (Borden)

Apparatus

1. Usual Babcock equipment for cream. (See page 187).
2. 9 gm 20 per cent Babcock ice cream test bottles.
3. Measuring burettes or pipettes for measuring ethyl alcohol, ammonium hydroxide, and normal butyl alcohol.

Reagents

1. Ethyl alcohol, pure, not denatured.
2. Ammonium hydroxide (NH_4OH) (28 to 29 per cent NH_3)
3. Butyl alcohol, b.p. 242.6 F (117 C)
4. Dilute H_2SO_4, sp. gr. 1.650 at 68 F (20 C). (Approximately 3½ parts conc. H_2SO_4, sp. gr. 1.820—1.825 to 1 part distilled water.)
5. Babcock sulfuric acid (H_2SO_4) sp. gr. 1.820-1.825 at 68 F (20 C).
6. Glymol, 0.85 at 68 F (20 C).

Procedure

1. Weigh 9 gm of sample into 9 gm 20 per cent ice cream test bottle.
2. Add 5 ml pure ethyl alcohol and shake thoroughly.
3. Add 2 ml NH_4OH and mix thoroughly.
4. Add 2.5 ml N butyl alcohol and mix for at least one minute.
5. Add 17.5 ml dilute H_2SO_4 (sp. gr. 1.650) and agitate with a rotary motion until digestion is complete.
6. Centrifuge for 5 minutes at the speed and temperature indicated for milk. (See page 183).
7. Fill test bottle to desired reading level with a freshly prepared hot mixture of sulfuric acid and water (1 part H_2SO_4 sp. gr. 1.820-1.825 to 2 parts H_2O). This mixture will be at an approximate temperature of 200 F (93 C) from the heat of reaction.
8. Centrifuge for 1 minute.
9. Temper test botles in a water bath at 135-140 F (57.2-60 C) for not less than 3 minutes.
10. Allow a few drops of glymol to run down the inside of the neck of the test bottle, and with the aid of dividers read the fat column from the bottom of the lower meniscus to the sharp line of demarcation between the glymol and the fat.

NOTE: The liquid in the neck of the test bottle below the fat column should be clear, light in color and quite transparent. If color of this liquid is too dark to permit easy reading of the fat column, decrease the strength or reduce the temperature (or both) of the acid in step #5.

This test may be used on condensed milk and cottage cheese by adding 5 ml of water to the weighed sample in the test bottle and shaking thoroughly before proceding as indicated above, except in step #7 where a mixture of 1 part H_2SO_4 and 3 parts water should be used. Cottage cheese is best prepared for sampling by mixing in a Waring blendor and weighed into a test bottle having the same type of side opening as the Paley butter test bottle.

Calibration of Babcock Glassware

The accuracy of glassware for the Babcock test is determined by using water, alcohol or mercury for determining the weight or delivery capacity. Various types of plunger displacement devices are also used for this work.

The neck of the Babcock bottle for testing milk for fat is graduated to read from 0 to 8 per cent with 0.1 per cent graduations and has a capacity of 1.6 ml. All methods for checking the accuracy of the bottle involve the determination of the range in capacity above or below 1.6 ml in the 8 per cent graduated section.

The limit of variability must be within the smallest graduation on the bottle, which represents 0.1 per cent in the milk-test bottle, and 0.5 per cent in the cream-test bottle. The brass plunger and the Nafis mercury insertion methods are most commonly used by the manufacturers and those responsible for officially checking the accuracy of Babcock test bottles.

1. Babcock Test Bottles

A. Mercury and Cork Method

Apparatus

Cork

Reagent

Mercury

Procedure

1. Test bottles must be clean and dry. Temperature of mercury, bottle, and room should be 68 F, 20 C.

Weigh 2.70942 gm of clean dry mercury into the bottle for each per cent of graduation in the neck of the bottle.

Weigh 27.0942 gm into a 10 per cent bottle and 21.6754 gm. into an 8 per cent bottle. For a 9 gm, 50 per cent cream bottle the charge is one half of 2.70942 or 1.3547 gm for each per cent or 50 x 1.3547 = 67.735 gm.

2. Insert cork or rubber stopper to exactly the highest graduation mark on the bottle.

3. Invert bottle and observe the level of the mercury column at the top of the meniscus.

Interpretation

If bottle is correct the mercury will be exactly at the zero mark.

NOTE: It is possible in this procedure to transfer the weighed charge of mercury from one bottle to another. This practice, however, requires great care to avoid losing any of the mercury when transferring.

B. MERCURY CALIBRATION METHOD

Apparatus

Analytical balance.

Reagent

Mercury

Procedure

1. Have room, mercury, and bottles at 68 F (20 C).

2. Fill clean, dry bottle with clean, dry mercury so that the meniscus is exactly on the zero mark.

3. Weigh in mercury, 13.5471 gm for each 5 per cent on a 10 per cent bottle; and 10.8377 gm for each 4 per cent on an 8 per cent bottle. Record the graduation level of the top of the meniscus. Bottles in each size should be checked at both the half way and top graduation mark.

NOTE: A special device made by the Nafis Division of Kimball Glass Co., delivers exact volumes of mercury at 68 F (20 C) into the neck of inverted test bottles as follows: 1.6 ml for 8 per cent milk-test bottle; 2.0 ml for 10 per cent milk-test bottle; 10.0 ml for 50 per cent 18 gm cream-test bottle; and 5.0 ml for 50 per cent 9 gm cream-test bottle. If the bottle is accurate it will exactly fill the space in the graduated portion of the neck.

Interpretation

Graduated scale should check within the limits of the smallest graduation mark.

C. ALCOHOL AND BURETTE METHOD

Apparatus

1. Burette, 10 ml graduated in 0.10 ml divisions.
2. Blotting paper.

Reagent

Alcohol colored with methylene blue.

Procedure

1. Fill test bottles with the alcohol so that the bottom of the meniscus is exactly on the zero mark.

NOTE: Bottles should be thoroughly cleaned so that water will run freely over their surfaces without leaving any adhering drops.

2. Dry neck of bottle with blotting paper if necessary. It is essential that the neck of bottle be dry.

3. Add sufficient colored alcohol from an accurately graduated burette to fill the neck of the bottle from the zero mark so that the bottom of the meniscus is exactly on the highest graduation. Record burette reading.

Interpretation

It should require 2 ml for the 10 per cent bottle and 1.6 ml for the 8 per cent bottle. If the bottle neck is in error to the extent of more than the volume occupied by the smallest division of graduation it is rejected.

NOTE: Cream bottles can be calibrated by applying the same principle. It requires 0.2 ml for each per cent on an 18 gm bottle and 0.1 ml for each per cent on a 9 gm bottle.

D. DISTILLED WATER METHOD

Apparatus

1. Analytical balance.
2. Blotting paper.

Reagents

1. Alcohol, colored with methylene blue.
2. Freshly boiled distilled water.

Procedure

1. Fill the thoroughly cleaned test bottle with the alcohol at 68 F (20 C) so that the bottom of the meniscus is exactly on the zero mark.
2. Dry neck of bottle with blotting paper. It is essential that the neck be dry.
3. Weigh bottle filled to zero mark with the colored alcohol.
4. Fill bottle so that the bottom of the meniscus is exactly at the top graduation, using freshly boiled and cooled distilled water at 68 F (20 C).
5. Reweigh, determine, and record the weight of water added.

Interpretation

If the bottle is accurate the difference in weight between the two weighings should be exactly 1.5971 gm at 68 F (20 C) for the 8 per cent test bottle. (1.9964 gm for the 10 per cent bottle.)

NOTE: 1. This method can also be applied to cream-test bottles. For a 50 per cent 9 gm bottle the difference in weight should be 4.991 gm; and the difference in 50 per cent 18 gm bottle should be 9.982 gm.

2. Weight of 1 ml of distilled water at 68 F (20 C) is 0.998203 gm.

E. BRASS PLUNGER METHOD

Apparatus

A brass plunger capable of displacing a given amount of liquid and of being inserted into the test bottle. As a special plunger is required for each type of bottle, specify type of bottle for which the plunger is to be used when ordering the plunger.

Reagent

Alcohol, colored with methylene blue.

Procedure

1. Fill thoroughly cleaned bottle with colored alcohol so that the bottom of the meniscus is exactly on the zero mark, and dry neck with blotting paper.
2. Insert special brass plunger to a point where the liquid level rises to the half-way mark on the plunger and record position of the bottom of the meniscus in neck of the bottle.
3. Continue displacement of the liquid until it rises to the upper mark on the plunger and record position of the bottom of the meniscus in the neck of the bottle.

Interpretation

When the brass plunger is inserted in the liquid to the half-way mark the liquid should be exactly at the 4 per cent graduation in the case of an 8 per cent milk-test bottle. Similarly when upper mark of the plunger is submerged the liquid should be exactly at the 8 per cent graduation.

NOTE: A tolerance of 1 division representing a variation of 0.1 per cent in the case of an 8 per cent milk-test bottle is allowed in most official work.

F. BABCOCK PIPETTES

A Babcock pipette is required to contain 17.6 ml at 68 F (20° C) when the bottom of the meniscus coincides with the mark on the suction tube. The Babcock pipette should deliver about 17.5 ml, but must be calibrated to contain 17.6 ml. Pipette must be clean and dry before calibrating.

2. Babcock Pipettes

A. DISTILLED WATER METHOD (VOLUMETRIC)

Apparatus

1. Burette 25 ml graduated in 0.10 ml divisions.

Reagent

1. Freshly boiled distilled water.

Procedure

Measure from a burette the volume of freshly boiled and cooled distilled water at 68 F (20 C) which the pipette holds up to the graduation mark. If pipette is accurate it will require exactly 17.6 ml of water.

B. MERCURY METHOD

Apparatus

Pipette of known accuracy.

Reagent

Mercury.

Procedure

Pipettes can be rapidly and accurately tested by using a pipette of known accuracy as a standard. The standard pipette should hold 238.428 gm (17.6 x 13.5471) of mercury at 68 F (20 C).

Mercury can then be transferred from the standard to other pipettes to determine their accuracy. The pipettes should be held almost horizontal to facilitate speed in measuring. Be sure that all the air trapped between the mercury and the glass is removed by jarring the pipette.

NOTE: A rubber pad on the finger which closes the tip of the receiving pipette is a great aid in transferring mercury from one pipette to another.

3. Acid Measures

Acid measures should be checked to determine that they hold about 17.5 ml. Extreme accuracy of acid measures is a matter of less importance than the accuracy of the pipettes.

GENERAL CAUTIONS ON CALIBRATING GLASSWARE

1. Always make certain that glassware to be tested is thoroughly cleaned so that drops of liquid used in testing will not adhere to sides of the glassware.

2. Carefully temper room and calibrating liquid to as near 68 F (20 C) as possible. Temper the distilled water, colored alcohol, or mercury by placing in a bath of water or ice and water as conditions may require.

3. Mercury becomes dirty from continued use. It should be kept clean. This is easily done by filtering through filter paper inserted in a funnel. A small pin hole should be in the apex of the filter paper to allow the mercury to pass through slowly.

4. When employing mercury to determine volume, be certain that no air is entrapped between the glass and the mercury.

CHAPTER 14

ROESE-GOTTLIEB FAT DETERMINATION

(Mojonnier Modification[1])

Mojonnier Tester

The Mojonnier method is based on the Roese-Gottlieb ether extraction method. The use of thermostatically controlled vacuum drying ovens and hot plates and temperature control of desiccators by water circulation has made the test much more rapid than the original Roese-Gottlieb ether extraction method. While the Babcock test, owing to its simplicity and greater speed, is largely used for milk and cream, the Mojonnier test is more generally used for ice cream, evaporated milk, condensed milk and other dairy products which are difficult to test by the Babcock method.

Milk, Skim Milk, Buttermilk and Whey

Apparatus

Complete Mojonnier milk tester with chemical balance (see figure 17, page 22) which includes a quantity of fat dishes, extraction flasks, and 1, 2, 5 and 10 gram pipettes. (Figures 76 and 77.)

Reagents

1. Distilled water. Free from oil or any kind of mineral residue.
2. Ammonium hydroxide, concentrated NH_4OH, c.p.
3. Grain alcohol, 95 per cent, 190 proof, 0.8164 sp. gr. at 60 F (15.6 C). It should leave no residue on evaporation.
4. Ethyl ether, USP quality; should contain not more than 4 per cent of water, sp. gr. 0.713-0.716 at 77 F (25 C), boiling point about 95 F (35 C); should leave no residue on evaporation.
5. Petroleum ether, purified, sp. gr. 0.638-0.660 at 77 F (25 C); boiling point 120-140 F (49-60 C); should leave no residue on evaporation.

CAUTION: Both ethyl and petroleum ethers are highly *inflammable* and *explosive,* and the utmost care must be used in handling. Store in a cool place, and do not permit smoking, open flames, or electrical equipment which causes sparks where these reagents are handled. Purification of these ethers by redistillation is of doubtful value, extremely dangerous even in experienced hands, and is considered inadvisable.

NOTE: If there is any doubt as to the purity of the reagents, blank determinations, using distilled water instead of milk, should be made and repeated as frequently as necessary. The best guarantee of purity is a reliable source.

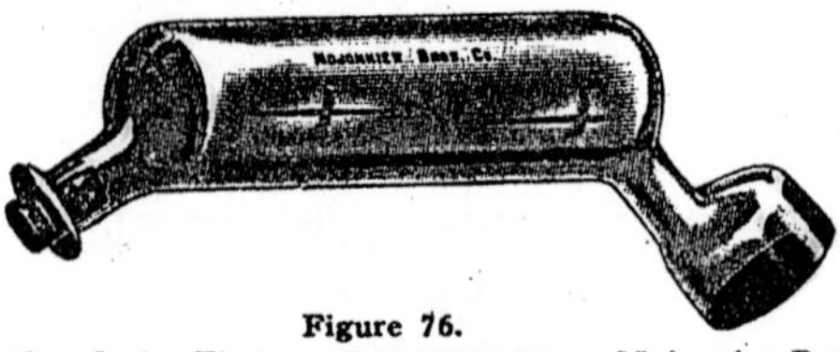

Figure 76.
Fat extraction flask. Figures 76 to 88 courtesy Mojonnier Brothers Co.

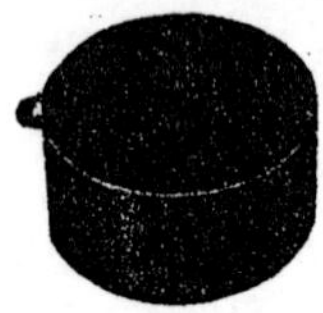

Figure 77.
Fat dish.

Procedure

1. Mix the sample thoroughly by pouring several times from one vessel to another. Should the fat be in a hardened condition or tend to adhere to the sample container, the sample should be slowly warmed to a temperature not exceeding 100 F (37.8 C), thoroughly mixed, then cooled to 60-70 F (15.6-21.1 C) before sampling.

Frozen or partially soured milk may be prepared for sampling by slowly warming to 100 F, vigorously mixing and quickly sampling. Such samples should be run in duplicate.

2. Weigh a 10 gm sample into a properly labeled extraction flask using a 10 gm pipette.

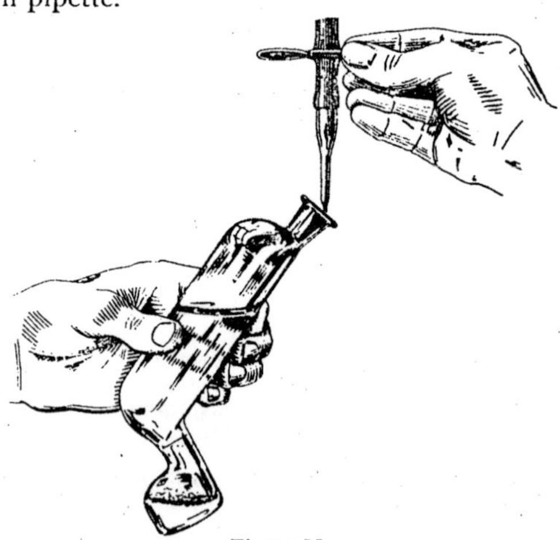

Figure 78.
Method of adding reagents.

Figure 79.
Correct position of flask holder and flasks for adding reagents when making four tests.

FOR FIRST EXTRACTION

1. Add 1.5 ml of NH_4OH and mix thoroughly. The NH_4OH neutralizes any acid present and dissolves the casein. (Figure 78.)

NOTE: A few drops of phenolphthalein indicator will help to sharpen the division line.

2. Add 10 ml of 95 per cent grain alcohol, insert the cork and shake for 1½ minutes. The alcohol prevents the formation of a jelly or colloid like substance, which always forms when ethyl ether is added without the presence of alcohol. (Figure 79.)

PRECAUTION: When shaking the contents of the extracting flask, keep the hand or finger over the cork to prevent leakage or spillage. Also keep the flask with the large bulb down so as to insure that the contents are all thoroughly mixed and shaken. (Figures 80 and 81)

3. Add 25 ml of ethyl ether, insert the cork and shake vigorously for 1½ minutes. The ethyl ether dissolves the fat.

4. Add 25 ml of petroleum ether, insert the cork, and shake for 1½ minutes. The petroleum ether takes out the last traces of moisture from the ethyl ether extract and also aids in dissolving the fat. (Figures 80 and 81.)

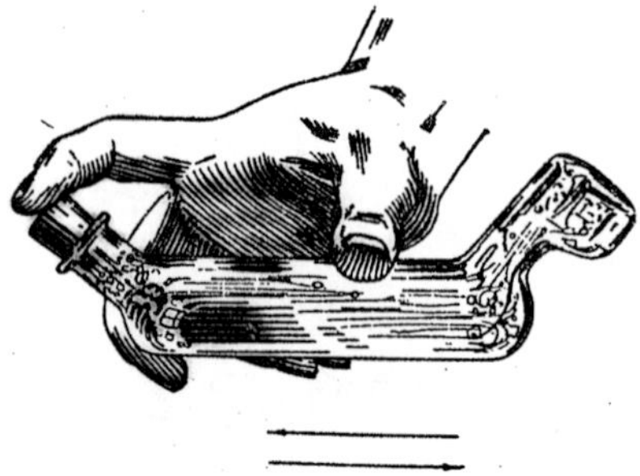

Figure 80.
Correct position of flask when shaking.

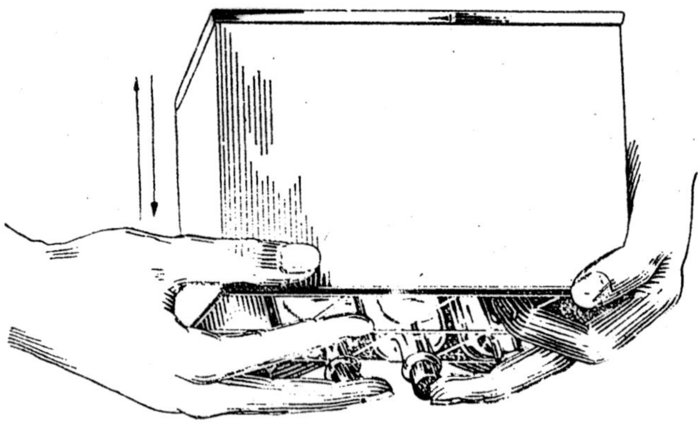

Figure 81.
Position in which to hold the four flasks that are being shaken at one time.

5. Centrifuge thirty turns, taking ½ minute.

6. Pour off the ether mixture containing the extracted fat into a fat dish which has been heated to 275 F (135 C) in the fat oven for 5 minutes, subsequently cooled in the fat cooling desiccator for 7 minutes, and weighed to the nearest 1/10 mg. (Figure 82.)

PRECAUTION: When decanting the ether solution into the dishes be very careful not to pour any suspended solids into the dish.

For Second Extraction

7. Add 5 ml of alcohol to the residue in the flask and shake.

8. Add 15 ml each of ethyl and petroleum ether, inserting the cork and shaking for 30 seconds after the addition of each reagent.

9. Centrifuge thirty turns, taking ½ minute.

Figure 82.
Correct procedure when pouring ether solution into dish.

10. Pour off the ether mixture into the same fat dishes as were used for the first extraction, taking care not to pour off any of the residue below the ether solutions.

NOTE: Before pouring off the ether mixture, if the level of the residual water layer is below the neck of the flask between the two bulbs, add sufficient distilled water to bring the level about half way on the neck so that after decanting, a minimum of ether mixture remains in the flask. (Figure 83)

11. Evaporate the ether from the fat dishes on the electric hot plate at 275 F (135 C).

PRECAUTION: Ether fumes are dangerous and must not be allowed to spread through the room. Always keep the cover over the hot plate when evaporating the ether, and have the proper exhaust pipe taking the fumes trom the tester to out doors. Do not permit smoking in the laboratory as ether fumes are very inflammable. (Figure 84)

FOR THIRD EXTRACTION

12. Make a third extraction in the same manner as the second, omitting the addition of alcohol.

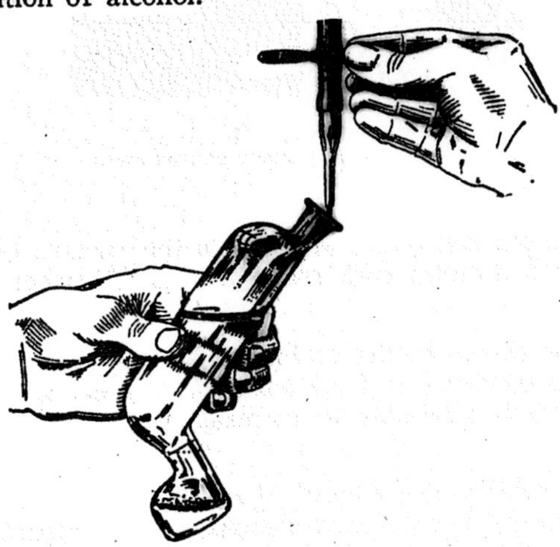

Figure 83.
Position of dividing line before and after raising.

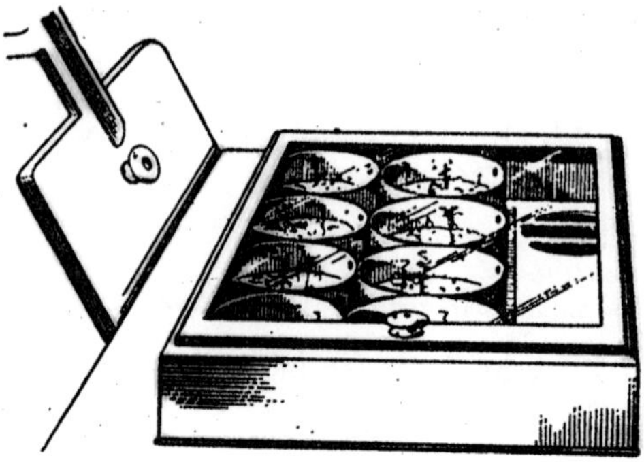

Figure 84.
Evaporating the ether.

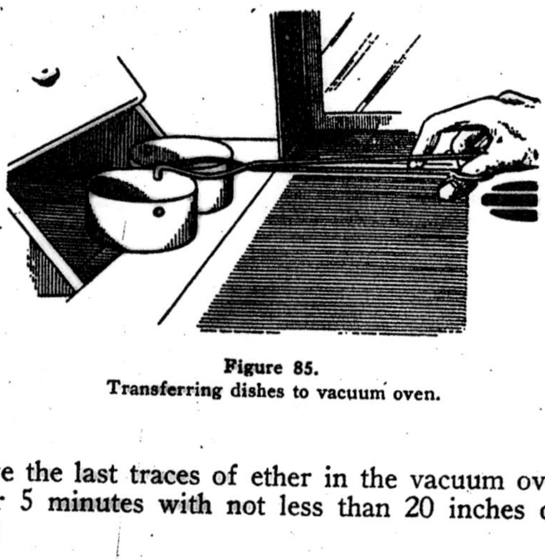

Figure 85.
Transferring dishes to vacuum oven.

13. Remove the last traces of ether in the vacuum oven at 275 F (135 C) for 5 minutes with not less than 20 inches of vacuum. (Figure 85.)

14. Cool the dishes in the cooling desiccator with the circulating water at room temperature for 7 minutes. If there are several dishes in the oven, it is advisable to increase the time of cooling to 10 minutes.

15. Weigh rapidly and record the weight of fat. All weights should be recorded to the fourth decimal place. (Figure 86.)

$$\text{Percentage Fat} = \frac{\text{Weight of fat}}{\text{Weight of sample}} \times 100$$

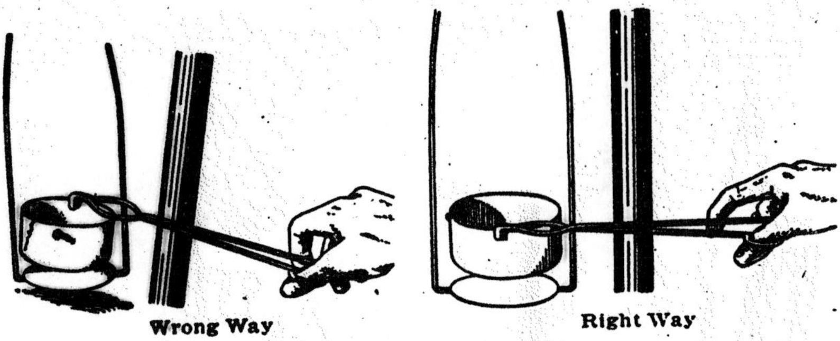

Figure 86.
Method of placing dish upon the balance pan.

Cream
(Mojonnier)

Apparatus

1. Same as given on page 209.

Reagents

1. Same as given on page 209.

Procedure

1. Adjust the sample to room temperature and mix thoroughly by pouring several times from one vessel to another.

PRECAUTION: Cream in which the fat has become hardened may be brought into the proper condition for mixing by slowly and thoroughly warming to a temperature not exceeding 100 F (37.8 C). Should the milk fat be separated or churned, heat the cream to 105-110 F (41-43.3 C); shake it thoroughly, and immediately weigh out the test portions. In such cases duplicate samples are imperative.

2. Weigh 2 gm of the well mixed sample into a properly labeled extraction flask. (See figures 87 and 88.)

NOTE: It is not necessary to use exactly 2 gm, but actual weights should be recorded to the fourth decimal place. In case of the cream containing over 25 per cent milk fat, a 1 gram sample is used.

3. Add 5 ml of distilled water to the sample in the extraction flask and mix well.

NOTE: Add 6 ml of distilled water for cream testing over 25 per cent milk fat.

4. Add 1.5 ml of NH_4OH and proceed with the remainder of the test as for milk except that 25 ml of ethyl and of petroleum ether must be added in making the second extraction.

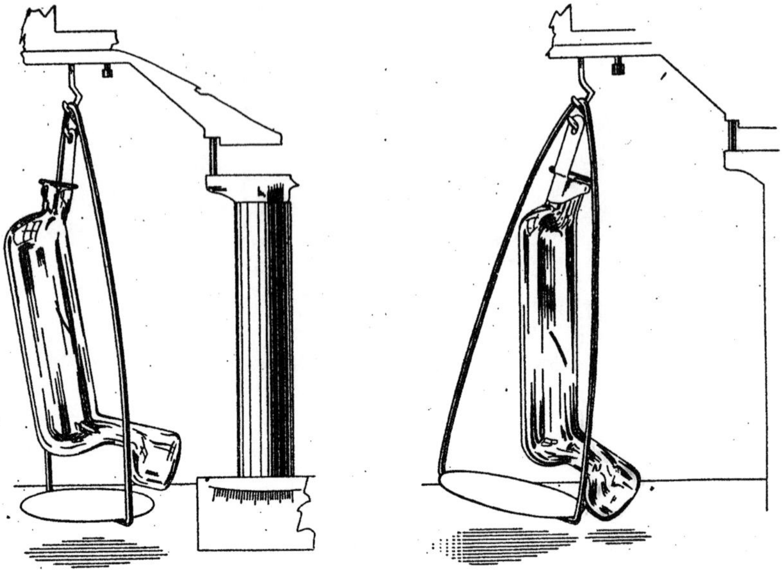

Figure 87.
Flask hanger with flask suspended to balance arm.

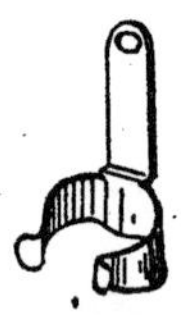

Figure 88.
Flask hanger.

Ice Cream
(Mojonnier)

Apparatus

1. Same as given on page 209.

Reagents

1. Same as given on page 209.

Procedure

1. Adjust the sample to room temperature and mix thoroughly by pouring several times from one vessel to another.

PRECAUTION: Fruited ice cream and samples containing nuts are difficult to sample. Care should be taken to get even distribution of fruits or nuts.

2. Weigh a sample of about 5 gm (recording the weight accurately) into a properly labeled extraction flask.

3. Add 5 ml of distilled water, mix well, and proceed as for milk except that 25 ml of ethyl and of petroleum ether must be added in making the second extraction.

Evaporated Milk, Condensed Buttermilk and Unsweetened Condensed Milk

Apparatus

1. Same as given on page 209.

Reagents

1. Same as given on page 209.

Procedure

1. Adjust the sample to room temperature and mix thoroughly by pouring several times from one vessel to another.

NOTE: In case of heavy superheated products and canned evaporated milk it may be necessary to warm the sample to 90-100 F (32.2-37.8 C). Should the milk fat be separated or churned, heat the sample to 100-110 F (37.8-43.3 C); shake it thoroughly and immediately weigh the 3 gm test portions. In such cases duplicate samples are imperative.

2. Weigh a sample of about 5 gm (recording the weight accurately) into a properly labeled extraction flask.

3. Add 6 ml of distilled water and proceed as for milk except that 25 ml of ethyl and of petroleum ether must be added in making the second extraction.

Sweetened Condensed Milk

Apparatus

1. Same as given on page 209.

Reagents

1. Same as given on page 209.

Procedure

1. Adjust the sample to room temperature and mix thoroughly.

NOTE: Sweetened condensed milk in which the sugar has settled out may be brought into proper condition for mixing by slowly and thoroughly warming to a temperature not exceeding 110 F (43.3 C).

2. Weigh a sample of about 5 gm (recording the weight accurately) into a properly labeled extraction flask.

3. Add 8 ml of hot distilled water and shake thoroughly.

4. Proceed with the remainder of the test in the same way as for milk except that 25 ml each of ethyl and of petroleum ether must be added in making the second extraction.

Butter, Cheese, Malted Milk, Chocolate and Cocoa

Apparatus

1. Same as given on page 209.

Reagents

1. Same as given on page 209.

Procedure

1. Thoroughly mix the sample. In the case of cheese and milk chocolate it is suggested that the sample be cut up into very small particles.

2. Weigh the sample portion directly into the extraction flask. In the case of malted milk, chocolate, and cocoa use a 0.5 gm sample. In the case of cheese and butter use a 1 gm sample.

3. Add 8 ml of hot distilled water.

4. Add 1.5 ml of NH_4OH (3 ml in the case of cheese) and proceed with the remainder of the test in the same way as for milk except that 25 ml each of ethyl and of petroleum ether must be added in making the second extraction.

NOTE: If greater accuracy is desired in determining the fat content of cheese, chocolate, and cocoa, the A.O.A.C. method may be used.[2]

Dry Skim Milk, Buttermilk Powder and Whole Milk Powder

Apparatus

1. Same as given on page 209.

Reagents

1. Same as given on page 209.

Procedure

1. Thoroughly mix a representative sample.

2. Weigh a 1 gm sample directly into the extraction flask.

3. Add 9 ml of hot distilled water. Shake vigorously until dissolved.

4. Add 1.5 ml of strong ammonium hydroxide (3 ml in the case of buttermilk) and proceed with the remainder of the test in the same way as for milk.

Causes for High Fat Tests

1. Not keeping the bottom of the dishes flat.

2. Improper shaking and centrifuging shown by nonfatty residue in the dish.

3. Impure reagents (if in doubt, run a test on the reagents substituting distilled water for milk).

4. Temperature in fat oven too low.

5. Contamination of the dish after the ether was poured into it.

6. Improper reading or posting of weights.

7. Weights becoming lighter because of wear.

8. Weighing the dish containing the fat at a lower temperature than prevailed when the dish was weighed empty.

NOTE: Weights should be calibrated at regular intervals with Bureau of Standards reference weights.

Causes for Low Fat Tests

1. Leaky corks. Use the best corks obtainable.

2. Insufficient shaking.

3. Adding too much water or too little alcohol.

4. Having the dividing line too low, so that too much ether is left behind. If such is the case, add more water to bring the line to the proper height before pouring off, or make a third extraction.

5. Too high temperature in the vacuum oven.

6. Insufficient water circulating through the cooling desiccator. The water tank must be kept filled and the circulating pump must be kept in good working order.

7. Improper reading or posting of weights.

8. Spattering of the fat in the oven due to transferring the dish to the oven before the ether solution has all evaporated, or if too high heat is carried in the vacuum oven.

9. Weighing the dish at a higher temperature than prevailed when the dish was weighed empty.

REFERENCES

1. Mojonnier, Timothy, and Troy, H. C.: *Technical Control of Dairy Products,* Bul. 101, Mojonnier Brothers Company, Chicago, 1925.

2. *Methods of Analysis,* A.O.A.C. 6th edition (1945), pp. 229, and 337.

Liquid Eggs, Frozen Eggs and Dried Eggs

A. FROZEN EGG YOLK

Apparatus

1. Same as given on page 209.

Reagents

1. Same as given on page 209.

Preparation of Sample

In order to get an accurate sample of frozen egg yolks, they should be removed from cold storage and allowed to temper, preferably in a cooler at 50 F (10 C). Application of heat or setting in hot water to hasten melting should be avoided as it may cause the yolks to thicken. After melting, beat thoroughly before sampling.

B. LIQUID EGGS

Acid Hydrolysis (A.O.A.C. Method[1])

From a well mixed sample, weigh accurately by difference into the fat extraction tube about 2 gm of yolks or 3 gm of whole eggs or 5 gm of whites. Add slowly, with vigorous shaking, 10 ml of HCl. Set tube in water bath heated to 158 F (70 C), bring to a boil, and continue heating at boiling point for 30 minutes, shaking tube with care at 5 minute intervals. Remove tube from water bath, add water merely to fill lower bulb of tube, and cool to room temperature.

C. DRIED EGGS

Transfer 1 gm of the well mixed sample to Mojonnier tube, add slowly, washing down any egg particles adhering to the sides of the tube, 10 ml of HCl (4 plus 1) ; and proceed as directed in liquid eggs.

Procedure

1. To the Mojonnier tube containing the treated sample, add 25 ml of ethyl ether, and mix. Add 25 ml of redistilled petroleum ether (b. p. not over 140 F (60 C)), mix and allow to stand until solvent layer is clear.

2. Decant as much as possible of ether-fat solution through filter consisting of pledget of cotton packed just firmly enough to stem of funnel to allow free passage of ether into weighed 125 ml beaker-flask containing some porcelain chips or broken glass. Before weighing beaker-flask, dry it and a similar flask as a counterpoise in drying oven at 212 F (100 C) and then allow it to stand in air to constant weight.

3. Re-extract liquid remaining in tube twice, each time with only 15 ml of each ether.

4. Draw off clear ether solutions through filter into same flasks as before, and wash tip of pledget, funnel, and end of funnel stem with a few milliliters of a mixture of the two ethers in equal volume free from suspended water.

5. Evaporate the ether slowly on steam bath, then dry fat in drying oven at 212 F (100 C) to constant weight (about 90 min).
6. Remove flask and counterpoise from oven, allow to stand in air to constant weight (about 30 min) and weigh. (Owing to size of flask and nature of material, there is less error by cooling in air than in desiccator.)
7. Correct this weight by blank determination on reagents used.
8. Report as percentage fat by acid hydrolysis.

D. Ovson Method for Fat in Egg Yolks[2]

Apparatus

1. Sampling Dipper.
2. 50 ml Erlenmeyer flask.
3. Röhrig extractor.
4. 125 ml Erlenmeyer flask.
5. Funnel fitted with pledget of cotton.
6. Distillation apparatus.
7. Steam bath.
8. Aspirator or vacuum oven.

Reagents

1. Distilled water.
2. Concentrated hydrochloric acid (Sp. Gr. 1.18).
3. Di-ethyl Ether.
4. Petroleum Ether (Re-distilled—fraction distilled under 140 F (60 C).

Procedure

1. Into a weighed 50 ml Erlenmeyer flask, weigh approximately 2 gm of egg yolk, and 2 ml of distilled water, and shake until dissolved.
2. Add 10 ml of concentrated hydrochloric acid, heat in steam bath for ½ hour or more, shaking at intervals or until all the proteins are hydrolyzed. Cool, add 7 ml of distilled water.
3. Transfer to Röhrig extractor, washing the lips of the flask with a portion of 25 ml of diethyl ether. Transfer the remainder of the ether into the extractor, stopper, and shake three times by inversion, holding the thumb over the vapor outlet and keeping the stopcock closed. Add similarly 25 ml of redistilled petroleum ether and shake as before. Allow to settle for at least 15 minutes. Repeat the above extraction twice but using 15 ml of each ether.
4. Drain the ether each time into the same weighed 125 ml Erlenmeyer flask through a small funnel containing a pledget of cotton.
5. Distill off ethers. Exhaust flask in vacuum oven or aspirate for ½ hour at 131 F (55C). Cool and weigh the flask. Report as percentage fat by acid hydrolysis.

References

1. *Methods of Analysis,* A.O.A.C., 6th edition (1945), p. 347.
2. Ovson Egg Co., Chicago, Ill., Personal Correspondence.

CHAPTER 15

GERBER TEST FOR FAT

The Gerber test, perfected by Dr. N. Gerber, a Swiss chemist and used extensively in Europe, employs sulfuric acid and amyl alcohol. The sulfuric acid dissolves the solids other than fat, liberates the fat, and creates heat to keep the fat in a liquid condition. The amyl alcohol prevents the charring of the fat and facilitates a clear fat column. The fat, after chemical liberation from the other milk solids is separated by centrifugal force as in the Babcock Test.

Milk, Plain or Homogenized

Apparatus

1. Gerber milk-test bottle (butyrometer) (see figure 89) with solid or nipple type rubber stoppers.

2. Gerber milk pipette, calibrated to deliver 11 ml of milk at 60 F (15.6 C).

3. Automatic acid measure for dispensing 10 ml charges (see figure 90).

4. Automatic amyl alcohol measure for dispensing 1 ml charges (see figure 91).

5. Gerber or Babcock centrifuge (see figure 92).

NOTE: Babcock centrifuges must be equipped with cup adapters to accommodate the Gerber test bottles, and must be heated as for Babcock test (see page 179).

6. Filling and shaking rack.

7. Tempering bath maintained at 135-140 F (57.2-60.0 C).

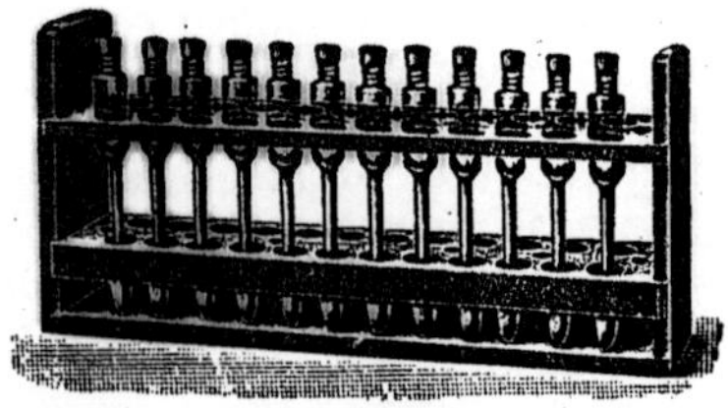

Figure 89.

Gerber test bottles, called butyrometers. Figures 89 to 94 courtesy Dairy Industry Supply Co.

Figure 90.
Acid measure.

Figure 91.
Tilt Measure for Amyl alcohol.

Figure 92.
Gerber electric centrifuge.

Figure 93.
Appearance of fat column in neck of Gerber test bottle.

Reagents

1. Sulfuric acid (H_2SO_4), sp. gr. 1.820-1.825 at 68 F. (20 C).

NOTE: Proper strength of H_2SO_4 since the test bottles hold, and must contain, a definite milk-acid-amyl alcohol volume which precludes the addition of a smaller or larger charge of acid to compensate for stronger or weaker concentration of acid.

2. Amyl alcohol[1] sp. gr. 0.815-0.818, b. p. 262 to 270 F (127.8-132.2 C), free from fat, furfural, or any substance that might appear in the fat column after centrifuging. (See figure 93.)

NOTE: Keep amyl alcohol in well-stoppered brown glass bottles, and in a cool dark place. Suitability of each new shipment of amyl alcohol for fat testing should be checked by testing a series of milk samples in duplicate and comparing results with tests obtained by use of amyl alcohol of known suitability.

Procedure

1. Transfer 10 ml of H_2SO_4 at 60-70 F (15.6-21.1 C) to the Gerber milk-test bottle.

2. Measure into the test bottle by means of the Gerber milk pipette, 11 ml of the well-mixed milk sample, allowing it to flow slowly down the side of the test bottle to form a distinct layer above the acid.

3. Add 1 ml of amyl alcohol to the test bottle and, if necessary, before mixing adjust the unmixed contents to 60-70 F (15.6-21.1 C) by partial immersion in a water bath.

NOTE: Always add amyl alcohol after the milk, never allowing it to contact the H_2SO_4 directly for this will cause an imperfect test.

4. Tighten the stopper and mix by shaking the bottle at a 45 degree angle until all visible curd has been dissolved (about 2 min.). Next, with the flat band in the verticle plane, invert the bottle several times to insure homogeneous mixing of the entire bottle contents.

NOTE: It is extremely important to shake the bottle at a 45° angle before inversion. This allows 8 of the 10 ml of H_2SO_4 to mix with the 11 ml of milk and brings about solution of the solids not fat without any tendency to "burn" the sample. Keeping the flat band of the bottle in the verticle plane during inversion will bring the acid-milk-amyl alcohol mixture to homogeniety in a minimum of time.

4a. For homogenized milk, shake as above for an interval sufficient to dissolve all visible curd, and then shake additionally for a period at least 50 per cent longer before inverting the bottles to mix the remaining contents.

NOTE: The additional shaking aids in liberating all of the small fat globules in homogenized milk thus to permit them to more readily agglomerate and rise.

5. Place the test bottle in the centrifuge and, with the machine balanced, whirl for 5 minutes at the proper speed.

NOTE: Proper speed for Gerber centrifuge is 1,000 r.p.m. Babcock centrifuge speed should be adjusted to speed recommendations on page 183 and heated as stated on page 180.

6. After centrifuging, temper the bottles in the water bath at 135-140 F (57.2-60.0 C) for 5 minutes and then, holding the bottle vertically, immediately adjust the bottom meniscus (flat meniscus) to zero or the nearest whole percentage graduation and observe the reading from that point to the lowest part of the upper meniscus. (See figure 94).

NOTE: It has been reported that homogenization of milk does not affect the Gerber test (Trout, G. M. and Lucas, P. S. A Comparison of the Babcock, Gerber, Minnesota, Pennsylvania, and Mojonnier Methods for Determining the Percentages of Fat in Homogenized Milk. *J. Dairy Sci., 30*: 145-159 (1947) (see page 185).

Optional Gerber Procedure for Testing a Large Number of Samples

When a large number of samples are to be tested and there is a possibility of considerable time elapsing between preparation of the first and last samples there may be a tendency for tests to "burn" because of prolonged contact between acid and milk. To prevent such difficulty it has been found feasible to transfer the 11 ml portion of milk directly to the test bottle and, holding the test bottle at a 45 degree angle with the flat band in the vertical plane to add the 10 ml portion of H_2SO_4 cooled to 40 F (4.44 C) by running it down the inside surface of the test bottle. The acid being heavier than the milk displaces it from the bulb and neck of the test bottle and the low temperature of the acid prevents any local action as it passes by the milk. The 1 ml of amyl alcohol may then be added to the milk layer and after adjusting the unmixed contents of the test bottle to 60-70 F (15.6-21.1 C) by partial immersion in a water bath the regular testing procedure may be followed to completion.

This procedure permits the use of the Gerber test bottle as a sample container for the 11 ml portion of composite milk sample measured at the receiving station and transported to the laboratory for testing.

Figure 94.
Position for Reading the Gerber Test Bottle.

Skim Milk and Buttermilk

Apparatus

1. Gerber skim milk test bottle.
2. Other apparatus the same as for milk. (See page 222).

Reagents

1. Same as for milk. (See page 223).

Procedure

The procedure is the same as for testing milk except that skim milk bottles are centrifuged twice for 5 minutes and between the centrifugings are placed in a water bath at 140 F (60 C) for a few minutes.

Chocolate Milk and Chocolate Drink

Apparatus

1. Same as for milk. (See page 222).

Reagents

1. Same as for milk. (See page 223).

NOTE: If excessive amounts of chocolate flavoring have been added, it may be necessary to use dilute H_2SO_4, which is made by slowly adding 94 ml of the Gerber acid (sp. gr. 1.820 to 1.825) to 6 ml of water.

Procedure

The procedure is the same as for testing milk except that the test bottle is immersed in a tempering bath at 140°F. (60 C) for a few minutes after the first 5 minute centrifuging, then centrifuged for an additional 5 minutes before final tempering and reading the test as directed on page 224.

Cream

Apparatus

1. Gerber cream-test bottles.
2. Torsion balance sensitive to 30 ml.
3. Cream-bottle support for weighing on the torsion balance.
4. Weight, 5 gm (tolerance ± 5 mg), clean and free from corrosion.
5. Pipette, 5 ml capacity for transferring cream.
6. Other apparatus same as for milk. (See page 222).

Reagents

1. Same as for milk. (See page 223).

Procedure

1. Weigh exactly 5 gm of well mixed cream into the cream-test bottle containing 10 ml of Gerber H_2SO_4 as it is balanced for weighing on the torsion balance.

2. Add 5 ml of water and 1 ml of amyl alcohol to the cream-test bottle and, if necessary, before mixing adjust the unmixed contents to 60-70 F (15.6-21.1 C) by partial immersion in a water bath.

3. Proceed as directed under procedure 4-6 for milk (page 224).

Ice Cream and Ice Cream Mix

Ice cream contains such a high percentage of sugar that charring by H_2SO_4 often results. In the Troy-Fucoma method[2], this difficulty is largely overcome by the use of dilute H_2SO_4 and amyl alcohol.

Apparatus

The same as for cream, except that special Gerber ice cream test bottles are used.

Reagents

Same as for milk, except that sulfuric acid is diluted by mixing commercial H_2SO_4 (sp. gr. 1.820-1.825) with water as follows: For ice cream mix and all flavors excepting chocolate, measure 13 ml of water and add slowly 87 ml of acid. For chocolate ice cream measure 6 ml of water and add slowly 94 ml of acid.

CAUTION: Always add acid to water, never vice versa.

Procedure

1. Transfer 10 ml of acid into the bottle.
2. Melt the sample. Mix and weigh 5 gm into a special ice cream test bottle on top of the acid.
3. Add 5 ml of water and 1 ml of amyl alcohol.
4. Insert the stopper. Shake, centrifuge for 5 minutes, temper at 140 F (60 C) for 5 minutes and make readings as for milk.

NOTE: Chocolate ice cream should be centrifuged twice (see procedure for chocolate milk on page 226).

REFERENCES

1. Herrington, B. L.: Amyl Alcohol as a Source of Error in the Gerber Test. Jour. Dy. Sc., Vol. 16, No. 6, (1933).

2. Fisher, R. C., and Walts, C. C.: A Comparative Study of Methods for Determining the Per Cent of Butterfat in Dairy Products. II. Ice Cream. Jour. Dy. Sc. Vol. 8, No. 3 (1925).

3. N. Y. State Dept. of Agric., circular 632 (revised May, 1943).

4. Henderson, Laton M.: Manual for Milk Testers in N. J., N. J. Agric. Exp. Sta. Bul. 647, (1938).

CHAPTER 16

MOJONNIER DETERMINATION OF TOTAL SOLIDS

Milk, Skim Milk, Buttermilk and Whey

Apparatus

Complete Mojonnier total solids tester with chemical balance (Figure 17, page 22) which includes a quantity of Mojonnier solids dishes (Figure 95) a dish cover, a dish contact maker (Figure 96) a quantity of 1 gm pipettes, and a pair of tongs.

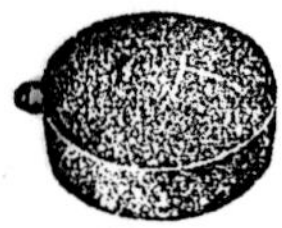

Figure 95.
Solids dish. **Figures 95 to 97, courtesy Mojonnier Brothers. Co.**

Reagent

Distilled water.

PRECAUTION: It is important that the distilled water contain no solids. A blank determination should be run whenever a new supply of distilled water is secured.

Procedure

1. Mix the sample thoroughly by pouring several times from one vessel to another.

2. Weigh with cover in place a 2 gm sample to the fourth decimal place into a solids dish which has been heated for 10 minutes in the solids oven at 212 F (100 C), subsequently cooled for 5 minutes in the solids cooling desiccator and weighed. Make all weighings to the four decimal place and with the cover on the solids dish, as shown in figure 97.

Record:

(a) Weight of dish.
(b) Weight of sample.

3. By gently tilting the dish, spread the milk in a thin film over the entire bottom.

4. Place the dish on the hot plate at 356 F (180 C) and heat until the residue begins to turn light brown. Insure uniform evaporation by using the dish contact maker.

5. Transfer the dish to the solids vacuum oven at 212 F (100 C). Heat for 10 minutes under not less than 20 inches of vacuum.

6. Cool in a cooling desiccator with water circulating for 5 minutes.

7. Weigh the dish and solids with the dish covered. Record the weight of the solids.

$$\text{Percentage of Solids} = \frac{\text{Weight of Solids}}{\text{Weight of Sample}} \times 100$$

Figure 96.
Dish contact maker, to press the dish against the hot plate.

Cream

Apparatus and Reagent

Same as given on page 228.

Procedure

1. Weigh a 1 gm sample directly into a tared and weighed dish.

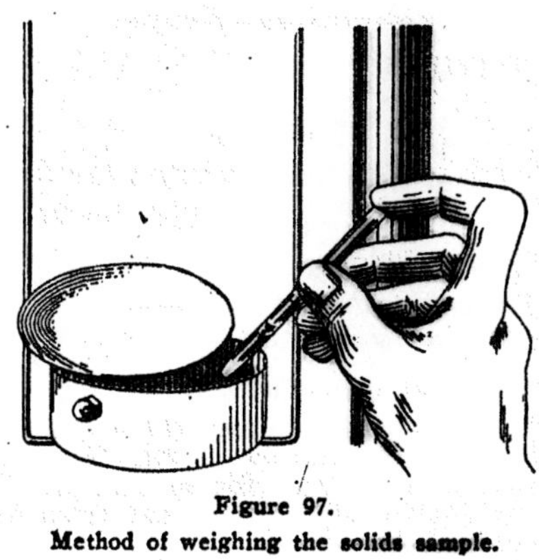

Figure 97.
Method of weighing the solids sample.

2. Add from 1 to 1.5 ml of distilled water so as to dilute the sample and spread it evenly over the bottom of the dish.

3. Place the dish on the solids hot plate and proceed as for milk.

NOTE: In case of high testing cream, over 30 per cent, use only a 0.5 gm sample.

Ice Cream

Apparatus and Reagent

Same as given on page 228.

Procedure

1. Weigh a 1 gm sample directly into a tared and weighed dish.
2. Add 1 ml of distilled water and spread the mixture evenly over the bottom of the dish.
3. Place the dish on the solids hot plate and proceed as for milk.

Unsweetened Condensed Milk and Condensed Buttermilk

Apparatus and Reagent

Same as given on page 228.

Procedure

1. Weigh a 1 gm sample directly into a tared and weighed dish.
2. Add 1 ml of distilled water to the sample in the dish and mix thoroughly with the sample.
3. Place the dish on a hot plate and proceed as for milk.

NOTE: In case of extra heavy superheated products use a 0.5-gm sample and add 2 ml of distilled water.

Sweetened Condensed Milk

Apparatus and Reagent

Same as given on page 228.

Procedure

1. Weigh a 0.25 gm sample directly into a tared and weighed dish.
2. Add 2 ml of hot distilled water to dissolve thoroughly the sugar and properly dilute the sample.
3. Evaporate on the hot plate as for milk.
4. Dry in the oven for 20 minutes.
5. Deduct 0.30 per cent from the per cent of total solids obtained.

NOTE: Sometimes deduction of 0.30 per cent is made because 20 minutes drying in the oven does not complete the drying. For all practical purposes it is as accurate as drying in the oven for 90 minutes. The diluted sample may also be placed in electric drying oven and dried to constant weight. (Approximately 4 hours will be required.)

Butter

Apparatus

Same as given on page 228.

Procedure

1. Weigh a sample about 1 gm directly into a dish.
2. Add no water and proceed as for milk.

Cheese

Apparatus and Reagent

Same as given on page 228.

Procedure

1. Weigh a sample of about 0.5 gm directly into a dish.

2. Weigh with the dish a blunt pointed glass rod that can be used to break up the cheese lumps.

3. Add 1.5 ml of hot distilled water and with the blunt pointed glass rod break up the lumps of cheese so as to insure an even film of the sample and added water over the entire bottom of the dish. The glass rod is left in the dish for the entire determination.

4. Proceed as for milk except that the sample is dried for 20 minutes in the vacuum oven.

NOTE: If the cheese is hand homogenized, the accuracy of this method will be appreciably increased.

Soft Cheeses

Apparatus and Reagent

Same as given on page 228.

Procedure

Use about 4 gm of cheese and enough hot distilled water (from 2 - 3 ml) to make a paste over the entire bottom of the dish; in other respects, proceed as for milk except that the sample is to be dried for 20 minutes in the vacuum oven.

Malted Milk, Chocolate and Cocoa

Apparatus

Same as given on page 228.

Procedure

1. Weigh about 0.3 gm of a sample directly into a dish.

2. Add 2 ml of hot distilled water and spread the sample and water in an even film over the bottom of the dish.

NOTE: In the case of chocolate and cocoa, the addition of 1 ml of ethyl ether aids in dispersing the sample.

3. Proceed as for milk except that the sample is to be dried for 20 minutes in the vacuum oven.

Dry Milk Powder, Whole Milk Powder and Buttermilk Powder

Apparatus

Same as given on page 228.

Procedure

1. Weigh a sample of about 0.3 gm directly into a dish.

2. Add 2 ml of hot distilled water and spread the sample with the water in an even film over the entire bottom of the dish.

3. Proceed as for milk.

NOTE: For greater accuracy dry to constant in a drying oven weight at 221 F (105 C) or the tolual distillation method given on Page 237.

Egg Yolk

Apparatus

Same as given on page 228.

Procedure

Follow the regular Mojonnier procedure as outlined under test for ice cream for total solids except that for fresh egg yolk or frozen egg yolk use 0.5 gm and for powdered egg yolk use 0.3 gm.

Gelatin

Apparatus and Reagent

Same as given on page 228.

Procedure

1. Weigh the sample directly into the dish on the balance pan. Use a sample of about 0.3 gm.

2. Add 2 ml of hot distilled water and spread the sample with the water in a thin film over the entire bottom of the dish.

3. Proceed as for milk.

4. Determine the percentage of moisture by difference.

NOTE: Drying to constant weight in a drying oven at 221 F (105 C) is more accurate for moisture determination of gelatin.

Causes for High Total Solids Tests

1. The bottoms of the dishes were not kept level.

2. Incomplete evaporation on the solids hot plate or evaporation at too low a temperature. Do not remove the dish until all visible moisture has been evaporated and the first trace of brown color appears.

3. Improper reading or posting of the weights. Weights light because of wear. Frequent calibration with Bureau of Standard reference weights is suggested.

4. Contamination of the dish after the sample had been weighed into it.

5. Low vacuum oven temperature.

6. The vacuum not up to standard.

7. Too large a sample was taken, rendering it impossible to remove all the water under standard conditions.

8. Weighing the dish with the solids in it at a lower temperature than prevailed when the empty dish was weighed.

Causes for Low Total Solids Tests

1. The sample was browned too much because of too long an exposure or the use of too high a temperature on the hot plate.

2. Operating vacuum oven above 212 F (100 C).

3. Milk spattered from the dish. This will not happen if the temperature is kept at 356 F (180 C).

4. Improper reading or posting of the weights.

5. Water was not running through the desiccator.

6. Weighing the dish with the solids in it at a higher temperature than prevailed when the dish was weighed empty.

REFERENCE

Mojonnier, Timothy, and Troy, H. C.: *Technical Control of Dairy Products*, **Bulletin** 101, Mojonnier Brothers Company, Chicago, Ill. (1925)

SUMMARY OF METHODS FOR MAKING FAT AND TOTAL SOLIDS TEST

Product to be tested	METHODS FOR MAKING FAT TESTS Keep sample in oven 5 min. at 135° C., and 7 min. in cooling desiccator								How long to centrifuge	Reagents to add, and how to shake—Second Extraction. Add neither water nor ammonia			How long to centrifuge	METHODS FOR MAKING TOTAL SOLIDS OR MOISTURE TESTS Summary of operations—Weigh directly into dish upon balance		
	How to prepare representative samples	How to weigh fat samples	Size of sample to take for fat test in grams	Reagents to add, and how to shake First extraction										Approximate size of sample to take for solids or moisture test in grams	Amount of water to add to sample in dish in cc	How long to keep sample in oven and cooling desiccator
				Water	Ammonia	Alcohol	Ethyl Ether	Petroleum Ether		Alcohol	Ethyl Ether	Pet'l'm Ether				
Fresh Milk	Mix thoroughly	Measure with 10 gram pipette. Drain 15 seconds	10	No water	1.5 cc Mix thoroughly	10 cc Shake half minute	25 cc Shake 20 seconds	25 cc Shake 20 seconds	30 turns	5 cc Shake 20 seconds	15 cc Shake 20 seconds	15 cc Shake 20 seconds	30 turns	2	None	10 min. in oven at 100°C. 5 min. in cooling desiccator at room temp.
Skim-Milk	″	″	10	″	″	″	″	″	″	″	″	″	″	2	″	″
Whey	″	″	10	″	1.5 cc 3 cc if whey is acid mix thoroughly	″	″	″	″	″	″	″	2	″	″	″
Buttermilk	″	″	10	″	″	″	″	″	″	″	″	″	″	2	″	″
Cream testing less than 25% B. F	Mix thoroughly Heat slightly if churned to melt fat	Weigh directly into flask. If necessary use butter boat	About 2	5 cc Mix thoroughly	″	″	″	″	″	″	″	″	″	1	1	″
Cream testing more than 25% B. F.	″	″	About 1	6 cc Mix thoroughly	″	″	″	″	″	″	25 cc Shake 20 seconds	25 cc Shake 20 seconds	″	1	1	″
Ice Cream Mix	″	Use weighing pipette or weigh directly into flask.	About 5	5 cc Mix thoroughly	1.5 cc Mix thoroughly	″	″	″	″	″	″	″	″	1	1	″

SUMMARY OF METHODS FOR MAKING FAT AND TOTAL SOLID TESTS (Continued)

Evaporated Milk	Shake in can or mix in bulk very thoroughly	"	About 5	4 cc Mix thoroughly	"	"	"	"	"	"	"	"	"	1	1	"
Bulk Unsweetened Condensed Milk	Mix very thoroughly	"	About 5	6 cc Mix thoroughly	"	"	"	"	"	"	"	"	"	1	1	"
Bulk extra heavy Unsweetened Condensed Milk	"	"	About 3	6 cc Mix thoroughly	"	10 cc Shake one minute	"	"	"	"	"	"	"	0.5	2	"
Sweetened Condensed Milk	Proceed without diluting. Mix very thoroughly	"	About 5	8 cc hot water. Mix until thoroughly dissolved	"	"	"	"	60 turns	"	"	"	60 turns	0.25	2	90 min. in oven at 100°C or 20 min. and deduct 0.30% from total. 5 min. in cooling desiccator at room temp.
Condensed Milk with Sugar and Chocolate	"	"	About 5	"	"	"	"	"	"	"	"	"	"	0.5	2	10 min. in oven at 100° C. 5 min. in cooling desiccator at room temp.
Condensed Buttermilk (semi-solid)	"	If fluid use weighing pipette otherwise use butter boat	About 3	6 cc hot water Mix thoroughly	3 cc Mix thoroughly	10 cc Shake half minute	"	"	30 turns	"	15 cc Shake 20 seconds	15 cc Shake 20 seconds	30 turns	0.5	2	"
Skim-milk Powder	Pulverize in close grained mortar. Transfer to sealed jar	Use butter boat	About 1	8 cc hot water. Mix thoroughly	1.5 cc Mix thoroughly	"	"	"	"	"	"	"	"	0.3	2	"
Whole Milk Powder	"	"	About 1	"	"	"	"	"	"	"	25 cc Shake 20 seconds	25 cc Shake 20 seconds	"	0.3	2	"
Buttermilk Powder	"	"	About 1	"	3 cc Mix thoroughly	"	"	"	"	"	15 cc Shake 20 seconds	15 cc Shake 20 seconds	"	0.3	2	"

SUMMARY OF METHODS FOR MAKING FAT AND TOTAL SOLID TESTS (Continued)

METHODS FOR MAKING FAT TESTS Keep sample in oven 5 min. at 135° C., and 7 min. in cooling desiccator														METHODS FOR MAKING TOTAL SOLIDS OR MOISTURE TESTS Summary of Operations—Weigh directly into dish upon balance		
Product to be tested	How to prepare representative samples	How to weigh fat samples	Size of sample to take for fat test in grams	Reagents to add, and how to shake First extraction					How long to centrifuge	Reagents to add, and how to shake—Second Extraction. Add neither water nor ammonia			How long to centrifuge	Approximate size of sample to take for solids or moisture test in grams	Amount of water to add to sample in dish in cc	How long to keep sample in oven and cooling desiccator.
				Water	Ammonia	Alcohol	Ethyl Ether	Petroleum Ether		Alcohol	Ethyl Ether	Pet'l'm Ether				
Malted Milk	''	Use butter boat	About 0.5	8 cc	1.5 cc Mix thoroughly	10 cc	25 cc	25 cc	''	5 cc	25 cc Shake 20 seconds	25 cc Shake 20 seconds	30 turns	0.3	2	20 min. in oven at 100° C. 5 min. in cooling desiccator at room temp.
Cocoa	''	''	About 0.5	''	''	''	''	''	''	''	''	''	''	0.3	2	''
Milk Chocolate	''	''	About 0.5	''	''	''	''	''	''	''	''	''	''	0.3	2	''
Cheese	''	''	About 1	''	3 cc Mix thoroughly	10 cc Shake one minute	''	''	''	''	''	''	''	0.5	2	''
B[illegible]	''	''	About 1	''	1.5 cc Mix thoroughly	10 cc Shake half minute	''	''	''	''	''	''	''	1 *	None	10 min. in oven at 100° C. 5 min. in cooling desiccator at room temp.

* Pulverize in close grained mortar, and transfer to sealed jar.

CHAPTER 17

TOTAL SOLIDS DETERMINATION WITHOUT MOJONNIER EQUIPMENT

Mojonnier equipment frequently is not available in milk plant laboratories. In such cases, the simple but much longer procedure of drying the sample at 212 F (100 C) to constant weight may be followed.

Milk, Skim Milk, Buttermilk and Whey

Apparatus

1. Drying oven or water bath.
2. Quantity of flat-bottom dishes and covers, 5 cm in diameter.
3. Chemical balance equipped with weights.
4. Desiccator containing anhydrous calcium chloride, concentrated sulfuric acid or other suitable desiccant.

Procedure

1. Dry the flat-bottom dish in the drying oven at 212 F (100 C) for 10 minutes.
2. Place the dish in the desiccator and allow it to cool to room temperature.
3. Quickly weigh the dish with the cover to the fourth decimal place.
4. Pipette into the dish 3 - 5 ml of a well mixed milk sample and weigh quickly, with the cover on the dish, to the fourth decimal place.
5. Heat the dish and sample at the temperature of 212 F (100 C) until it comes to constant weight (see note 2 to follow). The approximate time required is from 2 to 3 hours.
6. Cool in a desiccator to room temperature before each weighing.
7. Weigh rapidly, with the cover on the dish, to the fourth decimal place. The increase in weight of the dish over the empty dish is the weight of the solids in the sample.
8. Calculate the percentage of solids as follows:

$$\text{Percentage of solids} = \frac{\text{Weight of solids}}{\text{Weight of sample}} \times 100$$

Thus:

Weight of dish and milk........ 16.4235
Weight of dish empty........ 12.1345

Weight of milk........ 4.2890

$$\text{Percentage of solids} = \frac{0.5418}{4.2890} \times 100 = 12.39$$

Weight of dish and dried milk........ 12.6763
Weight of dish empty........ 12.1345

Weight of dried milk or solids........ 0.5418

NOTE: 1. The foregoing method is suitable for milk, skim milk, buttermilk and whey. For other products, such as ices, sherbets, ice cream, evaporated milk, plain condensed, sweetened condensed, cream, and dry milk containing more solids, reduce the size of sample in proportion and add hot distilled water to dilute to a milk-like consistency.

For example, in the case of condensed milk use only about 1 gm of sample and add about 1 ml of hot distilled water.

NOTE 2: After the first weighing the dish and sample should be placed in the oven again for at least 1 hour and reweighed. This process is repeated until two consecutive weighings differ by not more than 0.0002 gm, that is, constant weight.

Dried Milk

(American Dry Milk Institute Method)

Since dried milk solids absorb moisture readily, it is very important that all utensils and containers used for sampling be clean and dry, and that sampling be done as speedily as possible. Samples should always be taken at points beneath the exposed surfaces. For complete laboratory analysis one pound samples are required. Prepare a composite sample by taking at least three good sized equal portions and immediately transferring to a dry container in which they can be thoroughly mixed. The container should be larger than the final composite sample to be taken so that the sample can be readily mixed.

Important

1. Do not sample unless hands and clothing are clean and dry.

2. Do not sample in a humid atmosphere or in a room into which steam is being discharged.

3. Do not sample in a damp cold storage room or with utensils or containers which have just been brought from a cold room into a warm room.

4. Do not sample when your attention is being distracted by other duties.

Preparing the Sample for Analysis

With a suitably sized spoon, turn over the contents by lifting it from the bottom of the can, at the same time slowly rotating the can at an angle of 45 degrees. Mix the powder in this manner by turning over a spoonful every second for a period of one minute. It is important during mixing to take the powder out of the lowest edge of the can. Precautions should be taken to reduce to a minimum the possibility of absorption of moisture during the mixing; and as soon as mixing is completed, the sample should be returned to its original container and covered tightly.

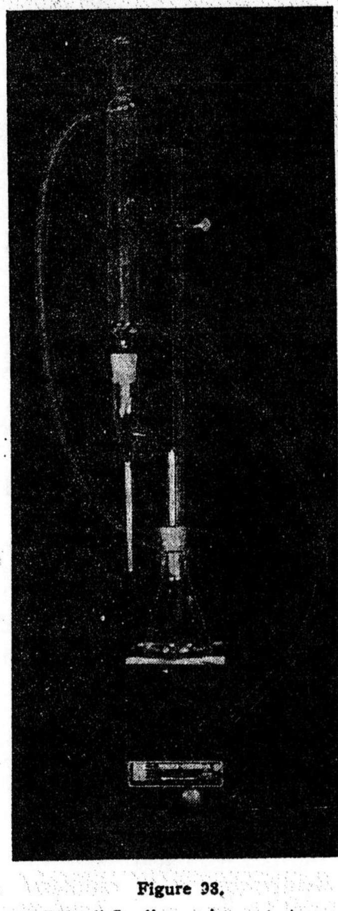

Figure 38.

Bidwell-Sterling moisture tester.

MOISTURE—TOLUOL DISTILLATION METHOD

Apparatus

1. Bidwell and Sterling moisture tester complete, including 300 ml Erlenmeyer flask, 2 one-hole cork stoppers, distillation tube and condenser. (See figure 98).

2. Precision electric heater, type RH with rheostat.

3. Rubber tubing, suitable for water connections, ¼ inch inside diameter.

4. Condenser brushes (bristle ½ inch in diameter, 3½ inches long, wire handle 24 inches long).

5. Tube brushes (brushes ¼ inch in diameter, 4 inches long, wire handle 13 inches long).

6. Ring stands and clamps suitable for holding equipment.

Reagent

Toluol, redistilled or moisture free.

Procedure

1. Transfer a 50 gm sample as quickly as possible to a clean, dry 300 ml Erlenmeyer flask. Immediately pour sufficient toluol into the flask to cover the sample. This requires about 75 - 100 ml.

2. Connect the flask with the condenser by means of a distillation tube.

NOTE: **The condenser and distillation tube must be clean and free of grease to prevent water adhering to the inner surface.**

3. When ready to start the distillation, fill the distillation tube by carefully pouring toluol through the top of the condenser. Bring to a boil rapidly by turning the heater on full. Shake frequently so as to prevent the powder from burning on the bottom of the flask, which would cause higher results. Just as soon as boiling has begun, reduce the heat so that the toluol will condense into the distillation tube at a rate of about 4 drops per second.

4. Thirty (30) minutes after distillation has begun, dislodge any particles of water in the condenser tube by means of a condenser brush, and wash down with 10 ml of toluol.

5. Twenty (20) minutes later repeat and observe the moisture reading.

6. Continue the distillation for an additional 10 minutes. Again dislodge any water particles and wash down the condenser tube as before and note if there has been any increase in the moisture reading.

7. If an increase has taken place, continue the distillation another 15 minutes and wash down as before. If no increase has taken place, continued distillation will be unnecessary. Spray powders do not usually require the additional distillation, although this may be necessary with roller process powder.

8. After the distillation tube has come to room temperature, dislodge any larger water particles in the toluol layer with a piece of wire, and finer particles by the use of a tube brush.

9. The water level in the graduated distillation tube is read after it has returned to room temperature. This reading multiplied by 2 equals the percentage of moisture in the sample.

NOTE: The foregoing method is recommended. It is recognized that the tentative method for the determination of moisture in dry milk [*Methods of Analysis*, A. O. C.; 6th edition (1945), p. 328] is equally satisfactory except that more time is required. This consists of weighing a 1 to 1½ gm sample into a suitable metal dish and drying to constant weight (approximately 5 hours), maintaining a pressure not to exceed 100 mm at the temperature of boiling water. During the drying, a slow current of air (about 2 bubbles per second), dried by passing through sulfuric acid, is admitted to the oven.

NOTE: Samples should check (a) within 0.2 per cent when determinations are made on the same sample in a given laboratory; (b) within 0.4 per cent when determinations are made on duplicate samples in different laboratories.

REFERENCES

1. *The Grading of Dry Milk Solids,* American Dry Milk Institute, 1942.
2. *J. Dairy Sci. 13:4* (1930).

Cheese

Apparatus

1. Drying oven controllable at a temperature of 212 F (100 C), using steam, water, or preferably electricity.

2. Flat-bottom moisture dishes covered with sand and provided with small glass rods for stirring the cheese.

3. Desiccator for cooling.

Procedure

1. Place the dish with the sand and the glass rod in the oven, and dry to constant weight (See Note 2, p. 238).

2. Weigh about 5 gm of the composite sample into the dish. In case of hard cheeses mix the sand and the cheese with the glass rod. Soft cheeses can be placed in the oven for a short time before mixing.

3. Leave the samples in the oven for 15 hours. Cool in the desiccator and weigh. In routine work the weighed samples are usually placed in the oven at 5 o'clock in the afternoon and removed at 8 o'clock the next morning.

CHAPTER 18

MOISTURE, SALT, AND FAT DETERMINATION IN BUTTER AND CHEESE

The following procedure is a rapid and fairly dependable method for determining moisture, fat, and salt in butter and cheese. Operators will find this test practical for control work in the plant. With careful manipulation, the test is sufficiently accurate for factory control work. When a greater accuracy is desired, use the Mojonnier test, or the A. O. A. C. method [*Methods of Analysis,* A. O. A. C., 6th edition (1945) pp. 331-336].

Butter

A. MODIFIED KOHMAN METHOD[1]

Apparatus

1. Electric hot plate, with three heat switch.
2. Special moisture and fat scale. A special scale, designed for this test, can be bought from the Torsion Balance Co. as No. 1735; also, the ordinary torsion balance, No. 1705, made by the same company, can be used. (Figure 99).
3. Aluminum beaker, 125 ml capacity.
4. Crucible tongs.
5. Spatulas, each with 4-inch blade of stainless steel.
6. Sample jars, of non-absorbant material, preferably with straight sides and tight-fitting non-absorbant closures, such as aluminum screw caps with Vinilite liners.
7. Weights. One 10 gm weight for use with the No. 1735 special scale, or a 10 gm, 9 gm, or 1 gm weight for use with the No. 1705 torsion balance.
8. Roll of tissue for wiping triers and spatulas.
9. Glass stirring rods, rubber tipped.
10. Several white cups.
11. Water bath.
12. Measuring flask, 250 ml.
13. Burette with stand, 50 ml.
14. Pipette, 25 ml.

NOTE: Mojonnier tester and apparatus can be adapted to moisture, fat, and salt test herein outlined.

Figure 99.
Torsion balance. *Courtesy Torsion Balance Co., New York.*

Reagents

1. Petroleum ether. Distillation between 95 and 176 F (35 - 80 C). Sp. gr. 0.634 - 0.660 at 77 F (25 C).

2. Distilled water.

3. Silver nitrate ($AgNO_3$) standard solution. Dissolve 29.0512 gm of pure silver nitrate in 1 liter of distilled water.

4. Potassium chromate indicator solution. Dissolve 5 gm of potassium chromate in 100 ml of distilled water. To be certain of the elimination of chlorides, add a few drops of silver nitrate until a brick red precipitate is formed and filter before using the indicator.

Procedure for Moisture

1. Obtain an accurate sample by taking plugs from various parts of the churn or package. Place these in a glass bottle of suitable size. Place the sample bottle in a water bath with water 90 - 95 F (32.5 - 35 C). During this operation the butter should be constantly stirred with the handle of a teaspoon or with a spatula until the butter softens into a uniform mass resembling thick or whipped cream.

PRECAUTION: Do not allow the butter to melt or oil off during this operation. This will not occur if the water in the water bath is not over 110 F (43.3 C) and if the sample is stirred constantly.

2. Place a thoroughly cleansed, dried, and cooled aluminum cup on the right hand pan of the moisture scale.

3. Properly tare the moisture scale by having the percentage riders on the two front scale beams set at zero, and balancing the scales with the empty cup on the right pan by using the two riders on the two rear beams.

PRECAUTION: After balancing the scale, do not change the position of the riders. The scale must be tared with each determination.

4. Weigh out exactly 10 gm of the sample prepared as described into a tared, cold aluminum beaker or cup.

PRECAUTION: Speed and accuracy are essential in weighing. The cup when weighed must be at room temperature, as temperature affects the weight.

5. Heat the weighed sample slowly over the electric hot plate, rotating the aluminum beaker constantly in a circular motion by means of the tongs to prevent spattering.

PRECAUTION: Do not allow the cup to rest directly on the hot plate or over flame without being rotated. When no electric hot plate is available an alcohol or gas flame may be used, but in such cases a wire gauze or asbestos mat should be used to prevent too rapid heating or overheating.

6. When the heating of the sample is complete, the foaming will have ceased and the curd in the bottom of the aluminum beaker will have a characteristic slightly brown coffee color. A whitish, yellowish color indicates insufficient heating and is productive of too low tests. A dark brown or black curd indicates overheating and will result in too high tests.

7. Allow the pan to cool in the air. For quick cooling place on a smooth and clean stone or iron surface. When cooled, replace the cup on the right hand pan of the scales.

8. Using the percentage rider on the lower front scale beam, again balance the scales and read the percentage of moisture direct.

NOTE: If moisture content only is desired, the test is completed with **(8)**. For milk fat and salt content, continue according to the next descriptions.

Procedure for Fat

1. Remove the beaker with the dried butter without disturbing the adjusted moisture counter balances.

2. Mix the dried butter sample with 100 ml of petroleum ether by stirring thoroughly with a glass rod. Remove the glass rod and leave the beaker with the contents undisturbed for three or four minutes, and decant the ether-dissolved fat.

PRECAUTION: The decanting of the ether should be done with extreme care to avoid pouring off any of the residue.

3. Repeat the extraction using 50 ml of petroleum ether.

4. Warm the aluminum beaker very slowly by holding it above the hot plate. (Do not allow the beaker to touch the hot plate as quick heating will cause the curd to pop out of the beaker.)

5. Heat until residue is dry, free from solvent odor, and powdery in appearance.

6. The beaker is then cooled, dried as before, and replaced on the scale and balanced, milk fat counter-balances being used.

7. Read the percentage of milk fat directly from the calibrated milk fat beams, the counter-balances of which were adjusted to balance the loss in weight of fat extracted by the petroleum ether.

Procedure for Salt

1. The beaker is then rinsed three or four times with hot distilled water into a 250 ml measuring flask. Such rinsings are made up to volume with distilled water and mixed thoroughly by inverting flask several times.

2. Pipette 25 ml of the water extract into a beaker or white porcelain cup. Add 2 or 3 drops of potassium chromate indicater and titrate with the standard silver nitrate solution to a permanent brownish red color. Each cubic centimeter of silver nitrate solution used is equivalent to 1 per cent of salt in the sample. Thus, if it required 3.5 ml of the silver nitrate solution the sample contained 3.5 per cent of salt.

NOTE: A more definite end point will be observed if dichlorofluorescein indicator is used in place of potassium chromate. From 2 to 3 drops of a 0.1 per cent solution of dichlorofluorescein is recommended. The change of color is from opalescent green to pink. (A 0.1 per cent solution may be prepared by dissolving 0.1 gm of dichlorofluorescein in 74 ml of 95 per cent alcohol and diluting to 100 ml with distilled water.)

Calculations

The percentage of curd is determined by subtracting the sum of moisture, fat, and salt (in the case of salt butter) percentages determined as has been outlined from 100.

Example: Total sample = 100%

Moisture	16.47
Fat	80.50
Salt	2.41
Total	99.38%
Curd	0.62%

REFERENCES

1. Laboratory Manual—Methods of Analysis, American Butter Institute, Chicago, Illinois (1947).

Kohman, E. F.: Rapid and Accurate Method for Butter Analysis for Factory Control Work, *Ind. Eng. Chem. 11*:36 (1919).

Overman, O. R., and Okimoto Saichi: A Comparison of the Decantation Method with Other Methods for the Determination of Fat in Butter, *J. Dairy Sci. 3*:425 (1920).

The Chemical Analysis of Butter, Report of Committee on Chemical Methods, *Ibid. 13*:380 (1930).

B. OFFICIAL A.O.A.C. METHOD

(This method has been edited to conform to the style of this publication.)

Sampling

The sample should be representative of the lot or shipment.

1. In the case of tub or cube butter, use a butter trier to extract three or more cores diagonally from top to bottom.

2. Transfer the cores with a spatula to a sample container, a 1-inch plug being left to plug the hole left by the core. The trier should be dried after each trierful. Do not include in the sample moisture adhering to outside of trier.

3. In the case of print butter, withdraw one print from each of a number of cases equal to the square root of the total numbers of cases in the lot. At least five samples should be taken. With prints of 1 pound or over, the print may be quartered and diagonally opposite quarters selected as samples. With prints of 4 ounces or 8 ounces, take the whole print as a sample.

Preparation of Sample

1. Soften the entire sample in a closed vessel by heating in a water bath kept at 102 F (39 C), shaking occasionally until the temperature of the entire mass is 95 F (35 C), or by heating in a constant temperature oven at 95 F until the temperature of the entire mass has reached 95 F.

2. Shake vigorously until a homogeneous semisolid mass is obtained.

3. Weigh the portion for analysis at once. If the sample is kept for any length of time, again soften and prepare as has been described before withdrawing portions for analysis.

Apparatus

1. Flat bottom dishes, 5 cm or more in diameter.
2. Gooch crucible with asbestos mat.
3. Vacuum apparatus.
4. Filter flask with crucible holder.
5. Volumetric apparatus, flasks, burettes, and pipettes.
6. Several various size beakers, 100 to 500 ml capacity.
7. Separatory funnel.
8. Electric drying oven.

Reagents

1. Absolute ether or petroleum ether.
2. Nitric acid (HNO_3), c.p., concentrated.
3. Silver nitrate ($AgNO_3$), c.p.
4. Potassium chromate (K_2CrO_4), c.p.

Procedure for Moisture

1. Weigh 1.5 - 2.5 gm of prepared sample into a flat bottom dish not less than 5 cm in diameter.

2. Dry to constant weight in oven at the temperature of boiling water.

3. Loss in weight is reported as moisture.

NOTE: Clean, dry sand or asbestos may be used if fat is not to be determined in the residue by the indirect method which is next given.

Procedure for Fat (Indirect Method)

1. Take up the dry butter obtained in the moisture determination just given, in which no absorbent was used, by macerating with 15 ml of absolute ether or petroleum benzin.

2. Transfer to a weighed Gooch crucible with the aid of a wash bottle filled with the solvent.

3. Wash free from fat with 100 ml of solvent.

NOTE: Last 25 ml of the 100 ml of solvent should pass through the crucible without the aid of suction.

4. Dry crucible and contents at the temperature of boiling water until the weight is constant.

5. Repeat the washing with 25 ml of solvent and dry to constant weight.

6. Repeat the operation until there is no loss of weight due to washing.

Procedure for Fat (Direct Method)

1. From the dry butter obtained in determination of moisture with or without the use of absorbents, extract the fat with anhydrous, alcohol-free ether or with petroleum benzin (b. p. below 149 F (65 C), receiving the solution in a weighed flask.

2. Evaporate the solvent.

3. Dry to constant weight at the temperature of boiling water.

4. Report the weight of residue in the flask as fat.

Procedure for Casein, Ash and Salt

1. Cover the crucible containing the residue from the fat determination by the indirect method and heat, gently at first, raising the temperature gradually to not over 932 F (500 C), remove cover, and continue heating until contents of crucible are white.

2. Loss in weight represents casein, and the residue in the crucible, mineral matter.

3. Dissolve this mineral matter in distilled water slightly acidified with HNO_3 and determine chlorine, either gravimetrically (*Methods of Analysis,* A.O.A.C., 6th edition (1945), p. 128), or volumetrically (*Ibid.,* 6th edition (1945), p. 128-129), and calculate the NaCl.

Procedure for Salt

1. Weigh in a counterpoised beaker 5 to 10 gm of sample.
2. Add about 20 ml of hot distilled water and after the butter has melted, transfer the whole to a separatory funnel.
3. Insert stopper and shake for a few minutes.
4. Let stand until all fat has collected on top of the water.
5. Draw off the water into a flask being careful to let none of the fat globules pass.
6. Add hot distilled water again, rinsing the beaker, and repeat the extraction 10 - 15 times, using 10 - 20 ml of water each time. The washings will contain all but a mere trace of the salt originally present in the butter.
7. Determine the quantity in the whole or in an aliquot part of the liquid by titration with standard $AgNO_3$, using K_2CrO_4 as indicator.

NOTE: For more complete analysis including the examination of fat microscopic examination, preservatives, and coloring, see *Methods of Analysis,* A. O. A. C., 6th edition, 1945, p. 335.

Cheese

MOISTURE

Apparatus

1. Mojonnier solids dish and cover or similar type dish.
2. Drying oven.
3. Desiccator.

Procedure

1. Dry and weigh or tare accurately a small dish and cover.
2. Place exactly 10 gm of a well ground composite sample in the dish and put on the cover.
3. Put dish and cover in a 212 F (100 C) drying oven for 24 hours (cover prevents loss of sample should spattering occur).
4. Place dish in a desiccator and when cool weigh accurately.
5. Loss in weight times 10 equals the percentage of moisture.

MODIFIED MOJONNIER METHOD FOR MOISTURE

Apparatus

1. Electric hot plate with temperature regulator or plate on Mojonnier fat side.
2. Total solids dish with cover, bottom must be perfectly flat.
3. Glass rod, diameter 4 mm or less.
4. Pipette with fine orifice, 1 ml.
5. Thermometer well with mercury.
6. Mason jar, pint size.
7. Laboratory homogenizer.

Sampling

Mix the cheese thoroughly in the carton or glass by stirring and turning the cheese over from the bottom of the container. The sample is then run through a laboratory homogenizer into the Mason jar, thoroughly mixed, and then sealed. The sample should be tested promptly to avoid any loss from evaporation.

Procedure

1. Weigh into a previously dried and cooled solids dish approximately 2 gm of cheese.
2. Add 1 ml of hot 149 F (65 C) distilled water and spread the cheese over the bottom of the dish, with the small glass rod.
3. By means of the 1 ml pipette, add 1 ml of hot water, wash off any particles of cheese from the glass rod into the dish, using the tip of the pipette to scrape the glass rod if necessary.

NOTE: Excessive water will cause spattering in the drying process.

4. Place dish and sample on hot plate previously regulated to a temperature of 279 F (135 C). Continue heating until the cheese turns to a light tobacco brown color. This usually takes from 12 to 15 minutes.
5. Cool in desiccator for 10 minutes and weigh. Express results in terms of per cent of total weight.

SALT

Apparatus

Erlenmeyer flask, 300 ml.

Reagents

1. Silver nitrate ($AgNO_3$), 0.1711 N.
2. Nitric acid (HNO_3) c. p.
3. Potassium permanganate, saturated solution.
4. Ferric ammonium sulfate.
5. Potassium or ammonium sulfocyanate, 0.1711 N.

Procedure

1. Weigh approximately 3 gm of a well ground sample in Erlenmeyer flask.

2. Add an excess of $AgNO_3$ (about 18 ml).

3. Add 15 ml of HNO_3 and 50 ml of distilled water and bring to boiling.

4. While boiling add 15 ml of saturated potassium permanganate in 5 ml portions.

5. Boil until cheese is digested.

6. Dilute to 100 ml with distilled water and let stand until all solid matter settles.

7. Decant clear solution, wash precipitate with 100 ml of distilled water, and again decant.

8. Combine the two portions of decanted liquid and add 3 ml of saturated ferric ammonium sulfate (as an indicator).

9. Titrate excess $AgNO_3$ with potassium or ammonium sulfocyanate until a reddish color appears, indicating the end-point.

10. Calculate per cent of salt, using the formula:

$$\frac{\text{Ml of } AgNO_3 \div \text{Ml of sulfocyanate}}{\text{Weight of sample}} = \%\ \text{Salt}$$

CHAPTER 19

TITRATABLE ACIDITY

Acidity

Titratable acidity generally has been presented as though the acidity arrived at by titration with 0.1N sodium hydroxide is due solely to lactic acid. This is not the case. What is actually measured is the quantity of alkali necessary to bring the pH of the milk to approximately 8.3 at which point phenolphthalein shows the characteristic pink color. This is a measure of the buffering capacity of milk. In general, freshly drawn milk does not contain lactic acid. Bacterial growth usually, but not always, results in a change of the titration value which may be due partially to lactic acid. For this reason the acidity test as customarily used is only of value when employed with a full appreciation of its meaning.

Since the titratable acidity of the milk is an arbitrary figure arrived at by following a definite procedure, such questions as which is more accurate 0.1 N or 0.02 N sodium hydroxide, how much water should be added, and strength of indicator are of paramount importance. Any variation in procedure is likely to give a slightly different result. Hence a standard procedure is adopted so that all acidity results will be on the same basis of comparison. For a more detailed discussion of the titration of milk, the technician is referred to Chapter 6, *Fundamentals of Dairy Science,* Associates of Rogers, 3rd Edition.

Milk and Cream

Apparatus

1. Babcock pipette, 17.6 ml, or 9 ml.
2. White porcelain cup or casserole.
3. Burette, 25 ml capacity, graduated in 0.1 ml divisions. (Figure 100.)
4. Burette stand and necessary clamps.
5. Glass stirring rods.
6. Cream test scales and weights.

Reagents

1. Sodium hydroxide (NaOH) 0.1 N.
2. Phenolphthalein indicator (see page 539 for preparation).

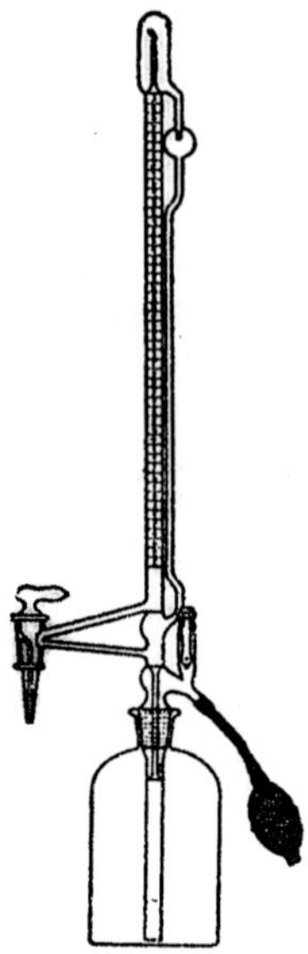

Figure 100.
All-automatic acidity burette. Courtesy Scientific Glass Apparatus Co.

Procedure

1. Thoroughly mix the milk or cream by pouring several times from one container to another.

NOTE: It is permissible and desirable to warm lumpy or frozen milk or cream to a temperature not exceeding 100 F (37.8 C) before mixing for testing. In this case the sample must be cooled back to 60 F (15.6 C) before pipetting. Duplicate tests are essential in all such cases.

2. *Milk.* Measure 17.6 ml into the white porcelain cup or casserole, add an equal volume of recently boiled distilled water by rinsing the measuring pipette. Or, using a 9 ml pipette, measure 9 ml into the cup and add an equal volume of water.

NOTE: Measuring with a pipette is the popular method but it is more accurate to weigh the sample.

2a. *Cream.* Weigh exactly 9 or 18 gm of cream into the cup and add an equal weight of distilled water.

3. Add 0.5 ml of the phenolphthalein indicator.

4. Titrate to the first definite change to a pink color which remains for 10 to 15 seconds by the slow addition of 0.1 N NaOH, taking care to stir constantly.

NOTE: Extreme care must be exercised to avoid adding an excess of 0.1 N NaOH. Accuracy demands the very sparing addition of 0.1 N NaOH when nearing the end-point. The color standards in the form of stirring rods are helpful in determining the end-point.

5. Record the number of milliliters of 0.1 N NaOH used.

$$\text{Per cent of acid} = \frac{\text{ml } 0.1\text{N NaOH} \times 0.009}{\text{gm sample used}} \times 100$$

Expression of Results

Results are recorded as per cent of lactic acid, but in speaking of the acidity it is customary to drop the decimal point and say 13 or 13.5 instead of 0.13 per cent or 0.135 per cent. When using a 9 gm sample, the per cent acidity can be read directly by multiplying the burette reading by 0.1. Should it be desired to use a greater quantity of sample, the following factors should be used:

Size of sample grams	Multiply ml used by following factor
9	0.10
10	0.09
18	0.05
20	0.045
25	0.036
50	0.018
100	0.009

Example 9 gram sample used: 1.3 ml 0.1 N NaOH used.
Then acidity = 1.3 × 0.10 = 0.13% acidity.

Skim Milk and Buttermilk

This procedure is the same as for milk except that in the case of buttermilk the sample should be weighed.

Ice Cream and Ice Cream Mix

Reagent and procedure same as for milk except:

1. The sample is usually weighed instead of measured and a 9 gm sample is generally used.

2. Twice the weight of distilled water is added to the sample.

Sherbets and Ices

The acid found in ices and sherbets is largely citric, lactic, or both. For this reason there is no point in calculating the acidity as lactic or citric acid since there is no way of knowing which acid is present by the titration alone. It is recommended that the results be expressed as ml of 0.1 N NaOH per 10 gm sample ("per cent acidity").

Method

1. Weigh 10 gm into the white porcelain cup.

2. Add 20 ml of distilled water and 0.5 ml phenolphthalein indicator.

3. Add slowly 0.1 N NaOH until a permanent change of color is obtained.

NOTE: 1. Ices and sherbets containing coloring matter which interfere with a sharp end-point can not be tested by the method given above. In this case the acidity can be determined by an electrometric titration. See page 258.

2. By using a 1 gm sample diluted to 100 ml with distilled water and titrating with 0.01 N NaOH, the interference of color can be eliminated.

Condensed Milk

Same as for milk except sample must be weighed and twice the weight of distilled water added.

Dry Whole Milk

(Reconstituted Basis)

Apparatus

1. Apparatus listed for milk.
2. Electric (malted milk type) mixer.
3. 18 or 20 oz tumblers.
4. Torsion balance.

Reagents

1. Phenolphthalein indicator solution (see page 539 for preparation).
2. Sodium hydroxide, 0.1 N NaOH.

Procedure

1. Dissolve and disperse 13 gm of the sample in 100 ml of neutral or distilled water, using the mixer and let stand for approximately 1 hour.
2. Stir gently and, with a pipette, withdraw 17.6 ml, and transfer to the white porcelain cup or similar container.
3. Rinse the same pipette with 17.6 ml of neutral or distilled water and add it to the liquid in the cup.
4. Add 0.5 ml of the phenolphthalein indicator.
5. Titrate with 0.1 N NaOH until a faint pink color persists for 30 seconds.
6. The milliliters of 0.1 N NaOH used divided by 20 indicates the percentage of acidity of the reconstituted sample in terms of lactic acid.

NOTE: When sediment is determined by the tumbler method, the reconstituted sample may be used for the acidity determination. After reading the sediment number, thoroughly agitate and withdraw 17.6 ml and follow the foregoing procedure.

REFERENCE

1. The Grading of Dry Whole Milk, Bulletin 913, American Dry Milk Institute, Chicago, 1947.

Non-Fat Dry Milk Solids

(Reconstituted Basis)

Proceed as directed for Dry Whole Milk except that 10 gm of sample is dispersed (Procedure 1) in 100 ml of distilled water.

Sour or Ripened Cream and Starter

Proceed as directed for cream except that the sample should be boiled and cooled before titrating to expel the carbon dioxide. In this case proceed as directed for cream. A pH determination of sour cream and high ripened starters is a much more reliable guide for control of quality.

Butter

The titratable acidity gives little information pertaining to butter and is therefore not reliable. A hydrogen ion determination in this product gives more reliable and comparable results.

Cream Cheese

To 10 gm of finely divided cheese add water at a temperature of 104 F (40 C) until the volume equals 105 ml; shake vigorously and filter. Titrate 25 ml portions of the filtrate, as representing 2.5 gm of the sample, with 0.1 N NaOH, using phenolphthalein indicator. Express the result in terms of lactic acid: 1 ml of 0.1 N NaOH solution = 0.0090 gm of lactic acid.

NOTE: Titratable acidity of cream cheese is of doubtful value. The pH determination is of more value in checking the manufacturing process and the keeping quality of the cheese.

CHAPTER 20

HYDROGEN ION DETERMINATION

Theory

The hydrogen ion concentration, or the actual acidity, in many cases affords more reliable information than a titration. The pH is closely related to the chemical changes in the milk proteins and is a valuable guide in the course of cheese manufacturing and processing. It is a determining factor in the bacteriological and enzymatic processes taking place in the ripening of cream, butter, cheese and other products.

A more detailed and technical explanation may be had by reference to Clarke and Lubs.

A brief nontechnical explanation of the meaning of pH may be helpful.

Any water solution contains some hydrogen ions and some hydroxyl ions. An acid is, strictly speaking, any substance which furnishes hydrogen ions. A base is any substance which furnishes hydroxyl ions. Consequently, any solution, no matter how basic, has a certain degree of acidity. Likewise, any solution, no matter how acidic, has a certain degree of basicity. The term pH is defined as the reciprocal of the logarithm of the hydrogen ion concentration. Therefore, pH is a measure of the degree of acidity of the solution. In general, we do not consider a solution to be an acid in the common sense of the word unless the hydrogen ion concentration is greater than the hydroxyl ion concentration.

Perhaps the following explanation will make the above discussion clearer. Pure distilled water is an example of a truly neutral substance. It has been found that water breaks up or, as we say, ionizes to a very slight extent in the following manner:

$$\underset{}{H_2O} \leftrightarrows \underset{\text{Hydrogen ions}}{(H+)} + \underset{\text{Hydroxyl ions}}{(OH-)}$$

For every particle or molecule of water ionized, there must be formed one hydrogen ion and one hydroxyl ion. Thus in pure water there are just as many hydrogen ions as there are hydroxyl ions.

It has been found that the product of the hydrogen ion concentration and hydroxyl ion concentration in any water solution is approximately constant at any particular temperature.

Thus: $(H^+) \times (OH^-) = K_{w_t} = 10^{-14}$ at 77 F (25 C).

In the case of pure water, in which all the hydrogen and hydroxyl ions come from the ionization of the water, the hydrogen and hydroxyl ion concentrations are equal and therefore the hydrogen ion concentration is:

$$H^+ = \sqrt{10^{-14}} = 10^{-7}$$

$$\text{Reciprocal is } \frac{1}{10^{-7}} = 10^{-7};\ \log.\ 10^{-7} = pH = 7$$

Now let us add hydrochloric acid, which is practically completely ionized in dilute solution. Suppose we add 3.65 gm of HCl per liter, which is what we call 0.1 mol. Then the hydrogen ion concentration is 0.1 mol. Since the ionization constant of water remains the same we have:

$$(H^+) \times (OH^-) = 10^{-14}$$
$$(0.1) \times (OH^-) = 10^{-14}$$
$$(10^{-1}) \times (OH^-) = 10^{-14}$$
$$(OH^-) = \frac{10^{-14}}{10^{-1}} = 10^{-13}$$

Thus the hydrogen ion concentration in the foregoing very acid solution is 10^{-1} and the pH is 1. The hydroxyl ion concentration is 10^{-13}.

Now let us take the case of an alkaline solution. If we add 0.1 mol of sodium hydroxide, which is almost completely ionized in dilute solution, we have:

$$(H^+) \times (OH^-) = 10^{-14}$$
$$(OH^-) = 10^{-1}$$
$$(H^+) \times (10^{-1}) = 10^{-14}$$
$$(H^+) = \frac{10^{-14}}{10^{-1}} = 10^{-13};\ \text{thus pH} = 13$$

Thus we see that in acid solutions the pH is less than 7 and in alkaline solution the pH is greater than 7. The pH scale is sometimes compared to the scale of a thermometer. On the basis of this analogy we can think of temperatures above 32 F as degrees of heat and temperatures below 32 F as degrees of cold. Likewise pH above 7 can be thought of as degrees of alkalinity and pH below 7 as degrees of acidity:

Thermometer Scale		pH Scale	
70	increasing heat	14	increasing alkalinity
65		13	
60		12	
55		11	
50		10	
45		9	
40		8	
32	water freezes	7	neutral point
30	increasing cold	6	increasing acidity
25		5	
20		4	
15		3	
10		2	
5		1	

It must be remembered, however, that the pH scale is a logarithmic one, therefore a decrease of 1 pH unit represents a 10 fold increase in the hydrogen ion concentration, whereas the thermometer scale is a straight line relationship and an increase of 1 thermometer unit represents a 1 unit increase in degree of heat.

The importance of pH as a measure of actual acidity can best be illustrated if we consider two acids in common use. Consider a 10 per cent hydrochloric acid solution and a 10 per cent acetic acid solution.

It is common knowledge that the hydrochloric acid solution is much stronger than the acetic acid solution. This is explained by the fact that hydrochloric acid is very largely ionized and forms a high concentration of hydrogen ions in solution. On the other hand, acetic acid is very slightly ionized and forms only a small concentration of hydrogen ions. A titration of the above solutions would only have revealed that both were 10 per cent acid solutions. However, pH determinations would show that the degree of acidity of the hydrochloric acid solution was much greater than that of the acetic acid solution.

Another term that is often used in connection with pH is buffer solutions and buffer capacity. A buffer is a substance which when added to a solution enables the solution to resist a change in pH when a small quantity of acid or alkali is added to the solution. Buffers are usually salts of strong bases and weak acids.

Colorimetric Method of pH Measurements

The colorimetric pH determination is used to some extent in dairy products in detecting over-neutralized cream, sour cream, or milk not suitable for pasteurizing or condensing. The method can also be used to detect abnormal milk from diseased cows.

Figure 101.
LaMotte colorimetric hydrogen ion block-comparator. Courtesy of LaMotte Chemical Products Co.

Colorimetric pH determinations should be used only as an aid for the detection of abnormal products but not as conclusive evidence of such abnormality. There are various sets on the market for the colorimetric determination of pH. Detailed instructions for making the test are furnished with each set. See illustrations of two popular sets in figures 101 and 102.

NOTE: A color chart of indicators, covering the range of pH values from 1.2 to 9.6, can be purchased from The Williams and Wilkins Co., Baltimore, Md.

All of these methods involve the addition of a definite quantity of indicator to a definite quantity of milk or other solution to be tested and the comparison of the color produced with standards representing various pH values.

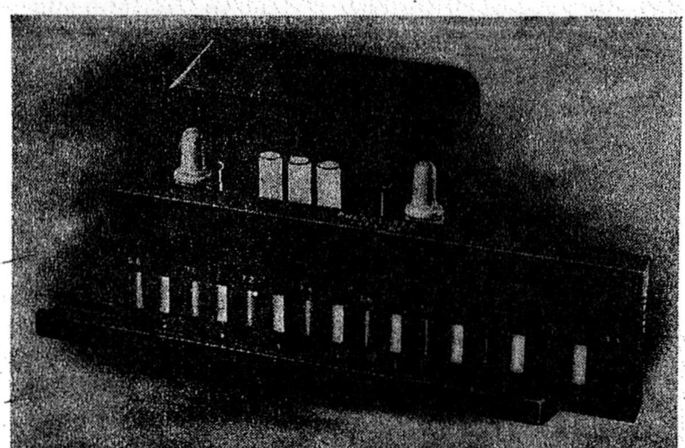

Figure 102.
Taylor colorimetric hydrogen ion slide comparator. Courtesy of W. A. Taylor & Co.

Potentiometric Method of Measuring pH

QUINHYDRONE ELECTRODE

The quinhydrone electrode has been extensively used in the past in milk control work. It is a reasonably reliable method in the hands of a competent technician and the equipment required is less expensive than that necessary for glass electrode measurements. It is not reliable in alkaline solutions much above pH 8.0 and can not be used in solutions which contain strong oxidizing or reducing substances.

A. MILK, CREAM, SKIM MILK, BUTTERMILK, ICE CREAM, ICE CREAM MIX, DRY MILK RECONSTITUTED, WHEY AND WHEY POWDERS

Method

1. Place in a beaker the solution to be tested and add quinhydrone crystals. About 0.1 gm to 50 ml of solution is sufficient, although an excess will not do any harm.

2. Immerse a plain platinum or gold electrode in the solution. The electrode should be prepared by immersing in boiling water for several minutes before use.

3. The sample is then connected to a saturated calomel half cell by means of a salt bridge or dripping contact.

4. The two cells are then connected to a potentiometer through a reversing switch.

5. Balance the potentiometer, make the reading, and note whether the calomel cell is positive or negative.

6. With the potential difference measured and the potential of the calomel half cell known, the pH can be calculated by means of the following formula: 77 F (25 C)

$$pH = \frac{453 - [\text{millivolts} + 0.77\ (t - 25)]}{59.1}$$

t = temperature of room and sample

Cleaning the Electrodes

The platinum electrode is rinsed thoroughly with a strong stream of distilled water from a wash bottle. A piece of absorbent cotton is then moistened with a chemically pure grade of acetone and the glass and wire are wiped free from fat and other adhering materials. The electrode is then rinsed free from acetone with distilled water and placed in a test tube filled with distilled water for future use. The calomel electrode glass tubing is rinsed in a similar way and placed in a test tube filled with a saturated solution of a chemically pure grade of potassium chloride. Before using it is rinsed with distilled water to remove the potassium chloride.

B. CHEESE

Procedure

1. Place a saturated solution of potassium chloride in a 150 ml beaker and the lower end of the calomel electrode is immersed as usual.

2. Place a composite sample of cheese, about 10 gm in weight into a small glazed porcelain mortar provided with a wide bottom pestle. It is important that the temperature of the cheese be the same as that of the room in which the apparatus is placed.

3. Add about 0.1 gm quinhydrone powder and the mass is ground to a uniform paste as quickly as possible.

4. Tamp or push tightly the cheese-quinhydrone paste into a glass tube.

5. Imbed the platinum electrode in the cheese paste and arrange the entire unit so that only the lower end of the glass tube is immersed in the saturated potassium chloride solution contained in the 50 ml beaker.

The glass tubes used are made of a good grade soda lime glass 7 mm outside diameter and cut in pieces of 4 cm lengths. By the proper use of a large nail and a flame, one end of each tube is flared out to facilitate filling.

If the cheese is too soft, the tube can be fixed to the platinum electrode by means of a rubber band.

C. BUTTER

Procedure

1. Fill a 250 ml centrifuge cup with a composite sample of butter (refer to moisture test for butter).

2. Heat the sample to 158 F (70 C) in boiling water and centrifuge for 5 minutes at 1,000 r. p. m.

3. Insert a 50 ml pipette to the bottom of the cup.

4. Remove the curd and the serum to a 50 ml Pyrex glass beaker.

5. Cool the solution in a refrigerator for approximately 2 hours to cause any fat to rise to the surface and harden.

6. After the fat forms a hard layer, plunge a pipette through it and draw off all serum possible.

7. Transfer to a clean 30 ml beaker, and warm to 77 F (25 C).

8. Determine the pH of the serum portion of the butter by the method used for milk.

If a centrifuge is not available, 500 gm of butter can be placed in a sealed Mason glass jar and the curd and serum separated by placing the jar in a heating oven at 140 F (60 C) for 1 hour. The curd and the serum are then removed, cooled to room temperature, thoroughly mixed by a motor stirrer and tested as has been stated.

D. CONDENSED MILK

Procedure the same as for milk except that it is diluted to a reconstituted basis.

E. SHERBETS AND ICES

Procedure same as for ice cream mix.

F. POWDERED EGG YOLK

Make a 10 per cent water solution and proceed as under milk.

THE GLASS ELECTRODE

This method of measuring the pH is extensively used. It is probably the most accurate method and has the advantage of being applicable over the entire pH range and is not appreciably affected by strong oxidizing and reducing substances..

There are several assemblies for pH determination by the glass electrode. Those in common use are, Coleman pH apparatus, Beckman pH apparatus, and Leeds and Northrup pH apparatus. Detailed instructions for operating these instruments accompany the equipment.

The preparation of the samples for pH measurement by the glass electrode is the same as for the Quinhydrone electrode except that the addition of quinhydrone is omitted.

The glass electrode pH measurement offers a convenient method of making acidity determinations on samples of dairy products that are highly colored or too opaque to distinguish the end point in an ordinary titration.

The method of titrating with a potentiometer to determine the end-point is called electrometric titration.

Method

1. Weigh the same quantity of sample as would be used for an ordinary titration, such as 9 grams in the case of chocolate milk or ice cream.

2. Add the same quantity of water as should be added in an ordinary titration.

3. Add 0.1 N NaOH from a burette in 0.5 ml portions and check the pH of the solution after each addition of 0.1 N NaOH.

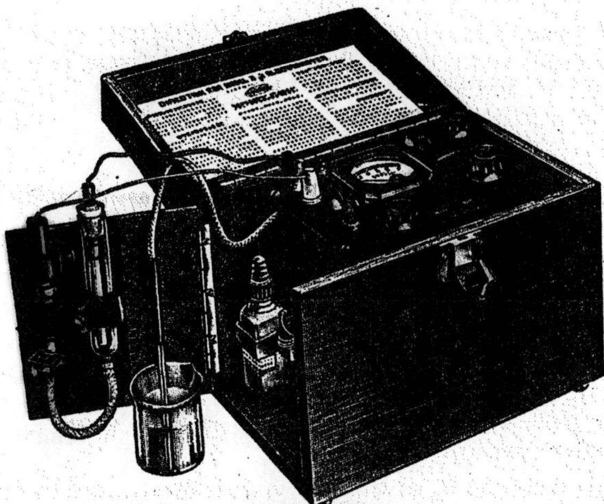

Figure 103.
Coleman pH electrometer, model 3D. Courtesy of Coleman Instruments, Inc.

4. When the pH approaches 8.0 add the 0.1 N NaOH in 0.05 ml portions until the pH of the solution reaches 8.2 - 8.4. This is the point at which phenolphthalein would change color if it could be seen.

5. From the number of milliliters of 0.1 N NaOH used and the weight of sample used, the quantity of alkali needed to bring the entire batch to the desired pH may be calculated.

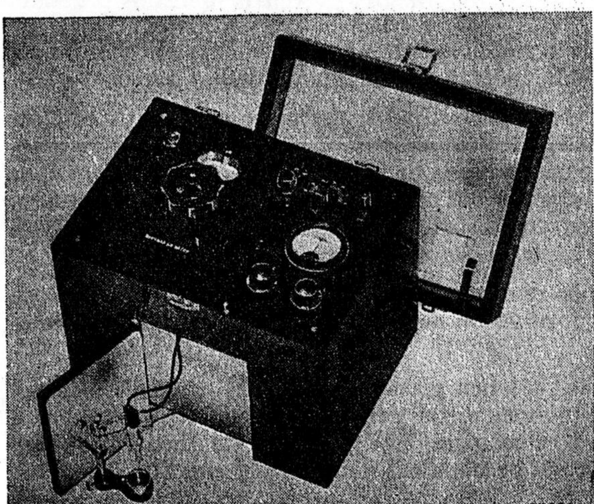

Figure 104.
Beckman pH meter. Courtesy of National Technical Laboratories.

For some products, especially ices and sherbets, there is a point at which a small addition of alkali will cause a large increase in the pH. The 0.1 N NaOH should be added in very small portions near this point in order to avoid going beyond the proper end-point.

Oxidation-Reduction Potential Measurements

A term which has come into use in recent years is called the oxidation-reduction potential and is usually designated as E_h. This is a measure of the oxidizing or reducing pressure or tendency of a solution.

Oxidation-reduction potentials are determined by immersing in the sample to be tested a noble metal electrode. This may be either platinum or gold, although gold is preferable since platinum electrodes tend to catalyze side reactions.

The single potential E_h of this electrode is determined by connecting the solution to a reference electrode and measuring the cell potential E_c existing between the two electrodes by means of a potentiometer and then compensating this value for the known reference potential E_r.

$$E_h = E_c - E_r$$

E_c is given a negative value if the noble metal electrode connects to the negative potentiometer terminal for balance.

The oxidation-reduction potential E_h varies with the pH in an orderly manner so that the pH as well as the E_h of the solution should always be determined.

For a given solution, increasing E_h at constant pH or increasing pH at constant E_h indicates increased oxidizing power.

E_h may be measured with the same equipment employed for the potentiometric determinations of pH by means of the glass electrode. All that is necessary is to substitute a noble metal for the glass electrode and express the results in millivolts. The potential measured is E_c and from this E_h can be calculated by using values of E_r given in the tables for the reference electrode in question.

It must be remembered that the noble metal electrode potential may be either positive or negative with respect to the calomel reference electrode.

Accordingly, when glass electrode pH meters are used, it is often necessary to reverse the connections between cell chain and instrument to effect balance. The necessity for this will be obvious when balance cannot be obtained with normal connections.

REFERENCE

[1]Clarke, W. M.: *The Determination of Hydrogen Ions,* Williams & Wilkins Co., Baltimore (1928).

CHAPTER 21

PHOSPHATASE TEST FOR PASTEURIZATION CONTROL

Research work by Kay and Graham[1] led to the discovery that the heat sensitive enzyme phosphatase, normally present in milk, is inactivated by exposure to the pasteurizing temperature 143.0 F (61.7 C) for 30 minutes or 160.0 F (71.1 C) for 15 seconds. When heat treatment is less than these specifications some of the phosphatase remains active and, in a laboratory test under carefully controlled conditions, will liberate phenol from disodium phenylphosphate in a buffered solution at a pH of approximately 9.5. The presence of free phenol is determined by a colorimetric test. A satisfactorily pasteurized milk yields a relatively colorless solution in this test, and a raw or incompletely pasteurized product shows a deep to a faintly blue solution dependent upon the degree or completeness of pasteurization.

Gilcreas[2] refined the method by a careful correlation of the depth of blue color with the pasteurizing time and temperature and the percentage of raw milk present. This method has been included as a tentative one in the *Official and Tentative Methods of Analysis* of the Association of Official Agricultural Chemists.

Gilcreas Method

Collection of Sample

Preservative should not be used in the sample and not more than 48 hours should elapse between the time of sampling and testing. Samples must be refrigerated but not allowed to freeze.

Apparatus

NOTE: All glassware and rubber stopper surfaces which samples and reagents may contact during performance of the phosphatase test must be rendered free of any trace of phosphatase, phenol, or other interfering substances before use for testing purposes. This requires that all test tubes, pipettes, funnels, reagent bottles, medicine droppers, dropper bulbs, and rubber stoppers be washed and rinsed free of all surface films (avoiding the use of disinfectant soaps) and, as an added precaution, soaked in hot water at a minimum of 180 F (82.2 C) for at least 10 minutes shortly before use. Avoid use of plastic closures containing phenolic compounds because of the danger of phenolic contamination.

1. Test tubes, 20 mm x 160 mm.
2. Test tubes, 13 mm inside diameter.
3. Paper glazed surface.
4. Water bath.

5. Incubator or bath at 93-99 F (33.9-37.2 C).

6. Funnels to take 11 cm filter paper.

7. Filter paper, free from phenol and other interfering substances, Whatman, No. 40, or Eaton and Dikeman "New Filt" Nos. 1 and 3.

8. Daylight lamp.

9. Light filter, turquoise blue, unglazed, opaque glass plate.

Reagents

1. Buffer substrate. Dissolve 1.09 gm of disodium phenylphosphate and 11.54 gm of sodium veronal (sodium diethyl barbiturate) in distilled water saturated with chloroform and dilute to 1 liter. Add 10 ml of $CHCl_3$ per liter and store the reagent in a refrigerator.

2. Folin-Ciocalteu phenol reagent. Dissolve 100 gm of sodium tungstate ($Na_2WO_4.2H_2O$) (according to Folin), and 25 gm of sodium molybdate ($Na_2MoO_4.2H_2O$) in 700 ml of distilled water in a 1,500 ml flask connected by a ground glass joint to a reflux condenser. Add 50 ml of 85 per cent H_3PO_4 and 100 ml of HCl, and reflux gently for 10 hours. Cool, add 150 gm of lithium sulfate, 50 ml of water, and 4 to 6 drops of liquid bromine. Boil (without condenser) for 15 minutes to remove excess bromine. Cool, transfer to a 1-liter volumetric flask, dilute to volume with distilled water, and filter. (The finished reagent should have a golden yellow color; reject the reagent if it has a greenish tint). Keep in a refrigerator and protect from dust. For use, dilute 1 volume of this stock reagent with 2 volumes of distilled water.

3. Sodium carbonate solution. Prepare a 14 per cent solution from anhydrous Na_2CO_3.

4. Chloroform ($CHCl_3$), reagent grade.

5. Permanent color standard solutions.

Preparation of Permanent Color Standard Solutions

1. Color solution, gray: Dissolve in distilled water 31.9 gm cobalt chloride ($CoCl_2.6H_2O$); 67.5 gm copper sulfate ($CuSO_4.5H_2O$); 75 gm nickel sulfate ($NiSO_4.6H_2O$); add 32 ml HCl and 45 ml H_2SO_4; dilute to 500 ml.

2. Color solution, red: Dissolve 476 gm of $CoCl_2.6H_2O$ in distilled water and filter. To the filtrate add 100 ml of HCl and dilute to 1 liter.

3. Color solution, blue: Dissolve 300 gm of $CuSo_4.5H_2O$ in distilled water, add 20 ml of H_2SO_4, and dilute to 1 liter. (Should crys-

tals appear when solution is cooled to below 68 F (20 C), warm slightly before using to insure complete solution).

Prepare a series of permanent color standards equivalent to phenol concentrations from 0.01 to 0.15 mg per 0.5 ml of sample by combining the quantities of color solutions (a), (b), and (c) as indicated in table 4, and diluting to 10 ml with distilled water in each case. For example, 0.3 ml solution (a), plus 0.106 ml solution (b), plus 0.96 ml solution (C), plus water to make a volume of 10 ml, which is equivalent to a phenol concentration of 0.01 mg in 0.5 ml of sample.

TABLE 4. PREPARATION OF PERMANENT COLOR STANDARD SOLUTIONS

Phenol values	Color solution		
	Gray (a)	Red (b)	Blue (c)
Mg./0.5 Ml.	Ml.	Ml.	Ml.
0.01	0.30	0.106	0.96
0.02	0.40	0.140	1.16
0.03	0.55	0.180	1.65
0.04	0.65	0.216	2.10
0.06	0.92	0.286	3.00
0.09	1.30	0.326	4.40
0.12	1.70	0.360	5.70
0.15	2.50	0.396	7.10

These color standards are suitable for use only in natural daylight. If, however, a turquoise blue, unglazed, opaque glass plate is used to deflect the light from a daylight lamp through the tubes of standards and sample, accurate color comparisons can be made in the absence of daylight. Since the standards are prepared for use only at 13 mm depth of color, tubes of different diameters cannot be used for accurate work.

Procedure

Transfer 10 ml of the buffer substrate (disodium phenyl phosphate) solution into a test tube 20 mm x 160 mm and add 0.5 ml of the milk to be tested. Add a few drops of $CHCl_3$, mix thoroughly by rotating the tube, and cover to protect contents from dust. (Do not use rubber or cork stoppers; glazed paper is satisfactory for placing over the open end of the tubes). Warm to 99-102 F (37.2-38.3 C) in a water bath and incubate at 93-99 F (33.9-37.2 C) for not less than 18 nor more than 24 hours. After incubation add 4.5 ml of the diluted Folin-Ciocalteu reagent. Mix and allow to stand for 3 minutes. Filter and transfer 5 ml of the filtrate to a test tube of 13 mm diameter. Add 1 ml of the 14 per cent Na_2CO_3 solution and mix thoroughly by rotating the tube. Place the tube in a boiling water bath for 5 minutes and filter. Cool and estimate the color of the filtrate by comparison with the permanent color standards.

Control Test

(To check deterioration of reagents and the presence of interfering substances in the milk sample.)

To 10 ml of the buffer substrate solution add 4.5 ml of the diluted Folin-Ciocalteu reagent and 0.5 ml of the milk sample. (Do not incubate.) Mix thoroughly, allow to stand for 3 minutes and filter. To 5 ml of the filtrate add 1 ml of the 14 per cent Na_2CO_3 solution, mix thoroughly by rotating the tube, heat in a boiling water bath for 5 minutes, and filter. Cool, and compare the color of the filtrate with the permanent color standards. If the phenol value obtained is greater than 0.02 mg (See table 4, page 267), subtract the excess from the phenol value of the incubated sample to obtain the phenol value indicative of pasteurization treatment.

Interpretation

A phenol value of 0.04 mg per 0.5 ml of sample generally indicates milk heated to 143 F (61.7 C) for 30 minutes. Greater values than this indicate progressively inadequate heat treatment. In reporting results, give the milligrams of phenol per 0.5 milliliter of sample as well as an interpretation as to whether the milk is pasteurized or under-pasteurized. According to more recent work of Gilcrease[3] a value of less than 0.05 mg of phenol per 0.5 ml of sample generally indicates milk heated to 143 F for 30 minutes.

This procedure yields very accurate results but requires from 18 to 24 hours to complete.

Scharer Methods[4]

The dairy industry required a phosphatase test which could be completed quickly so it could be used during plant operation to determine whether the milk was being adequately pasteurized. A perfect method for this purpose would be one that would yield results almost immediately. The nearest approach to it is the generally used field test of Scharer. This author has also developed two more sensitive laboratory procedures.

SCHARER LABORATORY TEST I[5]

Collection of Sample

In sampling bulk or bottled milk, thoroughly mix by pouring from one clean vessel into another three or four times. If this procedure is impractical, thoroughly stir the milk for at least 30 seconds with a suitable appliance long enough to reach to the bottom of the container. If cream has formed on the milk, continue the mixing until all of it is detached from the sides of the vessel and evenly distributed throughout the liquid. Care should be taken that the stirrer or mixing container used has not been contaminated with any raw or inadequately pasteurized milk as this will interfere with an accurate determination.

Place samples in nonabsorbent, airtight containers, and keep them cold but at a temperature above freezing until ready for examination. If necessary borax may be added as a preservative (0.8 gm per 100 ml of milk). Should samples become sour, neutralize to pH 6.6 with sodium carbonate before proceeding with the analysis.

Apparatus

Note: See note on page 265.

1. Pipette, 1 ml.
2. Test tube, pyrex 15 mm x 125 mm.
3. Test tubes, 16 mm x 150 mm and 10 mm x 100 mm.
4. Water bath.
5. Incubator or water bath maintained at 99-113 F (37.2-45 C).
6. Funnel.
7. Filter paper (any inexpensive paper of qualitative grade).
8. Brown glass dropping bottle, ¼ fluid ounce, with dropper calibrated to deliver 50 drops per ml of B. Q. C. solution with dropper in vertical position.
9. Light filter, double fluxed glass or sheet of Plastacele No. C-1605 HH, 0.015 in. thick.
10. Permanent colors standards.

Reagents

1. Borate buffer. Dissolve 28.427 gm of sodium borate ($Na_2B_4O_7.10H_2O$) in 900 ml of distilled water. Add 3.27 gm sodium hydroxide (81.75 ml of normal NaOH solution) and dilute to 1 liter with distilled water.

2. Gibbs phenol reagent. Dissolve 40 mg of 2.6—dibromoquinone-chloroimide (BQC) in 10 ml of methyl or of 95 per cent ethyl alcohol. Keep reagent tightly stopped and under refrigeration. Discard solutions that have begun discoloration.

3. Lead acetate solution. Dissolve 50 gm of lead acetate ($Pb(C_2H_3O_2)_2.3H_2O$) in 100 ml of distilled water.

4. Buffer substrate. Dissolve 0.5 gm crystalline disodium phenyl phospate in 5 ml of water in a small (10 x 100 mm) test tube. Add 0.5 ml borate buffer. Shake well and add 0.04 ml BQC solution. (Two drops from a dropper delivering approximately 50 drops per milliliter of the BQC solution.) Shake well. Allow 5 minutes for color development. If color is present, extract indophenol by shaking with 2 ml neutral n-butyl alcohol. Allow to stand until alcohol has separated completely. Remove supernatant alcohol layer with a pipette or medicine dropper and discard. Dilute remainder with 100 ml of borate buffer and sufficient water to make 1 liter. This buffer substrate is phenol free. Store under refrigeration. Because of pos-

sible decomposition *prepare quantities of this reagent sufficient for immediate needs only*. The pH of this solution is approximately 9.6 (blue to thymolphthalein solution 0.04 per cent in 50 per cent ethyl alcohol).

Tablets are available for convenient preparation of small quantities of the buffer substrate and BQC solution from Applied Research Institute, 2 East 23rd St., New York 10, N. Y.

NOTE: Do not keep tablets in the refrigerator. Packages of these tablets bare an expiration date when shipped, indicating that the tablets will remain stable for at least one year from the shipping date if kept at room temperature in a dark, dry place.

5. Sodium pyrophosphate solution ($Na_4P_2O_7.10H_2O$. 10 gm dissolved in 100 ml distilled water.

Permanent Color Standard Solutions

(a) *Acid Series*

1. Color solution, red. Dissolve 0.5 N cobalt chloride in 1 per cent HCl (59.59 gm $CoCl_2.6H_2O$ per liter).

2. Color solution, blue. Dissolve 30 per cent copper sulfate in 1 per cent HCl (300 gm $CuSO_4.5H_2O$ per liter).

3. Color solution, yellow. Dissolve 0.5 N ferric chloride in 1 per cent HCl (45.05 gm $Fe\ Cl_3.6H_2O$ per liter).

Combine quantities indicated in the following table and dilute to 5 ml with water.

	Color Solutions		
Units	Blue ml.	Red ml.	Yellow ml.
1	0.2	0.35	0.5
2	0.5	0.6	0.55
3.5	0.7	0.5	0.5
5	1.0	0.75	0.5
7.5	1.5	0.75	0.5
10	2.0	1.0	0.25

(b) *Ammoniacal Series*

The following ammoniacal color standards supplement the acid series above by affording a more extensive range.

1. Color solution, red. Dissolve 1.8 gm roseo (aquopentamine) cobaltic chloride ($[Co(NH_3)_5.H_2O]Cl_3$) per liter of 2.8 per cent ammonium hydroxide.

2. Color solution, yellow. Dissolve 0.84 gm ammonium chromate $(NH_4)_2CrO_4$) per liter of 2.8 per cent NH_4OH.

3. Color solution, blue. Dissolve 6.24 gm copper sulfate ($CuSO_4.5H_2O$) per liter of 2.8 per cent NH_4OH.

Combine the quantities indicated in the following table and dilute to 5 ml with water.

Color Solutions

Units	Blue ml.	Yellow ml.	Red ml.
1	0.25	0.4	0.5
2	0.5	0.5	0.75
3.5	0.8	0.7	1.0
5	1.0	0.55	0.6
7.5	1.25	0.75	0.75
10	1.5	0.5	0.25
15	2.5	0.5	
20	2.75	0.5	
25	3.0	0.5	
50	3.75	0.5	
100	4.0	0.5	
500	4.5	0.5	

Procedure

1. Transfer 1 ml of well-mixed milk or cream to be tested to a pyrex test tube (15 mm x 125 mm) and add 10 ml of the buffer substrate. Mix thoroughly.

2. Warm to about 104 F (40 C) and place in an incubator or water bath for 1 hour at 98.6-113.0 F (37.0-45.0 C). A temperature of 105.8 F (41.0 C) is preferred.

3. After incubation, place the tube in boiling water for 5 minutes.

4. Cool in ice water.

5. Add 0.1 ml lead acetate solution. Shake immediately and well. Ordinarily proteins will coagulate and separate sharply. If separation is incomplete, add an additional 0.05 ml of lead acetate.

6. Filter.

7. To 5 ml of filtrate in a test tube (16 mm x 150 mm) add 0.5 ml of borate buffer. The addition of a few drops of the pyrophosphate solution will clarify any turbidity of the filtrate.

8. Add 0.04 ml of BQC solution (2 drops from recommended dropper). Mix thoroughly by rotating the tube.

9. After 15 minutes, estimate the color by comparison with permanent color standards, holding plastacele light filter in back of tubes and viewing against light source.

NOTE: The permanent color standards are suitable for use in natural or artificial light provided, in the latter case, an opal double-fluxed glass or a sheet of Plastacele (see page 269, apparatus, number 9) is used as a light filter. Make color comparison of unknowns in tubes similar to those used for color standards.

Control Tests

To check reagents, add to 5 ml of buffer substrate, 0.04 ml (2 drops) of BQC solution. The development of a blue color after 15 minutes incubation indicates substrate decomposition. An alternate procedure is to incubate 10 ml of buffer substrate with 1 ml of a boiled milk sample, add BQC indicator and note the color. Indophenol development indicates substrate decomposition. Avoid contact of BQC solution with rubber. A darkening of this solution indicates decomposition.

Certain strains of micro-organisms, when grown in dairy products, produce microbial phosphatase (which is more thermostable than milk phosphatase) thereby causing "false positive" reactions. When there is some doubt as to whether a positive test is due to the presence of milk phosphatase, conduct a "double blank" determination as follows: heat 5 ml of the sample to at least 145 F and hold for 30 minutes or to 160 F for 2 or 3 minutes but not above 170 F (which might tend to destroy the microbial phosphatase). Make regular phosphatase test on the reheated sample and test a portion of the original sample as an unincubated control to demonstrate degree of phenol contamination. If there is no significant reduction in the phosphatase value as a result of the reheating, the original test may be regarded as a "false positive" and presumed not to be due to faulty pasteurization or to contamination with raw milk or cream. Under such circumstances the presence of microbial phosphatase may be assumed and can be established by boiling the sample momentarily (to destroy any microbial phosphatase), cooling and testing for phosphatase, in which case a negative test should result—if the test is still positive it indicates phenol contamination of the sample or decomposition of the reagents.

To check the presence of interfering substances in milk sample, add to 9 ml of water 1 ml of borate buffer and 1 ml of milk sample. Mix well, place in boiling water for 5 minutes, and follow the regular test procedure. The development of a blue color indicates the extent of indophenol color due to interfering substances in the milk sample.

Interpretation

"A phenol value of 2 units or more generally indicates inadequate heat treatment i. e., temperature or time, or both, less than 143 F for 30 minutes or 160 F for 15 seconds in the case of milk. To assure a value less than 2 units for cream it will be necessary to heat at least one degree higher than for milk or hold at least 35 minutes at 143 F or 17 seconds at 160 F. Commercially pasteurized milk and cream usually yields 0 to 1 unit because plant pasteurizing equipment, as customarily operated, treats the product somewhat in excess of the minimum time-temperature requirements".

SCHARER LABORATORY TEST II[6]

COPPER SULFATE MODIFICATION

This simplified modification of the Scharer Laboratory Test I, substitutes copper sulfate for lead acetate as the precipitant, omits the boiling and cooling steps, uses a simplified buffer, and is said to yield about twice the color gain per unit of enzyme activity, with the color formation being complete in about 2 minutes.

Apparatus

Same as listed on page 269 and 270.

Reagents

1. Neutral n-butyl alcohol, colorless, boiling range 239 to 244 F (115-117.8 C) highly purified and free from excessive acid or esters.

NOTE: If the alcohol is very old, it should be redistilled with NaOH. To test suitability of the alcohol, shake a sample with an equal quantity of neutral distilled water, allow it to separate, and test the water layer with bromothymol blue. A blue or green color should result. Errors are induced when the n-butyl alcohol does not meet the standard specified or when the extraction process is not conducted carefully. Too much or too little color may be obtained thereby inducing errors that result in a too severe or a too lenient determination.

2. Buffer—Dissolve 3 gm of twice recrystallized sodium sesquicarbonate dihydrate in 100 ml of distilled water.

3. Buffer substrate—see page 269, No. 4.

4. Gibbs phenol reagent—see page 270, No. 2.

5. Methyl alcohol or 95 per cent ethyl alcohol.

6. Copper sulfate solution, 5 per cent, dissolve 5 gm $CuSO_4.5H_2O$ in 100 ml distilled water.

Permanent Color Standard Solutions

1. Color solution, red—see page 270. Acid Series No. 1.

2. Color solution, yellow—see page 270. Acid Series No. 3.

3. Color solution, blue—Dissolve 62.425 gm of copper sulfate, $CuSO_4 5H_2O$ in 1 liter of 1 per cent HCl.

Combine quantities as indicated in the following table and dilute to 5 ml with distilled water.

	Color Solutions		
Units	Red ml	Yellow ml	Blue ml
2	0.6	0.15	3.0
3.5	0.8	0.1	4.0

Procedure

1. Transfer 1 ml of the well mixed (milk or cream) sample to a test tube (15 mm x 125 mm) add 10 ml of buffer substrate and mix thoroughly.

2. Heat to approximately 102 F (39C), temperature is not critical, and place in an incubator or water bath at 98.6 to 113 F (37-45 C), 102 F (39 C) preferred, for 30 or 60 minutes depending on the degree of sensitivity desired.

NOTE: The longer the incubation time the greater will be the distinction between a properly and improperly pasteurized sample.

3. Remove tube from incubator or water bath, add 0.5 ml of the 5 per cent $CuSo_4$ solution, shake briefly but vigorously by placing thumb over mouth of test tube and filter.

NOTE: If precipitation of proteins is not complete add a little more $CuSO_4$ solution.

4. To 5 ml of the water-clear filtrate in a test tube, add 0.5 ml of the 3 per cent solution of sodium sesquicarbonate buffer then add 0.04 ml of BQC solution (2 drops from recommended dropper), and mix thoroughly by rotating the tube.

NOTE: Indophenol color indicative of improper pasteurization appears in about 30 seconds and is completely formed in about 2 minutes.

5. Estimate the depth of color by comparing with permanent color standards. (See Note on page 271, No. 9).

This test is said to be sensitive to the presence of 0.1 per cent raw milk or 0.2 ppm. of phenol even with an incubation period of less than 30 minutes. A faint blue color on pasteurized milk is caused by the biuret reaction of the proteins and copper but it is reported that this color is so faint as to not interfere with the detection of as little as 1 gamma of phenol in 5 ml of solution.

Control Tests

When first employing this test it will be advisable to make a blank test by incubating the sample with 10 ml of the buffer solution diluted 1 to 20, or by heating a portion of the sample under test, momentarily, to a temperature in excess of 170F (77 C), so as to inactivate any phosphatase which might be present, and to incubate a 1 ml. portion of the heated sample with the buffer substrate. This pasteurized "control" procedure is particularly effective when testing chocolate ice cream or drink since it compensates for the interfering substances that might be present.

Interpretation

A value of 2 units or greater in commercially pasteurized milk indicates inadequate heat treatment. With the use of the improved reagents it has been demonstrated that milk commercially pasteurized at a minimum of 143F for 30 minutes or 160F for 15 seconds will yield less than 1 unit.

SCHARER FIELD TEST[4]

Apparatus

(See Note p. 265)

1. Test tubes 10 x 110 mm, calibrated at 5 ml, 5.5 ml, and 7.5 ml measured to top of meniscus. Gum rubber stoppers to fit.

2. Medicine droppers for use as pipettes.

3. Small (¼ oz) brown glass pharmaceutical dropping bottle with dropper calibrated to deliver 50 drops per ml of BQC solution with dropper in vertical position.

4. Portable water bath.

5. Color comparison rack, backed by light filter screen (apparatus 9, page 269). Obtainable from Applied Research Institute.

Reagents

1. Neutral n-butyl alcohol, colorless, boiling range 116-118 C (240.8-244.4 F), highly purified and free from excessive acid or esters. Obtainable from Applied Research Institute.

NOTE: If the alcohol is very old, it should be redistilled with NaOH before use. To determine suitability of the alcohol, shake a sample with an equal quantity of neutral distilled water, allow it to separate, and test the water layer with brom thymol blue. A blue or green color, indicating a neutral reaction, should result. Errors are induced when the n-butyl alcohol does not comply with the specifications or when the extraction process is not conducted carefully.

2. BQC solution. If tablets are used, one tablet is equivalent to 20 mg BQC and should be dissolved in 5 ml of 95 per cent ethyl or methyl alcohol.

3. Buffer substrate solution. If tablets are used, crush a buffer substrate tablet in a test tube and dissolve in 50 ml of distilled water.

NOTE: Because of climatic conditions the tablets may occasionally develop minute amounts of free phenol. To remove any trace of phenol dissolve one buffer substrate tablet in 5 ml distilled water, add 2 drops of BQC solution and shake well. Allow five minutes for color development. Extract any indophenol blue by shaking vigorously with 2 ml n-butyl alcohol. Allow to stand until alcohol separates completely. Remove alcohol layer with medicine dropper and discard. Dilute remainder of substrate solution to 50 ml with distilled water.

Permanent Color Standard Solutions

1. Color solution, red. Same as acid series on page 270.

2. Color solution, yellow. Same as acid series on page 270.

3. Color solution, blue. Dissolve 62.425 gm $CuSO_4.5H_2O$ copper sulfate in 1 per cent HCl.

Combine the quantities indicated in the following table, and dilute to 5 ml with water.

Color Solutions

Units	Red ml.	Blue ml.	Yellow ml.
2	0.4	1.5	0.5
5	0.2	2.2	0.5

Procedure

1. To 5 ml of buffer substrate, add 0.5 ml of the milk sample and mix.

2. Incubate for 10 minutes in a water bath at 97 - 111 F (36.1 - 43.9 C) (105.8F (41 C) preferred).

NOTE: Incubation for 30 minutes at this temperature increases the sensitivity of the test.

3. After incubation, add 6 drops BQC solution. Shake well immediately.

4. Allow to stand for 5 minutes.

5. The appearance of a blue color indicates inadequate pasteurization.

For greater sensitivity add 2 ml of the neutralized n-butyl alcohol. Extract the indophenol blue by quickly inverting the test tube 4 times through an arc of 180 degrees. Immediately lay the tube on a flat surface in the horizontal plane and allow about 2 minutes to permit separation of the n-butyl alcohol from the substrate milk mixture. Repeat the *extraction* and *separation* procedure, return the tube to the upright position for final separation of the supernatent alcohol layer.

NOTE: Avoid vigorous agitation when mixing n-butyl alcohol because a relatively opaque emulsion may result. Brief centrifuging will break the emulsion and clarify the alcohol layer sufficiently for color comparison.

6. Compare the color in the ¾ inch supernatant alcohol layer with the permanent color standards using white Plastacele sheet (see page 269) as a filter screen in back of the tubes and viewing against the light source. Report test as containing more or less of the blue color than the standards of 2 and 5 units.

Interpretation

A value of 2 units or more generally indicates faulty pasteurization.

Before extraction with n-butyl alcohol in the field test, pasteurized milk will yield a gray or brownish color, and cream a gray or white color. Appearance of even slight blue color before extraction indicates gross under-pasteurization (raw milk or cream shows an intense blue color). Upon extraction, adequately pasteurized homogenized milk frequently gives a slight yellow color. This is due to the more complete extraction of the butterfat and should not be confused with a positive result. Because of this effect an improperly pasteurized homogenized milk, (or one slightly contaminated with raw milk), may show a greenish blue color upon extraction.

In a positive test the depth of color will vary in proportion to the degree of failure in the pasteurizing process or the quantity of raw product with which the pasteurized product may have been contaminated. It is advisable to routinely extract the indophenol with n-butyl alcohol otherwise a trace of blue color may be difficult to distinguish.

CAUTION: (1) After use, thoroughly wash all equipment and then rinse in water at 180 F (82.2 C) before the next test. The BQC reagent is sufficiently sensitive to demonstrate the leaching of phenol by water from phenolic resin bottle closures, hence such closures should be avoided anywhere in the test.

(2) Although both buffer substrate and BQC tablets remain stable when kept at room temperature in a dark, dry place, the solutions made from them tend to decompose rapidly and should be stored under refrigeration or better still, prepared shortly before use.

(3) A reagent control should be run with each series of tests by adding 3 drops of BQC to 5 ml of substrate. If a blue color results the substrate solution should be discarded. If the butyl alcohol procedure is utilized this reagent blank should be extracted with the alcohol.

Common Causes of Error in Conducting the Test

1. Insufficient mixing of milk with substrate at the beginning of the test.

2. Insufficient shaking after the BQC indicator has been added.

3. Not waiting at least 5 minutes after the BQC indicator has been added before extracting the color.

4. Insufficient mixing when extracting with the alcohol.

5. Use of old or decomposed buffer substrate or BQC solutions which may give either false positive or false negative tests depending on the age of the respective solutions.

6. Contamination of glassware or rubber stoppers.

7. Cross contamination between positive and negative samples through contact with common pipettes or rubber stoppers.

General Precautions in Interpreting Phosphatase Tests

There are certain precautions that should be taken in interpreting the results obtained with the phosphatase test on dairy products.[4, 5]

Contamination with raw milk or cream in small quantity subsequent to pasteurization of milk or cream will very likely introduce sufficient phosphatase to yield a positive test even though the product had been subjected to adequate time-temperature treatment during the pasteurizing process. This means that the pasteurized product should not come in contact with any utensils or equipment that have contained raw milk or cream.

When cream is separated from milk there is a concentration of the phosphatase in it, consequently, if any raw cream is added to other products a correspondingly higher phosphatase reading will be obtained than if raw milk were added in the same volume. [4]

Pasteurization of milk at 143 F for 30 minutes or at 160 F for 15 seconds leaves a small residual of the phosphatase enzyme which tends to concentrate in the cream layer, (not enough to yield a positive test in the well mixed sample), and if such adequately pasteurized milk is subsequently separated it is possible that sufficient of the enzyme may be concentrated in the resulting cream to yield a positive test[9]. These facts point to the necessity for completely mixing pasteurized milk samples before testing for phosphatase and to the advisability of pasteurizing milk at higher temperatures or for longer periods of time if the milk is to be subsequently separated to produce pasteurized cream.

Experience has shown that a pasteurized cream with a negative phosphatase test after storage at refrigeration temperatures as low as 39 F (4 C) for several days may yield a positive reaction.

Barber and Frazior[7] discovered that the change was due to phosphatase produced by strains of the spore-forming species B. cereus and B. mesentericus. The positive reaction was obtained when the count on these organisms was above 1.5 million per ml. [8, 9, 11] Bacterial phosphatase withstood a heat treatment of 170 F (76.7 C) for 30 minutes whereas the phosphatase in milk was inactivated by 145 F (62.8 C) for 30 minutes. The presence of microbial phosphatase was distinguished from milk phosphatase by heating the sample of cream for 30 minutes at temperatures above those used for pasteurizing cream 150 F (66 C) but below 170 F and then testing for phosphatase activity in the usual way.

Hammer and Olsen [10] report the isolation of several strains of bacteria which produce phenol or microbial phosphatase in sufficient amounts to interfere with the testing of butter samples for milk phosphatase. In view of the possible formation of microbial phosphatase [7] or of phenol or both [10] by bacterial action in cream it is evident that a positive test on stored cream cannot be regarded as conclusive evidence of improper pasteurization.

The application of the phosphatase test in the control of butter is more complex than is the case with fresh milk and cream. Brown and Parfitt [11] found the test of value in determining the adequacy of pasteurization of the cream used in the manufacture of butter. However, they report the butter tended to give a more positive reaction than the cream from which it was made (concentration of enzyme residuals). Parfitt, Brown, and Shadwick [12] found no consistent relation between the phosphatase content of butter and its yeast and mold count, its total bacterial count, or its keeping quality. They also report that under the conditions normally associated with the marketing of butter a significant number of samples will react negatively when fresh, and positively after having been exposed to conditions prevailing during commercial distribution.

In testing butter Scharer[9] suggests melting the sample at a temperature below 100 F (37.8 C), centrifuging to effect complete separation of the water serum, pipetting 1 ml of this serum directly into 10 ml of the buffer substrate and proceding as for milk or cream. When present the enzyme tends to concentrate in the water serum and so yields higher results than the corresponding melted butter. He reports butter made from holder pasteurized cream showed only slight phosphatase activity, butter churned from H. T. S. T. (160 F for 15 seconds) pasteurized cream exhibited considerable enzyme activity, and butter made from raw cream retained high phosphatase activity even after prolonged cold storage.

The preparation of *ice cream mix* employs such a variety of products that special precaution has to be taken in the interpretation of phosphatase reading,[13,14] not only because of the ingredients may not have received enough heat treatment to destroy the enzyme present, but also because they may contain a sufficient quantity of a phenolic material to give a positive reaction.[15]

When *butter* is used in making mix, a butter made from an unpasteurized cream might unknowingly be employed. In such cases, a temperature sufficiently high to inactivate the phosphatase in the butter should be used to adequately pasteurize the mix. In case the cold mix method is employed, all the fluid dairy products entering into the make up of the mix should be tested for phosphatase.

Among some of the products which may contain both the enzyme and phenolic substances are fresh or cold packed *fruits.*[15] Cooked fruits and also synthetic fruit flavorings are apt to give a positive test due to phenols. Vanilla powder, vanillin, or coumarin, if present in sufficient quantity, may give high phenol values, but the cause for it may be readily determined by testing the flavoring or by making a blank test on the ice cream.[14]

Occasionally *coloring compounds* may introduce brown or reddish colors in the indophenol test which obscure the blue color. In this case, the indophenol may be extracted with n-butyl alcohol and compared with a similar extract from a sample known to be properly pasteurized.

In the Gilcreas method many non-phenolic compounds give a positive reaction with the Folin-Ciocalteu reagent.[8] Amino acids, tyrosine, tryptophane, cystine, I-leucine, and traces of copper give strong reactions. The hydrolysis of casein would form some of these products. It is found that the hydrolysis of bacterial nucleic acids to oxypurines also gives the reaction.[17]

The enzyme phosphatase may be present in unroasted *nut meats.* Uncooked lemon juice also contains the enzyme and in a particular lemon concentrate a positive reaction was obtained from a phenolic substance.[14]

The earlier phosphatase technic[9] while suitable for differentiating between cheese made from raw milk and cheese made from pasteurized milk, did not stress the importance of achieving optimum pH conditions, nor did it take into consideration the wide range of acidity present in the many different types of cheese, nor the complex changes which occur in cheese casein through the aging process, nor the variation in amounts of fat present in different types of cheese.[16] Any slight residual of phosphatase in pasteurized milk tends to become concentrated in the cheese made from that milk and thus it is possible for a "phosphatase-negative" milk to yield a cheese which will give a slight positive test.[16] On the other hand cheese made from raw milk and subsequently pasteurized usually shows no trace of phosphatase because the chemical environment provided by cheese makes the enzyme more thermolabile than it is in milk.

The ripening process in *cheese* does not seem to produce appreciable amounts of phosphatase. Surface growth of yeast and some molds will produce considerable phosphatase, but if this growth is removed before the cheese sample is prepared for testing, a false positive test is usually avoided.[16] In relish cream cheese and several other types of cheese the colored ingredients may interfere with the reading. The color due to indophenol may be extracted from the interfering color with n-butyl alcohol and compared with the color of a known negative test similarly extracted.[9]

Goat's milk contains less phosphatase than cows' milk, in the raw state, and the amount contained is destroyed at lower time-temperature levels than required for adequate pasteurization of cows' milk, hence, proper pasteurization of goats' milk cannot be positively determined by the phosphatase test. A positive phosphatase test on a sample of pasteurized goats' milk would indicate a gross failure in pasteurization or contamination with an appreciable amount of the raw product.

It is quite evident from the foregoing discussion that "double blank" determinations (see page 272) are essential if it is to be established beyond doubt that positive phosphatase tests are due to the presence of milk phosphatase alone or whether such positive results are due in whole or in part to the presence of microbial phosphatase, phenol, or other chemical factors that may give a color reaction with the phenol indicator.

Sanders and Sager Method

(A. O. A. C. Method, First Action, Adopted Oct., 1947)

Sanders and Sager[18] report modifications and improvements in the laboratory phosphatase test for pasteurization, and the application of the test to virtually all dairy products. In October, 1947, this laboratory procedure[19] was adopted as an official method, first action, by the Association of Official Agricultural Chemists for use with fluid milk and cream, cheddar type cheese, and the soft, unripened cheeses and as a tentative method for other types of cheese, ice cream mix, sherbet mix, chocolate-flavored milks and skim milk, butter, sweet-cream and cultured buttermilks, fermented milks, goat's milk, cheese whey, and concentrated milks.

Improvements in the Phosphatase Test Suggested by Sanders and Sager

1. Substitution of barium borate-hydroxide for sodium tetraborate as a buffer which is said to aid in precipitating the proteins and especially to precipitate phosphates and citrates, which interfere with the activity of the phosphatase enzyme. It is claimed that use of this buffer resulted in an average increase of approximately 12 per cent in the quantity of phenol liberated by the enzyme.

2. It is reported that the barium borate-hydroxide buffer has a high buffering capacity in the pH region in which the enzyme is most active. Concentration and the alkalinity of the barium buffer are specified for each type of product and are so adjusted as to yield the optional pH of approximately 10, or within a range of 9.85 to 10.2 (never below 9.5) during incubation, thus improving the quantitative accuracy. Sanders and Sager state that correct pH during incubation is apparently the most important condition for a successful, reliable test.

3. More uniformly clear filtrates are claimed by use of a protein precipitant containing zinc (for testing ripened products) or zinc and copper (for fresh products), instead of the lead precipitant used earlier. The zinc-copper precipitant enhances the development of the blue color and decreases the interfering yellow color produced by a given quantity of BQC indicator.

4. Sanders and Sager suggest the use of a color development buffer, a solution of sodium metaborate and sodium chloride in water which is added to the filtrate and serves to partially dilute the interfering compounds present and also adjusts the pH to 9.3 to 9.4, which was found to be optimal for the development of the blue color with a minimum of yellow color.

5. A dilution method has been introduced to make possible the full color development from all the phenol present, at the same time reducing the amount of off color which the authors claim increases the precision and makes the test quantitative even for raw samples.

Apparatus

1. Supply of chemical glassware—flasks, beakers, funnels, watch glasses, test tubes, stirring rods, graduates, etc.
2. Volumetric flasks, 100 ml, 250 ml, 1000 ml.
3. pH meter.
4. Photometer.
5. Balance.
6. Pipettes, 1 ml, 3 ml.
7. Thermometer, 0 to 110 C.
8. Water bath.
9. Brown glass, ¼ fluid oz. dropping bottle with dropper calibrated to deliver 50 drops per ml of BQC solution with dropper in verticle position.

Reagents

(In all cases, the term "water" means distilled water).

1. Buffers:

a. Barium borate-hydroxide buffer. Dissolve 25.0 gm C. P. barium hydroxide [$Ba(OH)_2 .8H_2O$—fresh, not deteriorated] in distilled water and dilute to 500 ml. In another flask or cylinder dissolve 11.0 gm C. P. boric acid (H_3BO_3) and dilute to 500 ml. Warm each to 50 C, mix the two together, stir, cool to approximately 20 C, filter, and stopper the filtrate tightly (pH 10.6).

The buffer thus prepared is designated as the 25-11 buffer, the figures indicating the grams per liter of each of the respective reagents. Modifications in the quantities of these two reagents, necessary in preparing the appropriate buffers for testing various products, are indicated in Tables 5 and 6.

b. Color development buffer. Dissolve 6.0 gm sodium metaborate ($Na\ BO_2$) and 20 gm sodium chloride in water and dilute to 1 liter with water (pH 9.8).

NOTE: $NaBO_2$ is obtainable from Applied Research Institute, New York, 10, N. Y.

c. Color dilution buffer. Dilute 100 ml of color development buffer (1-b) to 1 liter with water.

d. Standard borax buffer, 0.01-molar, for checking pH meter, pH 9.18 at 25 C. Dissolve 0.9603 gm pure borax (Bureau of Standards Sample 187) in distilled water (distilled recently or freshly boiled and cooled) and dilute to 250 ml. Keep stoppered tightly.

NOTE: Determine all pH values at 25°C (77°F) or correct to that temperature.

2. Buffer substrates:

a. For evaluating pasteurization. Dissolve 0.10 gm phenol-free crystalline disodium phenyl phosphate in 100 ml of the appropriate (Tables 5 and 6) barium borate-hydroxide buffer (1-a above).

NOTE: Phenol-free crystalline disodium phenyl phosphate is obtainable, relatively pure, from Applied Research Institute.

b. For quantitative results with raw milk and raw-milk products.—Dissolve 0.20 gm of the phenol-free crystalline disodium phenyl phosphate in 100 ml of the appropriate (Tables 5 and 6) barium borate-hydroxide buffer (1-a above).

3. Protein precipitants:

a. Zinc-copper precipitant for milk. Dissolve 3.0 gm zinc sulfate ($ZnSO_4.7H_2O$) and 0.6 gm copper sulfate ($CuSO_4.5H_2O$) in water and dilute to 100 ml with water. The precipitant thus prepared is designated as the 3.0-0.6 precipitant.

b. Zinc-copper precipitant for unripened cheese. Dissolve 6.0 gm zinc sulfate and 0.1 gm of copper sulfate in water and dilute to 100 ml with water. This precipitant is designated as the 6.0-0.1 precipitant.

c. Zinc precipitant for ripened cheese and for butter. Dissolve 6.0 gm zinc sulfate in water and dilute to 100 ml with water. This precipitant is designated as the 6.0 precipitant.

NOTE: The quantities of the respective reagents to use in preparing the precipitants for testing other products, not mentioned under "Protein precipitants" above, are indicated in Tables 5 and 6.

4. BQC or 2.6-dibromoquinonechloroimine solution (Gibbs' reagent)—Dissolve 40 mg of BQC powder in 10 ml of absolute ethyl or methyl alcohol and transfer to a dark-colored dropper bottle. This reagent remains stable for at least a month if kept in the ice tray of a refrigerator. Do not use it after it begins to turn brown.

NOTE: BQC powder may be obtained from Applied Research Institute.

5. a. Copper sulfate, 0.05 per cent, for standards.—Dissolve 0.05 gm copper sulfate in water and dilute to 100 ml.

b. Butyl alcohol, n-butyl alcohol, boiling point 116-118 C. To adjust the pH, mix 50 ml of the color development buffer (1-b) with 1 l of the alcohol.

6. Phenol standards:

a. Stock solution. Weigh accurately 1.0 gm pure phenol, transfer to a liter volumetric flask, dilute to 1 l with water, and mix. One ml contains 1 mg of phenol. Use this stock solution to prepare standard solutions. It is stable for several months in the refrigerator.

b. Preparation of standards.—Dilute 10.0 ml of the stock solution (6-a) to 1 l with water, and mix. One ml contains 10 micrograms

(0.00001 gm, 10 mmg, or 10 units) of phenol. Use this standard solution to prepare more dilute standard solutions; e.g., dilute 5, 10, 30, and 50 ml to 100 ml with water to prepare standard solutions containing 0.5, 1.0, 3.0, and 5.0 mmg or units of phenol per ml, respectively. Keep standard solutions in the refrigerator.

In a similar manner, prepare from the stock solution as many more concentrated standard solutions as may be needed, containing, for example 20, 30, and 40 units per ml.

Measure appropriate quantities of the phenol standard solutions into a series of tubes (preferably graduated at 5.0 and 10.0 ml) to provide a suitable range of standards as needed, containing 0 (control or blank), 0.5, 1.0, 3.0, 5.0, 10.0, etc., to 30 or 40 units. To increase the brightness of the blue color and improve the stability of the standards, add 1.0 ml of 0.05 per cent copper sulfate solution (5-a) to each.

Add 5.0 ml of color dilution buffer (1-c) and add water to bring the volume to 10.0 ml. Add 4 drops (0.08 ml) of BQC, mix, and allow to develop for 30 min. at room temperature. If the butyl alcohol extraction method is to be used in the test, extract the standards as described under "Conducting the test."

Read the color intensities with a photometer, subtract the value of the blank from the value of each phenol standard, and prepare a standard curve (straight line). When the standards are to be used for visual comparisons they should be stored in a refrigerator.

Sampling

1. Milk and other fluid products.—Mix the product well, pour several ml into a small tube, stopper the tube, and keep in a refrigerator.

2. Hard cheese.—Take a sample from the interior with a clean Roquefort trier, place in a small tube, stopper the tube, and keep in a refrigerator.

3. Soft and semi-soft ripened cheese.—Harden the cheese by chilling in the freezing chamber of a refrigerator. Taking special precautions to avoid contaminating the sample with phosphatase that may be present on the surface, use either of the following methods for sampling:

(a) Cut a portion from the end of the loaf or from the side of the cheese, extending inward at least 2 inches if possible or to a point somewhat beyond the center in the case of a small cheese. Cut a slit ¼ to ½ in. deep at least halfway around the portion and midway between the top and bottom. Break the portion into two parts, pulling it apart so that it breaks on a line with the slit, being careful not to contaminate the freshly exposed, broken surface. Remove the sample from the freshly exposed surface at or near the center of the cheese.

(b) Remove the surface of the area to be sampled—e.g., the end and the adjacent sides—with a clean knife or spatula, to a depth of ¼ in. Clean the instrument and hands with hot water and phenol-free soap and wipe them dry. Remove the freshly exposed surface to a similar or greater depth, and repeat the cleaning. Then take the sample from the center of the freshly exposed area, preferably at or near the center of the cheese in the case of a small cheese.

4. Process cheese, cheese spreads, butter, and other non-fluid products.—Take the sample from beneath the surface with a clean knife or spatula.

5. Ice cream and sherbet.—Melt the portion removed and allow it to remain melted for an hour or longer before testing, testing it as a fluid product.

NOTE: Avoid the use of samples contaminated with mold.

6. Preservation.—If a preservative is necessary, for liquid products add 1 to 3 per cent of chloroform; for solid products put 1 to 3 ml of chloroform in the container, cover with a plug of cotton, insert sample and stopper container tightly. Label preserved samples: "Poison, preservative added."

Procedure

The chemical principles involved in the detection and measurement of milk phosphatase activity are the same for all dairy products. Some modifications, described below, have been found necessary for different dairy products, because of their differences in physical properties, compositions, and especially buffering capacities.

Cheese:

1. Weigh, on a clean balance pan or watch glass, a 0.50-gm sample (preferably in duplicate) and place in a culture tube 16 or 18 x 150 mm. Similarly, weigh another sample and place in a tube as a control or blank. If the cheese is sticky, weigh the sample on a piece of wax paper about 1 x 1 inch and insert the paper with the sample into the tube. Macerate the blank and the test with a glass rod about 8x180 mm.

2. Add to the blank 1.0 ml of the appropriate (Table 5) barium buffer (1-a, without substrate added), macerate with the rod, leave the rod in the tube, heat for about a minute to at least 85 C (185 F) in a beaker of boiling water with the beaker covered so that the entire tube is heated to approximately 85 C, cool to room temperature, and macerate again with the rod.

3. Add to the test 1.0 ml of barium substrate (2-a or 2-b) (Table 5) and macerate.

NOTE: From this point, treat the blank and the test in a similar manner.

Add 9.0 ml of the appropriate barium buffer substrate (2-a or 2-b) (total, 10.0 ml added), and mix. The rod may be left in the tube during incubation; if it is removed at this point, cut a piece of filter paper about 1 x 1 in., wrap and hold it tightly around the rod, rotate the rod while withdrawing it from within the tube so as to wipe the rod clean, insert the paper with the adhering fat into the tube, and stopper.

4. Incubate in a water bath at 37-38 C for 1 hour, mixing or shaking the contents occasionally.

5. Place in a beaker of boiling water for nearly a minute, heating to approximately 85 C (use a thermometer in another tube containing the same volume of liquid), and cool to room temperature.

6. Pipette in 1.0 ml of the zinc precipitant (3-c) for ripened cheese or the zinc-copper precipitant (3-b) for unripened cheese, and mix thoroughly (pH of mixture, 9.0 to 9.1).

7. Filter and collect 5.0 ml of filtrate in a tube, preferably graduated at 5.0 and 10.0 ml.

NOTE: A 5 cm funnel with 9 cm Whatman No. 42 or No. 2 filter paper is recommended.

8. Add 5.0 ml of color development buffer (1-b) (pH of mixture, 9.3 to 9.4).

9. Add 4 drops of BQC, mix, and allow the color to develop for 30 min. at room temperature. NOTE: For merely detecting underpasteurization in testing unripened cheese, 2 drops are sufficient, provided the visual standards likewise are prepared with 2 drops.

10. Determine the intensity of blue color by either of two methods:

a. With a photometer: Read the color intensity of the blank and that of the test, subtract the reading of the blank from that of the test, and convert the result into phenol equivalents by reference to the standard curve described under "Phenol standards." The butyl alcohol extraction method ordinarily is necessary when using a photometer.

Photometric Determination: To read the color in aqueous solution, use a filter with maximum light transmission in the region of 610 mu wave length.

To read the color in butyl alcohol, extract the color as described above and centrifuge the sample for 5 min. to break the emulsion and to remove the moisture suspended in the alcohol layer. A Babcock centrifuge can be adapted for this purpose by making special tube holders as follows: Slice a section ¼ inch thick from a rubber stopper of suitable diameter to fit into the bottom of the centrifuge cup. Glue together two cork stoppers of appropriate diameter, bore through the center a hole of proper size to hold the tube snugly, and insert the double cork section into the cup. After centrifuging, remove nearly all of the butyl alcohol by means of a pipette with a rubber bulb on the top end. Filter the alcohol into the photometer cell and read with a filter with maximum light transmission in the region of 650 mu wave-length.

If more than approximately 4 ml of butyl alcohol is required for the photometer used, conduct the test in a larger tube and extract the color, in both the test and the standards, with the necessary quantity of butyl alcohol rather than with 5 ml specified above.

b. With visual standards: For quantitative results in borderline instances, e.g., tests yielding 0.5 to 5 units of color, extract with butyl alcohol (5-b). Add 5.0 ml of the alcohol and invert the tube slowly several times, centrifuge if necessary to increase the clearness of the alcohol layer, and compare the blue color with the colors of standards in the alcohol.

NOTE: With samples yielding more than 5 units, compare the colors in aqueous tests with those of aqueous standards.

11. Dilution method for quantitative results: In tests that are observed to be strongly positive during color development—e.g., 20 units or more—in which 4 drops of BQC may be much less than sufficient to combine with all of the phenol, pipette an appropriate proportion of the contents into another tube, make up to 10.0 ml with color dilution buffer (1-c) and add 2 drops more of BQC in the case of an unripened product or 4 drops in the case of a ripened product. With each test, dilute and treat the blank in the corresponding manner. Dilute each strongly positive test thus until the final color is within the range of the visual standards or photometer. Allow 30 min. for color development after the last addition of BQC, and make the reading at the end of the 30-min. period. To correct, multiply by 2 for a 5 + 5 dilution, by 10 for a 1 + 9 dilution, and by 50 for a 1 + 9 followed by a 2 + 8 dilution.

Alternatively, to reduce the amount of yellow off color, add 2 instead of 4 drops of BQC after each dilution, and allow the color to develop. Then test the completeness of color development by adding a third drop. Repeat the dilution procedure until the addition of an extra drop does not cause any further increase in the amount of blue color.

12. Calculation and evaluation of result: When using 0.5 gm of solid sample and adding a total of 11.0 ml of liquid, multiply the value of the reading by 1.1 to convert it to units of color or phenol equivalents per 0.25 gm of cheese. The result, if desired, may be converted

to phenol equivalent per 1 gm, by multiplying by 4.4. Evaluate the result by comparing it with the criteria of pasteurization in Table 5.

Milk and other fluid products:

1. Pipette a 1.0-ml sample (preferably in duplicate) into a tube and pipette 1.0 ml into another tube as a control or blank. In testing cream, the sample may be weighed (1.0 gm) if desired; in testing goats' milk, use a 3.0 ml sample (Table 6).

2. Heat the blank to the temperature indicated under "Cheese, No. 2," and cool to room temperature. From this point, treat the blank and the test in a similar manner.

3. Add 10.0 ml of barium buffer substrate (2-a or 2-b) (Table B), stopper the tube, and mix.

4 to 11, inclusive. Follow the directions given for the corresponding steps under "Cheese" above, substituting the appropriate precipitant (Table 6) in No. 6, and, for merely detecting under-pasteurization, using 2 rather than 4 drops of BQC in No. 9.

12. Calculation and evaluation of result: When using 1.0 ml of fluid sample and adding 11.0 ml of liquid (total liquid 12.0 ml, 5.0 ml of filtrate used), multiply the value of the reading by 1.2 to convert it to phenol equivalents per 0.5 ml of sample. If desired, the result may be converted to phenol equivalents per 1 ml by multiplying by 2.4. Evaluate the result by comparing it with the criteria of pasteurization in Table 6.

Butter:

1. Weigh, on a piece of wax paper about 1 x 1 inch on a balance, a 1.0-gm sample (preferably in duplicate) and insert the paper with the sample into the tube. Similarly, weigh another sample and place in a tube as a control or blank.

2. Heat the blank to the temperature indicated under "Cheese: No. 2" and cool to room temperature. From this point, treat the blank and the test in a similar manner.

3. Add 10.0 ml of barium buffer substrate (2-a or 2-b) (prepared with 18-8 barium buffer, Table 6), stopper the tube, and mix.

4 to 11, inclusive. Follow the directions given for the corresponding steps under "Cheese" above, mixing the contents frequently and thoroughly during incubation, substituting the appropriate zinc precipitant (Table 6) in No. 6, and, for merely detecting underpasteurization, using 2 rather than 4 drops of BQC in No. 9.

12. Calculation and evaluation of result: When using 1.0 gm of butter and adding 11.0 ml of liquid, multiply the value of the reading by 1.1 to convert the result to phenol equivalent per 0.5 gm of butter. Evaluate the result by comparing it with the criterion of pasteurization in Table 6.

MODIFICATIONS FOR DIFFERENT CHEESES AND OTHER DAIRY PRODUCTS

Different dairy products, and different kinds of cheese and cheeses of different ages, have different buffering capacities, and therefore some of them require modification of concentrations of the reagents. The modifications of the barium buffer needed to produce optimal pH conditions during incubation (9.85 to 10.20), and of the precipitant to yield uniformly clear filtrates and to minimize interference during color development under optimal pH conditions (9.3 to 9.4), are specified in Tables 5 and 6.

With some samples, especially cheese samples of unknown history, slight deviations from the optimal pH range may occur, but such deviations do not very materially affect the results. For example, pH values as low as 9.6 or as high as 10.35 during incubation have been found to result in an average decrease of not more than 20 per cent below the maximum in the quantity of phenol liberated. The use of the 25-11 buffer substrate with samples for which the 27-11 buffer substrate is specified yields pH values not lower than 9.8.

A trace of cloudiness in the filtrate, following the use of the precipitant as prescribed, indicates that the concentration of barium hydroxide in the buffer was not sufficiently great, i. e., that the buffer substrate was not sufficiently alkaline. For example, the 25-11 buffer diluted 5 + 5 with water, for use with fresh milk, may yield a cloudy filtrate if used with old milk having a pH below approximately 6.0, or with milk that has soured, and the test should be repeated with a more concentrated buffer and precipitant. Likewise, the 25-11 buffer, for use with unripened cheese, may yield a cloudy filtrate if used with ripened cheese, indicating that the concentration of the buffer used was not sufficient. Increasing the concentration of zinc sulfate in the precipitant also eliminates turbidity of the filtrate.

In testing cheese of unknown history or age, information as to the percentage of solids, especially the nonfat solids, is useful as an indication of the correct buffer to use; cheese with a relatively high percentage of nonfat solids generally requires the use of a relatively concentrated buffer to adjust the pH of the mixture correctly.

For precise quantitative results on unknown samples, adjust the pH to 10.0 to 10.05 for the incubation.

Cottage cheese curd is heated in the presence of considerable acid during manufacture, and therefore its phosphatase values are comparatively low. Alternatively, to increase the sensitivity of the test on cottage cheese, apply the following modifications: Use a 1.0 gm sample, 27-11 buffer substrate, 2-hour incubation, 6.0 to 0.1 precipitant, and a pasteurization criterion of 2 units per 0.5 g.

Phosphatase activity is much less in goats' milk than in cow's milk. The details of a modification designed to increase the sensitivity of the test applied to goats' milk are specified in Table 6.

To test concentrated milk products, reconstitute the product with water to its original concentration of milk solids and test in the manner specified for the original product.

To test for the presence of microbial phosphatase, which is indicated by blue color in the blank prepared as directed above, repeat the determination, adding 1 ml of the appropriate barium buffer (without substrate) to the blank and heat it for 5 min. in boiling water in a covered beaker. If the blank treated thus is negative, it indicates that the blue color in the original blank was due to microbial phosphatase, i.e., a "false positive" sample.

Precautions

The length of time that the crystalline disodium phenyl phosphate and the BQC powder will remain stable can be increased greatly by keeping them in the freezing chamber of a refrigerator. However, container must be warmed to room temperature before opening for use.

NOTE: Experience of many dairy laboratories indicate that crystalline disodium phenyl phosphate and BQC powder or the reagents in tablet form (as supplied by Applied Research Institute) will retain stability longer if kept at room temperature in a dark, dry place. (See note, page 270).

The glassware, stoppers, and sampling tools should be scrupulously clean, and it is desirable to soak them in hot, running water after cleaning.

The bottles containing solid barium hydroxide and the barium buffer must be kept stoppered tightly to prevent absorption of carbon dioxide.

Phenolic contamination from plastic closures on reagent bottles has been encountered, and therefore the use of plastic closures should be avoided. Rubber stoppers should not be used in flasks in which butyl alcohol is stored. Glass or cork stoppers should be used.

TABLE 5.—*Phosphatase test modifications for different kinds of cheese and cheese of different ages*

KIND OF CHEESE	Age or extent of curing; other details	Buffer for Optimum pH (9.85-10.20)	Precipitant	Criterion, Experimental, Phenol Equivalents[a]
Cheddar, granular, stirred curd, hard cheese	< 1 wk.	25-11[b]	6.0-0.1[c]	3
	1 wk.–1.5 mo.	25-11	6.0[d]	3
	1.5–4 mo.	26-11	6.0	3
	> 4 mo.	27-11	6.0	3
Washed curd, soaked curd, Colby	< 1 wk.	25-11	6.0-0.1	3
	1 wk.–2 mo.	25-11	6.0	3
	> 2 mo.	26-11	6.0	3
Swiss, Gruyère	< 1 wk.	25-11	6.0-0.1	3
	1 wk.–1 mo.	25-11	6.0	3
	1–3 mo.	26-11	6.0	3
	> 3 mo.	27-11	6.0	3
Brick, Muenster	< 1 wk.	25-11	6.0-0.1	3
	1 wk.–1 mo.	25-11	6.0	3
	1–2 mo.	25-11	6.0	3
	> 2 mo.	26-11	6.0	3
Edam, Gouda	< 1 wk.	25-11	6.0-0.1	3
	1 wk.–2 mo.	25-11	6.0	3
	2–4 mo.	26-11	6.0	3
	> 4 mo.	27-11	6.0	3
Blue mold, blue	< 1 wk.	25-11	6.0-0.1	3
	1 wk.–1 mo.	26-11	6.0	3
	1–4.5 mo.	27-11	6.0	3
	> 4.5 mo.	28-11	6.0	3
Camembert, Limburger	< 1 wk.	25-11	6.0-0.1	4
	1 wk.–1 mo.	25-11	6.0	4
	1–2 mo.	26-11	6.0	4
	> 2 mo.	27-11	6.0	4
Monterey	< 1 wk.	25-11	6.0-0.1	3
	1 wk.–2 mo.	25-11	6.0	3
	> 2 mo.	26-11	6.0	3
High moisture Jack	< 1 wk.	25-11	6.0-0.1	3
	1 wk.–2.5 mo.	25-11	6.0	3
	>2.5 mo.	26-11	6.0	3
Provolone, pasta filata	< 1 wk.	25-11	6.0-0.1	3
	1 wk.–1 mo.	25-11	6.0	3
	1–3 mo.	26-11	6.0	3
	> 3 mo.	27-11	6.0	3

TABLE 5.—*(continued)*
1000G (Each small scale division represents 2 microns).

KIND OF CHEESE	Age or extent of curing; other details	Buffer for Optimum p_H (9.85-10.20)	Precipitant	Criterion, Experimental, Phenol Equivalents
Parmesan, reggiano, monte, modena, Romano, asiago old	< 1 wk.	25-11	6.0-0.1	3
	1 wk.–2 mo.	26-11	6.0	3
	2–6 mo.	27-11	6.0	3
	6 mo.–1 yr.	28-11	6.0	3
	> 1 yr.	29-11	6.0	3
Asiago fresh	Same as Cheddar			
Asiago medium	< 1 wk.	25-11	6.0-0.1	3
	1 wk.–1 mo.	25-11	6.0	3
	1–3 mo.	26-11	6.0	3
	> 3 mo.	27-11	6.0	3
Gorgonzola	Same as blue			
Cottage,[e] cook cheese, koch kaese	Dry	25-11	6.0-0.1	1
	Moist	25-11 (8+2)[f]	4.5-0.1	1
Cream cheese		25-11 (7+3)	4.5-0.1	3
Semi-soft cheese	< 1 wk.	25-11	6.0-0.1	3
	1 wk.–1 mo.	25-11	6.0	3
	> 1 mo.	26-11	6.0	3
Soft ripened cheese	< 1 wk.	25-11	6.0-0.1	4
	1 wk.–1 mo.	25-11	6.0	4
	> 1 mo.	26-11	6.0	4
Nokkelost, kuminost, sage cheese	< 1 wk.	25-11	6.0-0.1	3
	1 wk.–1.5 mo.	25-11	6.0	3
	1.5–4 mo.	26-11	6.0	3
	> 4 mo.	27-11	6.0	3
Pasteurized process, past. proc. pimiento, past. proc. with fruits, meats, etc.	Soft, mild	25-11	6.0	3
	Medium firm	26-11	6.0	3
	Firm, sharp (incl. Swiss, Gruyère)	27-11	6.0	3
Past. proc. cheese foods; past. proc. cheese foods with fruits, meats, etc.	Same as pasteurized process			
Past. proc. cheese spreads; past. proc. cheese spreads with fruits, meats, etc.	Soft, high moisture, incl. cream spreads	25-11	6.0	3
	Less soft, incl. Blue	26-11	6.0	3

TABLE 5.—*(continued)*

KIND OF CHEESE	Age or extent of curing; other details	Buffer for Optimum *p*H (9.85-10.20)	Precipitant	Criterion, Experimental, Phenol Equivalent[a]
Cold pack, club; cold pack cheese foods; cold pack cheese foods with fruits, meats, etc.	Mild to medium flavored, soft	26-11	6.0	3
	Sharp, firm	27-11	6.0	3

[a] Values higher than those shown indicate under-pasteurization.
[b] Grams $Ba(OH)_2 \cdot 0H_2O$ and H_3BO_3 respectively, per liter.
[c] Grams $ZnSO_4 \cdot 7H_2O$ and $CuSO_4 \cdot 5H_2O$ respectively, per 100 ml.
[d] Grams $ZnSO_4 \cdot 7H_2O$ per 100 ml.
[e] See also more sensitive modification in text (p. 90), alternative.
[f] Eight parts of 25-11 buffer plus 2 parts of water.

TABLE 6.—*Phosphatase test modifications for various dairy products other than cheese*

Product	Quantity of Sample	Buffer for Optimum *p*H (9.85-10.20)	Precipitant	Criterion, Experimental, Phenol Equivalent[a]
				mmg
Milk: Fresh	1 ml	25-11[b] (+5)[c]	3.0-0.6[d]	2/0.5 ml
Old or slightly sour	1 ml	25-11	6.0[e]	2/0.5 ml
Cream: Fresh	1 ml or 1 gm	25-11 (5+5)	3.0-0.6	2/0.5 ml or 0.5 gm
Old or slightly sour	1 ml or 1 gm	25-11 (8+2)	4.5	2/0.5 ml or 0.5 gm
Ice cream mix	1 ml	25-11 (8+2)	4.5-0.1	2/0.5 ml
Sherbet mix	1 ml	25-11 (5+5)	3.0-0.6	2/0.5 ml
Chocolate drink	1 ml	25-11 (8+2)	4.5-0.1	2/0.5 ml
Butter	1 gm	18-8	6.0	2/0.5 gm
Sweet buttermilk	1 ml	25-11 (5+5)	3.0-0.6	2/0.5 ml
Cultured buttermilk and fermented drinks: Medium acid	1 ml	25-11	6.0	2/0.5 ml
Very acid, *p*H<4.5	1 ml	26-11	6.0	2/0.5 ml
Goats' milk[f]	3 ml	27-11	7.5-0.1	1/1.5 ml
Cheese whey	1 ml	25-11 (5+5)	3.0-0.6	2/0.5 ml

[a] Values higher than those shown indicate under-pasteurization.
[b] Grams $Ba(OH)_2 \cdot 0H_2O$ and H_3BO_3 respectively, per liter.
[c] Five parts of 25-11 buffer plus 5 parts of water.
[d] Grams $ZnSO_4 \cdot 7H_2O$ and $CuSO_4 \cdot 5H_2O$ respectively, per 100 ml.
[e] Grams $ZnSO_4 \cdot 7H_2O$ per 100 ml.
[f] Four-hour incubation period; use 7.0 ml of filtrate and add 3.0 ml of color development buffer (1-b).

REFERENCES

1. Kay, H. D., and Graham, W. R.: The Phosphatase Test of Pasteurized Milk. J. Dairy Research, Vol. 6, pp. 191-203 (1935).

2. Gilcreas, F. W., and Davis, W. S.: The Practical Value of the Phosphatase Test in Determining the Efficiency of Pasteurization. Int'l. Assn. Milk Dealers, Laboratory Section, p. 34 (1937).

3. Gilcreas, F. W.: Value of Phosphatase as an Indication of Pasteurization. A.J.P.H., 29:158 (1939).

4. Scharer, H.: A Rapid Phosphomoesterase Test for Control of Dairy Pasteurization. J. Dairy Sci., 21:21 (1938).

5. Scharer, H.: Improvements in the Rapid Phosphatase Test for Detection of Improper Pasteurization of Milk and Its Products. J. Milk Tech., Vol. 1:35 (1938).

6. Standard Methods for Examination of Dairy Products, 9th edition (1948), American Public Health Association, New York, N. Y.

7. Barber, Franklin W., and Frazier, W. C.: The Development of a Positive Phosphatase Test in Refrigerated Pasteurized Cream. J. Dairy Sci. 26:343 (1943).

8. Burgwald, L. H.: The Phosphatase Test—A Review of the Literature on its Application for Detecting Irregularities in the Pasteurization of Milk and Dairy Products. J. Dairy Sci. 22:854 (1939).

9. Scharer, H.: Application of the Rapid Phosphatase Test to Dairy By-Products. J. Milk Tech. 2:1, 16 (1939).

10. Hammer, B. W., and Olsen, H. C.: Phosphatase Production in Dairy Products by Micro-Organisms. J. Milk Tech. 4:83 (1941).

11. Brown, W. H., and Parfitt, E. H.: The Application of the Phosphatase Test to the Butter Industry. Abs. J. Dairy Sci. 21:149 (1938).

12. Parfitt, E. H., Brown, W. H., and Shadwick, G. W. Jr.: Application of the Phosphatase Test to Butter. A.J.P.H., 30:240 (1940).

13. Hahn, A. J., and Tracy, P. H.: Determining the Efficiency of Ice Cream Mix Pasteurization Through the Use of the Phosphatase Test. J. Dairy Sci. 22:219 (1939).

14. Caulfield, W. J., and Martin, W. H.: Certain Factors Affecting the Phosphatase Test When Applied to Ice Cream. J. Dairy Sci. 22:261 (1939).

15. Scharer, H.: Phosphatase Test for Gauging the Efficiency of Mix Pasteurization, Proc. 38th Ann. Conv. International Association of Ice Cream Manufacturers (1938). Production and Laboratory Council, pages 50-55.

16. Scharer, H.: Application of the Phosphatase Test to Cheese. A.J.P.H. Vol. 35:4, p. 358 (1945).

17. Leahy, Sandholzer and Woodside: Possible Errors in the Phosphatase Test Resulting from Bacterial Growth in Milk. J. Milk Tech., Vol. 4, 3:183 (1940).

18. Sanders, G. P., and Sager, O. S.: Present Status of the Phosphatase Test. J. Milk and Food Tech., Vol. 11, 2, 67 (1948).

19. Sanders, G. P., and Sager, O. S.: Phosphatase Test for Pasteurization for Fluid Milk, Cream, Cheddar Type Cheese, Soft Uncured Cheeses, and Tentative for Other Types of Cheese, Ice Cream Mix, Sherbet Mix, Chocolate-Flavored Milk and Skimmilk, Butter, Sweet Cream and Cultured Buttermilks, Fermented Milks, Goats' Milk, Cheese Whey, and Concentrated Milks. Jour. A.O.A.C., Vol. 31, No. 1, page 82 (1948).

20. Gilcreas, F. W.: Report on Tests for Pasteurization of Dairy Products. J. Assoc. Official Agr. Chem., 21:372 (1938).

21. Burgwald, L. H., and Giberson, E. M.: An Evaluation of the Various Procedures for Making Phosphatase Tests. Int'l. Assoc. Milk Dealers, Assoc. Bul., 31st yr., No. 9, page 211 (1939).

22. Roger, D. W.: The Practical Application of the Phosphatase Test. Int'l. Assoc. Milk Dealers, Assoc. Bul., 31st yr., No. 9, p. 233 (1939).

23. Paley, C.: A Study of the Effect of Growth of Same Organisms in Milk on the Phosphatase Test. J. Milk Tech., 2:251 (1939).

24. Gilcreas, F. W.: Report of Tests for Pasteurization of Dairy Products. Phosphatase Test in Examination of Milk and Cream. III. The Rapid Phosphatase Test. IV. Further Study of Tentative Method. V. Further Study of the Technic for Use in the Field. J. Assoc. Official Agr. Chem., 24:559-574 (1941).

25. Burgwald, L. H.: A Critical Review of the Phosphatase Test. Int'l. Ass'n. of Milk Dealers, Assoc. Bul., 33 yr., No. 14, p. 366 (1941).

26. Horwitz, W.: A Modified Scharer Phosphatase Test for Milk and Soft Cheeses. Jour. A.O.A.C., Vol. 29, No. 2, p. 129 (1946).

27. Kay, H. D., Aschaffenburg, R., Neave, F. K.: The Phosphatase Test for Control of Efficiency of Pasteurization. Tech. Comm. No. 1, Imperial Bureau of Dairy Science, October 1939.

28. Sanders, G. P., and Sager, O. S.: Heat Inactivation of Milk Phosphatase in Dairy Products. J. Dairy Sci., 31:845 (1948).

29. Hetrick, J. H., and Tracy, P. H.: Effect of Short-Time Heat Treatments on Some Properties of Milk. I. Inactivation of the Phosphatase Enzyme. J. Dairy Sci., 31, 867 (1948).

CHAPTER 22

NEUTRALIZER DETECTION

Hankinson and Anderson Method

Apparatus

1. Potentiometric titration set-up consisting of a pH meter (See page 263); 2 burettes, mechanical agitation and liquid container.
2. Glass funnels, 6 inch.
3. Erlenmeyer flask, 500 ml capacity.
4. Beakers, 150 ml.
5. Pipettes, 100 ml and 20 ml.
6. Balance.
7. Filter paper, 25 cm diameter, rapid filtering.
8. Facilities for making Babcock fat test. (See page 179.)

Reagents

1. Sodium hydroxide (NaOH), N/10. Standardized to 0.1000 N, ±0.0002 N.

2. Hydrochloric acid (HCl), N/10. Standardized to 0.1000N, ±0.0002 N.

3. Uranyl nitrate, approximately 0.2M with respect to the anhydrous salt. This solution is made by dissolving about 82 gm of $UO_2(NO_3)_2 \cdot 6H_2O$ in 1 liter of distilled water. The solution is filtered free from insoluble matter which is usually present, and is standardized by diluting with distilled water until exactly 10 ml of the solution requires 34.0 to 34.5 ml of 0.1N NaOH to neutralize to pH 8.40 at 77 F (25 C). The pH meter is used to adjust the end-point. The 10 ml sample is not diluted with water for the titration. It is found that titration of a solution of $UO_2(NO_3)_2$ with NaOH gives a typical acid-base reaction, so the solution is standardized arbitrarily in this manner. The titration shows a sharp rise in pH at about pH 6.00. A deep yellow precipitate will also appear at about pH 6.00.

4. Potassium acid phthalate, 0.05M. A solution containing 10.207 gm of the c. p. salt is used as a reference buffer for the purpose of standardizing the Beckman pH meter. Its pH is accepted as 3.92.

Procedure

1. Weigh 150 gm of cream into a 500 ml Erlenmeyer flask. Add 130 ml (most 100 ml graduates are tall enough to be graduated to 130 ml) distilled water and mix well.

2. Add accurately 20 ml of the standard $UO_2(NO_3)_2$ solution. The solution should be added slowly and with agitation of the diluted cream.

3. Place the mixture in a water bath at 140-143.5 F (60-62 C) for 15 minutes. Agitate until the mixture reaches the temperature of the bath.

4. At the end of 15 minutes, remove from the water bath and filter, while hot. Collect at least 110 ml of the filtrate and cool to 77 F (25 C).

5. Pipette 100 ml of the filtrate into a 150 ml beaker.

6. Immerse the electrodes of the potentiometer in the solution and start the mechanical agitator. Record the pH of the filtrate at 77 F when equilibrium is established with the electrode system (usually 30 seconds to 1 minute).

7. Add 0.1N NaOH from the burette until the reaction of the solution is pH 8.40. Record the amount of 0.1N NaOH used.

8. Add 0.1N HCl from the burette until the reaction of the solution is pH 2.80. Record the amount of 0.1N HCl used.

9. Divide the HCl titer by the NaOH titer to obtain the buffer ratio.

Calculations

A determination of fat (Babcock) is necessary to place cream samples of different serum contents on the same basis. The pH of the filtrate and the buffer ratio must each be adjusted, since each is increased by increases in the serum content. Experiments have been performed and tables made from which the adjustment is read directly from the fat percentage (table 7).

After the pH and buffer ratio are corrected by use of table 7, these values are compared, using figure 105 to determine if the sample has been neutralized. An indication of the amount of neutralization accomplished in a sample may be obtained by use of figure 106.

From the statistical analysis, and a visual analysis of the data, any sample which when plotted on the graph does not fall within plus or minus 3 standard error (S_y) measurements, or below pH 5.05, must be considered as abnormal. The detection of neutralizer in cream depends upon this fact.

REFERENCE

Hankinson, D. J., and Anderson, E. O.: A New Method for the Detection of Neutralized Cream Including a Comparison with Other Methods, *Storrs Agr. Exper. Sta.*, Storrs, Conn. Bul. 234 (1939).

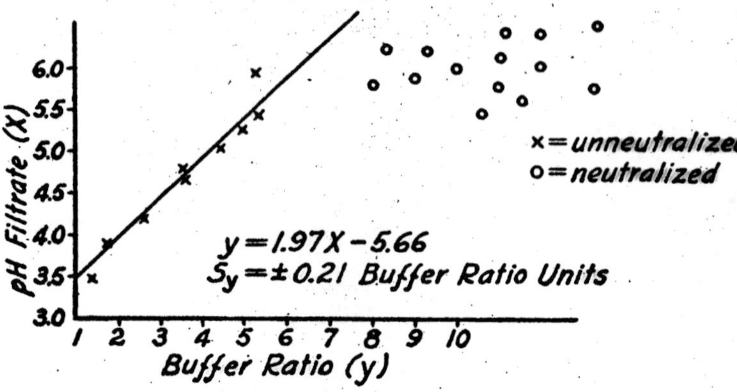

Figure 105.
Relation of buffer ratio to pH in neutralized and unneutralized cream.

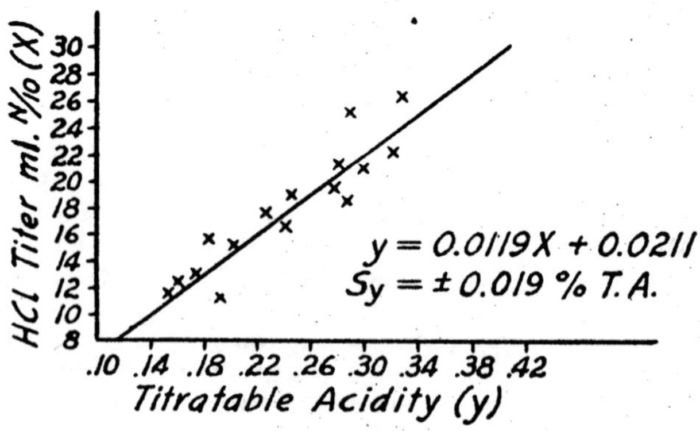

Figure 106.
Relation of titratable acidity to HCl titer to predict acidity of cream before neutralization.

TABLE 7

CORRECTION FACTORS WHICH MUST BE SUBTRACTED FROM THE EXPERIMENTAL VALUES TO CONVERT TO BASIS OF 40 PER CENT CREAM

	New Method				
Fat, %	pH, 25 C	HCl titer	Buffer ratio	Acidity	Buffer capacity 0.1 N NaOH, ml
0.	.75	10.82	1.25	.048	13.74
0.5	.74	10.73	1.24	.047	13.57
1.	.74	10.64	1.23	.047	13.40
1.5	.73	10.55	1.23	.046	13.23
2.	.72	10.46	1.22	.046	13.06
2.5	.72	10.37	1.21	.045	12.89
3.	.71	10.28	1.20	.044	12.71
3.5	.70	10.19	1.19	.044	12.54
4.	.69	10.10	1.19	.043	12.37
4.5	.69	10.01	1.18	.043	12.20
5.0	.68	9.92	1.17	.042	12.03
5.5	.67	9.82	1.16	.041	11.85
6.	.66	9.73	1.16	.041	11.68
6.5	.66	9.63	1.15	.040	11.51
7.	.65	9.53	1.15	.040	11.34
7.5	.64	9.44	1.14	.039	11.17
8.	.63	9.34	1.13	.038	10.99
8.5	.62	9.24	1.12	.038	10.82
9.	.62	9.14	1.12	.037	10.65
9.5	.61	9.05	1.11	.037	10.48
10.	.60	8.95	1.10	.036	10.31
10.5	.59	8.84	1.09	.035	10.14
11.	.58	8.73	1.08	.035	9.96
11.5	.58	8.63	1.08	.034	9.79
12.	.57	8.52	1.07	.034	9.62
12.5	.56	8.41	1.06	.033	9.45
13.	.55	8.30	1.05	.032	9.28
13.5	.54	8.19	1.04	.032	9.11
14.	.54	8.09	1.02	.031	8.93
14.5	.53	7.98	1.01	.031	8.76
15.	.52	7.87	1.00	.030	8.59
15.5	.51	7.75	1.01	.029	8.42
16.	.50	7.62	1.02	.029	8.24
16.5	.49	7.50	1.03	.028	8.07
17.	.48	7.37	1.04	.028	7.90
17.5	.48	7.25	1.05	.027	7.73
18.	.47	7.12	1.06	.026	7.56
18.5	.46	7.00	1.07	.026	7.38
19.0	.45	6.87	1.08	.025	7.21
19.5	.44	6.75	1.09	.025	7.04
20.0	.43	6.62	1.10	.024	6.87
20.5	.42	6.47	1.11	.023	6.70
21.	.41	6.32	1.11	.023	6.53
21.5	.40	6.16	1.12	.022	6.34
22.	.39	6.01	1.12	.022	6.18
22.5	.38	5.86	1.12	.021	6.01
23.	.37	5.71	1.12	.020	5.84
23.5	.36	5.56	1.12	.020	5.67
24.	.35	5.40	1.12	.019	5.50
24.5	.34	5.25	1.12	.019	5.32
25.	.33	5.10	1.12	.018	5.15

TABLE 7.—(*continued*)

	New Method				Buffer
Fat, %	pH, 25 C	HCl titer	Buffer ratio	Acidity	capacity 0.1 N NaOH, ml
25.5	.32	4.93	1.11	.017	4.98
26.	.31	4.76	1.10	.017	4.81
26.5	.30	4.59	1.08	.016	4.64
27.	.29	4.42	1.07	.016	4.46
27.5	.28	4.25	1.06	.015	4.29
28.0	.27	4.08	1.04	.014	4.12
28.5	.26	3.91	1.02	.014	3.95
29.	.25	3.74	.99	.013	3.78
29.5	.24	3.57	.97	.013	3.61
30.	.23	3.40	.95	.012	3.44
30.5	.22	3.24	.91	.011	3.26
31.	.21	3.07	.88	.011	3.09
31.5	.20	2.91	.84	.010	2.92
32.	.19	2.74	.81	.010	2.75
32.5	.18	2.58	.77	.009	2.58
33.	.17	2.41	.73	.008	2.40
33.5	.16	2.25	.69	.008	2.23
34.	.15	2.08	.66	.007	2.06
34.5	.14	1.92	.62	.007	1.89
35.	.13	1.75	.58	.006	1.72
35.5	.12	1.58	.52	.005	1.55
36.	.10	1.40	.46	.005	1.37
36.5	.09	1.23	.41	.004	1.20
37.	.08	1.05	.35	.004	1.03
37.5	.07	.88	.29	.003	.86
38.	.05	.70	.23	.002	.69
38.5	.04	.53	.17	.002	.52
39.	.03	.35	.12	.001	.34
39.5	.01	.18	.06	.001	.17
40.	.0	.0	.0	.0	.0
40.5	—.01	—.17	—.07	—.001	—.17
41.	—.03	—.34	—.13	—.001	—.34
41.5	—.04	—.51	—.20	—.002	—.52
42.	—.05	—.68	—.26	—.002	—.69
42.5	—.07	—.85	—.33	—.003	—.86
43.	—.08	—1.01	—.40	—.004	—1.03
43.5	—.09	—1.18	—.46	—.004	—1.20
44.	—.10	—1.35	—.53	—.005	—1.37
44.5	—.12	—1.52	—.59	—.005	—1.55
45.	—.13	—1.69	—.66	—.006	—1.72
45.5	—.15	—1.87	—.72	—.007	—1.89
46.	—.18	—2.05	—.77	—.007	—2.06
46.5	—.20	—2.23	—.83	—.008	—2.23
47.	—.22	—2.41	—.88	—.008	—2.40
47.5	—.25	—2.60	—.94	—.009	—2.58
48.	—.27	—2.78	—1.00	—.010	—2.75
48.5	—.29	—2.96	—1.05	—.010	—2.92
49.	—.31	—3.14	—1.11	—.011	—3.09
49.5	—.34	—3.32	—1.16	—.011	—3.26
50.	—.36	—3.50	—1.22	—.012	—3.44

pH Method

A determination of the pH of milk or cream suspected of being neutralized will detect only over-neutralized products. A hydrogen ion concentration of pH of 7.0 or over indicates over-neutralization. The pH may be determined either colorimetrically or by means of the potentiometer as described previously on pages 258 and 260.

PART IV

PHYSICAL CONTROL METHODS FOR DAIRY PRODUCTS

CHAPTER 23

SPECIFIC GRAVITY DETERMINATION OF MILK

The specific gravity of milk is usually determined for the purpose of calculating the solids content of the milk. When the specific gravity and fat content of a milk sample are known, it is possible to calculate the percentage of solids not fat. In routine practice specific gravity of milk is generally determined by the lactometer, with the milk being adjusted to 60 F (15.5 C). It has been pointed out by Sharp and Hart[1] that the lactometer reading will be affected by the lag in the change in physical state of the fat globules as the temperature is adjusted to that at which the specific gravity is determined, and that the lag in dissolving or solidifying of the fat globules is at a maximum at about 59 F (15 C). These investigators demonstrated that a sample of milk held cold for some time and then warmed to 59 F will have a greater specific gravity reading than the same milk held warm and then cooled to 59 F. The magnitude of the effect increases with the fat content. The determination of the specific gravity at 86 F (30 C) after previous warming to 113 F (45 C) for one-half minute is recommended as a method which will insure that the determination is made while the fat is in the liquid state. Directions for determining the specific gravity of milk by the conventional lactometer method and the procedure suggested by Sharp and Hart follow.

Lactometer Method

(Conventional)

Apparatus

1. Quevenne or New York Board of Health lactometer. (Figure 107).

NOTE: The lactometer should be calibrated by checking with a Bureau of Standards lactometer.

2. Cylinder or jar in which to float the lactometer.

3. Floating dairy thermometer, maximum graduation interval 2 degrees, accurate to 1 degree F, range 20 to 220 degrees F, not more than 40 degrees F per inch.

NOTE: A floating dairy thermometer made especially for this type of work has facilitated rapid reading of temperature. This thermometer is about 25 cm in length, has a scale graduated in single degrees with a range 30 to 90 degrees F. The scale is 8 cm long, starting at a point about 15 cm from the bottom of the mercury bulb.

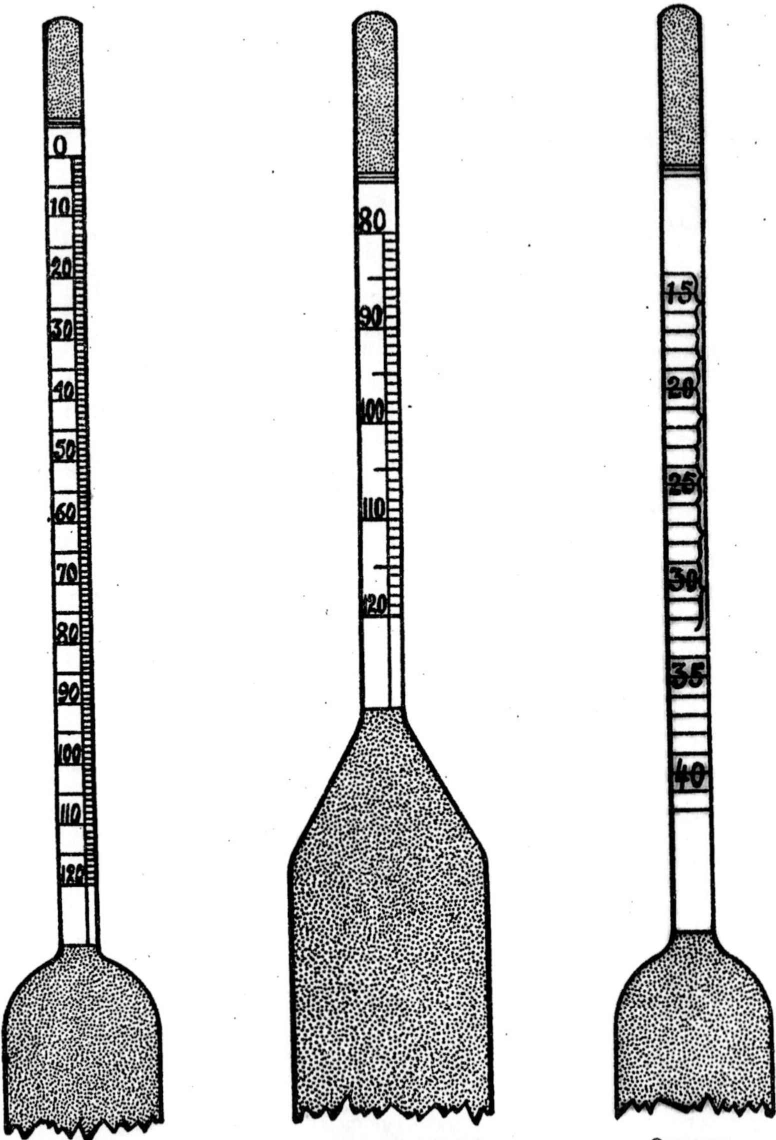

N. Y. Board of Health (regular type).

N. Y. Board of Health (special scale for accurate reading).

Quevenne

Figure 107.
Lactometer scales.

Procedure

1. Adjust the milk to as near 60 F as is consistent with reasonable speed of testing; never test below 50 F or above 70 F. Mix gently by pouring several times from one vessel to another, being careful to prevent incorporation of air.

NOTE: Specific gravity determination should not be attempted for at least one hour after the milk is drawn from the cow because of the presence of dissolved gases in freshly drawn milk.

2. Pour sufficient milk into the jar so that, with the lactometer floating, the milk level will be near to overflowing the jar.

3. Center the lactometer in the milk so the instrument floats freely from the sides of the jar and allow it to assume a constant level. Do not allow more than 30 seconds to elapse before reading.

4. Simultaneously read the temperature of the milk and the lactometer scale at the top of the milk meniscus.

5. Record the lactometer reading and the temperature of the milk and, when necessary, correct the reading to 60 F according to the following directions:

Quevenne Scale

Add 0.1 degree for each degree of Fahrenheit temperature above 60 F.

Subtract 0.1 degree for each degree of Fahrenheit temperature below 60 F.

New York Board of Health Scale

Add 0.3 degree for each degree of Fahrenheit temperature above 60 F.

Subtract 0.3 degree for each degree of Fahrenheit temperature below 60 F.

NOTE: These corrections apply only when the reading is made at temperatures not lower than 50 F and not higher than 70 F.

Calculation of Milk Solids from Lactometer Reading

The following formula yields approximately correct results for determining the percentage of total solids.

Per cent T. S. =

$$\frac{\text{Corrected Quevenne reading}}{4} + (1.2 \times \text{per cent of fat})$$

NOTE: New York Board of Health lactometer degrees may be converted to Quevenne degrees by multiplying by the factor 0.29.

The formula for the calculation of solids not fat is:

Per cent S. N. F. =

$$\frac{\text{Corrected Quevenne reading}}{4} + (0.2 \times \text{per cent of fat})$$

EXAMPLE: If the Quevenne lactometer reading is 32.5 at 55 F and the fat test is 3.9 per cent, then the corrected lactometer reading at 60 F is 32.5 minus 0.5, or 32.

$$\text{Per cent S. N. F.} = \frac{32}{4} + (0.2 \times 3.9)$$

Per cent S. N. F. = 8 + 0.78
Per cent S. N. F. = 8.78
Per cent total solids = S. N. F. + Fat
Per cent total solids = 8.78 + 3.9
Per cent total solids = 12.68

Lactometer Method

(Sharp and Hart Modification[1])

Apparatus

1. Equipment as listed on page 305.
2. Water bath facilities for heating and tempering milk samples as directed under procedure.

Procedure

1. Thoroughly mix the milk by pouring several times from one vessel to another, avoiding incorporation of air as much as possible.
2. Warm the milk to 113 F (45 C), hold at this temperature for 30 seconds, and cool to 86 F (30 C).
3. Float the lactometer in the jar with the well mixed milk so that the milk level will be near to overflowing the jar.
4. Make certain that the temperature of the milk is at 86 F (30 C) and read the lactometer scale at the top of the milk meniscus. Record the lactometer reading.
5. Calculate the total solids from the fat content and the specific gravity according to the following equation:

(according to correction by Herrington[2])

$$\text{Total solids} = 1.2537\ \text{Fat} + \frac{268.0\ (Q + 3)}{Q + 1000}$$

EXAMPLE: If Quevenne lactometer reading at 86 F (30 C) is 30, specific gravity of sample is 1.030, and the milk fat test is 3.90 per cent, then:

$$\text{Per cent total solids} = 1.2537 \times 3.9 + \frac{268.0\ (30+3)}{30+1000}$$

$$\text{Per cent total solids} = 4.889 + \frac{8844}{1030}$$

Per cent total solids = 4.889 + 8.586, or 13.475
S.N.F. = Percentage total solids − Percentage fat, or
S.N.F. = 13.475 − 3.90, or 9.575

REFERENCES

1. Sharp, P. F. and Hart, R. G. The Influence of the Physical State of the Fat on the Calculation of Solids from the Specific Gravity of Milk. *J. Dairy Sci., 19*:683 (1936).

2. Herrington, B. L. The Calculation of Total Solids by Means of the Sharp and Hart Equation. *J. Dairy Sci., 29*:87 (1946).

The Westphal Balance

Apparatus

1. Westphal balance.
2. Plummet weighing 15 gm and displacing 5 gm of distilled water at 60 F (15.6 C).
3. Set of weights or riders.
4. Glass cylinders.

Procedure

1. With the balance in readiness, adjust the milk sample to as near 60 F (15.6 C) as possible. Mix carefully by pouring several times from one vessel to another, then transfer the sample to the cylinder by pouring down the side in such a manner as to prevent the incorporation of excess air.
2. Suspend the plummet in the milk and add the various size riders to the beam until the buoyant force of the milk is conterbalanced, recording the temperature of the milk at the time the beam is balanced.

Interpretation

When the plummet is suspended in distilled water at 60 F (15.6 C), it is buoyed up by a force equal to 5 gm and this is counterbalanced by hanging the 5-gm rider on the hook with the plummet. In the case of liquids heavier than water, such as milk, the 5-gm rider indicates tenths, according to its position on the ten-notched beam, and the riders weighing 0.5, 0.05 and 0.005 gm indicate hundreths, thousandths, and ten thousandths of specific gravity according to their relative positions on the beam at the time the buoyant force of the liquid is counterbalanced.

The following example will indicate how the specific gravity is determined:

5.00 gram rider on notch 10:	$5 \times 1 =$	5.000
0.50 gram rider on notch 3:	$0.5 \times 0.3 =$	.150
0.05 gram rider on notch 1:	$0.05 \times 0.1 =$	.005
0.005 gram rider on notch 8:	$0.005 \times 0.8 =$	.004
		5.159, grams displaced by the plummet

Then, since the plummet displaces 5 gm. of distilled water and 5.159 gm of milk, $\frac{5.159}{5}$, or 1.0318, is the specific gravity of the milk sample.

NOTE: It is customary to read the specific gravity directly from the beam by observing the positions of the respective riders. Since the 5-gram rider indicates tenth, the 0.50-gram rider hundredths, the 0.05-gram rider thousandths, and the 0.005-gram rider ten thousandths, their positions on the beam as indicated above would signify a specific gravity of 1.0318.

Detecting Adulterated Milk—Watering, Skimming

Water added to milk lowers the specific gravity because water is lighter than milk. Added water decreases the percentage of all the milk solids in the same proportion. Removing milk fat from milk by skimming increases the specific gravity and the solids-not-fat content of the milk. Skim milk added to milk lowers the fat content but raises the solids not fat. Addition of water and skim milk lowers the fat content but may leave the solids-not-fat at a normal level. Low specific gravity may indicate added water but does not constitute definite proof of such adulteration. It is essential to supplement the specific gravity determination with a fat test when attempting to determine whether or not milk has been watered or skimmed. All suspicious results should be verified by obtaining representative samples from the original source. When such verification is not possible, added water may be determined by the application of the cryoscopic method, the acetic serum method, the sour serum method, or the copper serum method. (See Chapter 24).

CHAPTER 24

DETERMINATION OF ADDED WATER

Cryoscopic Method

Apparatus

1. Hortvet Cryoscope. (Figure 108.)

2. Control thermometer, a solid-stem instrument approximately 58 cm in length and having a scale range of +20 C to –30 C with scale graduations accurate to within 0.10 degree C.

3. Standard thermometer, designed especially for testing milk, a solid-stem instrument measuring a total length of 58 cm, with a scale portion measuring about 30 cm. The total scale range is 3 degrees C, from +1 degree C to –2 degrees C, and each degree division is subdivided into tenths and hundreths. The length of a degree division approximates one decimeter, thus making the smallest subdivisions of such magnitudes as to enable easy observation and reading estimated to 0.001 degree C.

4. Pressure and suctior pump, for forcing dry air into the apparatus at a suitable rate and conducting the mixed vapors out through the base of the drain tube into the sink.

Reagents

1. Sulfuric acid (H_2SO_4).

2. Ether.

3. Ethyl alcohol.

4. Recently boiled distilled water.

5. Sucrose solution—A: 7 gm of pure sucrose dissolved in distilled water and made up to 100 ml at 20 C (68 F).

6. Sucrose solution—B: 10 gm of pure sucrose dissolved in distilled water and made up to 100 ml at 20 C (68 F).

NOTE—A sample of pure sucrose may be obtained from the Director of the Bureau of Standards, Department of Commerce, Washington, D. C.

Standardization of the Thermometer

Make three freezing point determinations as directed under "Procedure" on each of the following:

a Recently boiled distilled water.

b Sucrose solution A.

c Sucrose solution B.

Tabulate the results in the following form:

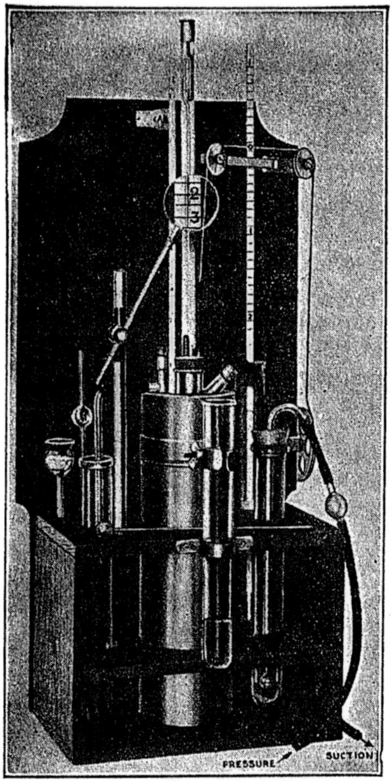

Courtesy of the American Public Health Association

Figure 108.

The Hortvet cryoscope.

Express the results as degrees freezing-point depression below the average of the observed freezing points obtained on the sample of pure water (±W), which may be above (+) or below (−) the 0-mark on the scale.

TABLE 8

Freezing point observations	Pure water	7-Gm Sucrose Solution (Solution A)		10-Gm. Sucrose Solution (Solution B)	
		Observed freezing point (—S)	Freezing point depression S—W (algebraic)	Observed freezing point (—S)	Freezing point depression S—W (algebraic)
1st					
2nd					
3rd					
Averages	± W	xxxxxx		xxxxxx	

Obtain each freezing-point depression of the sucrose solutions by the algebraic subtraction of the average of the freezing-point readings of pure water (±W) from each observed freezing point.

NOTE: Omit results which are in marked disagreement with other results obtained by carefully following instructions.

Apply the average of the freezing-point depressions obtained on the standard sucrose solutions for the purpose of correcting the thermometer readings obtained on samples of milk in the manner illustrated in the tables accompanying Figure 109.

Procedure

Freezing Point Determination of Water and Sucrose Solution

1. Insert funnel tube into vertical portion of the T-tube at one side of the apparatus and pour in 400 ml of ether previously cooled to 50 F (10 C) or lower. Close the vertical tube by means of a small cork and connect the pressure pump to the inlet tube of the air-drying attachment.

2. Adjust the pump so as to pass air through the apparatus at a moderate rate, as may be judged by the agitation of the sulphuric acid in the drying tube. Continuous vaporization of the ether will cause a lowering of the temperature in the flask from ordinary room temperature to 0 C in from 5 to 10 minutes. Continue lowering of the temperature until the control thermometer registers near –3 C.

3. At this stage, by lowering the gage tube into the ether bath then closing the top by means of the forefinger and raising to a suitable height, an estimate can be made as to the quantity of ether necessary to pour in for the purpose of restoring the 400 ml volume. When the volume of ether has been adjusted to 400 ml, an additional 10 to 15 ml is sufficient on an average for each succeeding determination.

4. Pour into the freezing tube sufficient distilled water (30-35 ml), boiled and cooled to 10 C or lower, to submerge the thermometer bulb.

Insert the thermometer, together with the stirrer, and lower the test tube into the larger tube. A small quantity of alcohol, sufficient to fill the lower space between the two test tubes, will serve to complete the conduction medium between the freezing bath and the liquid to be tested.

5. Keep the stirrer in steady up and down motion at a rate of approximately one stroke each 1 or 2 seconds, or even at a slower rate, provided the cooling proceeds satisfactorily. Maintain a passage of air through the apparatus until the temperature of the cooling bath reaches –2.5 C at which time the top of the mercury column in the thermometer usually recedes to a position near the freezing point of water.

6. Maintain the temperature of the cooling bath at –2.5 C and continue the manipulation of the stirrer until a super-cooling of the sample of 1.0–1.2 degrees C is observed. As a rule, at this time the liquid will begin to freeze, as may be noted by the rapid rise of the mercury. Manipulate the stirrer slowly and carefully three or four times as the mercury column approaches its highest point. By means of a suitable light-weight cork mallet, tap the upper end of the thermometer cautiously, a number of times until the top of the mercury column remains stationary for at least one minute. Observe the exact reading on the thermometer scale, taking necessary precautions to avoid parallax, and estimate to 0.001 degree C.

7. When the observation has been satisfactorily completed, make a duplicate determination; then remove the thermometer and stirrer and empty the water from the freezing tube.

Two Bureau of Standards tested thermometers gave intervals of 0.199° and 0.200°, respectively, between the freezing-point depression readings of the two sucrose solutions. One thermometer (illustrated) gave freezing-point depression —0.422° (+0.079 and —0.343) and —0.621° (+0.079 and —0.542), respectively, for the two sucrose solutions, while the other gave —0.422° and —0.622°, respectively.

Procedure

Freezing Point Determination of Milk

NOTE: Freezing point determination should not be made on samples of milk that show a titratable acidity of more than 0.18 per cent.

Sample Calculation:

Laboratory Thermometer No. 2

Water	7 grams sucrose to 100 ml	10 grams sucrose to 100 ml
Av. + 0.056°	-0.425°	-0.621°

Interval = 0.196
0.196 equiv. 0.199
Correction factor = 1.015

Laboratory Thermometer No. 24

Water	7 grams 100 ml	10 grams sucrose to 100 ml
Av. 0.000	-0.420°	-0.625°

Interval = 0.205
0.205 equiv. 0.199
Correction factor = 0.971

Example:

Laboratory thermometer No. 24
F.-pt. depression sample milk = 0.548
(0.548 − 0.420) 0.971 = 0.124
True F. pt. = 0.422 ± 0.124 = 0.546° below zero C.

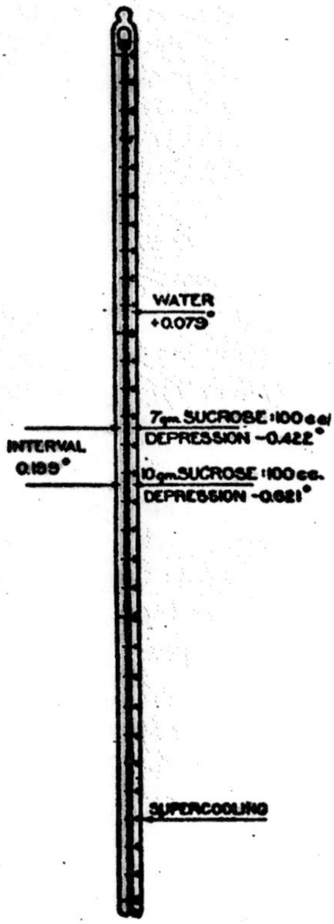

Figure 109

Bureau of Standards Tested Thermometer and Sample Calculation. Courtesy of the American Public Health Association.

1. Rinse the tube with about 25 ml of the sample of milk cooled to 10 C (50 F) or lower; measure into the tube 30-35 ml of milk, or enough to submerge the thermometer bulb, and insert the tube into the apparatus.

2. Maintain the temperature of the cooling bath at 2.5 degrees C below the probable freezing point of the sample. Make the determination on the milk, following the same procedure as that employed in determining the freezing point of water.

NOTE: It is usually necessary to start the freezing action in the milk by inserting the freezing starter (which has been kept in contact with ice for several minutes, and in the open end of which has been wedged a fragment of ice) at the time when the mercury column has receded 1.0 to 1.2 degrees C below the probable freezing point. A rapid rise of the mercury results almost immediately.

3. Remove the starter and manipulate the stirrer slowly and carefully two or three times while the mercury approaches its highest point. Complete the adjustment of the mercury column in the same manner as in the preceding determination; then, avoiding parallax, observe the exact reading on the thermometer scale and estimate to 0.001 degree C.

4. The algebraic difference between the average of readings obtained on the water and the reading obtained on the sample of milk represents the freezing point depression of the milk. To determine the true freezing point (T′) of the milk, subtract from the freezing point depression, the freezing point depression of 7 per cent sucrose solution as determined by the laboratory thermometer. Multiply the difference by the correction factor for the thermometer. Add to the product 0.422 (freezing point of 7 per cent sucrose solution by Bureau of Standards thermometer). See example page 315.

5. Ascertain the percentage of added water (W) corresponding to the determined freezing point from Table, pages 884-885, *Methods of Analysis*, A. O. A. C., (6th Edition), or by means of the following formula:

$$W = \frac{100\ (T - T')}{T}$$, in which

T = the average freezing point of normal milk (–0.550 C), and

T′ = the true freezing point on a given sample.

A tolerance of 3 per cent may be allowed on results for added water determined on the basis of an average freezing point depression of —0.550 C. Owing to the narrow variations found among market milks of genuine character it is not necessary to deduct the tolerance figure from results showing added water in excess of 3 per cent.

Acetic Serum Method

Apparatus

1. Supply of beakers, funnels, watch glasses.
2. Water bath.
3. Constant temperature bath to temper sample for refractometer reading.
4. Zeiss immersion refractometer.
5. Flat bottom platinum dish.
6. Electric muffle, with pyrometer attached.
7. Chemical balance.
8. Centigrade thermometer graduated in tenths of a degree. Range 0 to 50 C.

Reagents

1. Acetic acid—25 per cent, Sp. gr. 1.035,

Procedure

1. To 100 ml of the well mixed milk sample, measured at 20 C into a beaker, add 2 ml of the 25 per cent acetic acid. Cover the beaker with a watch glass and place in a water bath at 70 C for 20 minutes, then in ice water for 10 minutes, and separate the curd from the serum by rapid filtration through a small filter.

2. Transfer a portion of the clear serum to a refractometer beaker, place in the constant temperature bath, and take the refractometer reading when the temperature of the serum has been brought to exactly 20 C, as determined by a thermometer graduated in tenths of a degree. A refractometer reading below 39 indicates added water; between 39 and 40, the addition of water is suspected. If the reading is 40 or below, determine the ash in the serum as follows:

Ash Determination

1. Transfer 25 ml of the serum at 20 C to a weighed flat bottom platinum dish and evaporate to dryness on a water bath.

2. Heat over a low flame (to avoid spattering) until the contents are thoroughly charred, place the dish in an electric muffle, preferably with pyrometer attached, and ignite to a white ash at a temperature not greater than 500 C.

3. Cool and weigh. Express the result as grams per 100 ml. A result below 0.715 gm per 100 ml indicates added water.

The acetic serum ash multiplied by the factor 1.021 equals the sour serum ash (dilution of the acetic serum being 2 per cent).

Sour Serum Method

Apparatus

1. Same as listed for acetic serum method. (See page 314.)

Procedure

1. Measure 100 ml of the well mixed sample of milk into a beaker at 20 C and allow it to sour until it wheys.

2. Clarify the serum by filtration and determine the refractometer reading at 20 C as directed under procedure (2) on page 317.

A refractometer reading below 38.3 indicates added water.

Ash Determination

1. Transfer 25 ml of the serum at 20 C to a weighed flat bottom platinum dish and proceed with the ash determination as directed on page 314.

A result below 0.730 gm per 100 ml indicates added water.

Copper Serum Method

Apparatus

Same as listed for acetic serum method on page 317.

Reagents

1. Copper sulphate solution, 72.5 gm of $CuSO_4.5H_2O$ per liter, adjusted if necessary to read 36 at 20 C on the scale of the Zeiss immersion refractometer, or to a specific gravity of 1.0443 at 20 C.

Procedure

1. To 1 volume of the copper sulphate solution, add 4 volumes of milk. Shake well and filter.

2. Determine the refractometer reading of the clear serum at 20 C directed under Procedure (2) on page 317. A refractometer reading below 36 indicates added water.

When the refractometer reading is 36 or below, determine the ash of the sour serum as directed above, or of the acetic serum as directed on page 317.

NOTE: These methods are particularly useful in detecting the extent to which the first milk through equipment after sterilization may have been accidentally diluted with water. In such cases, comparisons should always be made between the refractive indexes of the milk in the raw storage supply and the same milk after it has passed through the equipment.

CHAPTER 25

SEDIMENT TESTS

Sediment tests on raw bulk milk and bottled milk ready for consumer use reveal the extent to which visible insoluble matter has been prevented from gaining entrance to the milk and the extent to which such material has been removed from milk by single service strainers on the dairy farm or by filtration or clarification at milk processing plants. High concentration of visible sediment in milk does not necessarily indicate excessive bacterial contamination because much of this insoluble foreign matter may be relatively free of bacteria.

Despite the limitations as to the interpretation that may be applied when visible sediment is or is not detected in milk the sediment test presents a simple, rapid, and quantitative means of indicating the cleanliness of milk with respect to visible dirt. One pint of milk is filtered through a white cotton lintine disc 1⅛ inches in diameter, and the disc containing the visible sediment makes an excellent exhibit with which to impress upon the dairyman or milk plant personnel the need for careful handling of milk to keep it clean.

Procedures for testing for sediment and extraneous matter in dairy products other than fresh fluid milk are somewhat more complicated because of the necessity of adjusting these products to such a condition that they may be filtered through the cotton lintine discs.

Milk as Received from Farm

Apparatus

1. Sediment tester filtering surface 1 inch diameter.

a. Pressure gravity or vacuum type for mixed sample (M-S) methods, see figure 110.

b. Suction (sampling), discharge (filtering) type for off-bottom (0-B) sample method, see figure 112.

2. White lintine cotton discs, 1¼ inch diameter, exposed filtration area 1⅛ inch diameter.
3. Can agitator.
4. Pint dipper for sampling from milk can or weigh vat.
5. Photograph of sediment disc ratings of A.P.H.A., see figure 111.

NOTE: Do not use photographs which have become faded, stained, soiled, or otherwise damaged.

Preparation of Standard Sediment Discs[1]

Since sediment encountered in different localities may vary in composition, specific gravity, color and other physical characteristics thus giving rise to variations in appearance of discs at the same concentration of visible sediment it is recommended[1] that standard sediment discs be prepared.

Make a uniform mixture of oven dried (100 C or 212 F) materials which meet the following screening specifications:

	Per cent
Cow manure, 40 mesh	53
Cow manure, 20 mesh, retained on 40 mesh	2
Garden soil, 40 mesh	27
Charcoal, 40 mesh	14
Charcoal, 20 mesh, retained on 40 mesh	4

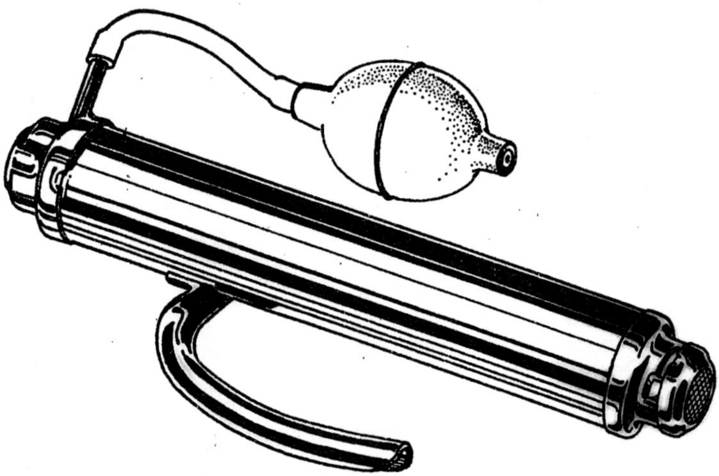

Figure 110.
Superior Sediment tester. Courtesy of Superior Manufacturing Co.

Moisten 2 gm. of the above mixture with 4 ml of 1 per cent aerosol solution or other suitable wetting agent, and 46 ml of 0.75 per cent gum solution such as gum arabic, and then make up to 100 ml with a 50 per cent by weight sucrose solution. Mix thoroughly, transfer to a flask a 10 ml portion (200 mg. of standard sediment) with a large-tip graduated pipette and make up to a liter with 50 per cent by weight sucrose solution. When thoroughly mixed, each milliliter contains 0.2 mg. of sediment. Add definite volumes of the sediment mixture to ¾ pt. of filtered milk. Mix thoroughly and pass the mixture through a standard sediment disc with a filtering area measuring

1⅛ inch in diameter. Wash the container promptly with 1¼ pint of filtered milk. For a permanent record, mount and spray the disc with 40 per cent formaldehyde, or with a solution containing 2.5 gm menthol and 2.5 gm thymol made up to 100 ml with 95 per cent alcohol. Following the above method, prepare a series of discs containing the sediment remaining from 0.0, 0.2, 0.5, 1.0, 2.0, 3.0, 4.0, 5.0, 6.0, 7.0, 8.0, 9.0, 10.0, 12.0, and 14.0 mg of the standard mixture. Identify each disc on the permanent record with the amount of sediment in milligrams used to prepare the pad. (See figure 111).

For the purpose of comparison with actual tests, the entire set of discs may be used. Usually it will be more convenient to mount 2, 3, or 4 of the prepared discs selected so as to classify the milk with respect to its sediment content in accordance with the investigation being made. Classification terms such as "Satisfactory," "Fair," "Unsatisfactory," and "Very Unsatisfactory," or "Reject Grade" may be used. To protect sets of discs which are to be used repeatedly, mount them under glass, transparent plastic sheets or other suitable materials. The discs prepared from the sample tested should be graded to the nearest standard sediment disc regardless of whether the actual amount of sediment is above or below the standard. Gross pieces of material, such as whole flies, large chunks of manure, etc., should be disregarded when grading the pads but should be plainly listed in the record describing the pad on which they occur. Where grading charts are prepared and where reports are made, the chart and report must indicate whether the mixed sample or the off-bottom method has been used.

Procedure

The procedure herein given is to be adhered to unless local municipal, or state authorities insist on some variation or modification; and if a deviation is practiced, the method used should be stated in the report.

1. a. Mixed sample (M-S) method:

Take a one pint sample from thoroughly agitated milk in a can or from milk freshly dumped into a weigh vat in which case no further agitation is necessary. Filter the milk through a properly adjusted, firm cotton lintine disc held in the sediment tester so that a filtration area of 1⅛ inch diameter is exposed.

NOTE: If the fat is not uniformly distributed throughout the mass it will tend to clog the sediment discs. The colder the milk the more difficult it will be to filter it through the cotton discs. A temperature of from 70-80 F (21.1-26-7 C) facilitates filtration, and in some localities during part of the winter warming of the sample may be necessary.

1. b. Off-bottom (O-B) method:

Insert the sediment tester into an unstirred can of milk being careful not to unduly agitate the milk in the act. Sample a pint of milk

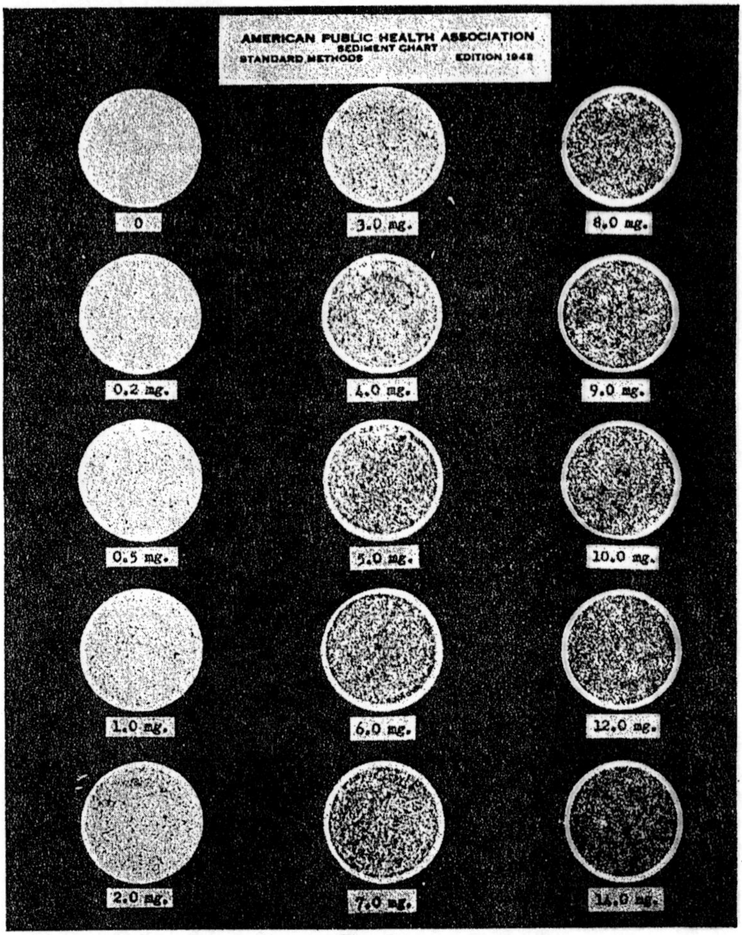

Fig. 111

Complete range of sediment obtained as described under Preparation of Standard Sediment Discs, page 320. This reproduction is not satisfactory for making comparisons. Photoprint may be obtained from the American Public Health Association, 1790 Broadway, New York 19, N. Y.

at a level not more than ¼ inch above the can bottom by simultaneously raising the plunger and pushing the head of the tester once across the bottom of the can.

NOTE: Any agitation just prior to sampling or sampling from cans with concave, convex, or dented bottoms will cause variability of results by the O-B method. Likewise samples from cans that have stood over night in farmers' cooling tanks will show comparatively more sediment than samples from morning's milk.

Figure 112. Carkuff Sediment Tester

After inserting the tester into the can of milk, the handle is pulled to draw the milk into the tester. The handle is then pushed to force the milk through the sediment disc.

With a downward movement of the plunger force the milk through the cotton lintine disc with the standard 1⅛ inch diameter, filter area exposed.

NOTE: The necessity for keeping foreign material, other than that contained in the sample out of the sediment tester before or after testing is self evident. It is good practice to rinse the tester with water after a sample of milk containing appreciable sediment in run through it.

2. Compare the sediment discs with the prepared standard sediment discs or with those shown in figure 111 on page 322 and

record the sediment score or cleanliness rating for each individual dairy indicating whether the M-S or O-B method was used and whether or not the sediment disc was graded in the wet or dry condition.

NOTE: It is customary to post the discs on a bulletin board, to return them to the milk producer, or to file them for the record. For posting on a bulletin board or filing, the wet discs placed on a cardboard sheet will adhere tightly by reason of the bonding power of the milk as it dries. If the discs are to be returned to the producer they may be dried and placed in a transparent glassine or cellophane envelope before mailing in a conventional envelope. To preserve the discs, spray with a solution of 5 to 10 parts 40 per cent formaldehyde to 95 to 90 parts of water or a disinfectant of equal effectiveness.

CAUTION: Be certain that the discs have adequate protection from dust while they are drying.

Interpretation

The presence of appreciable sediment in milk indicates careless or insanitary dairy farm practice. It should be borne in mind, however, that lack of sediment is not always indicative of ideal conditions, since visible sediment may be readily removed by straining at the dairy farm. Milk that has been divested of visible sediment by straining retains essentially the bacterial contamination incidental to the original sediment.

Milk After Processing (in Final Consumer Package)

The apparatus and procedure are the same as for raw bulk milk, except that mixing is accomplished by inverting the bottles several times. It is necessary to rinse the tester between the tests.

Interpretation

In this case the problem is not to measure the amount of visible sediment that may be present but rather to detect the presence of any sediment. Obviously, bottled milk should contain no visible sediment and samples showing any degree of visible sediment should be discriminated against accordingly.

Another simple procedure for checking sediment of bottled milk is to examine the bottom of the bottle for specks of sediment after the bottle of milk is allowed to stand undisturbed for 12 to 24 hours, so as to allow any sediment to settle to the bottom.

Fresh Fluid Cream (in Final Consumer Package)

Add to one pint of well mixed fresh cream in a suitable size vessel, one pint of filtered hot water, 158 to 194 F (70 to 90 C), and stir until the mixture reaches 113 to 140 F (45 to 60 C) and filter through a cotton lintine disc in a sediment tester which has been warmed to approximately 140 F (60 C) by passing hot filtered water through it. Make interpretation as directed above for milk after processing.

Sweet Cream (as Received at Creamery or Processing Plant)[2]

Place one quart of thoroughly mixed cream in a suitable container and warm to 185-194 F (85-90 C) with constant stirring. Place two ounces of the sample in a 400 ml beaker and add eight ounces of filtered water (distilled preferred) at 180 F (82 C) and stir thoroughly with a thermometer.

Pour the warm mixture into a sediment tester which has previously been thoroughly rinsed with warm filtered water. Force the mixture through the cotton lintine disc then rinse the sediment tester by forcing through at least four ounces of filtered, hot water (180 F) before removing the sediment disc. Score the sediment on the disc by comparing with an original photoprint obtained from American Butter Institute. (Figure 114).

Dry Whole Milk[3]

Apparatus

1. Balance, Torsion or similar type, approximately 500 gm capacity, sensitivity 0.1 gm.
2. Beakers, 500 ml, 1000 ml.
3. Electric mixer, malted milk type.
4. White cotton lintine sediment discs, 1¼ in. diameter.
5. Pressure type sediment tester.
6. Photoprint of A.D.M.I. Sediment Standards for Dry Milks (figure 113).
7. Tumblers, glass, 20 fl. oz. capacity having a flat, smooth, clear glass bottom. The inside diameter of the tumbler bottom should be between 45 and 48 mm to correspond with the diameter of the photo print disc of the standard chart. (Figure 113). The tumbler should be approximately straight sided. (Tumblers of this specified type are required for the Tumbler Method only.)

Procedure (Disc Method)

(Spray process samples can generally be tested satisfactorily by this method).

1. Place 500 ml sediment-free (distilled or filtered) water in a 1000 ml beaker, add 65 gm of sample and mix thoroughly with electric mixer until the product is dissolved and uniformly dispersed.
2. Pass the entire solution through the cotton disc in the sediment tester.

NOTE: If the reliquefied sample is allowed to stand before filtering, stir it vigorously just before pouring into the tester. Do not allow samples to stand uncovered.

3. Rinse the mixing container and tester with approximately 50 ml sediment-free water, passing the rinse through the disc.

4. Remove the filter disc, place it on a flat surface (sediment test cards manufactured for the purpose are convenient), and dry at low temperature in a dust-free atmosphere.

5. Compare the dry disc under uniform indirect or north light with the photoprint, disc method (figure 113). Any test falling between two standard discs should be assigned the lower disc number.

NOTE: Sediment testers often become worn so that reliquefied milk may pass around instead of through the sediment disc. To determine the efficiency of a tester, collect the filtered sample in a clean glass receiving vessel; cover the container and allow it to stand for several hours; then observe the bottom of the receiving vessel. The presence of sediment on the bottom indicates that the sediment tester has failed to filter the sample completely.

Procedure (Tumbler Method)

(Vacuum drum and roller proves samples usually require this method.)

1. Reliquefy 32.5 gm of sample in 250 ml sediment-free water with electric mixer.

2. Cover the sample and allow it to stand in the 20 oz. tumbler for 5 hours.

3. Compare the amount of sediment on the bottom of the tumbler with photoprint, tumbler method, (figure 113).

Nonfat Dry Milk Solids

(Scorched Particle Tests)

Apparatus

1. Waring Blendor
2. Aspirator or pressure type tester
3. Standard cotton discs
4. Scorched Particle Standard disc test card. (Test cards may be secured from American Dry Milk Institute, 221 North LaSalle Street, Chicago 1, Ill., or from Dairy Branch, Standards Section, U. S. Department of Agriculture, Washington 25, D. C.)
5. Graduated cylinder, beakers, stirring rod.

Procedure for Spray Process

(Water Disc Method)

1. Measure 250 ml sediment-free water into Waring Blendor jar, turn on mixer and add 25 gm of well-mixed sample.

2. Add 0.5 ml diglycol laurate S and mix for one minute.

3. Filter entire solution through standard cotton disc in tester.

NOTE: If reliquified sample is allowed to stand before filtering, stir vigorously just before pouring into tester. Do not allow samples to remain uncovered.

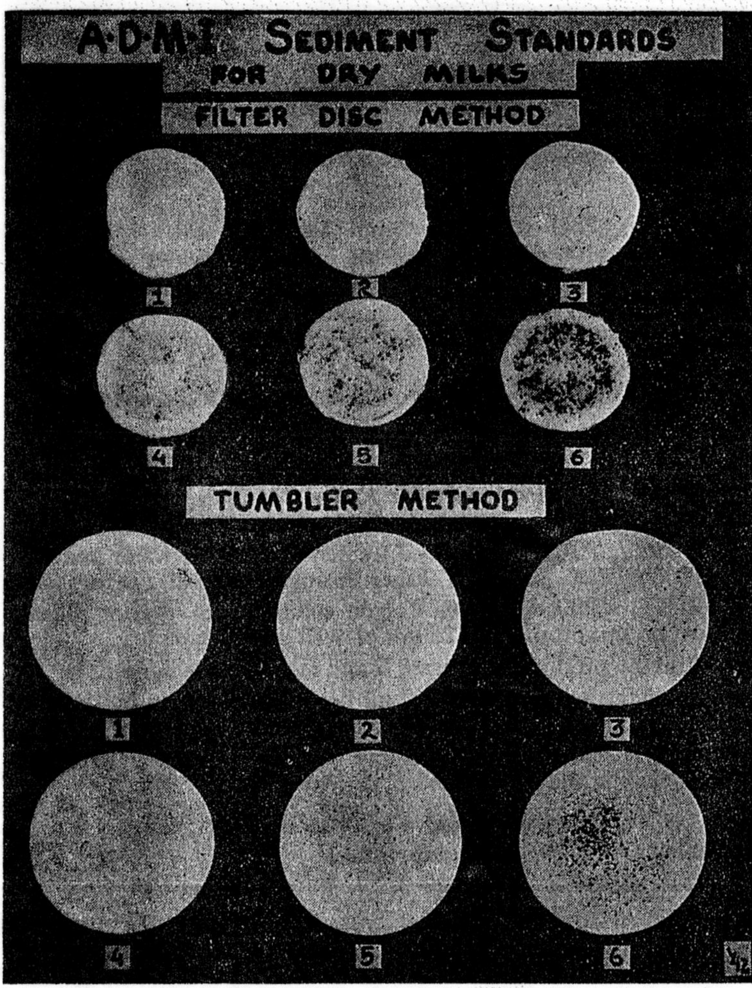

Figure 113.

American Dry Milk Institute sediment standards for dry milk. This reproduction is not satisfactory for making comparisons. Photoprint may be obtained from the American Dry Milk Institute, 221 N. La Salle St., Chicago 1, Ill.

4. Rinse mixing container and tester with 50 ml sediment-free water and pass through disc.

5. Remove filter disc, place in disc test card and dry at 86 to 104 F (30 to 40 C) in a dust-free atmosphere.

6. Compare the dry disc placed on a table and viewed from directly above with the standard test card under uniform indirect or

north light. Any test falling between two standard discs should be assigned the higher disc's letter. (Example: Disc showing more scorched particles than standard disc A but less than B should be assigned a B, and similarly for the other discs.)

Procedure for Roller Process
(Sodium Citrate Method)

1. Dissolve 100 gm sodium citrate in distilled water and make up to 1000 ml (10% solution). Filter through cotton disc before using.

2. Measure 200 ml of 10% sodium citrate solution at 176 to 194 F (80 to 90 C) into Waring Blendor jar, turn on mixer and add 25 gm of well-mixed sample.

3. Add 0.5 ml diglycol laurate S and mix for 30 seconds.

4. Proceed as directed for Spray Process, beginning with paragraph 3.

Procedure for Roller Process
(Calgon Method)

1. Dissolve 20 gm unadjusted Calgon in distilled water and make up to 1000 ml (2% solution). Filter through cotton disc before using.

2. Measure 250 ml of 2% Calgon solution into Waring Blendor jar, turn on mixer and add 25 gm of well-mixed sample.

3. Add 0.5 ml diglycol laurate S and mix for one minute.

4. Transfer solution to a 600 ml beaker, bring to a boil and immediately filter through standard cotton disc in tester.

5. Proceed as directed for Spray Process, beginning with paragraph 4, rinsing the beaker, and filter with 25 ml of water.

Sweetened Condensed Milk

Apparatus

Same as for milk. (See page 319.) Pint container.

Procedure

1. Weigh out 90 gm of the condensed milk in a pint container.

2. Fill container about three-fourths full with warm, filtered or distilled water 122 F (50 C) and stir until the condensed milk is in complete solution, then completely fill the container with the warm filtered water.

3. Proceed as directed under milk (see page 321), using a little clean warm water to rinse any sediment remaining in the pint container into the sediment tester.

Plain or Superheated Condensed Milk
(Whole or Skim)

Procedure

1. Dilute a sample of the condensed milk with filtered or distilled

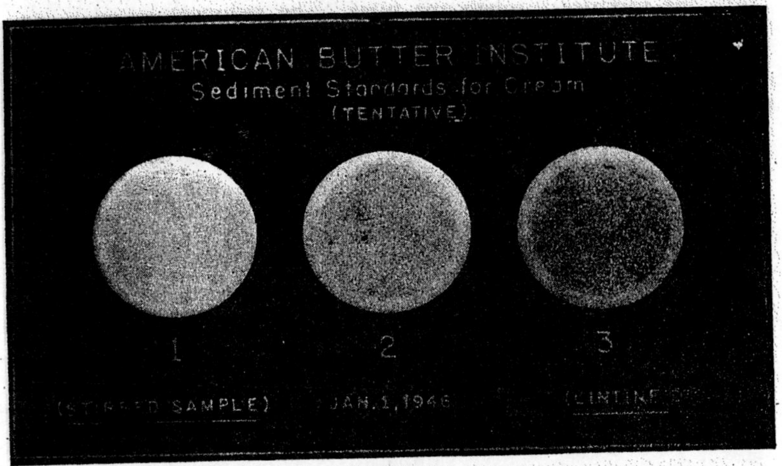

Figure 114.
American Butter Institute sediment standards for cream. This reproduction is not satisfactory for making comparisons. Photoprint may be obtained from the American Butter Institute, 110 N. Franklin St., Chicago 6, Ill.

warm water sufficient to give one pint of reconstituted milk. Proceed as directed under milk. (See page 321.)

Sour Cream (Am. Putter Inst. Methods)[2]

The three methods most commonly used are outlined. This does not preclude any modification which the individual operator may find more suitable to his conditions, provided that such method of modification will not destroy any of the extraneous matter.

(a) Acid Method (Recommended by Dr. B. E. Horall, Purdue University).

Apparatus

1. Sample jars, 2 oz.
2. Improved Parsons cream sediment tester.
3. Granite cup, glass beaker or pint fruit jar.
4. Reel of cream filtering material.
5. Water, free from sediment (filtered or distilled).

Reagent

1. Hydrochloric acid, 0.03 N (3.2 ml concentrated C.P. HCl in 1,000 ml distilled or filtered water).

Procedure

1. Mix 2 oz. of cream with 5 oz. of hot (180 F) acid solution and filter.
2. Pour warm mixture of cream and water into sediment tester

which has previously been thoroughly rinsed with filtered water. Force the mixture through the filter. The sediment tester is then thoroughly rinsed by forcing through at least 4 oz. of filtered, hot (180 F) water before removing the sediment sample. Score sediment on the disc by comparing with an original photoprint obtained from American Butter Institute (figure 114).

(b) Soda Solution Method (Generally known as the Illinois Method).

Apparatus

1. Same as Acid Method above.

Reagent

1. Baking Soda.

Procedure

1. Place 6 oz. water, free from sediment, in cup or pint jar. Add one-fourth of one-half teaspoon baking soda, (free from sediment), and a 2 oz. representative sample of cream. Heat while stirring to 160 F (71 C) by placing the cup in a kettle of hot water, providing a double boiler effect. Do not permit temperature to rise above 165 F (74 C) because of chemical changes. Allow two minutes to dissolve the curd and filter and score as described under Acid Method above.

(c) Parsons Method.

Apparatus

1. Same as for Acid Method above.

Reagent

1. A dry mixture consisting of 25 per cent powdered sodium hexametaphosphate (Calgon) and 75 per cent sodium bicarbonate, uniformly mixed.

Procedure

1. Place a 2 oz. sample of cream in a pint fruit jar. Add 2 to 4 oz. of hot water at 180 to 200 F (82 to 93 C) and one teaspoon of the reagent. Stir well and filter and score as described under Acid Method above.

Butter (Am. Butter Inst. Method)[2]

Apparatus

1. Buechner or Hirsch funnel fitted with a 40 mesh brass screen attached by rubber stopper to a suction flask.
2. Sharkskin filter paper, 7½ cm ruled.
3. Lintine J. & J. cream sediment test roll, 1¼ in.
4. Electric hot plate.
5. Water bath.
6. Fahrenheit thermometer (20-220 F).

7. Suction flask and aspirator.
8. Liter breaker.

Reagent

1. Hydrochloric Acid, 2 per cent solution.
2. Phosphoric Acid, 85 per cent solution.

Procedure

1. Heat to boiling 225 gm (½ lb.) butter in a one liter beaker with 200 ml of 2 per cent HCl. Stir during the latter part of the heating.

2. Filter at once through a 7½ cm ruled sharkskin filter paper supported on a 40 mesh brass section in a Buechner or Hirsch funnel. Do not allow mixture to accumulate on paper. When filtering is impeded, add hot water until the paper is cleared. Then add small amounts of the butter mixture alternately with the hot water treatment.

NOTE: Samples which foam strongly when heated to boiling usually filter slowly at first. Such samples are frequently, but not invariably, helped by the addition of 10 ml alcohol, with mixing, just prior to filtering. In some samples which are difficult to filter, continuous heating and stirring for five to ten minutes will be of assistance.

Alternate Method (Filtration on Sediment Pad): Weigh 225 gm of butter into a liter beaker. Add 250 ml hot water and 20 ml 85 per cent H_3PO_4. Bring to a boil stirring continuously. Filter with suction through 1¼ in Johnson sediment pad. Wash with hot water.

Examine filter paper with wide field microscope for foreign matter illustrated in Laboratory Manual, American Butter Institute, pages 34 to 37.

Butter

BORAX METHOD

Apparatus

1. Aspirator (filter pump).
2. Filter, 500 ml capacity.
3. Büchner funnel, 7 cm, fitted with rubber stopper.
4. Muslin circle, 7 cm diameter.
5. Filter paper, 7 cm, No. 86 Max Dreverhoff or other rapid flow paper.
6. Beaker, 500 ml.
7. Hand lens, 20x.

Reagents

1. Petroleum ether, b. p. 95-140 F (35-60 C).
2. Borax solution, 4 per cent, prepared by filtering a solution of 40 gm of 20-Mule Team Borax in 1,000 ml of water.

Procedure

1. Assemble filtering apparatus and attach to the aspirator. Wet the muslin circle and place it over the perforations in the Büchner funnel and cover with one piece of filter paper. Wet the paper and apply suction.
2. Place 4 oz butter in the pint measure or beaker.
3. Add 150-200 ml borax solution.
4. Carefully heat to boiling with stirring to melt thoroughly and distribute the butter.
5. Filter immediately through the filter paper and muslin which was prepared according to instructions given under (1).
6. Wash filter with about 25 ml petroleum ether.
7. Wash with a little hot water.
8. Remove filter paper from funnel and examine with hand lens for dirt, insects, foreign matter, or other undesirable materials.

Ice Cream and Ice Cream Mix

Apparatus

1. Usual sediment test material for milk (See page 319).
2. Suitable containers for melting or mixing samples of approxi mately, 1000 and 1500 ml capacity.

Procedure

Ice Cream

1. Place 1 pint sample in 1500 ml container, add 1 pint of hot filtered water, 58 F (70 C) and heat mixture on water bath to 149 F (65 C), stirring thoroughly to insure complete melting and mixing.
2. Pass entire solution through cotton disc in a sediment tester.

Ice Cream Mix

1. Place 0.5 pint sample in 1000 ml container, add 1 pint of warm filtered water, about 122 F (50 C) and stir thoroughly to insure complete mixing.
2. Pass entire solution through cotton disc in a sediment tester.

Interpretation

Consider as sediment all foreign matter deposited on disc, disregard any partially or wholly insoluble ingredient normally present in ice cream as a result of normal processing or manufacture. No appreciable amount of foreign material should be present as sediment.

Cheese

Microanalytical Division Food and Drug Administration

Introduction

The examination of cheese for extraneous material presents difficulties not encountered in the analysis of butter. Most of these are attributed to one or more of the following: (1) the use of rennet for precipitating the casein; (2) the variable methods used for manufacturing the different types likely to be encountered, including process cheese, wherein various organic or inorganic salts as well as gums may be incorporated; (3) the presence of an enormous number of bacteria that occur in most cheese.

Since the physical properties of different types of cheese vary considerably, some variation in the technique of determining the extraneous matter is necessary. Two methods are available, the Greene method, adapted for use on Limburger, soft Cheddar, and other soft types, and the Chernoff method, adaptable for use on certain hard varieties.

GREENE METHOD

Apparatus

1. Mortar and glass rods.
2. Beaker, 400 ml.
3. Beaker, 1,000 ml.
4. Electric hot plate or Bunsen flame with asbestos plate.
5. Büchner suction funnel with copper screen, 6 cm diameter, 50 mesh.
6. Filter paper, quantitative grade; Dreverhoff No. 86, 7 cm is recommended.
7. Field binocular microscope, 40x wide.

Reagents

1. Hydrochloric acid, c. p., sp. gr. 1.18, 36 per cent HCl.
2. Ethanol (ethyl alcohol) U. S. P., 95 per cent.

Procedure

1. Prepare a fresh filtered solution of 950 ml of 95 per cent ethanol and 50 ml of HCl (sp. gr. 1.18; HCl 36 per cent).
2. Weigh 50 gm of the cheese to be examined. If a hard rind is present on the cheese this should not be included.
3. Place the cheese in a suitable mortar and slowly add 100 ml of the above solution while stirring (not grinding) to a fairly smooth paste.
4. Pour the paste into a 400 ml beaker, rinsing the mortar with another 100 ml portion of the solution.
5. Heat slowly and allow to boil gently for 3 minutes. Filter at once through Dreverhoff filter paper placed in the Büchner suction funnel.
6. Wash the filter immediately with a large amount (½ liter) of previously filtered boiling water until the paper shows no cheese material left.

7. For the detection of extraneous material, the paper should be placed in a Petri dish and examined with the wide field binocular, using a 12.5 x ocular and a 3.4 objective, giving a magnification of about 40.

8. The transfer of debris found on the filter paper to a drop of glycerine on a microscopic slide and subsequent examination with a compound microscope may be necessary for the exact identification of most material.

NOTES: (a) The analyst must bear in mind that the foreign matter present is usually in very small particles; especially is this so in process cheese in which a grinding process is part of the method of manufacture. Fragments of insects, principally flies or their larvae (usually brown in color), rodent and other hairs, paper, wood, and nondescript debris have been found with this method.

(b) The substitution of water for alcohol in the solution will work with an occasional sample of certain kinds of cheese. The minute amount of ethyl chloride that may be formed probably bears no part in the process although the reactions involved in this method are not definitely known.

(c) If difficulty is encountered in mixing the solution with the cheese because of firmness of the latter, the cheese may be cut into cubes, placed in the beaker, and the solution added without mixing. The beaker is then heated until the solution boils. It should then be poured into the mortar for mixing, after which it is returned to the beaker and boiled slowly for a three minute period.

(d) With certain types of cheese, particularly unprocessed cheddar, there may be a tendency for the filter to plug. This can be relieved by the immediate addition of boiling water.

(e) A piece of 50 mesh copper screen cut to a diameter of 6 cm is essential to the support of the filter paper. The paper is centered over the screen and moistened with water to insure a tight seal before pouring the cheese solution.

CHERNOFF METHOD

Apparatus

1. One liter beaker and one 600 ml beaker.
2. A quantity of glass stirring rods.
3. One 500 ml graduated cylinder.
4. Bolting cloth, 60 mesh.
5. Bolting cloth, 80 mesh.
6. Hirsch porcelain funnel with perforated plate about 4.5-5.0 cm diameter. Place a circular piece of approximately 50 mesh copper screen (4 cm diameter) in the funnel and then place the bolting cloth over the copper screen.
7. Meat grinder, preferably with a nut-butter cutter.
8. Malted milk stirrer.
9. Electric hot plate.
10. Field binocular microscope, 40 x wide.
11. Filtering flask with vacuum apparatus.

Reagents

1. Normal butyl alcohol (technical grade is satisfactory).

2. Oxalate reagent. Dissolve 15 gm of sodium oxalate, c. p., in 500 ml of distilled water and filter if necessary. Add 10 ml of 10 per cent sodium hydroxide solution.

Procedure

1. Pass cheese through a meat grinder, preferably with a nut-butter cutter.

2. Place 8 ounces of ground cheese in a liter beaker and add 200 ml butyl alcohol. Stir a moment with a glass rod to break up large lumps; then place under a malted milk stirrer. Stir until a homogeneous creamy mass is obtained. This should take only about a minute or two.

3. Without stopping the stirrer, add 400-500 ml of hot boiling oxalate solution. Keep stirring about ½-1 minute and then filter immediately through 60 mesh silk bolting cloth on a Hirsch funnel by suction. A brass screen placed under the bolting cloth aids in filtration. Wash with hot oxalate solution or hot water. Examine as soon as possible under the microscope. The filtrate may be passed through 80 mesh silk bolting cloth.

NOTES: (a) Some chemists have found that ethanol (ethyl alcohol U. S. P.) in many instances may be substituted for butyl alcohol. However, butyl alcohol appears to be preferable because of the greater solubility of the fat in this reagent.

(b) A metal malted milk cup may be used in place of the 600 ml beaker, but glass is preferable because of visibility.

(c) Avoid long stirring with the butyl alcohol because sometimes the casein may coagulate to a lump, making it difficult to dissolve completely in the oxalate solution.

2. Transfer the measured sample to a pint milk bottle and add sufficient warm, 120 F (49 C), distilled or filtered water to fill the bottle.

3. Stir with a glass rod and proceed as detailed under milk. The test pad may be rinsed in the tester with 5 - 10 ml of clean water.

4. Large particles should be examined under a hand lens and classified as wood, paper, burlap, wire, or whatever the material appears to be.

Sugar

Apparatus

1. Same as for milk. (See page 319.) Pint milk bottle.

Procedure

1. In case of dry sugar weigh out 25 gm of the well mixed sample and transfer to a clean pint milk bottle.

2. Fill the bottle half full with warm filtered or distilled water and dissolve the sugar.

3. Fill the bottle with the warm clean water and proceed as under milk.

Salt

Apparatus

1. Aspirator (filter pump).
2. Büchner funnel (7 cm) fitted with rubber stopper to fit filter flask.
3. Filter flask, 500 ml.
4. Two feet of pressure tubing.
5. Rapid filter paper, 7 cm.
6. Magnifier (20 x) or low power microscope.
7. Hotplate or gas burner.

Procedure

1. Dissolve 100 gm of salt in 800 ml of cold distilled water. The solubility is only slightly increased by heating.
2. While the salt is dissolving assemble the filtering apparatus and attach to aspirator.
3. Place a filter paper over the perforations of the Büchner funnel, wet with a little water, and apply the suction.
4. Pour the salt solution on the Büchner funnel, discarding the filtrate as the filter flask becomes full.
5. Rinse beaker several times with a little hot water to clean down the side walls and put rinsings through the filter.
6. Disconnect the filter pump and remove the filter paper, putting it on a piece of white paper for examination.
7. Examine the filter paper with the 20 x magnifier or low power microscope.

Stabilizers

Apparatus

1. Same as for milk. (See page 319.)
2. Watch glass.

Procedure

1. Weigh out 1 gm of the well mixed sample and transfer to a 250 ml beaker containing approximately 150 ml of filtered or distilled water.
2. Let s[illegible]d at room temperature until the gelatin or stabilizer is completely w[illegible]ted. Do not stir during wetting procedure, which may take half an hour or longer.

3. When the sample appears to be completely wetted, place the beaker in hot water or on a steam bath, never over a direct flame or on a hot plate.

4. Heat the solution to 140-150 F (60-65.6 C) and stir with a glass rod, being careful not to get the stabilizer on the rod. Continue stirring and heating the mixture until complete solution is effected.

5. Fill the beaker to approximately the 225 ml level and proceed as under milk.

NOTE: Some types of stabilizers may be very difficult to dissolve. If so, proceed to step 2 and allow it to soak over night or longer at room temperature in a covered container.

REFERENCES

1. Standard Methods for the Examination of Dairy Products, 9th Edition, American Public Health Association, 1948.

2. Laboratory Manual, American Butter Institute, 1947.

3. The Grading of Dry Whole Milk, American Dry Milk Institute Bulletin 913, 1947.

4. The Grading of Nonfat Dry Milk Solids, American Dry Milk Institute Bulletin 911, second revised edition, 1944.

CHAPTER 26

CREAM VOLUME DETERMINATION

Milk Industry Foundation Method*

Apparatus

1. M. I. F. cream volume tubes with corks to fit (flat bottom tube with graduation mark at a point 200 mm from the inside bottom of the tube.) (Figure 115).

2. A suitable rack for holding cream volume tubes in ice water bath.

3. An ice water bath of sufficient capacity to hold the tube contents at constant temperature for twenty hours under favorable conditions.

4. Two or more accurate Fahrenheit thermometers.

5. Accurate millimeter scale.

Procedure

After deciding on a logical sampling sequence for the pasteurizing system under investigation, proceed to treat each sample as follows:

1. Adjust the sample to 60 F (15.6 C) and mix thoroughly by pouring several times from one vessel to another. If it should prove impracticable to cool the pasteurized milk to 60 F (15.6 C) before mixing, full consideration should be given to the fact that hot milk will appreciably shrink in volume on being immersed in ice water.

NOTE: Hot milk samples may be readily cooled to 60 F (15.6 C) or below by running the milk directly from the equipment through spiral tubing of suitable length sweated into a container for holding iced water or other cooling medium. The use of such a quick cooling apparatus is imperative when investigating the creaming ability of milk pasteurization in high-temperature short-holding systems.

2. Set aside a representative portion of the sample for duplicate milk fat determination or make fat test on contents of tubes after all cream volume observations have been made.

3. Fill duplicate tubes up to the mark with the properly mixed sample, cork tightly, and immerse the entire milk column in the ice water bath recording the time at which immersion took place.

*Formerly known as the International Association of Milk Dealers Method.

Figure 115.
Cream volume tubes ready for reading.

NOTE: Foam will not constitute a source of inaccuracy of measurement if the cream volume tube is overflowed to float off the foam and the level of the milk column is adjusted to the mark by pipetting off the excess milk. The cork will prevent water or ice from falling into the tube and minimize evaporation.

4. Hold the filled cream volume tubes undisturbed for twenty hours in the ice water bath. It is important that the bottom of the upper meniscus of the milk column be held level with the 200 mm mark.

5. Holding the cream volume tube in a vertical position, read the cream volume from the lower extremity of the layer to the 200 mm mark on the tube and record the results in terms of millimeters. (See figure 116).

NOTE: It is assumed that extreme care has been exercised in measuring out the milk so that the bottom of the upper meniscus will be on exact level with the 200 mm mark on the tube at the time of reading.

Interpretation

The length of the cream layer, in millimeters, divided by 2 yields the creaming ability of the milk in terms of percentage since the total length of the milk column in each tube is 200 mm.

CREAM VOLUME STUDY DATA SHEET

Date...

Equipment

(1) Type and size of the raw milk tank.
(2) Type and capacity of the heater.
(3) Is the heater equipped with automatic temperature control?
(4) Is the milk filtered or clarified?
(5) Type and capacity of the holder.
(6) Type and location in milk flow of the pumps used (show by a sketch on the opposite side the arrangement of the equipment and the location of the milk pumps).
(7) Type and capacity of coolers.

RESULTS

Source of samples	Time sample was taken	Temp.	Milk fat		Cream volume after 20 hours	
			A	B	A	B
1. Raw milk in storage tank						
2. Leaving filter						
3. Leaving heater						
4. First discharge from holder						
5. Last discharge from holder						
6. From cooler when holder is two-thirds empty						

Milk Bottle Gage Method

Apparatus

1. The cream volume gage is made of metal, plastic, or compressed paper. This gage is so constructed as to conform to the shape of the neck and shoulder of a quart or pint milk bottle. The gage is calibrated to read the bottle volume directly in percentage from the cap seat down to a point at the shoulder of the bottle, representing 25 per cent of the capacity. (See figure 117).

Procedure

The cream volume is determined by placing the gage in position and reading the percentage in volume directly at the point where the cream line meets the gage. (See figure 118).

NOTE: The gage must be specially constructed for each individual type bottle, and any one gage will not be adaptable to different shaped bottles. The bottle gage method is at best a rough measurement of cream volume. A much more accurate means of measuring the cream volume in a bottle is by the use of the burette method.

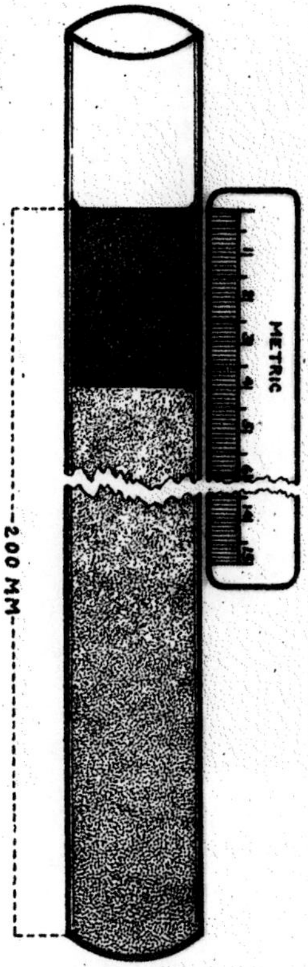

Figure 116.
Method of reading cream volume portion of cream volume tube. (M.I.F. method.)

Plant Method

(This method merely serves as a means of comparing the cream volume of the finished product from day to day or of contrasting the depth of the cream line on samples in bottles of like conformity.)

Apparatus

1. Wooden rack to hold six quart bottles in a row equipped with a taut, horizontal wire the height of which may be adjusted.

Procedure

1. Place the bottles of milk in the rack.
2. Adjust the wire to the height of the average cream line.
3. Compare the cream lines by observations.

Figure 117.
Method of holding gage against bottle.

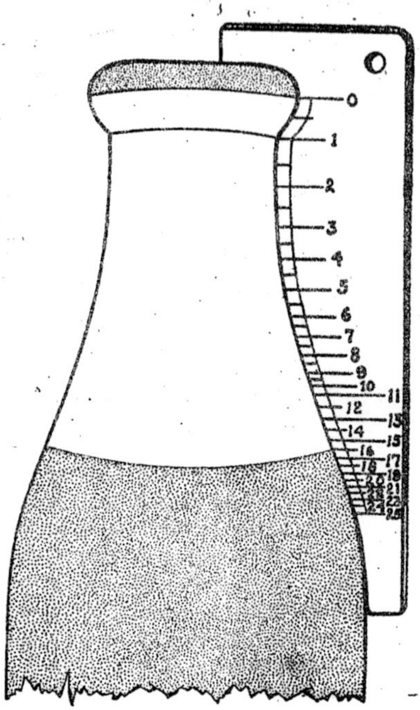

Figure 118.
Milk bottle gage showing scale.

Burette Method

(This method has been developed as a reasonably rapid and accurate method of measuring the percentage of cream volume in a quart sample of milk.)

Apparatus

1. Cream volume burette. (Figure 119).

Procedure

A sample of the milk to be tested is held below 50 F (10 C) in a quart milk bottle for twenty-four hours. A mark is then placed on the bottle at the upper and lower level of the cream column. These marks may be made with a glass marking pencil or with a piece of adhesive tape or similar adhesive material. Care should be taken that the marking is accurately done.

The milk is then removed, the bottle rinsed with water, inverted, and drained free of water. Water is run into the bottle from a burette until the bottom of the meniscus is just even with the top of the lower cream volume mark. The burette reading will indicate the volume of

Figure 119.
Cream volume burette, one quart capacity, calibrated in milliliters.

milk less the creamed portion. Record this figure and allow water to run into bottle to the top of the mark, the point to which the bottle was originally filled. This reading is the total volume. Subtract the first reading from the second to obtain the actual cream volume.

NOTE: The percentage cream volume is obtained by dividing total volume of the sample into the cream volume portion.

$$\text{per cent cream volume} = \frac{\text{Cream volume}}{\text{Total volume}} \times 100$$

CHAPTER 27

CURD TENSION DETERMINATION

American Dairy Science Association Method

Investigations in infant feeding have shown that soft-curd milk is sometimes more readily tolerated than milk of average curd tension.

Some cows produce milk with a curd tension of 30 gm or less, which is natural soft curd, but most of the milk of low tension on the market is produced by homogenization or other mechanical or chemical treatments.

Apparatus

1. A tempering bath that will not change more than 1 degree F in 10 minutes.

2. A pipette to deliver by gravity 100 ml in approximately 45 seconds.

3. A special curd-knife meeting the following specifications: Knife should consist of eight radial blades each having 9/16 in. of lineal cutting edge and being 0.02 in. thick, spaced equally and enclosed by a circular or ring blade, 1¾ in. outside diameter, 3/16 in. high and 0.031 in. thick. The radial blades should be attached to the inside of the circular or ring blade and extend upward from it a distance of 2 1/16 in., being reduced to a width of 5/32 in. above the circular or ring blade. Their upper ends must be curved inward and attached to a central spindle 5/16 in. in diameter. The lower or cutting edge of the circular or ring blades must be tapered from the outside at an angle of 30 degrees to the knife axis, to a dull knife edge. The lower or cutting edges of the radial blades are to be tapered on each side at an angle of 15 degrees to the knife axis, to a dull knife edge. The cutting edges of the ring blade and the radial blades must lie approximately in the same plane, deviating not more than 1/32 in. The total linear cutting edge of the knife must be 9.8 in. ± 0.10 in. All joints are to be mortised and soldered smooth and the knife constructed of a noncorrosive metal.

4. A tensiometer, imparting automatic movement to the knife or receptacle, should travel at the rate of 1 in. in 7-8 seconds. The sensitivity of the instrument must be such that readings can be made to an accuracy of ± 1.0 gm.

5. A heavy-wall jar, 3 to 4 in. high (inside) and having an inside diameter of 2⅜ in ± ⅛ in. The jars herein listed have been found to

meet these specifications satisfactorily and undoubtedly others will also be found which conform:

W. M. Welch Scientific Company, Chicago—Museum jar No. 4612A (8 oz.).

Central Scientific Co., Chicago—Museum jar No. 10396 (7½ oz.).

Owens-Illinois Glass Co., Toledo—Mayonnaise jar (8 oz.).

Reagent

1. Coagulant, 0.08 N hydrochloric acid to which U. S. P. (1 : 3,000) dry pepsin has been added at the rate of 450 mg per 100 ml, freshly prepared.

Procedure

1. Temper 10 ml of the coagulant in the jar for at least 3 minutes at 95 F (35 C). Adjust the undiluted sample of milk to the same temperature. The milk should be tempered in a 5 minute period immediately preceding the pipetting of the sample. The temperature of the latter should not exceed 100 F (38 C).

NOTE: A convenient alternative is to measure the milk and the coagulant into separate containers, temper, simultaneously, and rapidly pour the milk into the container holding the coagulant.

2. Add the sample of milk to the tempered coagulant from the 100 ml pipette, the milk to be blown from the pipette held vertically over the center of the beaker. There should be no rotation of contents.

3. Hold the sample at 95 F (35 C) for 10 minutes to coagulate. This temperature should not vary more than 1 degree F.

4. The curd tension should be determined at the end of 10 minutes ± 30 seconds from the time the sample was placed in the coagulating vessel.

5. Readings recorded are to be maximum obtained as the knife penetrates the surface. A sufficient number of tests should be made so that at least two or more will check within 10 per cent. The average of these readings is taken.

REFERENCES

1. Milk Ordinance and Code Recommended by the U. S. Public Health Service.

2. Hill, R. L.: The Physical Curd Character of Milk and Its Relationship to the Digestibility and Food Value of Milk for Infants. *Utah Agr. Exp. Sta.* Bul. 207 (1928).

3. Final Report of Committee on Methods of Determining the Curd Tension of Milk. *J. Dairy Science* 24:825-7 (1941).

CHAPTER 28

VISCOSITY DETERMINATION OF DAIRY PRODUCTS

The consuming public judges the quality of such products as cream and buttermilk largely by the way they pour from the container. It is very desirable, therefore, to have some simple, dependable method by which to measure the viscosity or body of such products and to check their uniformity from day to day.

A few tentative methods are submitted, by which the relative viscosity of fluid products can be compared.

Borden Method for Cream

Apparatus

1. Equipment illustrated in Figs. 120 and 121. *A* is a brass head fitted with a rubber gasket *B*, which in turn fits into the neck of the specially calibrated bottle *C*. The head is held in place by the special metal clip *I*. *D* is an air inlet tube. *E* is a rubber stopper closing the inlet of tube *D* at all times except when cream is flowing from the bottle. *F* is a carefully calibrated brass tip which screws into the cream outlet tube *G*. *H* is a two-hole stopper which supports the entire apparatus when inverted over a quart milk bottle.

2. Supply of 56.5 mm ½ pt. milk bottles, common sense finish.

Procedure

1. A ½ pint milk bottle (235 ml capacity) should be completely filled with the cream under test. The temperature of the cream in all cases before testing should be below 50 F (10 C). For light cream the temperature is adjusted to exactly 50 F (10 C) by placing the bottle in a dish of water of slightly higher temperature and stirring the contents gently with an accurate thermometer. For heavy cream the temperature is raised to 59 F(15 C) and removed from the water bath. Actual practice shows that by the time the test is made the temperature of the cream rises to 60 F (15.6 C). The final temperature should be carefully controlled and should not vary more than ½ degree F from the standard temperature.

2. Any foam on the surface of the bottle should be removed, after which the apparatus, which has previously been cooled in a stream of running water to 50 or 60 F (10 or 15.6 C), is inserted in the neck of the

Figure 120.
Apparatus for measuring viscosity of cream (Borden method).

bottle in such a fashion as to exclude air bubbles. The stopper *E* should be in place during this operation. Any excess of cream will rise in tube *G* and flow or spurt out of the tip *F*. When the head *A* is firmly in place, a metal clip *I* is slipped over it so as to clamp it in position.

3. The bottle is then inverted over a quart milk bottle into the neck of which stopper *H* fits so as to support the entire assembly. With a stop-watch in the right hand and the instrument supported with the left, the stopper *E* is removed, whereupon the cream starts to flow out of the tip. The exact second of time at which this begins is noted. When the cream ceases to flow out of the tip in a stream, the exact time is again noted, and the elapsed time is accurately reported in seconds. This value is called the flow time.

4. After use the flow meter should be cleaned by rinsing with warm water at 130-140 F (54.5-60 C) to wash out any fatty particles clinging to the tubes. It is important that no adhering particles remain in the outlet tube, since such particles will increase the flow time and prevent accurate measurement. The tip should be cleaned from time to time by running a small string through it.

5. Two tips are used with the flow meter. The one with smaller bore is designed for cream with milk fat content less than 30 per cent. The tip with the larger bore is for use with cream having a milk fat content of 30 per cent or higher. With water, the time required for

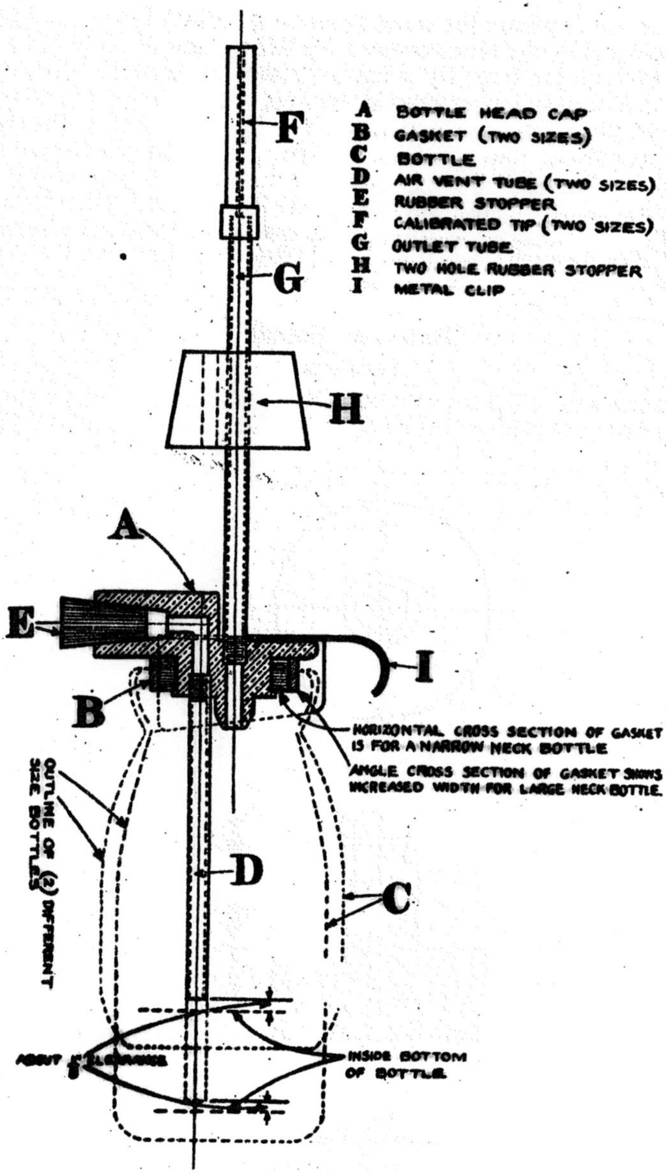

Figure 121.
Diagram of apparatus for (Borden) flow viscometer.

½ pint to flow through the small tip at 50 F (10 C) is approximately 50 seconds, while the time required for this volume at 60 F (15.6 C) to flow through the large tip is approximately 25 seconds. Cream requires considerably longer time, depending on how heavy the body is. If desired, the method of expressing results used by C. J. Babcock[1] may be employed. For instance, a 40 per cent cream which requires 200 seconds flow time at 60 F would have a value of 8, provided the constant of the instrument with water was 25 seconds (200÷25=8).

NOTE: For further information on this method the technician is referred to the Borden Research Laboratory, North Franklin St., Syracuse, New York, or to the *J. Dairy Sci.* *16*:1-10 (1933).

Babcock Method

Apparatus

An instrument similar to the standard Saybolt viscosimeter except that the orifice should be 0.07 inch in diameter and removable. (Figure 122.)

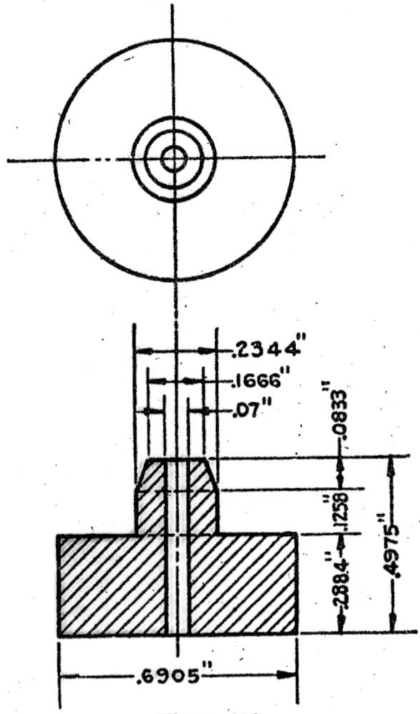

Figure 122.
Detail of the orifice for Babcock viscometer. Courtesy of U. S. Department of Agriculture.

Procedure

1. Place 70 ml of liquid in the viscosimeter and allow it to come to constant temperature.

2. Record in seconds the time required for the discharge of the first 60 ml.

3. For the maintenance of constant temperatures surround the viscosimeter with a water bath.

4. Determine the temperature by a thermometer passed through the lid of the viscosimeter into the center of the liquid within.

5. Record the results as relative viscosity, which may be defined as the relative internal friction of a liquid, and determined by taking that of water at the same temperature as equal to 1. The average of several determinations should be taken.

REFERENCE

1. Babcock, C. J.: Factors Affecting the Viscosity of Cream. *U. S. Dept. Agr.* Tech. Bul. 249 (1931).

Saybolt Viscosimeter Method

(Scales Modification)

Apparatus

1. Modified Saybolt Viscosimeter with 0.078 inch orifice in tube. (See Figure 123.)

Figure 123.
Saybolt viscosimeter (Scales modification).

Modified Saybolt Viscosimeter designed by F. M. Scales of Sealtest Laboratory, Sheffield Farms Company, Incorporated, New York, may be purchased from the Will Corporation, Rochester, New York.

2. Saybolt-like flask, 100 ml.

3. Stop watch.

4. Thermometer, range of 10 to 150 F (–12.2-65.6 C), accurate to 0.5 degree.

Procedure

1. Before a determination the tube and orifice of the viscosimeter should be inspected to see that they are clean. Hot water and soft tissue paper, if necessary, may be used to remove any residue. The tube should be rinsed with about 10 ml of the sample if it is wet.

2. Fill the bath of the viscosimeter with cold water and adjust to desired temperature with cracked ice or hot water.

NOTE: It is essential that the temperature be accurately controlled (± 0.5 F) during the test.

Product and Temperatures:

Light Cream .... 50 F
Heavy Cream .... 60 F

3. The cream in a half-pint bottle filled to within a half inch of the cap seat is brought to the correct temperature by immersing for 30 min. in a water bath adjusted to the desired temperature. At the end the first 15 min. the bottle is taken out of the bath and slowly inverted three times to mix the contents. At the end of 30 min. the mixing procedure is repeated. The cream is then ready for use.

4. Fill the viscosimeter tube flush to the line where it joins the shelf.

5. Place the dry 100 ml flask in the ring on the base directly beneath the orifice and check the temperature of the sample and water bath.

6. When the sample in the tube and the water bath are at the desired temperature, raise the orifice rod quickly and turn it slightly so that the pin rests on the top of the sleeve. Simultaneously start the stop watch.

NOTE: If the flask has been properly set within the ring on the base, the stream of cream will fall on the side of the cup at the top of the flask and run into it without forming bubbles or foam.

7. Stop the watch at the instant the cream reaches the level of the 100 ml mark on the neck of the flask. Record the time to the nearest two-tenths of a second.

Pipette Method

Apparatus

1. Measuring pipette, 100 ml. (See Figure 124.)

2. Thermometer.

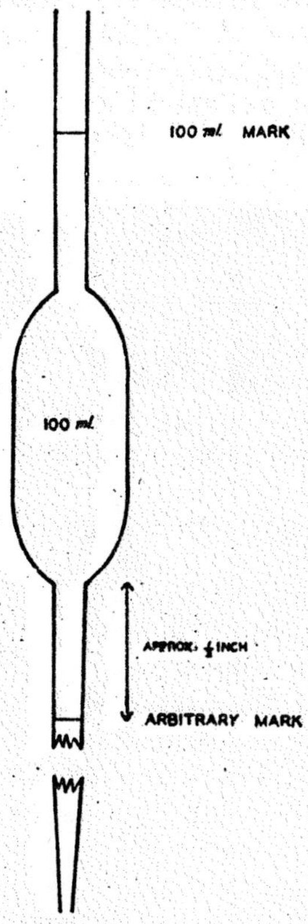

Figure 124.
Measuring pipette.

Procedure

1. Bring the sample to be tested to exactly 60 F (15.6 C).
2. Rinse the pipette with the sample two or three times.
3. Fill the pipette with the sample to the 100 ml mark and check the number of seconds required to discharge to an arbitrary fixed point about one-half inch below the place where the small discharge tube begins.
4. Record the results and compare from day to day.

NOTE: a. Only those results obtained from the same identical pipette are suitable for comparison.

b. If it is desired to compare results obtained from different pipettes, the viscosities should be converted to relative viscosities by determining the viscosity of water at the same temperature and dividing the time of discharge for cream by the time of discharge for water at the same temperature.

Falling Ball Method for Sweetened Condensed Milk

Apparatus

1. Viscosity apparatus, consisting of calibrated test tubes, calibrated steel balls, and special test tube rack, furnished complete by National Dairy Research Laboratories, Oakdale, Long Island, N. Y. (Figure 125.)

Figure 125.
Falling ball apparatus.

2. Stop-watch.
3. Water bath, deep enough to hold test tubes in vertical position.
4. Thermometer 10-150 F (−12.2-65.6 C), accurate to 0.5 degree F (0.3 degree C).
5. Rubber stoppers to fit test tubes.

Procedure

NOTE: It is very important that samples for viscosity determination be handled with the minimum amount of agitation. Do not stir to mix samples

1. Pour sample into a dry calibrated test tube, filling it to about 1 inch above the upper graduation.

2. Stopper the test tube and place it in the water bath, the temperature of which has previously been brought to 70 F (21.1 C).

3. Leave the test tube in the bath for exactly 30 minutes checking the temperature occasionally to see that it does not vary over plus or minus 0.5 degree F. (0.3 degree C.)

4. After 30 minutes take the test tube from the bath, remove stopper, and place test tube in the rack at the proper angle and immediately roll bearing down the wall of the tube.

5. Start the watch when the center of the ball passes the upper graduation and stop the watch when the center of the ball passes the lower graduation.

6. Record the elapsed time in seconds and immediately repeat the determination on the same tube with two more ball bearings without removing tube from rack.

7. Report the average of the three determinations as the viscosity in seconds.

NOTE: Each successive determination will run slightly lower than the preceding one.

NOTE: The modified Stormer viscosimeter is probably one of the best means for determining the viscosity of sweetened condensed milk. Reference: *Chem. Engineers' Handbook,* McGraw-Hill Book Co., Inc., New York (1934), p. 1273.

REFERENCES

1. Hening, J. C.: The Standardization of the Borden Flow Meter for Determining the Apparent Viscosity of Cream. *J. Dairy Sci. 18*:763 (1935). (Note: This page reference deals with the falling sphere method of measuring the viscosity of condensed milk.)

2. Hening, J. C.: Same paper, page 465, deals with the Borden flow meter.

3. Leighton, A., Leviton, A., and Williams, O. E.: The Apparent Viscosity of Ice Cream. I. The Sagging Beam Method of Measurement. II. Factors to be Controlled. III. The Effects of Milkfat, Gelatin, and Homogenization Temperature. *J. Dairy Sci. 17*:639-650 (1934).

4. Goriaew, M.: The Viscosity of the Fatty Matter in Butter. *Ibid. 20*: Abst. No. 224, p. 20 (1937).

CHAPTER 29

HOMOGENIZATION EFFICIENCY DETERMINATION

Determination of the U.S.P.H.S. Index of Homogenized Milk[1]

The U.S.P.H.S. defines homogenized milk as milk which has been treated in such manner as to insure break-up of the fat globules to such an extent that after 48 hours quiescent storage, no visible cream separation occurs on the milk and the fat percentage of the top 100 ml of milk in a quart bottle, or of proportionate volumes in containers of other sizes (50 ml per pint and 25 ml per one half pint), does not differ by more than 10 per cent of itself from the fat percentage of the remaining milk as determined after thorough mixing.

By way of explanation of the definition of homogenized milk the U. S. P. H. S. states:

> "According to the above definition, if the top milk in the quart bottle contains 4.0 per cent fat, the remainder of the milk must test at least 3.60 per cent. The difference, 0.4 per cent, is 10 per cent of the figure 4.0 per cent."

Apparatus

1. Babcock or Gerber equipment for testing milk for fat. (See pages 180 and 222.)
2. Graduate for measuring top 100 ml (50 or 25 ml) of homogenized milk.
3. Aspirator or suction pump and appropriate two-hole rubber stopper, glass tubing, and rubber tubing for collecting top 100 ml portion of milk in the measuring graduate when this method of sampling is preferred over direct pouring of the milk into the graduate. (See Figure 126.)

Reagents

1. Babcock sulfuric acid (H_2SO_4), sp. gr. 1.82–1.825 at 68 F (20 C), stored in tightly stoppered containers.
2. When the Gerber method is used for testing for percentage of fat, Amyl alcohol sp. gr. 0.815–0.818, b. p. 263 to 270 F (127.8–132.2 C), free from fat, furfural, or any substance that might appear in the fat column after centrifuging.

Procedure

1. Collect the homogenized milk sample in the finished package and after proper agitation store the sample at a temperature between 40 and 50 F (4 to 8 C) and hold in a quiescent state for 48 hours.

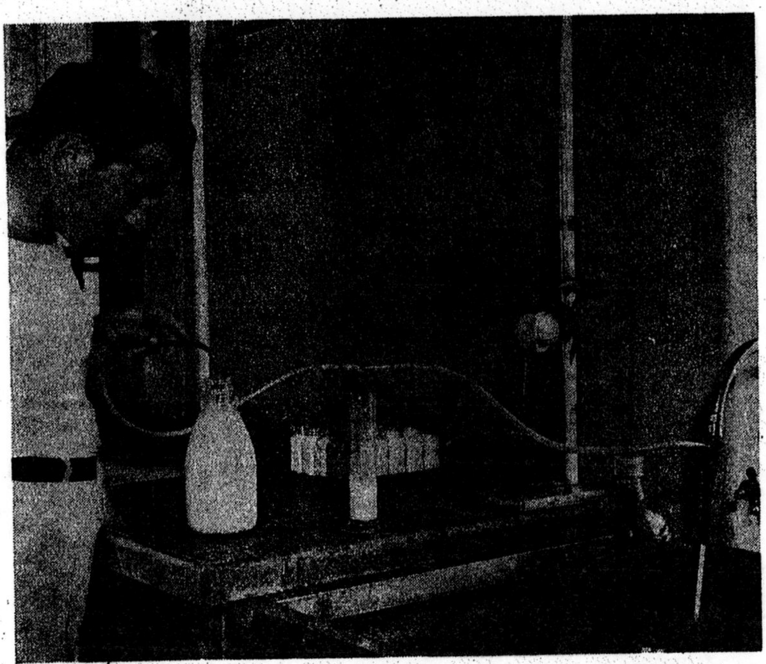

Figure 126.

Suction method of removing the top 100 ml of homogenized milk from a quart bottle in the U. S. Public Health Service index testing procedure.

NOTE: When homogenized milk samples are stored for more than 3 hours after filling it will be advisable to agitate by pouring the sample from one bottle to another 4 or 5 times. When samples are stored within 3 hours of filling time sufficient mixing of the bottle contents may be effected by slowly inverting the bottle 8 to 10 times.

2. After the 48 hour period of quiescent storage, carefully remove bottle cap and pass a stiff wire around the inner surface of the bottle neck to loosen any adhering butterfat. Observe critically for any visible sign of cream separation or cream plug. Remove the top 100 ml from the quart sample (proportionate amounts from samples of other sizes) by pouring directly into the graduate or by the suction method as illustrated in figure 126.

3. Thoroughly mix the 100 ml portion by pouring 3 or 4 times from one vessel to another, quickly measure the test portion and procede with the fat test for homogenized milk as directed on page 186 (Babcock) or page 222 (Gerber). Repeat this process with the remainder of the homogenized milk sample.

4. Calculate U.S.P.H.S. Index as follows:

Per cent top test — Per cent bottom test = difference in tests

$$\frac{\text{Difference in tests}}{\text{Per cent top test}} \times 100 = \text{U.S.P.H.S. Index (in terms of percentage)}$$

Example:

Per cent top test	=	4.00
Per cent bottom test	=	3.60
Difference in tests	=	0.40

$$\text{thus: } \frac{0.40}{4.00} \times 100 = 10.00\%$$

NOTE: Some states specify calibration of test bottles to ±0.10 per cent tolerance, and the use of a pair of test bottles having tolerances at the extreme limit of this specification would induce a possible 0.20 per cent difference between top and bottom tests due entirely to the variation in the accuracy of the graduations. When there is doubt as to the matched accuracy of test bottle graduations, it is advisable to test the bottom portion of homogenized milk in the same test bottle that has been used for testing the top portion.

REFERENCE

1. U. S. Public Health Service Ordinance and Code.

Microscopic Method

Apparatus

1. Microscope fitted with 16 mm and 4 mm objectives and 10 eyepiece.
2. Microscope slides.
3. Hanging drop slide and cover glasses.
4. Graduated cylinder, 100 ml.
5. Pipette, 1 ml.
6. Test tubes, 150 ml.

Procedure

1. Routine Examination

a. Dilute the sample to be tested with distilled water so that the sample contains from 0.1 to 0.2 per cent of fat.

The following table gives the approximate dilutions to use:

Milk	dilute 1 ml with 25 ml of distilled water
20% cream	dilute 1 ml with 150 ml of distilled water
30% cream	dilute 1 ml with 225 ml of distilled water
40% cream	dilute 1 ml with 300 ml of distilled water
10% ice cream mix	dilute 1 ml with 75 ml of distilled water
20% ice cream mix	dilute 1 ml with 150 ml of distilled water

b. Immediately prepare a regular or hanging drop slide and examine under low power (16 mm) objective. The light passing through the slide must be subdued owing to the transparency of the fat globules. The low power objective which produces a magnification of 100 diameters gives the observer a general idea of the efficiency of homogenization and the presence of fat clumps. For more detailed observation the high power (4 mm) objective (magnification 440 diameters) is used.

2. Measurement of Size of Fat Globules

The beginner will find it valuable to examine samples of milk (diluted 1-25 ml), unhomogenized cream (diluted according to fat content), etc., to get and idea of the general size and appearance of fat globules under the microscope. Further ability to note the efficiency of homogenization will come with the examination of a large number of samples produced under varying conditions. Measurement of the fat globules and clumps is of value to the trained microscopist and research worker but is unnecessary for routine examination of ice cream mix samples. (See illustration in figure 127.)

NOTE: It is suggested that examination for efficiency of homogenization be made at least once a week on all products regularly homogenized. Examination under the microscope is considered the most practical and reliable method to determine whether or not milk and other dairy products have been homogenized.

Figure 127. *Courtesy of Creamery Package Mfg. Co.*

Microscopic appearance of fat globules of homogenized and unhomogenized milk, about 1000X (each small scale division represents 2 microns.)

PART V

MISCELLANEOUS AND SPECIAL TESTS OF DAIRY PRODUCTS

CHAPTER 30

MISCELLANEOUS TESTS

The following simple and rapid tests are often useful for the identification of milk from cows with abnormal udder conditions.

Brom Thymol Blue Test

Procedure

1. Place 3 ml of milk in a small test tube.
2. Add 5 drops of a concentrated alcoholic solution of bromthymol blue (1 gm in 200 ml of 68 per cent ethyl alcohol) and mix.
3. A color ranging from yellow-green to greenish blue usually denotes a normal reaction.

The development of a bluish green color is supicious, and a blue color very suspicious indication of a mastitis infection or other abnormal conditions. A blue color may also indicate that the product has been neutralized. A dirty yellow to a straw yellow color indicates high acidity. It may also indicate acute mastitis. Bromthymol blue test papers furnished by laboratory supply companies provide a quick field test for this examination.

Chloride Test

(Sharp-Struble Method)

Procedure

1. With a Babcock pipette place 18 gm (17.6 ml) of milk in a glass beaker (about 400 ml capacity) in which the milk can be rotated rapidly.
2. Add 2 ml of a 10 per cent solution of potassium chromate.
3. Add from a burette silver nitrate solution containing 8.6245 gm of silver nitrate per liter of solution, constantly rotating the beaker, until the first slight reddish orange color is obtained.
4. Multiply the number of milliliters of silver nitrate solution used by 0.01 to obtain the percentage of chloride.

NOTE: The chloride content of milk from normal udders rarely exceeds 0.14 per cent (by the Sharp-Struble method) while abnormal udders may secrete milk with a much greater amount. The chloride test is useful in detecting mastitis. However, during the first few weeks after freshening and during the last month of lactation, the chloride content of milk is usually higher than normal. These facts should be taken into consideration when using the chloride test as a means of detecting mastitis. Milk having a high chloride content usually has a high lipase activity. Lipase is a normal constituent of milk, but its activity is often abnormally high during late stages of lactaton.

Blood in Milk

Chemical tests have been devised for detecting the presence of blood in milk. Blood in milk insufficient to cause an appreciable change in the color may be detected as follows:

Procedure

1. Place 15 ml of the sample in a conical-bottom centrifuge tube.
2. Insert the bottom of the tube into the small end of a No. 7 one-hole rubber stopper.
3. Place in centrifuge, taking care to have the machine properly balanced.

NOTE: The tube should be inserted about one half of the distance through the stopper. Cotton may be packed under and around the tube if stoppers are not available.

4. Centrifuge for 10 minutes or longer at the maximum speed for the centrifuge used.

The presence of blood is indicated by a red sediment in the bottom of the tube.

Alcohol Test for Determining Coagulability of Milk

The alcohol test is of practical value in determining the susceptibility of milk to coagulation by heat. It is largely used in evaporated milk plants to detect milk which has a tendency to curdle during the sterilizing process. The test will also aid in detecting abnormal milk such as colostrum, stripper milk, or milk in which the mineral balance has been disturbed in such a way as to make it more subject to coagulation than normal milk. Although the test is often used to differentiate high and low acid milk, it is reliable for this purpose.

Apparatus

1. Supply of test tubes, with a mark at 5 and 10 ml points.
2. Two 5 ml pipettes or dippers.

Reagent

1. Alcohol solution prepared by diluting 72 ml of 95 per cent grain alcohol to volume of 100 ml with distilled water.

Procedure

1. Place 5 ml of milk in a test tube and add an equal quantity of the alcohol solution.
2. Mix the contents of the test tube by inverting several times slowly.

NOTE: The temperature of the milk and alcohol should be about 70 F (21 C).

Interpretation

A positive test is indicated by large, medium, or small flakes of curd appearing on the side of the test tube. No change in the milk is a negative test and may be taken as partial evidence of heat stability. Large flakes generally indicate instability to heat, an acidity of over 0.20 per cent or other abnormal condition.

Catalase Test for Butter

(Hesse)

Apparatus

1. Test tube, 20 ml, graduated in 0.5 ml.
2. Funnel.
3. S-shaped tube.

Reagents

1. Hydrogen peroxide, 3 per cent H_2O_2.
2. Sterile water.

Procedure

1. Melt the butter at 105 F (40.6 C), shake, and measure 50 ml in a separatory funnel containing 20 ml of sterile water. Shake thoroughly and let it stand for a while.

2. Run 15 ml of the milky butter serum into the graduated test tube; fill with 3 per cent H_2O_2, insert a rubber stopper provided with an S-shaped glass tube, invert the apparatus and let it stand over night.

3. Measure in milliliters the amount of gas formed.

A high grade butter made under sanitary conditions (properly pasteurized cream, scalded churns, rinse water of a low bacterial count, etc.) should not produce gas. Contaminated butter, or butter of poor keeping quality, will show 20 ml or more of gas.

REFERENCES

Marting, B.: *Berlin Molk Zeitung 1*:166 (1890).

Sommer, H. H., and Hart, E. B.: The Heat Coagulation of Milk. *J. Biol. Chem. 40*:137 (1919).

Dahlberg, A. O., and Garner, H. S.: The Alcohol Test as Means of Determining Quality of Milk for Condensers. *U. S. Dept. Agr. Bull.* 944 (1921).

Sommer, H. H., and Binney, T. H.: A Study of the Factors That Influence the Coagulation of Milk in the Alcohol Test. *J. Dairy Sci. 6*:176 (1923).

Detection of Coloring Matter

ANNATTO

Apparatus

1. Flask, glass stopper, 100 ml.
2. Centrifuge, and centrifuge tubes.
3. Pipette, with rubber tubing.
4. Beaker, small.
5. Filter paper, Whatman No. 3, 4 cm.
6. Funnel, 1 inch.

Reagents

1. Ethyl alcohol, 95 per cent.
2. Gasoline, (73 octane, without lead).
3. Formaldehyde, 40 per cent.
4. Sodium hydroxide solution, 5 per cent.
5. Stannous chloride solution, 10 per cent, aqueous, containing sufficient conc. HCl to make the solution acid, and a small piece of tin to keep it reduced.

Procedure

1. To 10 ml of ethyl alcohol (95 per cent) in a glass stopper flask, add 10 ml of the prepared sample of cream. Shake vigorously for 20 seconds.

2. Add 20 ml of aviation gasoline (73 octane) without lead. Shake vigorously for 20 seconds.

3. Add 0.5 ml of formaldehyde and shake a few seconds.

4. Transfer to a centrifuge tube and centrifuge in an unheated machine for 30 minutes at the speed recommended for the Babcock test (page 183).

5. By means of a pipette, withdraw the alcohol-serum layer and transfer it to a small beaker. A piece of rubber tubing attached to the pipette will enable the operator to hold the tube before the eyes and thus observe when all the serum has been drawn into the pipette.

6. Fold a 4 cm Whatman No. 3 filter paper and place in a 1 inch funnel. Immediately before use, saturate the paper with a 5 per cent sodium hydroxide solution.

7. Filter the alcohol-serum solution through the prepared paper. Do not allow the filter to become empty until filtration is complete, otherwise filtration will be slowed by increased viscosity and by gumming of the filter.

8. Remove the filter and wash it while open with a gentle stream of water.

9. Dry the filter with or without heat.

10. Treat the paper with a few drops of a 10 per cent aqueous solution of stannous chloride (containing sufficient concentrated hydrochloric acid to make the solution acid and a small amount of metallic tin to keep the solution reduced.

11. Dry the paper without heat, and examine by transmitted light.

12. The appearance of any pink coloration confirms the presence of annatto.

The test requires two hours to complete, and is sensitive to 1 part in 500,000.

COAL TAR DYES

The curd of an uncolored milk is perfectly white after complete extraction with ether, as is also that of milk colored with annatto. If the extracted fat-free curd has a distinct orange or yellow color, the presence of coal tar dye is indicated. In many cases, on treating a lump of a fat-free curd in a test tube with a small amount of concentrated hydrochloric acid, the color changes to pink, indicating the presence of a dye similar to aniline yellow or butter yellow or perhaps one of the acid azo yellows or oranges.

In some cases the presence of coal tar dyes can be detected by mixing about 100 ml of the milk with an equal volume of HCl (sp. gr. 1.18) in a porcelain casserole, the dish being given a slight rotary motion. In the presence of some dyes the separated curd acquires a pink coloration.

For the separation and identification of coal tar dyes used in foods, see *Methods of Analysis*, A. O. A. C., 6th edition, 270-302 (1945).

REFERENCES

1. England, C. W., and Yongue, N. E.: The Use of Annatto as a Tracer in Cream for Manufacturing Purposes and Its Detection, *J. Dairy Sci. 23*:331 (1940).

2. *Methods of Analysis*, A. O. A. C., 6th edition (1945).)

3. *Standard Methods for the Examination of Dairy Products*, Amer. Pub. Health Assoc., 9th edition (1948).

Copper Determination in Milk

Apparatus

1. Pipette, 50 ml.
2. Evaporating dish, 100 ml.
3. Erlenmeyer flask with rubber stopper, 200 ml.
4. Gooch crucible and suction flask.
5. Nessler tubes, 12.
6. Hot plate.

7. Water bath or drying oven.
8. Asbestos fiber suspension in water.

Reagents

1. Hydrochloric acid, 1:1 dilution (HCl).
2. Hydrochloric acid, 1:4 dilution.
3. Nitric acid, fuming (HNO_3).
4. Nitric acid, 2 : 3 dilution.
5. Sulfuric acid, concentrated (H_2SO_4).
6. Sodium hydroxide solution, 25 per cent (NaOH).
7. Potassium ethyl xanthate, 0.1 per cent.
8. Methyl orange indicator.
9. Litmus paper.
10. Hydrogen sulfide (Kipp generator or cylinder).
11. Sodium thiosulfate solution ($Na_2S_2O_3$), 0.1N.

Preparation and Standardization of Copper Sulfate Standards

Weigh accurately 3.93 gm copper sulfate c. p. ($CuSO_4.5H_2O$) and make up to 1 liter in copper-free distilled water.

This contains 1 mg copper per ml. Standardize with standard sodium thiosulfate solution. Call this solution I.

Ten ml of solution I diluted to 100 ml with copper-free distilled water makes solution II.

Ten ml of solution II diluted to 100 ml with similar water makes solution III.

Solution III is used in preparing the standards and contains 0.01 mg copper per ml. Upon standardizing it will probably be found that solution I does not contain exactly 1 mg copper per milliliter. It is not necessary to try to adjust this, if the factor is taken into consideration in calculating the milligrams of copper per milliliter of solution III. For example the copper sulfate solution may be found to have 0.9853 mg copper per ml (solution 1). Solution III, therefore, instead of containing 0.01 mg per ml will contain 0.0098 mg per ml.

As this dilute solution tends to hydrolyze upon standing and so reduce the copper content by precipitation of basic copper sulfate, it is necessary to add 2 ml dilute sulfuric acid per 100 ml. of solution as a preventive.

A set of 9 standards in the following range is convenient:

No. 1 0.10 ml $CuSO_4$ solution 0.0010 mg Cu (if factors = 1)
No. 2 0.15 ml $CuSO_4$ solution 0.0015 mg Cu
No. 3 0.20 ml $CuSO_4$ solution 0.0020 mg Cu
No. 4 0.25 ml $CuSO_4$ solution 0.0025 mg Cu
No. 5 0.30 ml $CuSO_4$ solution 0.0030 mg Cu
No. 6 0.35 ml $CuSO_4$ solution 0.0035 mg Cu

No. 7 0.40 ml $CuSO_4$ solution 0.0040 mg Cu
No. 8 0.45 ml $CuSO_4$ solution 0.0045 mg Cu
No. 9 0.50 ml $CuSO_4$ solution 0.0050 mg Cu

Intermediate amounts may be estimated. For example, if a sample is deeper in color than the No. 4 standard but lighter in color than the No. 5 standard, it may be said to contain 0.0026, 0.0027, 0.0028, or 0.0029 mg of copper according to the intensity of the color.

Smaller amounts than 0.0010 mg copper cannot be accurately estimated by the above procedure but may be designated merely as less than this amount. If amounts larger than 0.0050 mg copper are present then an aliquot may be taken and diluted to 15 ml. The reading should fall within the given range.

Procedure

1. Pipette 50 ml of the milk to be tested into a small evaporating dish and heat on a water bath or in a drying oven at 100 C (212 F) until the moisture has all evaporated.

2. Burn over a free flame and finally incinerate to a white ash.

3. Dissolve ash in 1 : 1 HCl, raising the temperature to boiling for an instant to be sure all the ash has dissolved.

4. Make up to a known volume with distilled water (about 100 ml).

5. Neutralize this solution with 25 per cent NaOH, using litmus as an indicator. A heavy precipitate forms which should be dissolved by adding 1 : 4 HCl drop by drop. (It is very important to have the solution of exactly the right acidity before the potassium ethyl xanthate is added. That is the reason it is necessary to make the solution neutral to litmus, and then to adjust slightly to the acid side.)

6. Transfer the solution to a Nessler tube, make up to 50 ml, and add 10 ml of 0.1 per cent potassium ethyl xanthate. The degree of yellow color which develops is proportional to the copper content. It is necessary to make standards containing known quantities of copper in order to estimate the quantity of copper in the unknown solution. These standards should cover a range of 0, 0.01, 0.03, 0.05, 0.07, 0.09, 0.1, 0.2, and 0.3 mg copper per 50 ml respectively. Add one drop of 1 : 4 HCl to each of the standard tubes. If preliminary determinations or familiarity with the sample indicate that it contains copper in excess of the limit established by the standards, it is then necessary to make accurate dilutions appropriate to bring the sample within the established range.

The color changes on standing, hence it is necessary to make the color comparisons within ten minutes after preparation. This obviously necessitates the preparation of fresh standards each time a determination is made. It is a good practice to have on hand a solution containing 1 mg of copper per milliliter, which may be used for the preparation of standards.

Diacetyl and Acetylmethylcarbinol (Acetoin) Determination in Butter and Butter Starters

The following tests are very sensitive and capable of detecting minute quantities of diacetyl and acetylmethylcarbinol. The diacetyl in butter or butter starter may be distilled by steam. In the determination of acetylmethylcarbinol a separate sample is first treated with ferric chloride which oxidizes the acetylmethylcarbinol to diacetyl; the latter is then distilled as described.

Apparatus

1. Distilling flask, 2-liter.
2. Beakers, 200 ml, tall form.
3. Glass tubing.
4. Condenser.
5. Rubber tubing and stoppers.
6. Adapters.
7. Sintered or fritted glass Gooch crucibles (A. H. Thomas No. 4142, Medium Jena 1-G-3).
8. Clamps, clamp holders, and ring stand.
9. Crucible tongs.
10. Desiccator.
11. Drying oven, 120 C (248 F).
12. Analytical balance.

Reagents

1. Hydroxylamine hydrochloride, special quality (Eastman Kodak Company), 20 per cent solution.
2. Sodium acetate, c. p. (20 per cent solution).
3. Nickelous chloride ($NiCl_2$), c. p. (10 per cent solution).
4. Ferric chloride ($FeCl_3$), technical (40 per cent solution).

Procedure

A. DIACETYL DETERMINATION

1. Weigh 400 gm of butter or 200 gm of starter into a 2-liter distillation flask and steam distill into 10 ml of the following freshly mixed reagents:

To 4 ml of the 20 per cent hydroxylamine hydrochloride solution add 4 ml of the 20 per cent c. p. sodium acetate solution and 2 ml of the 10 per cent c. p. $NiCl_2$ solution. A quantity of this reagent may be made up, filtered, and kept in a glass-stopper reagent bottle. Use 10 ml for each test.

2. Continue the steam distillation until 100 ml have been collected. The glass tube at the end of the container must be submerged

in the reagent. After completing the distillation, heat the distillate to 80 C (176 F) and allow to stand over night.

3. Filter into a weighed dried Gooch crucible. Set the filtrate aside for another 24 hours and again filter out any precipitate that may have formed.

4. Dry the precipitate for 2 hours at 120 C (248 F) with not less than 20 inches of vacuum, or 1 hour in the solids oven of the Mojonnier tester.

5. Weigh, and express the results as milligrams of the nickel salt (nickel-dimethyl-glyoxime) per 400 gm of butter or 200 gm of starter.

B. ACETYLMETHYLCARBINOL DETERMINATION

1. Add 40 ml of a filtered solution of 40 per cent $FeCl_3$ to 400 gm of butter or 200 gm of starter in a distillation flask, and proceed as in the diacetyl determination.

The results of this distillation will give the combined diacetyl and acetylmethylcarbinol contents. This is the preferred method for analyzing butter and starter, as only one distillation is required.

To determine the amount of acetylmethylcarbinol present, subtract the results of the diacetyl determination from that for acetylmethylcarbinol. Express results as milligrams of the nickel salt per 400 gm of butter or per 200 gm of starter.

REFERENCES

1. Michaelian, M. B., Farmer, R. S., and Hammer, B. W.: The Relationship of Acetylmethylcarbinol and Diacetyl to Butter Cultures, *Iowa Agr. Exp. Sta.* Research Bul. 155 (1933).

2. Michaelian, M. B., and Hammer, B. W.: Studies on Acetylmethylcarbinol and Diacetyl in Dairy Products, *Ibid.* Research Bul. 179 (1935).

3. Hammer, B. W., Stahly, E. L., Werkman, C. H., and Michaelian, M. B.: Reduction of Acetylmethylcarbinol and Diacetyl to 2, 3-Butylene Glycol by the Citric Acid Fermenting Streptococci of Butter Cultures, *Ibid.* Research Bul. 191 (1935).

Differentiation of Oleomargarine, Butter and Renovated Butter

Procedure

A. FOAM TEST

Heat a small piece of the suspected material in a spoon or dish over a flame. Boil vigorously. Oleomargarine and renovated butter sputter and crackle nosily, and form little foam, whereas butter makes little noise and much foam.

B. WATERHOUSE TEST

1. Fill a 100 ml beaker or cup about half full with skim milk.
2. Heat to approximately 180 F (82 C).

3. Add from 3 to 5 gm of suspected material and stir with a small wooden spatula until the sample is melted.

4. Place the beaker in ice water and stir the sample constantly with the spatula until the fat hardens.

PRECAUTION: It is important to stir the melted fat continuously until it congeals.

Oleomargarine will collect in a sticky lump whereas butter and renovated butter will granulate.

The following table conveniently summarized the results.

Sample	Foam Test	Waterhouse Test
Butter	Foams.	Granulates
Renovated butter	Sputters	Granulates
Oleomargarine	Sputters	Collects in lumps

REFERENCES

Patrick: Household Tests for the Detection of Oleomargarine and Renovated Butter, U. S. Dept. Agr. *Farmers' Bul. 131* (1905).

Cochran, C. B.: The Detection of Renovated Butter, *Analyst 24*:88 (1899).

Sommer, H. H.: The Household Foam Test for Butter and Oleomargarine, *J. Dairy Sci. 2*:142 (1919).

Egg Yolk Determination in Dairy Products

The Smith method[1] for determining egg yolk in dairy products consists of the determination of the lipoid-phosphorus content of the egg ice cream mix, and the computation of the value obtained in terms of egg yolk, in which the quantity present varies only slightly under normal conditions.

Milk products in ice cream mix normally contain phospholipoids for which a correction must be made in estimating the quantity of actual egg lipoid-phosphorus present. Smith used the constant correction factor of 2.56 mg of lipoid-phosphorus per 100 gm of mix, which he obtained as an average on 21 mixes containing from 1.78 mg to 3.09 mg per 100 gm. It is thus obvious that the lipoid-phosphorus content of egg-free ice cream mixes is not as uniform as might be hoped for; this being further substantiated by other results tabulated in his work, showing lipoid-phosphorus contents of from 1.77 mg to 4.15 mg per 100 gm in egg-free mixes of various compositions ranging in fat content from 10.0 to 18.5 per cent. This indicates that the natural phospholipoid content is roughly in proportion to the fat content. In table 9 are given the results of analysis of several egg-free mixes which seem to bear this out reasonably well.

TABLE 9. FAT AND LIPOID-PHOSPHORUS CONTENTS OF EGG-FREE ICE CREAM MIXES

Sample	Fat per cent	Lipoid-phosphorus, mg per 100 gm	Lipoid-phosphorus in extract, mg per 100 gm
1	12.76	2.47	19.4
2	12.82	2.66	20.7
3	13.63	2.40	17.6
4	14.05	2.67	19.0
5	14.96	2.79	18.6
6	15.46	3.34	21.0
7	16.62	3.14	18.9
8	17.43	2.99	17.2
Average	14.72	2.81	19.1
Smith (average)	14.52	3.07	20.9
General average	14.62	2.94	20.0

The total fat present in an ice cream containing eggs is not entirely milk fat. It includes an appreciable quantity of egg yolk fat or lipoids. This would tend to give an erroneous milk fat value unless a correction is made. The correction is necessary also if the milk fat content is to be utilized in computing the natural milk lipoid-phosphorus to be subtracted from the total lipoid-phosphorus present in the egg mix.

TABLE 10. RATIO OF TOTAL LIPOIDS TO LIPOID-PHOSPHORUS IN EGGS[1]

Material	Analyst	Number of samples	Average total lipoids, per cent	Average total lipoid-phosphorus, per cent	Lipoids / Lipoid-phosphorus
Miscellaneous yolks	Smith	9	46.46	0.524	88.7
Whole eggs	J. L. P.	7	13.99	0.136	85.8

[1]Smith's values represent fresh and dried yolk products, including commercial material. J. L. Pearlman's values represent only fresh whole eggs which had a minimum manipulation.

In table 9 the lipoid-phosphorus values have also been computed on the basis of the total fatty extracts of the mixes, and an average value of 20.0 mg per 100 gm of fat is indicated. This figure is used for obtaining the variable correction factor in computing the true milk fat and egg yolk contents (0.2 mg phosphorus for each 1 per cent of milk fat).

The analytical procedure as recommended by Smith and others, with slightly modifications, constitutes a determination of the phosphorus in the fatty extract of the mix as obtained by the Mojonnier method of extraction.

Reagents

1. Chloroform ($CHCl_3$).

2. Magnesium nitrate [Mg $(NO_3)_2$] 50 per cent alcoholic solution.

3. Sulfuric acid (H_2SO_4) 1:1.

4. Sodium bisulfite (15 per cent) containing 0.5 per cent hydroquinone.

5. Ammonium molybdate; according to Griggs: dissolve 25 gm NH_4 molybdate c. p. in 300 ml distilled water plus 200 ml H_2SO_4 (3 : 5).

6. Potassium dihydrogen phosphate, standard solution containing 0.2193 gm per liter.

Procedure

1. Weigh 10 gm sample.

2. Make three thorough extractions and remove the solvents, heating on a steam bath with the aid of a gentle current of air from an electric fan. The extract is further dried to constant weight in an oven at 100 C (212 F) and the percentage of fat content calculated.

3. Dissolve the fatty extract in a small portion of $CHCl_3$, and transfer carefully, by means of a wash bottle containing $CHCl_3$, through a closely packed cotton pledget in a small funnel to a 50 ml volumetric flask, and make up to volume.

4. Transfer an aliquot part containing about 0.1 mg of phosphorus to a small flat bottom platinum dish, add 1 ml of 50 per cent $Mg(NO_3)_2$ (alcoholic), and evaporate to dryness on a steam bath with the aid of a gentle current of air from an electric fan.

5. Place the dish with contents on a hot plate and gradually increase the heat until maximum charring occurs. Complete ashing to a perfectly white residue in a muffle furnace at 500 C.

6. Moisten the residue thus obtained with a small quantity of distilled water and complete the solution by the addition of 2 ml of dilute H_2SO_4 (1 : 1).

7. Transfer the solution, which should be perfectly clear and colorless, to a test tube calibrated at 20 ml, washing the dish with small quantities of distilled water and adding the washings to the tube until the volume totals 20 ml.

8. Add the reagents as follows: 2 ml of sodium bisulfite (15 per cent, containing 0.5 per cent hydroquinone), and 2 ml of ammonium molybdate according to Griggs. A standard is prepared at the same time in an identical tube by treating 2 ml of a standard potassium dihydrogen phosphate solution (0.2193 gm per liter = 0.05 mg phosphorus per ml) with the acid, making up to volume with distilled water, and adding the reagents as specified above.

9. Mix well the contents of the tubes, immerse in a boiling water bath for 15 minutes, cool in running water, and make up to 50 ml volume prior to comparison in a colorimeter.

To ascertain the quantity of true milk fat in an egg ice cream, the egg-lipoid value must be subtracted from the total fat value obtained. The former can be obtained in close approximation from the total lipoid-phosphorus content, using Smith's fixed correction factor of 2.56 by means of the following formula:

$$L = \frac{(X - 0.02)\ 85.8E}{100}$$

in which:

L = Percentage of egg lipoids.
X = Percentage of lipoid-phosphorus in fatty extract.
E = Percentage of fatty extract in ice cream.
0.02 = Percentage of (average) lipoid-phosphorus in fatty extracts of egg-free mix.

In table 11 are given the analyses of known egg ice creams containing known quantities of milk fat showing the recovery from each formula.

To determine the amount of dry egg yolk solids Smith applies the formula:

$$E = \frac{(A - 2.56)\ 100}{777.7 - 2.56}$$

in which:

E = percentage of dry egg yolk.
A = lipoid-phosphorus in sample expressed in milligrams per 100 grams.
E = (A — 2.56) 0.129, in which
777.7 = milligrams of phosphorus per 100 gm of dry egg yolk, based on Hertwig's value for lipoid (P_2O_5) in commercial yolk (1.78 per cent).

To apply the variable correction factor the above formula must be changed thus:

$$E = \frac{(A - 0.2MF)\ 100}{777.7 - 0.2MF},$$

in which:

A = milligrams lipoid-phosphorus per 100 gm ice cream,
0.2 = milligrams lipoid-phosphorus per 1 per cent of actual milk fat present, and
MF = percentage of calculated milk fat.

The analyses of known egg ice creams and the recovery of yolk solids by each of the foregoing formulas are reported in table 11.

TABLE 11. RECOVERY OF LIPOID PHOSPHORUS ADDED TO ICE CREAM MIX IN FORM OF FRESH EGG YOLKS—PERCENTAGE OF EGG YOLK CALCULATED BY FORMULA BASED ON LIPOID PHOSPHORUS CONTENT

Lab. No.	Description	Fat and lipoids, per cent	Lipoid phosphorus (mg. per 100 grams)			Recovery of lipoid phosphorus, per cent	Calculated egg yolk by formula		Dry egg yolk actually present, per cent	Differences between* actual and calculated egg yolk	
			Found	Average	Added		Dry, per cent	Fresh, per cent		Fresh, per cent	Dry, per cent
562	Ice Cream Mix; egg-free; "A" Ice Cream Company	....	2.65 2.58	2.62	....	....	....	....	....	....	....
562-A	Ditto, containing 1 per cent fresh egg yolk, No. 564	12.86	6.48	6.48	3.96	98.50	0.50	1.00	0.50	0.0	0.0
562-B	Ditto, containing 2.91 per cent fresh egg yolk, No. 564	13.24	14.13	14.13	11.52	99.93	1.49	2.98	1.46	−0.07	−0.03
563	Ice Cream Mix; egg yolk-free; "B" Dairy Company	....	2.55 2.48	2.52	....	....	....	....	....	....	....
563-A	Ditto, containing 1 per cent fresh egg yolk, No. 564	11.34	6.46	6.46	3.96	99.67	0.50	1.00	0.50	0.0	0.0
563-B	Ditto, containing 2.91 per cent fresh egg yolk, No. 564	11.87	13.54 14.02	13.78	11.52	98.15	1.44	2.88	1.46	+0.03	+0.02
564	Fresh egg yolk prepared by separation of yolk from market eggs	34.52	402.0 389.3	395.6	....	....	....	....	50.01	....	....
695	Ice Cream Mix; egg-free; "A" Ice Cream Company	11.14	....	2.65	....	....	....	....	....	....	....
695-A	Ditto, containing 4.83 per cent fresh egg yolk, No. 696	12.27	20.71	20.71	20.06	91.20	2.34	4.68	2.47	+0.15	+0.13
695-B	Ditto, containing 1.12 per cent fresh egg yolk, No. 696	11.26	6.43	6.43	4.65	88.08	0.50	1.00	0.57	+0.12	+0.07
695-C	Ice Cream Mix No. 695, containing 7.92 per cent fresh egg yolk, No. 696	13.00	31.41	31.41	32.88	88.36	3.72	7.44	4.05	+0.48	+0.33
696	Fresh egg yolk prepared by separation of yolk of market eggs	33.09	412.7 418.0	415.3	....	....	....	....	51.11	....	....

695-D	Ice Cream Mix No. 695, containing 1 per cent fresh egg yolk, No. 696	11.50	7.02 6.05	6.54	4.15	[1]103.30 [1]90.30	[2]0.57 [2]0.45	[3]1.14 [3]0.90	0.51	−0.02	0.0
695 D	Ditto, after pasteurization for 40 min. at 145° F.	11.52	6.64	6.64	4.15	97.65	0.52	1.04	0.51	−0.04	−0.1
695-E	Ice Cream Mix No. 695, containing 3 per cent fresh egg yolk, No. 696	11.95	13.57	13.57	12.46	89.81	1.43	2.84	1.53	+0.16	+0.10
695-E	Ditto, after pasteurization for 40 min. at 145° F.	11.98	14.01	14.01	12.46	92.72	1.48	2.96	1.53	+0.04	+0.05
695-F	Ice Cream Mix No. 695, containing 5 per cent fresh egg yolk, No. 696	12.44	22.94	22.94	20.77	97.95	2.63	5.26	2.55	−0.26	−0.08
695-F	Ditto, after pasteurization for 40 min. at 145° F.	12.43	21.74	21.74	20.77	92.83	2.47	4.94	2.55	+0.06	+0.08
695	Base sample (mix); no egg; after pasteurization for 40 min. at 145° F.	11.25	2.73	2.73	......	......	......	......	......	......	......
848	Experimental egg-free mix, prepared in laboratory from milk, cream, condensed skim, sugar and gelatin	11.07	2.51 2.66 2.49	2.55	......	......	......	......	......	......	......
848-A	Mix No. 848, containing 3.83 per cent fresh egg yolk, No. 849	11.84	18.10 18.20	18.15	15.40	101.10	2.01	4.02	1.93	−0.19	−0.08
848-B	Mix No. 848, containing 1.14 per cent fresh egg yolk, No. 849	11.07	7.04 7.15	7.10	4.58	99.58	0.58	1.16	0.58	−0.02	0.0
848-C	Mix No. 848, containing 7.55 per cent fresh egg yolk, No. 849	12.54	31.37 31.66	31.52	30.36	95.78	3.73	7.46	3.81	+0.09	+0.08
849	Fresh egg yolk prepared by separation of yolks from market eggs	......	394.0 404.6 407.7	402.1	......	......	......	......	50.48	......	......
	Averages of all mixes (16 in all)	......	......	15.100	13.353	95.507	1.616	3.230	1.657	+0.034	+0.041

NOTE—Average percentage fresh egg yolk actually present in samples equals 3.263.

*Differences equal figures to be added to or subtracted from percentages found by formula to equal actual percentages of egg yolk.

[1] Average 96.80.

[2] Average 0.51.

[3] Average 1.02.

Samples 19 and 20 contained quantities unknown to the analyst at the time of analysis.

Owing to the narrow range of milk fat content of commercial ice cream as well as the small quantity of lipoid-phosphorus present therein, the use of the sliding scale for estimating the lipoid-phosphorus attributable to milk alone does not give values which vary widely from those obtained by using Smith's average value of 2.56 mg per 100 gm of mix.

[1]Smith, N. C.: Determination of Egg in Ice Cream. *California Agriculture Department Bulletin 19*:692-703 (1930).

Gelatin Detection in Dairy Products

Apparatus

1. Test tubes, 50 ml.
2. Glass funnel.
3. Filter paper.

Reagents

1. Acid mercuric nitrate solution (mercury dissolved in twice its weight of nitric acid, sp. gr. 1.42, and this solution diluted to 25 times its volume with water).

2. Reagent B—Saturated aqueous picric acid solution.

Procedure

1. To 10 ml of the sample, add an equal volume of reagent A; shake the mixture.

2. Add 20 ml of distilled water, shake again, allow to stand 5 minutes, and filter. If much gelatin is present, the filtrate will be opalescent and cannot be obtained quite clear.

3. To a portion of the filtrate in a test tube, add an equal volume of Reagent B. A yellow precipitate will be produced in the presence of a considerable amount of gelatin, while smaller amounts will be indicated by cloudiness. In the absence of gelatin, the filtrate will remain perfectly clear.

Heated Milk (Over 172 F.) Detection

There are occasions when it is desirable to learn whether or not milk has been heated higher than the pasteurization temperature. The following tests are applicable only in cases in which the milk has been heated over 172 F (77.8 C).

STORCH TEST

Reagents

1. Acidified hydrogen peroxide, 0.2 per cent. Dilute one part of 3 per cent H_2O_2 with 14 parts of distilled water, and add 0.1 per cent of H_2SO_4 by volume.

2. Paraphenylendiamine hydrochloride, 2 per cent aqueous solution. Dissolve in hot distilled water and filter. Do not use after 2 days.

Procedure

1. Pipette 10 ml of a sample into a test tube.
2. Add 2 drops of the acidified H_2O_2 and mix.
3. Add 2 drops of the paraphenylendiamine hydrochloride solution and mix thoroughly.

An intense blue indicates that the milk was not heated above 172 F (77.8 C). If the milk remains white, it is an indication that it has been heated to this temperature or higher.

NOTE: The principle of the test depends upon the destruction of the enzyme peroxydase.

ARNOLD TEST

Reagents

1. Guaiac, 10 per cent solution in acetone (prepared two or three days before use).

Procedure

To 10 ml of milk in a test tube, add about 8 drops of the 10 per cent solution of guaiac by allowing to run down the side of the tube. Milk not heated above 172 F (77.8 C) will show a blue color band at the point of contact between the two liquids. Milk heated above this temperature does not show any coloration.

REFERENCES

Jensen, C. O.: *Essentials of Milk Hygiene,* translation by Pearson, Lippincott, Philadelphia (1909).

Storch, V.: Chemical Method of Ascertaining Whether Milk or Cream Has Been Heated to at Least 80°. *Bied. Centr. 27*: 711-14 (1898); from *40 Beret. kgl. Vet. Landbohöjskoles Lab. landökon Fors.* Copenhagen 1-16 (1898).

Barthel, C.: *Methods Used in the Examination of Milk and Dairy Products* translated by Goodwin, Macmillan Co., New York (1910).

Lactic Acid Determination in Milk

Apparatus

1. Volumetric flask, 50 ml.
2. Graduated pipettes, 10 ml.
3. Funnels and filter papers.
4. Test tubes.
5. Screen, 24 mesh.

Reagents

1. Copper sulfate ($CuSO_4 \cdot 5H_2O$), 25 per cent solution.

2. Calcium hydroxide [$Ca(OH)_2$] suspension, prepared by slaking 300 gm of the best quality calcium oxide, using 1,400 ml water and working through a 24 mesh screen after it has stood 30 minutes.
3. Sulfuric acid (H_2SO_4), arsenic free, concentrated.
4. Solution of veratrole in water, 0.1 per cent (1, 2-dimethoxybenzene).
5. Lithium lactate, purified.
6. Tropeolin.
7. Fuchsin, 0.01 per cent solution.

Preparation of Standards

Natural standards may be prepared according to Hillig.[3] Dissolve 106.6 mg of purified lithium lactate in 100 ml of distilled water. One ml of this solution contains the equivalent of 0.1 per cent lactic acid. Consequently, to prepare a 0.01 per cent lactic acid standard, 5 ml of this solution is diluted to exactly 50 ml and mixed. One ml of this dilution is then placed in a test tube and 5 ml of concentrated H_2SO_4 added. The mixture is held in boiling water 5 minutes and cooled in ice water 3 minutes. Four drops of a 0.1 per cent solution of veratrole is then added and the color observed after 60 minutes. Additional standards may be made by diluting the original lithium lactate solution as required.

Artificial standards which are fairly satisfactory may be prepared according to the method of Nordbö.[6] Prepare just prior to use 25 ml of a 0.01 per cent solution of tropeolin 000 (Orange I). One-tenth ml of this solution diluted to 50 ml is equivalent to approximately 0.005 per cent lactic acid; 0.25 ml diluted to 50 ml is equivalent to approximately 0.01 per cent, and 0.9 ml diluted to 50 ml is equivalent to 0.02 per cent. Nordbö's standards are suitable only for concentrations below 0.03 per cent. Above this percentage the yellow color increases in intensity. Consequently, for a 0.05 per cent lactic acid standard, it is necessary to add 0.4 ml of 0.01 per cent tropeolin and 1.8 ml of 0.01 per cent fuchsin to 47.8 ml of water; for a 0.07 per cent standard, 1 ml of tropeolin and 2 ml of fuchsin to 47 ml of water; for a 0.10 per cent standard, 2 ml of tropeolin and 2 ml of fuchsin to 46 ml of water. These artificial standards should be checked against natural standards before using.

Procedure

1. Weigh 5 gm of milk in a 50 ml volumetric flask. Add 30 ml cold water and mix. Dilute a portion of a 25 per cent solution of $CuSO_4$ (1 : 3) and add drop by drop with agitation 2 to 8 drops of the dilute solution to the water and milk mixture until precipitation occurs.

2. Warm the flask in hot water until the contents reach 45 C (113 F). Add 6 ml of 25 per cent $CuSO_4$ solution and mix thoroughly by rotation. Hold the contents of the flask at 45-47 C (113-117 F) for 8 to 10 minutes.

3. With a bent wire, loosen any curd particles which may adhere to the flask.

4. Add 5 ml of $Ca(OH)_2$ suspension. Before mixing the added lime with the other constituents in the flask, water should be added to bring the contents to exactly 50 ml. Then mix until the contents are homogeneous and allow to stand at 45-47 C for 10 minutes.

5. Cool to 25 C (77 F) and filter, discarding the first few milliliters of filtrate.

6. Place 1 ml of the filtrate in a thoroughly clean and dry test tube. Hold the test tube in ice water and allow 5 ml of cold concentrated H_2SO_4 (arsenic free) to run down the side wall of the test tube. Mix gently while cooling.

7. Place the mixture in boiling water for exactly 5 minutes. Cool in ice water for 3 minutes. Add 4 drops of 0.1 per cent solution of veratrole in water and mix. Hold in ice water for 60 minutes and compare colors with artificial or natural standards.

The test is sensitive to about 0.005 per cent and is accurate to ± 0.004 per cent between the ranges of 0.005 and 0.12 per cent lactic acid.

REFERENCES

1. Derviz, G. V.: The Colorimetric Lactic Acid Determination According to Mendel and Goldscheider, *Zhur. Exptl. Biol. Med. 12*: 147-150 (1929), cited by *Chem. Abstr. 24*: 1880 (1930).

2. Fuchs, Hans J.: Einige Verbesserungen der kolorimetrischen Milchsäurebestimmung nach Mendol und Goldscheider, *Biochem. Ztschr. 217*:405-408 (1930).

3. Hillig, Fred: The Colorimetric Determination of Lactic Acid in Milk and Milk Products. *J. Assoc. Offic. Agr. Chemists 20*:130-140 (1937).

4. Mendel, Bruno: Zur Methode der kolorimetrischen Milchsäurebestimmung, *Biochem. Ztschr. 202*:390 (1938).

5. Mendel, Bruno, and Goldscheider, Ingeborg: Eine kolorimetrische Mikromethode zur quantitativen Bestimmung der Milchsäure in Blut, *Ibid. 164*:163-174 (1925).

6. Nordbö, Ragnvald: Zur Methode der Milchsäurebestimmung nach Mendel und Goldscheider, *Ibid. 271*: 213-215 (1934).

7. Troy, H. C., and Sharp, P. F.: Quantitative Determination of Lactic Acid in Dairy Products, *Cornell Univ. Agr. Exp. Sta. Memoir*, 179 (1935).

NOTE: Taken from article by B. Heinemann, Chief Chemist, Producers' Creamery Co., Springfield, Mo., entitled: A Rapid Colorimetric Method for the Determination of Lactic Acid in Milk. *J. Dairy Sci. 23*:969 (1940).

Oiling Off Test for Cream

(Massachusetts State College Method)

Apparatus

1. Babcock skim milk test bottle, of type with stem extending down into bottle proper to within ½ inch of bottom and stem should have no side hole in base of neck as in some styles.
2. Babcock centrifuge.
3. Pipette, 1 ml.

Procedure

1. Pipette 1 ml of cream into a skim milk test bottle.
2. Add water at least 200 F (93 C) to within ½ inch of the base of the neck. Mix thoroughly while adding.
3. Centrifuge for not longer than 10 seconds at the standard Babcock centrifuge speed (see page 183).
4. Remove the bottle from the centrifuge and tap the side of the bottle with the finger to break the very thin film of cream which has collected on the top. Do not agitate.

NOTE: If the machine is whirled longer than ten seconds a thick film will form on top which will be difficult to break and the pieces will clog the graduated neck.

5. Add water at 200 F (93 C) or above and bring the column up to the top of the neck and centrifuge for 5 minutes.
6. Read the oil layer as soon as the bottle is removed from the centrifuge. Each small division of the graduated scale should be read as 1.

Interpretation: Creams yielding a reading of 3 or more with this test will show a noticeable separation when used in coffee.

REFERENCE

Jenkins, Hubert, and Mack, Merrill J.: Study of Oiling Off of Cream in Coffee. *J. Dairy Science 20*:723 (1937).

Preservative Detection

FORMALDEHYDE

Leach Test[1]

Mix about 10 ml of milk with an equal volume of concentrated hydrochloric acid containing 1 ml of 10 per cent ferric chloride solution to each 500 ml of acid. Heat slowly directly over a gas flame for about 5 minutes to 80-90 C (176-194 F). Rotate the casserole to break up the curd. A violet color indicates formaldehyde. Confirm the test with a control sample free from formaldehyde.

Hehner Test[1]

To about 10 ml of milk in a wide mouth test tube, add about half the volume of concentrated commercial sulfuric acid, pouring the acid carefully down the side of the tube so that it forms a layer at the bottom without mixing with the milk. A violet or blue color at the junction of the two liquids indicates formaldehyde. The test is given only in the presence of a trace of ferric chloride or other oxidizing agents. This test may be combined with the Babcock test for fat, by noting whether a violet color forms on addition of the commercial H_2SO_4 to the milk in the test bottle. Confirm the test with a control sample free from formaldehyde. This test is sensitive to one part in 10,000.

SALICYLIC ACID

Ferric Chloride Test[1]

1. Acidify 100 ml of milk with 5 ml of hydrochloric acid (1 : 3). Shake until curdled.
2. Filter.
3. Extract with from 50-100 ml of ether.
4. Wash the ether layer with two 5 ml portions of water.
5. Evaporate the greater portion of the ether in a porcelain dish on a steam bath, allow the remainder to evaporate spontaneously.
6. Add a drop of 0.5 per cent neutral ferric chloride solution.
7. A violet color indicates salicylic acid.
8. Confirm with a control sample free from salicylic acid.

Jorissen Test[1]

1. Acidify, filter, and extract a portion of the milk as directed under the ferric chloride test.
2. Dissolve the residue from the ether extract in about 1 ml of hot water.
3. Cool 10 ml of the ether extract in a test tube, add 4 or 5 drops of a 10 per cent potassium nitrite solution, 4 or 5 drops of 50 per cent acetic acid, and 1 drop of a 10 per cent cupric sulfate solution; mix thoroughly.
4. Boil for 30 seconds and allow to stand 1 or 2 minutes.
5. In the presence of salicylic acid a blood red color will develop.
6. Confirm with a control sample free from salicylic acid.

BENZOIC ACID

1. Acidify 100 ml of milk with 5 ml of hydrochloric acid (1 : 3). Shake until curdled.
2. Filter.
3. Extract with 50-100 ml of ether.
4. Wash the ether layer with two 5 ml portions of water.
5. Evaporate the greater portion of the ether in a porcelain dish on a steam bath and allow the remainder to evaporate spontaneously.
6. If benzoic acid is present in considerable quantity, it will crystallize from the ether in shining leaflets and have a characteristic odor on heating.

7. Dissolve the residue in hot water, divide into two portions, and test as directed below. The residue may also be purified by sublimation and the melting point determined.

a. Make the solution alkaline with a few drops of ammonium hydroxide, expel the excess of ammonia by evaporation, dissolve the residue in a few milliliters of hot water, filter if necessary, then add a few drops of a neutral 0.5 per cent ferric chloride solution. A salmon colored precipitate of ferric benzoate indicates the presence of benzoic acid.

b. Add to the water portion 1 or 2 drops of a 10 per cent solution of sodium hydroxide and evaporate to dryness. To the residue add from 5 to 10 drops of concentrated sulfuric acid and a small crystal of potassium nitrate. Heat for 10 minutes in a glycerol bath at 120-130 C (248-266 F) or for 20 minutes in a boiling water bath. The temperature must not exceed 130 C (266 F). After cooling add 1 ml of water, make distinctly ammoniacal, and boil the solution to decompose any ammonium nitrate that may have been formed. Cool, pour into a test tube, and add a drop of fresh colorless ammonium sulfide, without allowing layers to mix.

A red-brown ring indicates benzoic acid. On mixing, the color diffuses through the whole liquid, and on heating, finally changes to greenish yellow. This differentiates benzoic acid from salicylic acid or cinnamic acid. The last two form colored compounds, which are not destroyed by heating. Confirm the test with a control sample free from benzoic acid.

CARBONATES

To about 10 ml of milk in a test tube, add 10 ml of alcohol and a few drops of a 1 per cent aqueous solution of rosalic acid, and mix. If a carbonate is present, a rose red color appears, whereas pure milk shows only a brownish coloration. Confirm with a control sample free from carbonates.

BORIC ACID[1]

Preliminary Test

1. Immerse a strip of turmeric paper in a sample of milk previously acidified with hydrochloric acid in the proportion of 7 ml of strong acid to each 100 ml of milk. Allow the paper to dry spontaneously. If borax or boric acid is present, the paper will acquire a characteristic red color. The addition of ammonium hydroxide will change the paper to a dark green, but the red color may be restored by HCl. Confirm the test with a control sample free from boric acid.

Confirmatory Test

1. Make about 25 ml of the sample strongly alkaline with lime water.

2. Evaporate to dryness on a water bath.

3. Ignite the residue at a low red heat to destroy organic matter.

4. Digest with about 15 ml of water, add concentrated HCl drop by drop until the ignited residue is dissolved, and then add 1 ml in excess.

5. Saturate a piece of turmeric paper with this solution, and allow the paper to dry without the aid of heat.

6. In the presence of borax or boric acid, the color change will be the same as described under the preliminary test.

REFERENCES

1. For details of official procedures, see *Official and Tentative Methods of Analysis* of the Association of Official Agricultural Chemists, 6th edition (1945), pp. 523-537; *Standard Methods for the Examination of Dairy Products*, 9th edition.

2. Richmond and Boesley: Note on the Detection of Formalin, *Analyst*, *20*:154 (1895).

3. Hehnert: Note on the Detection of Formalin, *Ibid.* 155 (1895).

4. Leach, A. E.: *Food Inspection and Analysis*, John Wiley and Sons, New York, 1913.

5. Gettler, A. O.; Critical Study of Methods for the Detection of Methyl Alcohol, *J. Biol. Chem.*, *42*: 311 (1920).

6. Woodman, A. G.: *Food Analysis*, McGraw-Hill Book Co., New York, 1941, p. 127.

7. Rupp, P.: The Detection of Hypochlorites and Chloramines in Milk and Cream, *U. S. Dept. Agr.* Bull. 1114 (1922).

Solubility Index of Dry Whole Milk

Apparatus

1. Conical sediment tubes, 50 ml, specially graduated in 0.1 ml divisions from 0 to 1 ml, in 0.2 ml divisions from 1 ml to 2 ml, in 0.5 ml divisions from 2 ml to 10 ml, and in 1 ml division from 10 ml to 20 ml. The tubes should contain a 50 ml mark at least ½ inch from the top.

2. Suitable glass siphon tubes.

3. A.D.M.I. solubility index mixer, with special mixing jars, available from Central Scientific Co., Chicago, Ill.

4. Wood adapter or rubber pad to hold the conical sediment tube in the centrifuge cup.

5. Centrifuge, unheated, and not operated by steam turbine.

Procedure

1. Prepare sample as described on page 238.

2. Add 13 gm of dry whole milk to 100 ml of distilled water at a temperature of 24 C (75 F) in special mixing jar.

3. Place the jar in the mixer and stir for exactly 90 seconds. If there is an immediate need for reuse of the mixing jar the entire mix sample may, at this point, be transferred to a suitable tumbler.

4. Allow the sample to stand until the foam has separated sufficiently to permit its complete removal by a spoon. The period of standing after mixing shall not exceed 15 minutes.

5. After removal of the foam, mix the sample thoroughly with a spoon for five seconds and immediately fill the conical tube to the 50 ml mark.

6. Centrifuge the tube 5 minutes at the required r.p.m. (see page 183).

7. Immediately siphon off the supernatant liquid to within 2 ml of the surface of the sediment level, using care not to disturb the sediment layer.

8. Add about 25 ml of distilled water at a temperature of 75 F (24 C) and shake the tube to disperse the sediment, dislodging it, if necessary, with a wire.

9. Fill the tube to the 50 ml mark with distilled water at a temperature of 75 F (24 C); invert and shake to mix the contents thoroughly and again centrifuge at the required r.p.m. for five minutes.

10. Hold the tube in a vertical position with the upper level of the sediment on a level with the eye and report the milliliters of sediment in the tube to the nearest graduated scale division. The sediment is easily distinguished when the tube is held between the eye and a strong source of light.

NOTE: It is important that the water used in determining the solubility be free from sediment and always at a temperature of 75 F (25 C). Periods of stirring and centrifuging must also be exactly observed if duplicate results are to check. Do not use a steam turbine centrifuge. Frequently check the speed of the electric centrifuge.

Solubility Index of Nonfat Dry Milk Solids

Apparatus and procedure is identical with that given for Dry Whole Milk except for step 2 of procedure where a 10 gm sample of Nonfat Dry Milk Solids is liquefied in 100 ml of distilled water at 75 F (24 C).

Stiffness and Stability Determination of Whipped Cream

The methods usually employed for measuring vicosity are as a rule too complicated to admit of practical application for detecting the relative stiffness of different samples of whipped cream.

PRESSURE METHOD

NOTE: The following simple method will enable any one to test the stiffness of different whipped creams very satisfactorily. It is, however, purely a relative measure intended for direct comparison of different creams and the determinations are not comparable with those obtained on the basis of other standards.

Apparatus

The apparatus is in the form of a balance. The right arm of the balance is divided into 60 equal spaces, zero being at the end, and is provided with a sliding weight. To the left arm a metal disc is suspended of such weight as to form a balance when the sliding weight is at zero. For each space that the sliding weight is moved the pressure exerted increases approximately 1 gm per square inch. The scale reading used may be interpreted as the approximate pressure per square inch in grams required to displace the cream. The pressure required to force the disc into the whipped cream is determined and used as a measure of the stiffness or quality of the whip. (Figures 128 and 129.)

Figure 128.
Balance for determining the relative stiffness of whipped creams. Courtesy of U. S. Department of Agriculture.

The following description may be applied to the various scale readings:

Scale Reading	*Description*
Below 6	failed to whip
7 to 12	very poor whip
13 to 15	poor whip
16 to 20	fair whip
21 to 24	satisfactory whip
25 to 30	good whip
31 to 35	very good whip
36 and above	excellent whip

REFERENCE

Babcock, C. J.: The Whipping Quality of Cream, *U. S. Dept. Agri.* Bull. 1075 (1922).

DRAINAGE METHOD

Apparatus

1. Glass funnel, large enough to hold about 1 pint of whipped cream.

2. Copper screen, funnel shaped, to fit into the size of funnel mentioned above. The screen should be about nine mesh to the inch.
3. Graduated cylinder.
4. The usual equipment for whipping cream.

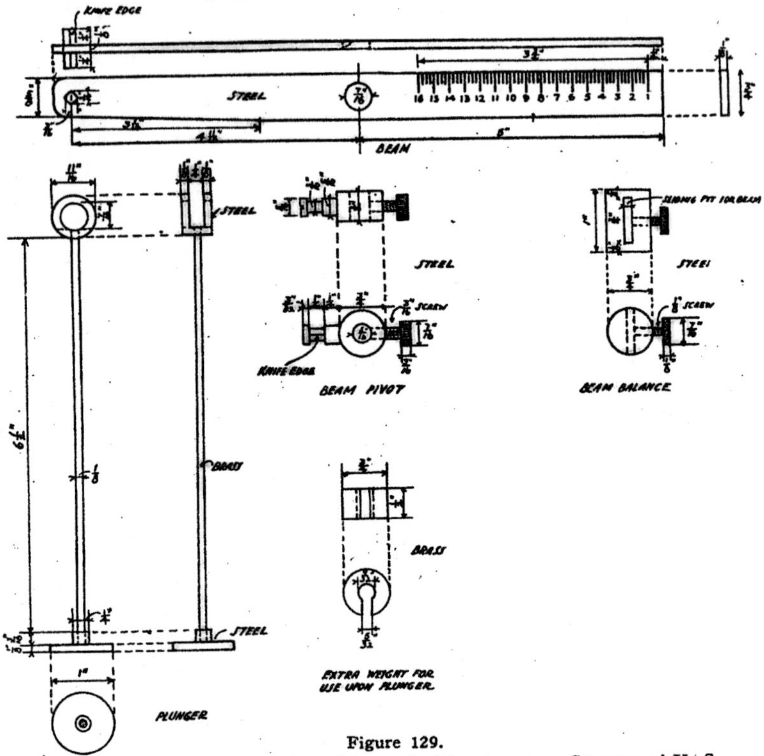

Figure 129.
Apparatus used for measuring stiffness of whipped cream Courtesy of U. S. Department of Agriculture.

Procedure

1. Place 1 pint of the whipped cream into the funnel with the copper screen in place. The test should be made at a temperature of about 40 F (4.4 C).
2. Collect the leakage from the funnel into the graduated cylinder.
3. Measure the stability of the whipped cream by comparing the number of milliliters collecting in the graduated cylinder for a given period and by comparing the apparent stiffness of the cream in the funnel.

REFERENCE

Dahlberg, A. C., and Hening, J. C.: Viscosity, Surface Tension, and Whipping Properties of Milk and Cream. *New York Agr. Exper. Sta. Tech. Bul.* 113 (1925).

Sucrose and Lactose in Dairy Products

COPPER REDUCTION METHOD (FOR SIMULTANEOUS DETERMINATIONS)

Apparatus

1. Gooch crucible.
2. Asbestos mat prepared as follows:

NOTE: The asbestos used in Gooch crucible should be of the amphibole variety. It must have been treated with dilute hydrochloric acid (1:3) for two or three days, washed free from acid, digested with 10 per cent sodium hydroxide solution for same time, then treated with hot alkaline tartrate solution, washed free from alkali, digested several hours with dilute nitric acid (1:3), washed free from acid, and kept in distilled water. Have a mat of this asbestos about ¼ in. thick on the bottom of Gooch crucible.

3. Volumetric flask, 250 ml.
4. Funnel.
5. Filter paper.
6. Beaker, 400 ml.
7. Ringstand.
8. Wire gauze.
9. Bunsen burner.
10. Wash bottle.

Reagents

1. Fehling's copper sulfate solution (see page 391).
2. Sodium hydroxide solution, (0.5 N NaOH).
3. Hydrochloric acid, (HCl) diluted 1 : 1.

Procedure

1. Weigh 10 gm of the well mixed sample (20 gm in the case of ice cream) into a 250 ml volumetric flask.

2. Add about 125 ml of boiling distilled water, rinsing down all material adhering to the neck.

3. Rotate briskly two or three minutes and cool to 20 C (68 F).

4. Add gradually, with vigorous shaking, 10 ml of Fehling's copper sulfate solution (see page 391) followed by 6 ml of 0.5 N NaOH.

5. Fill to 250 ml mark with distilled water and then add 1.5 ml more water (3.2 ml in the case of ice cream) to allow for the volume of precipitated protein and fat.

6. Mix well and filter, rejecting the first few milliliters of the filtrate. The liquid should have a faint blue color, indicating a slight excess of copper.

7. Determine reducing sugars in 50 ml of the filtrate by the regular Munson-Walker procedure, beginning with step 6 on page 391.

8. Filter rapidly into a 250 ml volumetric flask and wash well with about 80 ml of hot water.

NOTE: Use a tared Gooch filter and a clean suction flask if lactose is to be determined. For condensed milk, use the "1 lactose, 4 sucrose" column in Munson & Walker Table, *Methods of Analysis,* A.O.A.C., in 6th edition (1945), p. 830. Use the correction shown by the graphs in determining lactose in ice cream, remembering that the correction is to be subtracted from the gross weight of cuprous oxide.

9. Add 34 ml of 1 : 1 HCl and place the flask immediately in an actively boiling water bath.

NOTE: The contents must be at such a temperature as will give one of 21.1 C (70 F) after the addition of the acid. With fast work very little adjustment will be necessary.

10. Leave in the bath for exactly 5 minutes, then remove and cool rapidly.

NOTE: The safest plan is to hold for ½ minute, with constant rotation, under the hot water tap, and then an equal time under the cold top.

11. Neutralize with 50 per cent NaOH solution, taking care not to have either a local or a general excess of alkali in the warm solution lest some of the sugar be destroyed.

12. Cool to room temperature, make up to mark, shake well and run a Munson-Walker reducing sugar determination (beginning with step 6, page 391) on a 50 ml aliquot (equal to 0.4 gm of condensed milk or 0.8 gm of ice cream).

13. Filter through a tared Gooch filter, wash four or five times with hot water and once with alcohol, and dry 30 minutes at 100 C (212 F).

14 Add 1.6 mg to the weight of the cuprous oxide, calculate sucrose from the tables and divide by 0.004 to obtain the percentage of sugar. In the case of ice cream the amount to add is 1.0 mg, and the percentage of sucrose is found by dividing by 0.008.

NOTE: It is evident, that if invert sugar, or some reducing sugar other than lactose, is present it will be calculated as lactose. This would be the case of ice cream containing fruit juices, malt sugar, dextrose, invert syrups, corn syrups, or cornmeal syrups. It might also be true in condensed milks which had undergone slight decomposition.

NOTE: Excellent information on many ways of determining lactose in dairy products, with a lengthly bibliography, together with a description of a method for the simultaneous determination of sucrose and lactose in dairy products by copper reduction, can be found in a paper by White, as follows:

White, W. B.: The Simultaneous Determination of Sucrose and Lactose in Dairy Products by Copper Reduction. *Department of Agriculture and Markets, State of New York, Agr. Bull.* 234 (1930).

LACTOSE
(Munson and Walker Method)

Apparatus

Same apparatus as under copper reduction method (see page 389).

Reagents

1. Fehling solution: Prepare by mixing immediately before use, equal volumes of solutions (a) and (b).

Solution (a): Dissolve 34.639 gm of copper sulfate ($CuSO_4.5H_2O$) in water. Dilute to 500 ml and filter through prepared asbestos.

Solution (b): Dissolve 173 gm of Rochelle salt and 50 gm of sodium hydroxide in water. Dilute to 500 ml, allow to stand for two days, and filter through prepared asbestos.

2. Sodium hydroxide, 0.5 N NaOH.
3. Nitric acid, (HNO_3), diluted 1 : 3.

Procedure

1. Prepare a tared Gooch crucible.

2. Weigh 20 gm of well mixed ice cream into a 250 ml volumetric flask and add about 125 ml of boiling water, rinsing down all material adhering to the neck.

3. Rotate briskly two or three minutes, cool to 20 C (68 F) and add gradually, with vigorous shaking, 10 ml of Fehling's $CuSO_4$ solution followed by 6 ml of 0.5 N NaOH solution.

4. Make up to mark, and add 3.2 ml of water to allow for the volume of precipitated protein and fat.

5. Shake well and filter, rejecting the first few milliliters of the filtrate. The liquid should have a faint blue color showing a slight excess of copper.

6. To 50 ml of this filtrate in a 400 ml alkali resistant beaker add 25 ml each of the $CuSO_4$ and alkaline tartrate solution ("a" and "b" of Fehling's solution).

7. Heat the beaker on asbestos gauze over a Bunsen burner. Regulate the flame so that the boiling begins in 4 minutes and continues boiling for exactly 2 minutes. (This is very important.) Keep the beaker covered with a watch glass during the heating.

8. Filter the hot solution at once through an asbestos mat in a porcelain Gooch crucible using suction. Wash the precipitate of cuprous oxide thoroughly with water at 60 C (140 F), then with 10 ml of alcohol, and finally with 10 ml of ether.

9. Dry the precipitate for 30 min. in a water oven at 100 C (212 F), cool, and weigh. Calculate the weight of metallic copper. using the factor 0.8882. Obtain from Munson and Walker tables the weight of lactose equivalent to the weight of copper found.

NOTE: Munson and Walker tables are found in *Methods of Analysis*, A.O.A.C., 6th edition (1945), p. 830; or in Woodman, *Food Analysis*, 4th ed. (1941), pp. 271-3; Leach, *Food Inspection and Analysis*, and many other books on food analysis and sugar.

LACTOSE
(Polariscopic Method)

Apparatus

1. Polariscope.
2. Flask, calibrated, 102.6 ml.
3. Quevenne lactometer.

Reagent

1. Acid mercuric nitrate solution. Dissolve mercury in twice its weight of strong nitric acid and dilute with an equal volume of water; or,

2. Mercuric iodide solution. Dissolve 33.2 gm of potassium iodide and 13.5 gm of mercuric chloride in 20 ml of glacial acetic acid and 640 ml of water.

Procedure

1. Determine the specific gravity of the milk.

2. Measure the quantity of milk indicated in table 12 into a flask graduated at 102.6 ml. The volume of milk should be measured at the same temperature as that at which the lactometer reading is made.

TABLE 12. VOLUMES OF MILK CORRESPONDING TO A LACTOSE DOUBLE NORMAL WEIGHT

Specific gravity of milk	Volume of milk for lactose double normal weight (Ventzke scale) ml	Specific gravity of milk	Volume of milk for lactose double normal weight (Ventzke scale) ml
1.024	64.25	1.030	63.90
1.025	64.20	1.031	63.80
1.026	64.15	1.032	63.75
1.027	64.05	1.033	63.70
1.028	64.00	1.034	63.65
1.029	63.95	1.035	63.55

3. Add 1 ml of the acid mercuric nitrate solution or 30 ml of the mercuric iodide solution and fill to the mark (an excess of these reagents does no harm).

4. Shake, and clarify by passing through a dry filter.

5. Adjust to 20 C (68 F).

6. Polarize. If a 200 mm tube is used, divide the polariscope reading by 2 (or if a 400 mm tube is used, by 4) to obtain the percentage of lactose in the sample.

LACTOSE DETERMINATION WITH PICRIC ACID

This procedure is a modification of the method described by Folin and Denis.[8] A correction table (table 13) is included to overcome the necessity of having more than one standard solution. Futhermore, the procedure is adaptable to some dairy products as well as for milk.

TABLE 13. FACTORS FOR CONVERTING COLORIMETRIC READING TO PERCENTAGE OF LACTOSE

A. Colorimeter reading of unknown compared with standard color at 20.
B. Factor to use in calculation of results.

A	B	A	B	A	B	A	B	A	B
13.0	1.465	17.0	1.164	21.0	0.959	25.0	0.832	29.0	0.733
13.1	1.455	17.1	1.158	21.1	0.956	25.1	0.829	29.1	0.731
13.2	1.446	17.2	1.152	21.2	0.952	25.2	0.826	29.2	0.728
13.3	1.437	17.3	1.146	21.3	0.948	25.3	0.823	29.3	0.726
13.4	1.428	17.4	1.139	21.4	0.945	25.4	0.821	29.4	0.724
13.5	1.419	17.5	1.133	21.5	0.941	25.5	0.818	29.5	0.722
13.6	1.410	17.6	1.127	21.6	0.937	25.6	0.815	29.6	0.720
13.7	1.401	17.7	1.121	21.7	0.934	25.7	0.812	29.7	0.718
13.8	1.392	17.8	1.115	21.8	0.930	25.8	0.810	29.8	0.716
13.9	1.384	17.9	1.109	21.9	0.927	25.9	0.807	29.9	0.715
14.0	1.375	18.0	1.103	22.0	0.924	26.0	0.805	30.0	0.713
14.1	1.367	18.1	1.097	22.1	0.921	26.1	0.802	30.1	0.712
14.2	1.359	18.2	1.091	22.2	0.917	26.2	0.800	30.2	0.710
14.3	1.351	18.3	1.085	22.3	0.914	26.3	0.797	30.3	0.708
14.4	1.343	18.4	1.079	22.4	0.910	26.4	0.795	30.4	0.706
14.5	1.335	18.5	1.073	22.5	0.907	26.5	0.792	30.5	0.704
14.6	1.328	18.6	1.067	22.6	0.904	26.6	0.790	30.6	0.702
14.7	1.320	18.7	1.062	22.7	0.901	26.7	0.787	30.7	0.700
14.8	1.312	18.8	1.056	22.8	0.898	26.8	0.785	30.8	0.698
14.9	1.305	18.9	1.051	22.9	0.895	26.9	0.782	30.9	0.696
15.0	1.298	19.0	1.045	23.0	0.892	27.0	0.780	31.0	0.694
15.1	1.291	19.1	1.040	23.1	0.889	27.1	0.777	31.1	0.693
15.2	1.284	19.2	1.035	23.2	0.886	27.2	0.775	31.2	0.691
15.3	1.277	19.3	1.030	23.3	0.883	27.3	0.773	31.3	0.689
15.4	1.270	19.4	1.025	23.4	0.880	27.4	0.771	31.4	0.687
15.5	1.263	19.5	1.021	23.5	0.877	27.5	0.768	31.5	0.686
15.6	1.256	19.6	1.016	23.6	0.874	27.6	0.766	31.6	0.684
15.7	1.249	19.7	1.012	23.7	0.871	27.7	0.764	31.7	0.683
15.8	1.242	19.8	1.008	23.8	0.868	27.8	0.762	31.8	0.681
15.9	1.235	19.9	1.004	23.9	0.865	27.9	0.759	31.9	0.679
16.0	1.229	20.0	1.000	24.0	0.862	28.0	0.757	32.0	0.677
16.1	1.222	20.1	0.995	24.1	0.858	28.1	0.755	33.0	0.622
16.2	1.215	20.2	0.991	24.2	0.855	28.2	0.752	34.0	0.647
16.3	1.208	20.3	0.986	24.3	0.852	28.3	0.750	35.0	0.633
16.4	1.201	20.4	0.982	24.4	0.849	28.4	0.748	36.0	0.619
16.5	1.195	20.5	0.978	24.5	0.846	28.5	0.746	37.0	0.606
16.6	1.189	20.6	0.974	24.6	0.843	28.6	0.744	38.0	0.594
16.7	1.183	20.7	0.970	24.7	0.840	28.7	0.741	39.0	0.582
16.8	1.176	20.8	0.966	24.8	0.837	28.8	0.738	40.0	0.571
16.9	1.170	20.9	0.963	24.9	0.834	28.9	0.735		

A saturated solution of picric acid overcomes the error, otherwise present, by using a table of factors. This table not only makes the procedure simpler but shortens the calculation of the results.

Apparatus

1. Volumetric flask, 100 ml.
2. Colorimeter cup.
3. Westphal balance.
4. Pipette, 2 ml.
5. Water bath.
6. Mojonnier weighing cross.

Reagents

1. Picric acid, saturated solution in distilled water. Picric acid goes into solution very slowly, so a supersaturated solution is prepared by heating, and the excess acid is allowed to crystallize out at room temperature before the solution is used.

2. Sodium carbonate, pure anhydrous, dissolved in distilled water to the strength of 22 gm per 100 ml. Heating facilitates complete solution.

3. Standard lactose solution. Dissolve exactly 1 gm of pure dry lactose monohydrate in 1 liter of distilled water (or proportionate amounts to other volumes). For exacting work, a determination should be made on the lactose as a check on its purity before making up the standard.

NOTE: A standard may be made up in saturated picric acid and used for six or seven weeks with safety.

Preparation of Fresh Standard Color

1. Transfer 10 ml of the standard lactose solution to a 100 ml volumetric flask, add 20 ml of saturated picric acid (10 ml when the standard is made up in saturated picric, plus 10 ml of water) and 10 ml of 22 per cent sodium carbonate.

2. Mix, insert cork loosely, and heat in a boiling water bath for 15-25 minutes.

NOTE: The unknown solutions of the product to be tested must be prepared as subsequently described and heated simultaneously in the water bath with the standard color solutions.

3. Cool in cold water, simultaneously with unknowns, and dilute to 10 ml.

4. Transfer a portion of the standard color solution to a colorimeter cup and place on the left side of the colorimeter, setting the instrument at 20. (The table of factors is based on the standard color being set at 20.)

5. Compare the unknown solutions with the standard color and record the readings.

NOTE: Owing to the slight fading in color, it is advisable to change the standard color in the cup with each unknown and reset at 20. It is necessary to prepare a new standard color with each set of determinations, and for accurate results it is essential to prepare the standard color along with the unknown under identical conditions.

Procedure for Milk

1. Determine the specific gravity with a Westphal balance or an accurate lactometer.
2. Pipette exactly 2 ml of milk into a 100 ml volumetric flask, half filled with saturated picric acid solution.
3. Fill the flask to the 100 ml mark with saturated picric acid, mix, and filter.
4. Transfer 10 ml of the filtrate to another 100 ml volumetric flask; add 10 ml distilled water, 10 ml saturated picric acid, and 10 ml of 22 per cent sodium carbonate.
5. Mix, insert cork loosely, and heat simultaneously with standard color solution in a boiling water bath for 15-25 minutes.
6. Cool in cold water simultaneously with the standard color; dilute to 100 ml and mix.
7. Transfer a portion of the colored solution to the colorimeter cup and place on the right side of the colorimeter. The color is matched with the standard color in the usual manner. The average of five readings constitutes the color value.

Calculation

The percentage of lactose is found by obtaining the factor corresponding to the color value of the unknown solution from table 13 (page 393) and using either of the following formulas.

$$\frac{\text{Factor} \times 10}{\text{Weight of sample}} = \text{per cent lactose}$$

$$\frac{\text{Factor} \times 5}{\text{Sp. gr. of milk}} = \text{per cent lactose}$$

Procedure for Cream

Using a Mojonnier pipette and weighing cross, accurately weigh about 2 ml of cream. Transfer into a 100 ml volumetric flask previously half filled with saturated picric acid. Fill to the mark with saturated picric acid, shake, and filter. The subsequent procedure and calculation is then the same as for milk.

Procedure for Whey

Approximately 2 gm, accurately weighed, or exactly 2 ml of known specific gravity of sample is used, with procedure the same as that given for milk.

TABLE 14. COMPARISON OF LACTOSE CONTENT OF VARIOUS PRODUCTS BY THE COLORIMETRIC AND GRAVIMETRIC METHODS

Product	Sample No.	Colori-metric method (A) per cent	Fehling's method (B) per cent	Variation A-B per cent	Colori-metric method (ml.) per cent
Milk	1	4.75	4.82	—0.07	4.75
	2	4.70	4.76	—0.06	
	4	4.81	4.80	0.01	4.75
	5	5.69	6.73	—0.04	5.74
	6	3.75	3.70	0.05	3.81
	13	5.32	5.26	0.06	5.31
	Average			0.05	
Cream	10	3.67	3.72	—0.05	3.68
	19	4.88	4.88	0.00	4.89
	20	3.71	3.72	—0.01	3.72
	Average			0.02	
Whey	16	4.65	4.65	0.00	4.70
	17	5.28	5.35	—0.07	5.22
	Average			0.04	
Evaporated milk	7	9.93	10.14	—0.21	
	9	7.86	7.72	0.14	7.86
	18	9.59	9.45	0.14	9.62
	24	9.34	9.30	0.04	9.41
	Average			0.12	
Powdered milk	3	49.90	50.48	—0.58	50.05
	14	49.40	49.56	—0.16	49.75
	Average			0.37	
Ice cream	11	4.39	4.41	—0.02	
	12	3.87	3.96	—0.09	
	22	4.46	4.52	—0.06	4.45
	23	6.69	6.89	—0.20	6.64
	Average			0.09	
Condensed milk	8	10.08	9.80	0.28	
	15	8.99	8.58	0.41	
	21	12.35	12.45	—0.10	
	Average			0.26	

Procedure for Evaporated Milk

Approximately 1 gm of evaporated milk is accurately weighed and transferred to a 100 ml flask as for cream and the same subsequent procedure followed.

Procedure for Powdered Milk

Powdered milk is often lumpy and it is advisable to grind samples before weighing a representative portion. This applies especially to whole milk powders.

Weigh quickly and accurately about 0.2 gm of the fine powder on a small watch glass. Transfer to a 100 ml flask with the aid of a small funnel, washing in the powder with saturated picric acid. Shake vigorously until no lumps of powder are noticeable and fill the flask to the 100 ml mark with saturated picric acid, mix and filter, proceeding as under milk.

Because of the small size of the sample and the large dilution, this method may be in error to the extent of plus or minus 0.3-0.4 per cent of the lactose in the powder. Other methods, however, encounter the same difficulty because of the amount of dilution used.

Procedure for Ice Cream

It has been found that the sucrose in ice cream does not interfere with the determination of lactose by this procedure, provided the sucrose is not hydrolyzed during the time the product is in picric acid solution. Sucrose will invert comparatively quickly in the presence of cold saturated picric acid and produce high results, whereas uninverted it gives no reduction.

Weigh accurately, using Mojonnier pipette and weighing cross, approximately 1.8 gm of melted ice cream or ice cream mix and transfer to a 100 ml volumetric flask. Quickly fill to the 100 ml mark with saturated picric acid, mix, and filter. Rapidly transfer the first 10 ml of clear filtrate to another 100 ml volumetric flask, and proceed as for milk. It is advisable to transfer the sodium carbonate solution to the flask within 15 minutes of the time the sample is first introduced into the saturated picric acid to avoid inversion of some of the sucrose.

Procedure for Sweetened Condensed Milk

Sweetened condensed milk has more sucrose present than does ice cream, so that precautions to prevent inversion are even more important. The procedure is identical with that for ice cream except that only 1 gm of sample is used. It is important to obtain a representative sample from the can or batch. In cases of "sugar down," warm the sample, dilute with equal volume of water, and weigh accurately a 2 gm sample for analysis. Proceed as for milk.

It is advisable, however, not to have any picric acid in the flask when the sample is transferred to it, since this will increase the time the sample is exposed to the picric acid and accelerate inversion of the sucrose.

REFERENCES

1. Dehn, W. M., and Hartman, F. A.. The Picrate Colorimetric Method for the Estimation of Carbohydrates, *J. Amer. Chem. Soc. 36*:409 (1914).

2. Lewis, R. C., and Benedict, S. R.: A Method for the Estimation of Sugar in Small Quantities of Blood, *J. Biol. Chem. 20*:61 (1915).

3. Folin, O., and Denis, W.: The Determination of Lactose in Milk, *J. Biol. Chem. 33*:521 (1918).

4. Pacini, A., and Russell, D.: A Method for the Colorimetric Determination of Lactose in Milk, *Ibid. 24*:505 (1918).

5. Boch, J.: Notes on the Colorimetric Determination of Lactose, *Ibid.*, Scientific Proc. Sec. XIII, p. XIV (1920).

6. Benedict, S. R.: A Modification of the Lewis-Benedict Method for the Determination of Sugar in the Blood, *Ibid. 34*:203 (1918).

7. Lisk, H.: A Quantitative Determination of the Ammonia, Amino Nitrogen, Lactose, Total Acid, and Volatile Acid Content of Cows' Milk, *J. Dairy Sci.. 7*:79 (1924).

Vitamin C Determination in Dairy Products

1. MILK

The Vitamin C content of dairy products is determined as the sum of the ascorbic acid and the dehydroascorbic acid content. Ascorbic acid is determined by titration with 2-6 dichlorophenol-indophenol solution according to the method of Sharp.[1, 2] Dehydro ascorbic acid is determined in the same manner after conversion to ascorbic acid by means of a suspension of *Escherichia coli*, according to the method of Gunsalus and Hand.[3]

Apparatus

1. Pipette, 10 ml.
2. Beaker, 200 ml.
3. Cylinder, 25 ml.
4. Burette, 10 ml, graduated in 0.05 ml divisions.
5. Water bath.

Reagents

1. Sulfuric acid (H_2SO_4), 0.1 N.
2. Standard 2-6 dichlorophenol-indophenol solution.
3. Cucumber juice concentrate. (See page 399.)
4. *Escherichia coli*. (See page 400.)

Preparation of Standard 2-6 Dichlorophenol-Indophenol Solution

1. Dissolve 0.135 gm sodium 2-6 dichlorobenzenone-indophenol in)0 ml of hot distilled water. Pour through a filter into a one-liter

volumetric flask and add more hot water until the blue colored material has passed through the filter. Adjust to room temperature and make up to the volume of one liter with distilled water.

2. Standardization of 2-6 dichlorophenol-indophenol solution:

a. Weigh accurately, approximately 50 mg of ascorbic acid. Place in a one-liter volumetric flask, add 50 ml of 3 per cent metaphosphoric acid and make up to the volume of one liter with distilled water.

b. Immediately pipette 10 ml of the standard solution of ascorbic acid (prepared in 2 a) into a 200 ml beaker.

c. Add approximately 15 ml of tenth normal sulfuric acid and 10 ml of distilled water.

d. Titrate with the indophenol dye solution to a light pink color, permanent for at least 30 seconds.

e. Subtract from this titration value a blank determination made on the acid-water mixture brought to the same color.

NOTE: A blank of approximately 0.3 ml is usually obtained.

Calculations

The amount of ascorbic acid in the 10 ml aliquot, divided by the number of milliliters of dye solution required, gives the number milligrams of ascorbic acid corresponding to 1 ml of the dye solution. This value is called the "dye factor."

EQUATION 1

$$\frac{A}{100\,(d - b)} = \text{dye factor}$$

in which:

A = mg of ascorbic acid per liter
b = blank titration
d = ml dye to titrate blank

NOTE: (1) The metaphosphoric acid is used to delay the oxidation of ascorbic acid and may be omitted if the standardization procedure is performed with great rapidity.

(2) The dye solution should be restandardized at 2 to 4 day intervals. Dye solutions over two weeks old do not give sharp end-points and therefore should not be used.

Preparation of Cucumber Juice Concentrate

Puncture several large green frozen cucumbers with 2 or 3 small holes deep in one end of each. Suspend hole end downward in a piece of cheese cloth. Allow the juice to drain over night. Place the juice in lengths of ¾ inch cellophane tubing and suspend horizontally over a low temperature hot plate. Force a rapid current of air over the cellophane surface by means of a fan to produce rapid evaporation until the juice is concentrated to approximately 1/12th the original volume. The concentrated juice is filtered and stored cold with a drop of toluene as a preservative.

Preparation of Esch. coli—First Method

a. Add an 18 to 24 hour culture of *Esch. coli* to 5 liters of sterile broth containing 1.0 per cent peptone, 0.1 per cent glucose and 0.3 per cent secondary potassium phosphate.

b. Allow to grow 14 to 24 hours at 37 C (98.6 F).

c. Centrifuge the culture in a continuous centrifuge, wash the cells once with a sterile neutral phosphate buffer, and suspend in 50 ml of the sterile neutral phosphate buffer.

NOTE: (1) Neutral phosphate buffer is prepared by dissolving 7.4 gm disodium phosphate ($Na_2HPO_4 \cdot 12H_2O$) and 1.6 gm potassium dihydrogen phosphate (KH_2PO_4), and diluting to one liter with distilled water. Sterilize by autoclaving.

(2) Certain strains of *Esch. coli* are much more active than others in their ability to reduce dehydroascorbic acid. One strain, knows as "Crookies," is particularly active. This strain may be obtained from American Type Culture Collection, Georgetown University Medical School, Washington, D. C.

(3) The *Esch. coli* suspension may be stored cold for one month with no appreciable deterioration in activity.

Preparation of Esch. coli, Second Method

Explanation: If a continuous centrifuge is not available, the *Escherichia coli* may be cultured on a nutrient agar surface.

a. Add 80 ml of nutrient agar to each of 15 one-pint flat flasks.

NOTE: Formula of the nutrient agar is:

Ingredients per liter	
Beef extract	3 grams
Peptone	5 grams
Agar	15 grams
Glucose	3 grams

b. Sterilize and allow to harden with the flasks lying flat.

c. Add 2 ml of a 6 hour broth culture of *Esch. coli* to each flask. Allow to grow 18 hours at 37 C (98.6 F).

d. Pipette 10 ml of sterile neutral phosphate buffer into each flask and gently rock to remove the bacteria from the agar surface.

e. Remove the bacterial suspension from the flasks, place it in sterile centrifuge tubes and centrifuge.

f. Re-suspend the *Esch. coli* in 100 ml of sterile neutral phosphate buffer.

Determining Activity of Esch. coli Suspension

a. Incubate a mixture of 5 ml *Esch. coli* suspension, and 5 ml of 10 per cent glucose for 15 minutes at 40 C (104 F).

b. Add 4 mg of ascorbic acid to 100 ml of pH 6.2 phosphate buffer.

NOTE: To prepare phosphate buffer dissolve 1.36 gm of KH_2PO_4 in distilled water, add 17.2 ml. 0.1 N NaOH and dilute to one liter.

c. Immediately remove a 10 ml aliquot of the buffered ascorbic acid solution, add 25 ml 0.1 N H_2SO_4 and titrate to a stable (30 seconds) pink color with indophenol dye solution.

d. Add 0.2 ml of cucumber juice concentrate to the remaining 90 ml of ascorbic acid solution and allow to stand until all ascorbic acid is oxidized to dehydroascorbic acid.

NOTE: This oxidation should require not over 5 minutes. A 10 ml sample will titrate to a pink color with approximately 0.3 ml indophenol dye when the oxidation is complete.

e. Immediately pipette 10 ml samples of the dehydroascorbic acid solution into test tubes containing 1 ml, 2 ml, 3 ml, and 4 ml of the incubated *Esch, coli*-glucose mixture.

f. Incubate for 15 minutes at 40 C (104 F) in a water bath and titrate the entire contents of each tube with indophenol dye after adding 25 ml of the 0.1 N H_2SO_4. The amount of *Esch. coli*-glucose mixture required completely to reduce dehydroascorbic acid to ascorbic acid, as indicated by a titration value equal to that of the original ascorbic acid titration, is the amount which should be used for the analysis of each 10 ml sample of milk.

Procedure for the Determination of Reduced Ascorbic Acid

1. Pipette 10 ml of milk into a 200 ml beaker.

2. Add 25 ml of 0.1 N H_2SO_4.

3. Immediately titrate with indophenol dye solution until a definite pink color remains for at least 30 seconds.

4. Add sufficient cucumber juice concentrate to 100 ml milk completely to oxidize the ascorbic acid to dehydroascorbic acid in 5 minutes. This should require about 0.2 ml of the concentrate.

5. Pipette 10 ml of the milk treated with cucumber juice concentrate into a 200 ml beaker, add 25 ml of 0.1 N H_2SO_4 and titrate to a pink color as before. This titration is used as a blank and should be approximately 0.4 ml.

Calculations

The mg ascorbic acid per liter of milk is calculated from the dye titration of the milk after subtracting the blank titration.

EQUATION 2

$$(m - b) \cdot f \cdot 100 = \text{mg. ascorbic acid per liter of milk}$$

in which:

m = titration of milk
b = blank titration
f = dye factor

Procedure for the Determination of Dehydroascorbic Acid and Total Vitamin C

1. Pipette 10 ml of milk into a test tube and add the required amount of *Esch. coli*-glucose mixture (see activity of *Esch. coli*, page 400.)

2. Incubate in a water bath at 40 C (104 F) for 15 minutes.

3. Transfer the entire contents of the tube to a 200 ml beaker, add 25 ml of 0.1 N H_2SO_4 and titrate with indophenol dye to a stable (30 seconds) pink color.

Calculations

The number of milligrams of total vitamin C per liter is calculated from the dye titration of milk incubated with *Esch. coli* after subtracting the blank titration. The blank titration is that obtained by treating milk with cucumber juice concentrate as given in the procedure for determination of ascorbic acid. The number of milligrams of dehydroascorbic acid per liter of milk is calculated by subtracting the ascorbic acid content from total vitamin C.

EQUATION 3

$$(M - b) \cdot f \cdot 100 = \text{mg. total vitamin C per liter of milk}$$

in which:

M = titration of milk and Esch. coli.
b = blank titration
f = dye factor

EQUATION 4

$$C - a = \text{mg. dehydroascorbic acid per liter of milk}$$

in which:

C = mg total vitamin C per liter
a = mg reduced ascorbic acid per liter

2. DRIED MILK AND EVAPORATED MILK

The effect of heat on milk is to produce reducing groups which titrate with indophenol dye solution. Milk products which are heated during processing should be analyzed by the following procedure to determine the amount of these reducing groups. The amount of indophenol dye required to titrate the reducing group produced as the result of heat can then be subtracted from the total titration to give the correct value for ascorbic acid.

Procedure

1. Reconstitute milk powder 62.5 gm to 250 ml total volume with distilled water. If evaporated milk is to be analyzed, dilute 100 ml to a total volume of 200 ml with distilled water.

2. Pipette a 10 ml aliquot into a 200 ml beaker.

3. Add 25 ml of 0.1 N H_2SO_4.

4. Immediately titrate with indophenol dye solution until a definite pink color remains for at least 30 seconds.

5. Pipette a 10 ml aliquot into a test tube and add the required amount of *Esch. coli*-glucose mixture (see activity of *Esch. coli*, page 400.)

6. Incubate in a water bath at 40 C (104 F) for 15 minutes.

7. Transfer the entire contents of the tube to a 200 ml beaker, add 25 ml of 0.1 N H_2SO_4, and titrate with indophenol dye as before.

8. While the sample treated with *Esch. coli* is incubating, add sufficient cucumber juice concentrate to a 100 ml aliquot of the reconstituted milk powder or evaporated milk completely to oxidize the ascorbic acid to dehydroascorbic acid in 5 minutes. This should require about 0.2 ml of the concentrate.

9. Pipette a 10 ml aliquot of the sample treated with cucumber juice concentrate into a 200 ml beaker containing 25 ml of 0.1 N H_2SO_4. While the pipette is draining, pipette another 10 ml aliquot into a test tube containing the required amount of *Esch. coli*-glucose mixture (see activity of *Esch. coli*, page 400.)

10. Immediately titrate the aliquot added to the H_2SO_4 with indophenol dye solution to a stable (30 seconds) pink end-point.

11. Add acid and titrate the aliquot added to the *Esch. coli*-glucose mixture after incubation at 40 C (104 F) for 15 minutes.

NOTE: Cucumber juice concentrate, in addition to bringing about the oxidation of ascorbic acid to dehydroascorbic acid, slowly oxidizes the heat-produced groups present. This oxidation is usually complete in 5 minutes in the case of whole milk powder and a blank titration of approximately 0.7 ml is obtained. In the case of evaporated milk, this oxidation of heat-produced groups is usually incomplete in 5 minutes and a blank titration of over 3 ml is not uncommon.

Calculations

Total vitamin C is calculated from the dye titration of milk treated with cucumber juice concentrate and subsequently incubated with *Esch. coli*, after subtracting the blank titration obtained from the titration of milk treated only with cucumber juice concentrate. Dehydroascorbic acid is calculated from the increase in titration of the sample of milk after addition of *Esch. coli*. Ascorbic acid is obtained by subtracting the value for dehydroascorbic acid for total vitamin C.

EQUATION 5

$(j - b) \cdot d \cdot 100 =$ total vitamin C per liter of 2 × diluted evaporated milk

or

$(j - b) \cdot d \cdot 50 =$ mg total vitamin C per 125 gm milk powder

in which:

j = titration of milk and cucumber juice, and *Esch. coli*.
b = blank titration
d = dye factor

EQUATION 6

(M — m) · d · 100 = mg dehydroascorbic acid per liter of 2 × diluted evaporated milk

or

(M — m · d · 50 = mg dehydroascorbic acid per 125 gm milk powder

in which:

M = titration of milk and *Esch. coli.*
m = titration of milk
d = dye factor

EQUATION 7

C — a = mg ascorbic acid per liter of 2 × diluted evaporated milk, or per 125 gm milk powder

in which:

C = mg total vitamin C per liter of 2 × evaporated milk or per 125 gm milk powder
a = mg reduced dehydroascorbic acid

REFERENCES

1. Sharp, Paul F.: Rapid Method for the Quantitative Determination of Reduced Ascorbic Acid in Milk, *J. Dairy Sci. 21*:85-88 (1938).

2. Sharp, Paul F., Hand, D. B., and Guthrie, E. S.: Quantitative Determination of Dissolved Oxygen—Ascorbic Acid Oxidase Method, *Ind. Eng. Chem. (Anal. Ed.), 13*:593-597 (1941).

3. Gunsalus, I. C., and Hand, David B.: The Use of Bacteria in the Chemical Determination of Total Vitamin C, *J. Biol. Chem. 141*:853-858 (1941).

4. Stewart, A. P. 1942): *The Determination of Vitamin C in Milk in the Presence of Reducing Substances Produced by Heat.* (Unpublished.)

PART VI

MICROBIOLOGICAL, CHEMICAL, AND PHYSICAL TESTS FOR NON-DAIRY PRODUCTS

CHAPTER 31

CHEMICAL CONTROL PROCEDURES FOR WASHING AND STERILIZING SOLUTIONS AND BRINE

Total Hardness of Water

The determination of the total hardness of water by the soap method roughly approximates the amount of calcium and magnesium present. The procedure actually measures the soap-consuming power of a water.

Apparatus

1. Volumetric flask, 1 liter.
2. Glass stoppered bottle, 250 ml.
3. Burette, 50 ml, graduated in 0.1 ml.

Reagents

1. Pure calcite (calcium carbonate).
2. Pure, dry, white castile soap.
3. Ethyl alcohol, 80 per cent by volume.
4. Hydrochloric acid (HCl), diluted 1 : 9.
5. Distilled water.

Preparation of Standard Solutions

1. *Standard Soap Solution.*

Dissolve 100 gm of dry white castile soap in 1 liter of 80 per cent alcohol and allow this solution to stand several days before standardizing.

2. *Standard Calcium Chloride Solution.*

Dissolve 0.2 gm of calcite in a little dilute HCl, being careful to avoid loss of solution by spattering; a tall beaker or an Erlenmeyer flask should be used. Evaporate the solution to dryness and dissolve with distilled water. Repeat this process several times to expel excess of acid which is indicated when the residue is white without a trace of yellow color. Dissolve the residue in distilled water and dilute to 1 liter. One ml of this solution is equivalent to 0.2 mg of calcium carbonate.

3. *Standardization of Solutions.*

Dilute 20 ml of the calcium chloride solution in a 250 ml glass-stoppered bottle to 50 ml with distilled water which has been recently boiled and cooled. Add the soap solution from a burette, 0.2 to 0.3 ml at a time, shaking the bottle vigorously after each addition until the lather persists for 5 minutes over the entire surface of the water while the bottle lies on its side.

Adjust the strength of the stock soap solution with 80 per cent ethyl alcohol so that the resulting diluted solution will give a permanent lather (stable for 5 minutes) when 6.4 ml is properly added to 20 ml of a standard calcium chloride solution diluted to 50 ml. Usually from 75 to 100 milliliters of the stock soap solution is required to make 1 liter of the standard soap solution.

TABLE 15. TOTAL HARDNESS EQUIVALENT OF STANDARD SOAP SOLUTION

Ml of soap solution	Total hardness in ppm.	Grains per U.S. gallon	Ml of soap solution	Total hardness in ppm.	Grains per U.S. gallon	Ml of soap solution	Total hardness in ppm.	Grains per U.S. gallon
0.1	----	------	2.8	29.9	1.733	5.4	65.7	3.810
0.2	----	------	2.9	31.2	1.809	5.5	67.1	3.891
0.3	----	------	3.0	32.5	1.885	5.6	68.6	3.978
0.4	----	------	3.1	33.8	1.960	5.7	70.0	4.060
0.5	----	------	3.2	35.1	2.025	5.8	71.4	4.141
0.6	----	------	3.3	36.4	2.111	5.9	72.9	4.228
0.7	0.0	0.000	3.4	37.7	2.186	6.0	74.3	4.309
0.8	1.6	0.092	3.5	39.9	2.314	6.1	75.7	4.390
0.9	3.2	0.185	3.6	40.3	2.337	6.2	77.1	4.471
1.0	4.8	0.278	3.7	41.6	2.412	6.3	78.6	4.558
1.1	6.3	0.365	3.8	42.9	2.488	6.4	80.0	4.640
1.2	7.9	0.458	3.9	44.3	2.569	6.5	81.4	4.721
1.3	9.5	0.551	4.0	45.7	2.650	6.6	82.9	4.808
1.4	11.1	0.643	4.1	47.1	2.731	6.7	84.3	4.889
1.5	12.7	0.736	4.2	48.6	2.818	6.8	85.7	4.970
1.6	14.3	0.829	4.3	50.0	2.900	6.9	87.1	5.051
1.7	15.6	0.904	4.4	51.4	2.981	7.0	88.6	5.138
1.8	16.9	0.980	4.5	52.9	3.068	7.1	90.0	5.220
1.9	18.2	1.055	4.6	54.3	3.149	7.2	91.4	5.301
2.0	19.5	1.131	4.7	55.7	3.230	7.3	92.9	5.388
2.1	20.8	1.206	4.8		3.311	7.4	94.3	5.469
2.2	22.1	1.281	4.9	58.6	3.398	7.5	95.7	5.550
2.3	23.4	1.357	5.0	60.0	3.480	7.6	97.1	5.631
2.4	24.7	1.432	5.1	61.4	3.561	7.7	98.6	5.718
2.5	26.0	1.507	5.2	62.9	3.648	7.8	100.0	5.800
2.6	27.3	1.582	5.3	64.3	3.729	7.9	101.5	5.887
2.7	28.6	1.658				8.0	102.9	5.968

TABLE OF EQUIVALENTS FOR WATER HARDNESS DETERMINATION

	Parts per million	Grains per gallon	Degrees (Clark)
Parts per million	1.0	0.058	0.07
Grains per gallon	17.0	1.0	1.20
Degrees—Clark	14.3	0.83	1.0

Procedure

1. Measure 50 ml of the water to be tested into a 250 ml bottle.

2. Add standard soap solution 0.2 to 0.3 ml at a time, shaking vigorously after each addition until the lather persists for 5 minutes over the entire surface while the bottle lies on its side. If more than 7 ml of soap solution are required for 50 ml of the water, repeat, using a smaller sample, diluted to 50 ml with distilled water which has been recently boiled and cooled. Correct the final reading in proportion to the amount of sample.

Determine the hardness of the water by reference to table 15, which shows the total hardness in part per million as calcium carbonate for each tenth of a milliliter of soap solution when 50 ml of the sample are titrated.

REFERENCE

Standard Methods for the Examination of Water and Sewage, American Public Health Association, 9th edition, 1947.

Determination of Strength of Washing Solutions

1. LABORATORY METHOD

(Caustic Alkali and Total Alkali as NaOH)

Apparatus

1. Pipette, 50 ml.
2. Burette, 50 ml, graduated in tenths.
3. Pipette, 5 ml.
4. Erlenmeyer flask, 200 ml.
5. Volumetric flask, 250 ml.

Reagents

1. Sulfuric acid, 0.1 N.

2. Phenolphthalein indicator, 0.5 per cent solution in 50 per cent alcohol.

3. Methyl orange indicator, 0.05 per cent solution in distilled water.

Procedure

1. Samples of bottle-washing solutions may be taken from the petcocks on the glass gages showing the liquid level in the tanks by discarding about 4 ounces from the gage and then collecting a 4 to 6 ounce sample.

2. Transfer 5 ml of the bottle-washing solution to a 250 ml volumetric flask and make up to the mark with distilled water. After mixing well, transfer 50 ml of this solution to an Erlenmeyer flask.

3. Add a few drops of phenolphthalein indicator, and titrate with 0.1 N sulphuric acid to a colorless solution. Record the quantity of acid used. This is titration A.

4. Add to this same solution in the Erlenmeyer flask a few drops of methyl orange indicator and continue the titration until a slight pink color appears. The amount of acid used in this second titration is recorded as titration B.

Calculations

The percentage of free caustic NaOH equals the number of milliliters of titration A minus the Number of milliliters of titration B multiplied by 0.4. The percentage of total alkali in terms of NaOH equals the number of milliliters of titration A plus the number of milliliters of titration B multiplied by 0.4.

For example, if A is 5.5 ml and B is 0.5 ml, the difference, 5.0 times 0.4 equals 2.0, the per cent of free caustic. Likewise, the sum 5.5 and 0.5 and 0.5 is 6.0 times 0.4 equals 2.4, the per cent of total alkalinity.

2. DIRECT READING TITRATION METHOD

A convenient method for determining the strength of alkali solutions is to prepare a standard acid solution of such strength that the burette readings may be interpreted directly as percentage of alkali when a given amount of sample is employed. Such a standard acid solution may be prepared by diluting 66.63 ml (122.6 gm) of sulphuric acid, sp. gr. 1.84, to exactly 1 liter volume with distilled water. This solution is 2.5 N, and should be kept in a tightly glass stoppered bottle when not in use.

Procedure

1. Measure exactly 10 ml alkali solution to be tested into a white porcelain evaporating dish.
2. Add 5 drops phenolphthalein solution.
3. From a burette, graduated in tenths, add 2.5 N H_2SO_4 drop by drop until the solution becomes clear.

The number of milliliters of acid used equals the percentage of alkali (calculated as NaOH) in the sample.

NOTE: (1) Further titration with methyl orange as the indicator to determine the total alkalinity may be conducted as described on page 412.

(2) This procedure will give approximate results, but if greater accuracy is desired, the 2.5 N H_2SO_4 should be precisely standardized.

3. THE *A.B.C.B.* CAUSTIC TEST

Apparatus

1. Graduated cylinder, 10 ml.
2. Glass tumbler.
3. Stirring rod.

Reagents

A.B.C.B. Caustic test tablets (available from American Bottlers of Carbonated Beverages).

Procedure

1. Withdraw a sample of the solution from the soaker into any convenient container. Do not take the sample until the soaker has been an operation for at least five minutes. If very dirty, allow to stand until the sediment settles, or filter and cool.

2. Measure accurately 10 ml into the graduate and transfer it to a clean glass tumbler. If the solution is dark, add 1 ounce of distilled water.

3. Add one preliminary tablet (tablet No. 1) and crush with the stirring rod.

4. When the preliminary tablet is thoroughly dissolved, add one caustic test tablet (tablet No. 2) and dissolve with the aid one of the stirring rod. If the color is brick red, the strength of the solution is greater than 0.5 per cent caustic. Add caustic test tablets until a lemon yellow color appears.

Each caustic test tablet required to produce lemon yellow color represents 0.5 per cent caustic strength. To determine the total caustic strength in percentage, divide the number of total caustic tablets used by 2.

EXAMPLE: Two tablets used indicate 1 per cent caustic, four tablets indicate 2 per cent caustic, etc.

NOTE: This test was developed by a Fellowship of American Bottlers of Carbonated Beverages, 1128 16th St., N.W., Washington 6, D. C.

Determination of Strength of Washing Powders

1. PHENOLPHTHALEIN ALKALINITY OF CAUSTIC WASHING POWDERS

Dissolve 10 gm of the washing powder to be tested and make up to 100 ml with distilled water. After the powder is completely dissolved, transfer 10 ml of the solution to a flask and add 5 drops of phenolphthalein solution. Proceed with the test as outlined under 1 or 2 (pages 409-410). This will give the percentage of caustic soda and other alkalis calculated as NaOH in the washing powder.

2. FREE CAUSTIC IN THE PRESENCE OF CARBONATE, PHOSPHATES, AND OTHER ALKALIS

Apparatus

1. Burette, 50 ml graduated in tenths.
2. Pipette, 10 ml.
3. Erlenmeyer flask.
4. Stirring rod.

Reagents

1. Barium chloride solution ($BaCl_2$), saturated.

2. Sulfuric acid (H_2SO_4), 2.5 N (see page 410).

3. Phenolphthalein indicator, 0.5 per cent solution in 50 per cent alcohol.

4. Methyl orange indicator, 0.05 per cent solution in distilled water.

Procedure

1. Transfer another 10 ml portion of the solution prepared in (a) (see page 410, paragraph 3, top of page) to an Erlenmeyer flask.

2. Add 10 ml of saturated $BaCl_2$ solution to the sample, stirring thoroughly.

3. Add 3 drops of phenolphthalein. It will become a deep pink.

4. Titrate to the end-point of phenolphthalein with the H_2SO_4.

5. Add 1 or 2 drops more of the H_2SO_4 to cause the color to disappear. The number of milliliters of acid used multiplied by 10 gives the percentage of NaOH present.

CAUTION: Do not add the acid too rapidly, as it will destroy the carbonates or other alkalis locally.

6. Add 3 drops of methyl orange.

7. Continue slowly the addition of the H_2SO_4 to the end-point of the methyl orange. The number of milliliters of acid used multiplied by 5 gives the percentage of sodium carbonate present, and the sum of the two readings gives the percentage of total alkalinity, expressed as NaOH.

NOTE: Be sure to rinse out the 10 ml pipette after adding the saturated $BaCl_2$ solution and before sampling any detergent solutions. Never sample a detergent solution while hot as it will affect materially the accuracy of the test.

$BaCl_2$ solution is necessary only when trisodium phosphate and other alkali salts are present.

REFERENCES

1. Phillips, A. W., Mack, M. J., and Frandsen, J. H.: Washing Powders for Dairy Use, *Proc. 28th Ann. Conv. I. A. M. D. (1927), Lab. Sec.* pp. 78-92.

2. Zoller, H. F.: Certain Fundamental Factors Involved in Cleaning Dairy Equipment, *Proc. 21st Ann. Conv. Ibid.* (1928) pp. 35-53.

3. Baumann, L. A.: Recent Developments and Data in the Soaker Type Washer, *Proc. 23rd Ann. Conv: Ibid.* (1930) pp. 25-37.

4. LeFevre, P. E.: Cleaning of Milk Bottles, *Milk Plant Monthly 19*: 21-4 (1930).

5. Irwin, R. E.: Present Requirements and Procedure for Washing Milk Bottles, *Amer. J. Pub. Health 18*: 737 (1928)

6. Shere, L. and Hoeft, G.: Sterliziation of Milk Bottles in Soaker Type Washers. *Amer. J. Pub. Health 19*: 747 (1929).
7. Parker, M. E.: The Character of a Dairy Detergent and What It Means to the Ice Cream Manufacturer.
8. Parker, M. E.: Note on Dairy Detergency, *Creamery 8*: May, p. 62, June 54, and July p. 38 (1929).
9. Johnson, A. H.: Efficiency of Various Washing Compounds. *Proc. 23rd Ann. Conv. I. A. M. D. (1930), Lab. Sec.* pp. 38-44.
10. Parker, M. E.: Temperature Control Efficiency, *Ibid.* pp. 48-51.
11. Mudge, C. S. and Lawler, B. M.: Effect of Alkali Solutions on Bacteria Found in Unwashed Milk Bottles, *Ind. Eng. Chem. 20*: 378 (1928).
12. Levine, M. and co-workers: Germicidal Efficiency of Sodium Hydroxide, Sodium Carbonate, and Trisodium Phosphate at the Same Hydrogen Ion Concentration, *Ibid. 19*: 1338 (1927).
13. Myers, R. P.: The Germicidal Properties of Alkaline Washing Solutions, with Special Reference to the Influence of Hydroxyl Ion Concentration, *J. Agr. Research 38*: 521-63 (1929).
14. Mudge, C. S.: Cleaning and Sterilizing with Alkaline Detergents, *Food Industries, 1*: 613 (1929).
15. Buchanan, J. H., and Levine, M.: Bottle Washing and Its Control in the Carbonated Beverage Industry, *A. B. C. B. Educational Bulletin No. 1*, August 1929.
16. Mitchell, R. W.: The Cleaning of Metal, *Metal Cleaning and Finishing 2*: 13, 111, 207, 759 (1930).
17. Johnson, A. H.: Washing and Sterlizing Solutions. *Proc. 23rd Ann. Conv.* I. A. M. D. (1930), pp. 38-48.

Phosphoric Acid Determination

Official Gravimetric Method (A.O.A.C.)

Apparatus

1. Volumetric flask, 200 ml.
2. Funnels, 65 mm diameter.
3. Filter paper, 11 cm.
4. Erlenmeyer flask, 500 ml.
5. Pipette, volumetric.
6. Beakers, 250 ml.
7. Cylinder, graduated 100 ml.
8. Bunsen Burner.
9. Tripod and asbestos mat.
10. Wash bottle.
11. Electric furnace.
12. Desiccator.

Reagents

1. Ammonium hydroxide (NH_4OH) conc.
2. Nitric acid (HNO_3) conc.
3. Molybdate solution—dissolve 100 gm of molybdic acid (MoO_3) in a mixture of 144 ml of NH_4OH and 271 ml of distilled water. Pour this solution slowly and with constant stirring into a mixture of 489 ml of HNO_3 and 1148 ml of distilled water. Keep the final

mixture in a warm place for several days or until a portion heated to 40 C (104 F) deposits no yellow precipitate of ammonium phosphomolybdate. Decant the solution from any sediment and preserve in glass-stopper vessels.

4. Ammonium nitrate (NH_4O_3)—dissolve 100 gm of phosphate-free NH_4NO_3 in distilled water and dilute to 1 liter.

5. Ammonium hydroxide (NH_4OH).

6. Hydrochloric acid (HCl) conc.

7. Litmus paper or bromthymol blue solution.

8. Magnesia mixture—(1) Dissolve 11 gm of MgO in HCl (1 : 4), avoiding an excess of the acid; add a little MgO in excess; boil a few minutes to precipitate iron, aluminum, and phosphorus pentoxide; and filter. To the filtrate add 140 gm of NH_4Cl and 130.5 ml of NH_4OH and dilute to 1 liter. Or, (2) dissolve 55 gm of crystallized $MgCl_2 \cdot 6H_2O$ in distilled water, add 140 gm of NH_4Cl and 130.5 ml of NH_4OH, and dilute to 1 liter. Or, (3) dissolve 55 gm of crystallized $MgCl_2 \cdot 6H_2O$ in distilled water, add 140 gm of NH_4Cl, and dilute to 870 ml Add NH_4OH to each required portion of the solution just before using, at the rate of 15 ml per 100 ml of solution.

9. Ammonium hydroxide (NH_4OH) 1–9 for washing.—Should contain not less than 2.5 per cent of NH_3 by weight.

Procedure

1. Pipette an aliquot of the prepared solution corresponding to 0.25 gm, 0.50 gm, or 1 gm, into a 250 ml beaker; add NH_4OH in slight excess, and barely dissolve the precipitate formed with a few drops of HNO_3, stirring vigorously. If HCl or H_2SO_4 has been used as a solvent, add about 15 gm of crystalline NH_4NO_3 or a solution containing that quantity.

2. To the hot solution add 70 ml of the molybdate solution for every one-tenth gram of P_2O_5 present. Digest at about 65 C (149 F) for 1 hour, and determine whether or not the P_2O_5 has been completely precipitated by adding more molybdate solution to the clear supernatant liquid.

3. Filter, and wash with cold distilled water or preferably with the NH_4NO_3 solution.

4. Dissolve the precipitate on the filter with NH_4OH (1 : 1) and hot distilled water and wash into a beaker to a volume of not more than 100 ml.

5. Neutralize with HCl, using litmus paper or bromthymoll blue as an indicator, cool, and from a burette add slowly (about 1 drop per second), stirring vigorously, 15 ml of the magnesia mixture for

each one-tenth gram of P_2O_5 present. After 15 minutes add 12 ml of NH_4OH.

6. Let stand until the supernatant liquid is clear (usually 2 hours), filter, wash the precipitate with NH_4OH (1 :9) until the washings are practically free from chlorides, dry, burn at a low heat, and then ignite to constant weight, preferably in an electric furnace, at 950–1000 C, cool in a desiccator, and weigh as $Mg_2P_2O_7$.

7. This weight may be converted to a percentage of P_2O_5 by the formula:

$$\text{Wt. } Mg_2P_2O_7 \times \frac{0.6379}{\text{Wt. sample}} \times 100 = \text{Percentage of } P_2O_5$$

REFERENCE

Methods of Analysis, A.O.A.C., 6th edition (1945) p. 21.

Polyphosphate Determination in the Presence of One Another

As the use of one or more of the complex phosphates as ingredients of cleaning powders is becoming more general, a method for their estimation in cleansers should be available.

The procedures described furnish a method for the estimation of orth-, pyro-, meta-, and polyphosphates in the presence of each other.[1] Where the greatest possible accuracy is desired, gravimetric procedures should be employed.

Reagents

1. C.P. sodium hydroxide, N and 0.1 N NaOH, carbonate free.

2. Hydrochloric acid, N and 0.1 N HCl, c.p.

3. Ammonium molybdate solution. Prepared as described on page 413.

4. Barium chloride solution. Dissolve 25 gm of c.p. $BaCl_2 \cdot 2H_2O$ in 1 liter of distilled water.

5. Potassium nitrate solution. Dissolve 10 gm of c.p. KNO_3 in 1 liter of water.

6. Manganous chloride solution. Dissolve 100 gm of c.p. $MnCl_2 \cdot 4H_2O$ in 1 liter of distilled water.

7. Phenolphthalein indicator, 0.5 per cent solution. Dissolve 0.5 gm of c.p. phenolphthalein in 60 ml of 95 per cent ethyl alcohol, and dilute to 100 ml with previously boiled and cooled distilled water. Neutralize with 0.1 N NaOH.

8. Methyl orange indicator, 0.1 per cent solution.

9. Bromocresol green indicator, 0.4 per cent solution. Dissolve 0.4 gm of bromcresol green in 5.75 ml of 0.1 N NaOH and a few milliliters of alcohol. Dilute with water to 100 ml.

10. Sodium iodide solution.

Preparation of Test Solutions

1. Disodium orthophosphate. c.p. Recrystallize disodium orthophosphate three times from distilled water and dry at 110 C overnight. The sample, analyzed by magnesia mixture, should indicate the presence of 100 per cent disodium hydrogen orthophosphate with pyro- and metaphosphates (2, 3) absent.

2. Tetrasodium pyrophosphate. c.p. Recrystallize tetrasodium pyrophosphate three times from distilled water and dry to constant weight at 250 C. The sample, analyzed by magnesia mixture after conversion to orthophosphate in HNO_3, should indicate the presence of 100 per cent tetrasodium pyrophosphate, with ortho- and metaphosphates (3, 4) absent.

3. Sodium hexametaphosphate. c.p. Heat monosodium dihydrogen orthophosphate to about 900 C for 3 hours in a platinum dish and cool rapidly. Different batches of the resulting glassy product contain from 90 to 95 per cent sodium hexametaphosphate and, since ortho-, pyro-, and polyphosphate are absent, the balance of the phosphorus pentoxide must be present as trimetaphosphate. Prepare a solution of sodium hexametaphosphate with only 0.10 per cent of trimetaphosphate by precipitating the hexametaphosphate from the trimetaphosphate with $AgNO_3$ in a HNO_3 solution, acid to methyl orange. Wash the soluble trimetaphosphate out of the silver hexametaphosphate precipitate by decantation until the filtrate shows only a trace of phosphate. The silver hexametaphosphate is then suspended in water and stirred while slowly adding sodium iodide solution. The filtrate from the removal of silver iodide contains the hexametaphosphate.

4. Sodium trimetaphosphate. c.p. Heat monosodium dihydrogen orthophosphate for 1 hour at 300 C, increase to and hold at 610 C for 3 hours and then slowly decrease to 390 C where it is held for 14 hours. The phosphorus pentoxide determination indicates that the sample must be 100 per cent $(NaPO_3)x$. The sample yields no precipitate to $AgNO_3$ or $BaCl_2$ at any pH, thereby indicating the absence of hexametaphosphate, pyrophosphate, orthophosphate (5), tripoly- and tetraphosphate precipitate.

5. Sodium tripolyphosphate. Heat quantities equivalent to 1 $NaH_2PO_4 \cdot H_2O$ and $2Na_2HPO_4 \cdot 12\ H_2O$ in a platinum dish in an oven at 110 C. Dissolve the salts in the water of hydration, and after evaporation to dryness transfer to a muffle furnace at 250 C for 2

hours and then increase the temperature to approximately 800 C. Allow the mixture to remain at this temperature until the entire mass becomes molten, then lower the temperature to about 650 C and maintain at this temperature for 18 hours. At the end of this time lower the temperature from 650 C to 250 C over a period of 8 hours, and then transfer to a desiccator to cool. The white crystalline solid mass disintegrates into a fine white powder upon cooling to room temperature. Analysis of the material by the procedure given below indicates the following composition: $(NaPO_3)_3$, 4.4 per cent; Na_2HPO_4, 1.7 per cent; $Na_4P_2O_7$, 16.0 per cent; $Na_5P_3O_{10}$, 77.9 per cent.

A sample of a commercial sodium tripolyphosphate has been found to have the following composition: $(NaPO_3)_3$, 0.4 per cent; Na_2HPO_4, 3.0 per cent; $Na_4P_2O_7$, 10.6 per cent; $Na_5P_3O_{10}$, 86.0 per cent.

6. Sodium tetraphosphate. A commercial sodium tetraphosphate when analyzed by the current procedure indicates a composition of $(NaPO_3)_3$, 3.8 per cent; NaH_2PO_4, 3.8 per cent; and $Na_6P_4O_{13}$, 92.4 per cent.

Procedure

1. PREPARATION OF SOLUTION FOR ANALYSIS. Run a rough preliminary test by the method given below for the determination of the approximate total phosphorus pentoxide. Then prepare a solution of the sample in a standard volumetric flask such that a convenient size of aliquot will contain about 0.05 gm of total P_2O_5.

2. CONVERSION OF PHOSPHATES TO THE ORTHO FORM. Boil gently for 15 minutes the acidified phosphate solution containing a 25-ml excess of c.p. HNO_3 per 100 ml of solution. Before adding the ammonium molybdate, cool the solution to room temperature, and neutralize the excess acidity with NaOH so that the solution is just acid to litmus paper.

NOTE: This conversion of the phosphates, isolated by the various analytical steps, makes possible application of the volumetric molybdate method for determination of their phosphorus pentoxide content.

3. TOTAL PHOSPHORUS PENTOXIDE.

a. Transfer a suitable aliquot of the prepared solution to a beaker and dilute to 75 to 100 ml.

b. Convert all phosphates to the ortho form as directed in (2) and precipitate the phosphorus pentoxide with 50 ml of ammonium molybdate.

c. Wash the precipitate with a minimum volume of 1 per cent KNO_3 wash solution cooled to 10-12 C (50-53 F) until 5 ml of the wash solution turn alkaline to phenolphthalein indicator with not over 3 drops of 0.1N NaOH. One hundred ml of wash solution should be sufficient.

d. Determine the phosphorus pentoxide content of the phosphomolybdate volumetrically, using N or 0.1N reagents depending on the amount of precipitate.

4. HEXAMETAPHOSPHATE. This procedure should be carried out soon after the solution has been prepared, so that the metaphosphate ion, which undergoes hydration particularly in strongly acid or alkaline solutions, will not have time for conversion.

a. Transfer an aliquot of the prepared solution of the size used for total phosphorus pentoxide to a beaker and dilute to 50 ml with distilled water.

b. Add 1 drop of methyl orange indicator, make the solution just acid to the indicator with N HCl, then add 0.5 ml of the acid in excess. If the solution is already acid, make slightly alkaline to the indicator, then make final acid adjustment as indicated above.

c. Add slowly with constant stirring 15 ml of $BaCl_2$, and let the precipitate settle before filtering. If the solution becomes cloudy without the precipitate separating within a minute, stir vigorously for 15-second intervals until a precipitate commences to separate when stirring is stopped.

d. Wash the precipitate by decantation with a small amount of cold $BaCl_2$ (1 gm per liter).

e. Transfer the precipitate to the filter paper and wash four or five times with small portions of the cold $BaCl_2$ wash solution. Save the filtrate for the determination of trimetaphosphate.

f. Dissolve the precipitate in 1 : 1 HNO_3 and convert the metaphosphate to orthophosphate as above and determine the phosphorus pentoxide as for total phosphorus pentoxide.

TABLE 16. CONVERSION OF PYRO-, META-, AND POLYPHOSPHATES TO ORTHOPHOSPHATE

Phosphate		P_2O_5 Molybdate After 15 min. boiling	P_2O_5 Molybdate After 30 min. boiling	P_2O_5 by Mg Mix. After 2 hours boiling
Hexameta	95%	0.0472 gm	0.0473 gm	0.0472 gm
Trimeta	5%			
Pyro		0.0482 gm	0.0481 gm	0.0483 gm
Hexameta	60%	0.0798 gm	0.0798 gm	0.0799 gm
Trimeta	2%			
Pyro	33%			
Ortho	5%			
Tetra		0.0624 gm	0.0624 gm	....
Tripoly		0.0576 gm	0.0576 gm	....

5. TRIMETAPHOSPHATE.

a. To the filtrate from the separation of hexametaphosphate add N NaOH with stirring until the mixture remains definitely alkaline to phenolphthalein indicator.

b. Filter off the precipitated barium phosphates.

c. Determine the phosphorus pentoxide of the trimetaphosphate on the resulting filtrate as for total phosphorus pentoxide.

6. Orthophosphate.

a. Transfer a suitable aliquot of the prepared solution to a beaker, dilute to 100 ml, make just acid to methyl red indicator with N HCl.

b. Add 25 ml of $BaCl_2$ and filter off any barium phosphate precipitate.

c. To the filtrate cooled to 20-25 C (68-77 F) add 8-10 gm of N HNO_3 before precipitating the orthophosphate radical with ammonium molybdate and determining it volumetrically as above. Allow only 15 minutes for precipitation. Phosphates other than orthophosphate are not precipitated under these conditions.

7. PYROPHOSPHATE.

a. Transfer a suitable aliquot of the freshly prepared solution to a beaker and make just acid to methyl orange indicator with N HCl.

b. Add 10 per cent $BaCl_2$ in excess of the estimated amount needed for the hexametaphosphate present. In the absence of hexametaphosphate add no $BaCl_2$. Stir vigorously to hasten separation and settling of the barium metaphosphate

c. Filter.

d. Dilute the filtrate to approximately 125 ml with distilled water, and add 5 ml of $MnCl_2$ solution.

e. Adjust the pH of the solution with the aid of a glass electrode or other suitable pH-meter by adding 0.1 N NaOH drop by drop with stirring until the pH remains 4.1.

f. Add with stirring 7-8 ml of acetone and allow to stand 12-16 hours at 20-30 C (68-86 F) to permit the manganous pyrophosphate to precipitate.

g. Filter and wash the precipitate with a $MnCl_2$ solution at a strength of 1 gm per liter. Reserve filtrate for polyphosphate qualitative test.

h. Dissolve precipitate in 1 : 1 HNO_3, convert to orthophosphate, and determine the phosphate as for total phosphorus pentoxide.

8. POLYPHOSPHATE. Qualitive test.

NOTE: **This test must be performed on the filtrate immediately after removal of manganous pyrophosphate which has been previously precipitated from a freshly prepared solution.**

Add 25 ml of $BaCl_2$ solution to the filtrate from the pyrophosphate determination. The formation of a cloudiness or precipitate within 15 minutes (a behavior found to be characteristic of tetraphosphate) is considered indicative of the presence of this compound. Tripolyphosphate gives no visible change under these conditions.

The phosphorus pentoxide of the polyphosphates is taken as the difference between the total phosphorus pentoxide and the sum of the phosphorus pentoxide values of ortho-, pyro-, and the metaphosphates, and is calculated to tripoly- or tetraphosphate as dictated by the qualitative test.

NOTE: The calculation of polyphosphate is subject to an accumulation of additive errors; however, results from numerous experiments indicate that they are satisfactory for the purpose for which the method is intended. For greater accuracy, the total phosphate may be determined gravimetrically by precipitating with magnesia mixture, igniting the precipitate, and weighing as magnesium pyrophosphate. The barium hexametaphosphate and manganous pyrophosphate precipitates may be fused with sodium carbonate, the fusion extracted with cold water, and the phosphorus pentoxide in the filtrate determined gravimetrically by the magnesium pyrophosphate method. Before adding magnesia mixture, the filtrates must be acidified with nitric acid and boiled to eliminate carbon dioxide and convert the meta- and pyrophosphate radicals to orthophosphate.

REFERENCES

1. Jones, Loren T.: Estimation of Ortho-, Pyro-, Meta-, and Polyphosphates in the Presence of One Another. *Ind. Eng. Chem. Analyt. Ed. 14*: 536 (1942).

2. Kiehl, S. J., and Coats, H. P.: The Hydration of Sodium Monometaphosphate in Alkaline Solution at 75°. Phosphate V. *J. Am. Chem. Soc. 49*: 2180 (1927).

3. Kiehl, S. J., and Wallace, G. H.: The Dissociation Pressures of Monopotassium and Monosodium Orthophosphates and of Dipotassium and Disodium Dihydrogen Pyrophosphates. Phosphate IV. *Ibid. 49*: 375 (1927).

4. Boratynski, K.: Uber die colorimetrische Bestimmung des Orthophosphates in Gegenwart von meta- und pyrophosphaten. *Z. Anal. Chem. 102*: 421 (1935).

5. Mellor, J. W.: *Comprehensive Treatise on Inorganic and Theoretical Chemistry,* Vol. VIII, p. 987, Longmans, Green & Co., New York (1928).

Alkyl Benzene Sulfonate Determination

Apparatus

1. Nessler tubes, 50 ml (or 50 ml graduates).
2. Pipettes.
3. Burette, 10 ml, graduated in 0.05 ml divisions.

Reagents

1. o-Tolidine Solution. Add 1.0 gm of o-tolidine to 5 ml of 20 per cent hydrochloric acid (100 ml of HCl, sp. gr. 1.18-1.19, to 500 ml of distilled water) and grind to a thin paste. Add 150-200 ml of distilled water to produce a solution. Transfer to a 1 liter graduate and make up to 505 ml with distilled water, then make to 1 liter by adding 20 per cent HCl. Store in an amber bottle, out of direct sunlight.

2. Sodium Hypochlorite. Prepare a solution of sodium hypochlorite by slurrying 5 gm of high test calcium hypochlorite and 4 gm of anhydrous sodium carbonate in 334 ml of distilled water. Let the insoluble matter settle, then decant the solution through a filter Analyze it for available chlorine content according to the A.O.A.C. arsenious oxide titration method. Prepare a standard solution containing 300 p.p.m. of available chlorine.

3. Alky Benzene Sodium Sulfonate, or Unknown. Prepare a 0.05 per cent solution of the alkyl benzene sulfonate by suitably diluting a stronger solution with distilled water. Dilute a wash solution 1 to 5 or more, depending upon the approximate concentration of the alkyl benzene sulfonate. This solution must be as dilute as indicated, or if stronger must be measured accurately from a suitably calibrated burette or pipette.

Procedure

1. Transfer 40 ml of distilled water to each of two Nessler tubes. Add 1 ml of 300 p.p.m. sodium hypochlorite solution to each tube, and invert to mix. Add 2 ml of o-tolidine solution to each, and invert to mix.

2. Make the blank up to 50 ml with distilled water and invert several times to mix thoroughly.

3. Add the unknown solution of alkyl benzene sulfonate of definite dilution carefully in known amounts to a fresh mixture of reagents, until a color change just appears. Make the tube up to volume and compare with the blank. From the volume of unknown added, and the fact that 1 p.p.m. of alkyl benzene sodium sulfonate is required for a color change, calculate the strength of solution.

Used in the manner described the method is suitable for determining the strength of dilute solutions by visual observation where photometric equipment is not available.

REFERENCES

Harris, Joy. C., Colorimetric Determination of Alkyl Benzene Sulfonates, *Ind. Eng. Chem. Anal. Ed. 15*: 254 (1943).

Scales, F. M. and Kemp, Muriel. The Fundamental Principles of Chlorine Sterilization and a New and Positive Germicide. *Internat. Assoc. Milk Dealers* Association Bulletin 31 (1939), pp. 187-208.

Chlorine Solution Strength

A. TITRATION METHODS

When it is necessary to determine the concentration of available chlorine in a strong solution which may contain as much as 15 per cent of sodium hypochlorite, the most accurate procedure is by titration.

Some titration methods for the determination of available chlorine in both high and low concentrations are described.

THIOSULFATE METHOD

Apparatus

1. Beaker, 250 ml.
2. Stirring rod.
3. Graduated cylinder, 2.5 ml.
4. Pipette, 1 ml.
5. Burette, 50 ml, graduated in tenths.

Reagents

1. Sodium thiosulfate ($Na_2S_2O_3$) 0.1 N solution.
2. Hydrochloric acid (HCl), concentrated.
3. Potassium iodide (KI), 10 per cent solution.
4. Starch solution indicator, 2 per cent.

Procedure

1. Transfer by means of the graduate 20 ml of a 10 per cent solution of KI into a 250 ml beaker.
2. With the graduate, measure 5 ml of concentrated HCl and slowly pour into the beaker. Stir well during the addition.
3. While stirring, add slowly from the pipette 1.0 ml of the sodium hypochlorite solution to be tested.
4. Continue stirring and add 150 ml of distilled water.
5. Add, with stirring, 0.1 N $Na_2S_2O_3$ solution from the burette until the brown color changes to a pale yellow.
6. Add 2 or 3 ml of starch solution and continue the titration until the solution becomes colorless.

Calculation

The milliliters of 0.1 N $Na_2S_2O_3$ used in the titration, $\times$ 0.003546 $\times$ 100 = per cent of available chlorine by volume. One per cent is equal to 10,000 p.p.m. One ml of 0.1 N $Na_2S_2O_3$ = 0.003546 gm of available chlorine.

TITRATION METHOD FOR WEAK SOLUTIONS

The above procedure may be employed for the accurate determination of available chlorine in dilute sterilizing solutions.

Procedure

Add 50 ml of the dilute solution to those of the KI and HCl in the beaker. The titration may then be followed according to the procedure just described.

Calculations

The milliliters of 0.1 N $Na_2S_2O_3$ used in the titration $\times$ 0.003546 $\times$ 2 $\times$ 10,000 = p.p.m. The usual procedure in figuring parts per

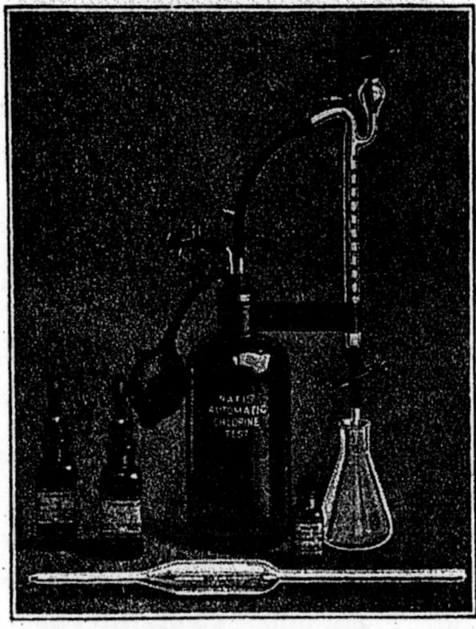

Figure 130.
Nafis automatic titration chlorine set. Courtesy Louis F. Nafis, Inc.

million is to move the decimal point four places to the right in the weight of available chlorine per milliliter and omit the final multiplication by 10,000.

NAFIS AUTOMATIC CHLORINE TEST SET

A titration set for the determination of available chlorine is sold under the name of the Nafis Automatic Chlorine Test by the Kimball Glass Co. of Vineland, New Jersey. (Figure 130.) The set includes the following materials and procedure.

Apparatus

1. Reservoir bottle.
2. Burette attached.
3. Pipette.
4. Erlenmeyer flask.

Reagents

1. Solution A, 99.5 per cent acetic acid.
2. Crystals B, potassium iodide crystals.
3. Indicator 3, 2 per cent starch solution.
4. Neutralizer, 0.0141 normal sodium thiosulphate solution.

Procedure for Available Chlorine

1. Assemble the apparatus as shown in Figure 130. Do not drop any material into the reservoir bottle. The neutralizer is supplied ready for use. Keep all openings of the apparatus closed with the pinchcocks when it is not in use.

NOTE: To refill the burette hold the short projecting rubber vent tube shut and compress the rubber bulb. The air pressure in the bottle causes the solution to rise into the burette; when it rises into the overflow bulb at the top of burette, release the rubber vent tube and the zero point is instantly and automatically adjusted.

In order to obtain accurate measurement and to fill the rubber tube and burette tip with the solution, draw out into the Erlenmeyer flask a small portion. Be sure to wash out the Erlenmeyer flask and to fill the burette to zero before beginning a test.

2. Shake into the Erlenmeyer flask ¼ to ½ gm of crystals B and add a little distilled water to dissolve.

3. Add two full droppers of solution A and thoroughly mix.

4. Pipette 50 ml of the chlorine solution to be tested. If chlorine is present, a pale yellow to an amber color will appear, depending on the amount.

5. Add neutralizer from the burette until a very faint yellow color remains.

6. Add one-half dropperful of indicator 3. A blue color will result.

7. Continue to add neutralizer from the burette very slowly until the solution is colorless. Note the amount of neutralizer used. The point on the burette level with the meniscus gives the parts per million of available chlorine.

Procedure for Sodium Hypochlorite Solution

1. Measure or pipette 1 ml of the solution to be tested into a 1,000 ml flask, rinse out the measure or pipette and add the rinsings to a flask. Make up to 1,000 ml with distilled water.

2. Put approximately 1 gm crystals B in the Erlenmeyer flask, dilute with 10 ml distilled water, mix thoroughly and proceed as for available chlorine. Multiply the result by 1,000.

NOTE: Keep the bottles containing the indicator and solution A corked and the apparatus openings closed with pinchcocks when not in use.

B-K CHLORINE TEST SET

The General Laboratories of the Pennsylvania Salt Company, 1000 Widener Building, Philadelphia, Pa., market a three bottle B-K testing set which employs the recognized starch-iodide-sodium thiosulphate reaction. The set consists of two bottles of reagents (Solutions Nos. 1 and 2) and an empty graduated bottle in which the test is made. Each drop of solution 2 dispensed from the dropper in the cap is equivalent to 10 parts of available chlorine per million parts of water.

Procedure

1. Fill the empty test bottle to red mark with the chlorine solution to be tested.

2. Add a dropperful of solution from bottle 1 and mix by rotating bottle.

3. Add solution from bottle 2 one drop at a time, holding dropper in vertical position with the tip in neck of the test bottle. Count drops accurately. Mix after each drop is added.

4. Continue to add solution drop by drop until blue color disappears. This is the end-point of the reaction.

Each drop of solution in bottle 2 is equivalent to 10 parts available chlorine per million parts of water (10 p.p.m.). For example, if 9 drops of solution from bottle 2 did not decolorize and 10 drops did, the solution contains between 90 and 100 p.p.m.

NOTE: The droppers in bottle 1 and 2 should be kept in their own bottles and not interchanged. Keep the tip of the dropper in bottle 2 clean. Dirt will change the size of the drop.

If solution in bottle 1 becomes discolored, add a drop from bottle 2 and shake, repeating until it is decolorized. The color of the solution in bottle 1 does not interfere with test.

Protect solutions from freezing.

Rinse test bottle thoroughly after each test.

B. COLORIMETRIC METHODS

The available chlorine in dilute sterilizing solutions may also be determined by one of the colorimetric methods. These depend on the depth of yellow color produced when an orthotoluidin reagent is added to a relatively dilute solution of sodium hypochlorite.

Orthotoluidin is a p-2-diamido-m-2-dimethyl-diphenyl compound, a dye intermediate similar to benzidine. When oxidized in the presence of hydrochloric acid by chlorine, a yellow dye substance, probably an HCl-addition compound, is formed. The color so formed is fairly proportional to the amount of chlorine present.

This reaction is not specific for chlorine; any oxidizing substance will produce a color with orthotoluidin in HCl solutions. Nitrates, chlorates, and ferric and manganic salts act as interfering substances, but since these are not frequently met with, at least not in appreciable amounts in sterilizing solutions, the method is a very convenient one for use in milk plants.

The use of the reagent in testing hypochlorites in chlorine sterilizing solutions for milk and ice cream equipment has found application in several chlorine test sets. The general method is essentially the same as for water supplies which contain very small chlorine concentrations. By the use of artificial color standards expressed in parts of chlorine per million parts of solution, a colorimetric comparison may be made. These give a range varying from 25 to 250 parts per million of chlorine. However, colors obtained with very high chlorine concentrations are not readily compared owing to the possibility that oxidation of the orthotoluidin structure is carried too far Most color comparison methods lend themselves best to dilute solutions and to a range in which varying color depths are most easily recognized by the eye. Standard color sets representing known amounts of chlorine are used to match the color of the sterilizing solution.

CHLORINE SLIDE COLORIMETER TESTING SET

This set may be obtained from W. A. Taylor & Company, 872 Linden Avenue, Baltimore, Md.

Procedure

1. After removing the cover from the base, fill three of the test tubes to the mark (5 ml) with the sample to be tested and place in the holes back of the slots in the base.

2. To the middle tube add 5 drops of colorimeter reagent 1 and mix well.

3. Add 0.5 ml of colorimeter reagent 2 and mix well. This order of adding the reagents must be strictly followed and each reagent well mixed with the sample.

4. Place the slide containing the color standards in position on the base and, with the back of the comparator toward a window or other source of daylight, move the slide back and forth in front of the test samples until a color match is obtained.

Read the chlorine content directly from the values on the front of the slide. All readings should be made within five minutes after the reagents have been added. On longer standing a greenish color may develop which will not match the standards.

NOTE: By means of dilution, solutions containing as high as 4-5 per cent available chlorine and 1-2 per cent free caustic may be tested with an accuracy sufficient for most practical purposes.

Sample	CHLORINE BY COLORIMETER			Chlorine by Titration
	Dilution	Reading p. p. m.	g/100 ml	g/100 ml
1	5:1000	225	4.50	4.25
2	10:1000	200	2.00	2.11
3	10:1000	175	1.75	1.73
4	10:1000	125	1.25	1.18
5	10:1000	75	0.75	0.82
6	10:1000	37.5	0.38	0.38

DIVERSEY CHLORINE TESTING SET

This set consists of a small graduated cylinder, a bottle of orthotoluidin solution, a test tube, and three tubes of color standards arranged in a small box. The values of the standards are 15, 50, and 100 p.p.m. The directions on the inside cover of the box are as follows:

"For testing wash and rinse solutions: Fill the graduated cylinder with chlorine rinse or wash solution to the line R, then add tap or distilled water to the line marked W. To this add toluidin solution to the line T. Shake the mixture and pour it into an empty test tube, and match the color of the solution with one of the three color tubes. After obtaining the closest match, read the side of the standard color tube which will tell you the strength of solution represented by that color and this will tell you how much available chlorine your rinse or wash solution contains."

LA MOTTE TESTING SET

This set consists of four chlorine color standards with values of 20, 30, 50, and 100 parts of chlorine per million, a 5 ml test tube, and a vial of orthotoluidin reagent, arranged in a case.

Rinse the test tube well with the solution to be tested and fill the tube to the mark with the chlorine solution. Add 10 drops of the orthotoluidin reagent by means of a dropper and mix well. Let stand for about 5 minutes for the color to develop. Using a block comparator, compare the color with the standards by placing the tube which contains the unknown solution in the center hole and placing consecutive color standards, such as 20 and 30, or 30 and 50, or 50 and 100 on each side of it.

Hold the block toward the daylight and compare the colors through the slots in the block. Change the position of the color standards until the color of the test sample either matches one of the color standards or lies between those of two consecutive standards. If the test solution is stronger than the highest standard, dilute the unkown in equal parts and fill to the mark with distilled water, when the color can be matched with one of the standard color. Use the dilution factor to obtain the strength of chlorine in the unknown solution.

LEHMKUHL SET

In this set, one standard is used instead of a number of standards, the range being obtained by varying the amount of the chlorine solution according to its concentration. Thus one standard may be employed for testing any solution from 3 to 500 p.p.m. of available chlorine by the simple expedient of diluting solutions stronger than 3 p.p.m. to that concentration.

A pipette which has been calibrated to show markings of from 500 to 50 p.p.m., and a tube (which also serves as a mixing and comparison tube) of from 30 to 3 p.p.m. As much of the solution to be tested is taken as parts per million are presumed to be present and this is introduced into the tube and made up to the top mark, with unchlorinated water, the stopper put in place, and shaken. Thus, if the chlorine solution is presumed to contain 100 p.p.m., fill the pipette to the 100 p.p.m. mark (the low point of the meniscus, coinciding with the line), introduce the sample into the rubber-stopper tube, blow out, and fill the tube to the 3 p.p.m. mark. The 100 p.p.m. solution having been diluted to 3 p.p.m. may be compared with the standard after treatment with the reagent. In the case of solutions ranging from 30 to 3 p.p.m. the pipette is not required.

It will be seen that this presumes a knowledge of the strength of the solution to be tested, but this is usually known, since the strength of stock chlorine compounds are known. In plant practice we do not deal with unknowns but solutions of known strength and the purpose of the test is that a constant concentration of chlorine be used at all times.

Directions for Hypochlorite Preparations

Measure an amount of the solution to be tested according to the strength it is presumed to be. Introduce this amount into the mixing tube, fill almost to the top line (3) with unchlorinated water, shake, add 5 drops of orthotoluidin reagent, replace the stopper and shake again. Place the stoppered tube in the case, close and after five minutes compare the color. If the color so formed is lighter, the chlorine solution is below the stated strength, and if darker, it is stronger.

Directions for Chloramine T. Preparations

Proceed exactly as for hypochlorites, except that after filling the tube almost to the 3 mark, introduce two 15½ grain salt tablets. Dissolve the salt tablets by shaking and then add the orthotoluidin reagent. In this case ten minutes must elapse before comparing the solution with the standard.

The solution in every case must be brought up to the top mark 3.

REFERENCES

1. What Chemicals Are Best Adapted for Sterilization of Dairy Equipment? *Proc. 23rd Ann. Conv., Lab. Sec.,* I.A.M.D. (1930).
   a. Myers, R. P.: Milk Bottles, p. 75-85.
   b. Hunziker, O. F.: Metal Surfaces (Coolers, Cans and Vats), pp. 85-96.
   c. Scales, F. M., and Russell, Harriet E.: Glass Lined Tanks, pp. 96-102.
2. Johnson, A. H.: Practical Tests for Determining the Strength of Chlorine Solutions, *Ibid.,* pp. 103-106.
3. Prucha, M. J.: Sterilization of Dairy Equipment, *Proc. 21st Ann. Conv., Lab. Sec.,* I.A.M.D. (1928), pp. 53-74.
4. Powers, A. J.: Chlorine Sterilization in the Market Milk Field, *Proc. 22nd Ann. Conv. Sec.,* I.A.M.D. (1929), pp. 39-43.
5. Scales, F. M., and Kemp, Muriel: Fundamental Principles of Chlorine Sterilization and a New and Positive Germicide. *Internat. Assoc. Milk Dealers, Assoc. Bul.* 31, 187-208 (1939).

APPLIED RESEARCH INSTITUTE CHLORATOR TEST
(Scharer Colorimetric Method)

This new test is designed to permit the rapid determination of available chlorine in concentrations of from 5 to 1,000 parts per million. It is exceedingly simple and will differentiate between a hypochlorite and a chloramine preparation. This test avoids the difficulty of matching yellow colors by producing distinctly clear and brilliant colors such as blue, green, yellow, orange and red, depending on the concentration of chlorine.

Two procedures are available—a titration technic and a quick field test. In the titration technic, one ml of the chlorator reagent is added to a test tube or flask and diluted with approximately 10 ml of distilled or unchlorinated tap water. Using a 10 ml pipette (graduated in 0.1 ml) the chlorine solution under test is added drop by drop with constant mixing until the color is a definite yellow.

$$\text{Calculation:}\ \frac{100}{\text{ml of chlorine solution used}} = \text{p.p.m. available chlorine}$$

In the quick field test, use is made of a pipette calibrated directly in parts per million of available chlorine. One ml of the chlorator reagent is diluted to approximately 10 ml with distilled or unchlorinated tap water. Using the calibrated pipette the hypochlorite solution under test is added in the amount indicated by the presumed or required strength. Shake briefly. A brilliant color forms at once when testing hypochlorites and the concentration is approximated from the following color scheme.

Blue = Approximately 50% weaker than desired concentration.
Green = Approximately 25% weaker than desired concentration.
Yellow = Correct or desired concentration.
Orange = Approximately 15% stronger than desired concentration.
Red = Approximately 25% stronger than desired concentration.

These colors are stable for at least 30 minutes. However, when testing chloramine T solutions the color formation is slower; initially a blue color is formed in all instances which gradually (in about two minutes) shifts through the range of other colors according to the amount of available chlorine present before becoming stable at the color representing the strength of the solution.

Determination of Strength of Quaternary Ammonium Solutions

HARTLEY-RUNNICLES METHOD (DU BOIS MODIFICATION[1])

This method is based on the qualitative alteration of the color of bromphenol blue from purple to pure blue by cations of high molecular weight.[2] This color change is used as the end point for the direct titration of surface active anions against surface active cations, leading to the formation of more or less insoluble salts. Removal of the surface active cation in this fashion, restores the alkaline purple color of the bromphenol blue.

Apparatus

1. Volumetric flasks, 100 ml, 1 l.
2. Microburette, 5 ml, graduated in 0.01 ml divisions.
3. Pipette, 2 ml, graduated in 0.10 ml divisions.
4. Pipette, 25 ml.
5. Supply of small test tubes.

Reagents

1. Duponol P C (sodium dodecyl sulfate), 1:2500 solution. Weigh accurately 0.40 gm of powdered Duponol P C, transfer to a one liter volumetric flask and adjust the volume to the one liter mark with distilled water. Heating may be necessary to bring into solution and the solution should be perfectly clear.

NOTE: DUPONOL solution may become cloudy upon standing and such a solution should not be used for titrations. Since Duponol solutions may contain impurities it is necessary to calibrate the 1:2500 solution against a sample of a high molecular quaternary ammonium compound of known composition and concentration. For this purpose, crystalline Hyamine 1622 (p-tert. octyl phenoxyethoxyethyl dimethyl benzyl ammonium chloride, monohydrate), and CTAB (cetyl trimethyl ammonium bromide) have been found most suitable.

2. Bromphenol blue, 0.04 per cent solution. Pipette 25 ml of 0.04 per cent solution into a 100 ml volumetric flask and making up to 100 ml with distilled water, adding a few drops of aqueous ammonia to make the solution slightly alkaline (a decided purple color).

NOTE. This indicator solution has a tendency to lose some of its ammonia on standing and, if it is not a decided purple color add a few drops of aqueous ammonia before use, otherwise the proper end-point will not be reached during the titration.

Procedure

a. Standardization of 1:2500 Duponol Solution.

1. Weigh exactly 0.40 gm of crystalline Hyamine 1622 or CTAB. Transfer to a one liter volumetric flask and make up volume to one liter mark with distilled water. This provides a 1:2500 concentration or 400 p.p.m. solution of the standard quaternary.

2. Place 2 ml of this 1:2500 standard quaternary solution in a test tube, add 0.1 ml of bromphenol blue indicator, and titrate with the 1:2500 Duponol solution, adding it drop by drop from a microburette and shaking after each addition.

3. At the end point the color changes from blue to purple, and is best observed under artificial light.

Theoretically 2 ml of the standard 1:2500 quaternary solution should be exactly equivalent to 2 ml of the 1:2500 Duponol solution and if such is the case the Duponol solution is of the proper strength. Duponol P C is a commercial product and may require standardization.

Assume it required 1.8 ml of the Duponol solution to titrate 2 ml of the standard quaternary, then, to adjust the one liter volume of Duponol solution to the standard 1:2500 strength, calculate the amount of distilled water to add by applying the formula shown under Standard Solutions, page 483.

$$\text{Distilled water} = \frac{998.2\ (2.0 - 1.8)}{1.8} = \frac{199.64}{1.8} = 110.9 \text{ ml}$$

b. Determination of the Strength of Quaternary Ammonium Solutions.

1. Place 2 ml of the quaternary ammonium solution of unknown strength in a test tube and proceed as directed in steps 2 and 3 above. Record the ml of 1:2500 standardized Duponol solution required to reach the end point.

NOTE: Since the relationship between the various types of quaternary ammonium compounds and the quaternary used for standardization of the Duponol solution is not quite linear it is advisable to establish an individual calibration curve for each quaternary ammonium compound to be tested. This may be done by preparing a series of known dilutions of each of the various types of quaternary ammonium compounds, titrating with 1:2500 standard Duponol solution, and listing the titration values as for the two types of quaternaries illustrated in table 17.

TABLE 17

Germicide	Concentration	ml. Standardized 1:2500 Duponol PC
Alkyl dimenthyl benzyl ammonium chlorides	1:1000	4.01
(Zylium, Roccal, Rodalon, B.T.C.)	1:2000	2.00
	1:3000	1.53
	1:4000	1.13
	1:5000	0.93
	1:7500	0.68
	1:10000	0.44
	1:20000	0.23
9-Octadecenyl dimethyl ethyl ammonium bromide	1:1000	4.77
(Onyxide)	1:2000	2.25
	1:3000	1.45
	1:4000	1.11
	1:5000	1.01
	1:7500	0.62
	1:10000	0.49
	1:20000	0.21

These values are the average of 5 determinations.

c. For routine laboratory determination of quaternary concentration a titration factor may be applied when the Duponol solution is stronger or weaker than the standardized 1:2500 strength.

Assume 2 ml of the standard quaternary solution = 1.8 ml of Duponol solution, then 1.1 ml of the standard quaternary solution = 1.0 ml of Duponol solution, therefore 1.1 times the number of ml of strong Duponol solution used will represent the equivalent in ml of standardized 1:2500 Duponol solution to be used in determining concentrations of quaternaries from the established graphs representing the relationship between concentration of quaternary and ml of standard 1:2500 Duponol solution.

Should the Duponol solution be so weak as to require 2.2 ml to titrate 2 ml of the 1:2500 standard quaternary, then the titration factor would be 0.81.

Since 2 ml of the standard quaternary solution = 2.2 ml of Duponol solution, then 0.81 ml of the standard quaternary solution = 1.0 ml of Duponol solution, therefore 0.81 times the number of ml of weak Duponol solution used will represent the equivalent in ml of standardized 1:2500 Duponol solution to be used in connection with the established graphs for determining the concentrations of the various quaternaries.

The DuBois modification of the Hartley-Runnicles Method is reported to give satisfactory assay results on cetyl pyridinium bromide, lauryl pyridinium chloride, cetyl trimethyl ammonium bromide, alkyl dimethyl 3.4—dichlorobenzyl chloride, p-tert. octyl phenoxyethyl ethoxy dimethyl benzyl ammonium chloride as well as the groups referred to in the table above.

Since this method is based on the interaction between a cationic and an anionic surface active agent, any soap or synthetic anionic surface active agent will interfere with the titration. Such organic substances as milk, beer, serum, etc., will also interfere and it is sometimes possible to filter out precipitates caused by these substances and to titrate the filtrate. Although these substances cause some interference with the test it has been shown [3] that it is possible to carry out a titration in the presence of milk.

NOTE: Solutions of quaternary ammonium compounds freshly prepared for sanitizing purposes should be at a minimum strength of 1:5000 or 200 p.p.m. of active quaternary. During the sanitizing treatment the active quaternary strength should not be permitted to drop below 1:10,000 or 100 p.p.m.

REFERENCES

1. Du Bois, A. S., Testing the Quaternary Ammonium, Soap and Sanitary Chemicals, November, 1946.

2. Hartley, G. S., and Runnicles, G. F., Proc. Roy. Soc. of London, Series A, No. 934, *168* 424, 1938.

3. Du Bois, A. S., and Dibblee, D. D. The Influence of Surface Active Cationic Germicides on the Bacterial Population of Milk, J. Milk Tech., Vol. 9, No. 5, 1946.

HARPER, ELLIKER, AND MOSELEY, QUATERNARY TEST[1]

This method is claimed to be quantitatively accurate for quaternary solutions ranging from 10 to 300 p.p.m. in concentration. It is based on the fact that eosin yellowish, essentially tetrabromfluorescein, reacts with quaternary ammonium compounds to form a bright red or pink precipitate and that addition of solutions of certain anionic surface-active agents replace the eosin in the salt formed with resultant loss of red color in the solution. Under controlled conditions the quantity of anionic compound required to remove the red color may be made to indicate the concentration of quaternary ammonium compound present.

Apparatus

1. Microburette, 5 ml, graduated in 0.01 ml divisions.
2. Pipettes, 1 ml, graduated in 0.01 ml divisions.
3. Supply of small test tubes.

Reagents

1. Indicator solution:

a. Dissolve eosin yellowish dye (dye concentration of about 90 per cent) in acetone (analytical grade) at the rate of 0.1 mg of dye to 1 ml of acetone.

b. Add acetone-eosin solution to tetrachlorethane at the rate of 1 ml of acetone-eosin to 9 ml of tetrachlorethane.

c. Remove the reddish color from the solution by adding citric acid crystals (analytical grade) at the rate of 5 mg of crystals to 1 ml of dye solution.

d. Shake for one minute.

e. Filter through Whatman No. 1 or equivalent grade of filter paper.

2. Buffer:

a. Prepare a 5 per cent solution of citric acid (analytical grade) and adjust to pH of 3.5 with N/10 NaOH. About 15 ml of NaOH are required per ml of citric acid. Add 1 per cent of tetrachlorethane to prevent mold growth.

3. Anionic solution:

a. Make up a 0.01 per cent solution of active anionic compound from 10 per cent Fisher Laboratory Aerosol (10 per cent di-octyl sodium sulfosuccinate) 1:1000 dilution of the 10 per cent Aerosol). Duponol paste has also been found satisfactory for this purpose.

Procedure

1. Place 1 ml of the solution to be tested in a test tube.
2. Add 0.1 ml of buffer solution.

3. Add 1 ml of dye solution. Two distinct layers will form.

4. Shake until a pink color appears in the lower layer. The upper layer will be clear and should show no color. About three to eight seconds of shaking are required.

5. Titrate with anionic solution. Add slowly and shake until color fades out and lower layer becomes white. Quantity of quaternary ammonium compound is indicated by the quantity of anionic compound required to remove red color from the lower layer. Standards may be prepared from solutions of known concentration and a standard reference curve may be established.

For evaluation purposes it will be necessary to prepare solutions of known strength of each quaternary to be tested and to record the titration values of the anionic solution as illustrated for the compounds shown in table 18.

TABLE 18

Compound	Milliliters of anionic solution required to remove pink color from solutions with following concentrations of quaternary ammonium compounds expressed in parts per millions:								
	10	20	30	40	50	75	100	150	200
	ml	ml	ml	ml	ml	ml	ml	ml	ml
Alkyl	.11	.23	.35	.43	.58	.83	1.30	1.62	2.43
Dimethyl	.10	.21	.34	.44	.54	.80	1.29	1.63	2.44
Benzyl	.11	.24	.33	.42	.56	.82	1.30	1.62	2.43
Ammonium	.10	.22	.34	.43	.57	.80	1.31	1.60	2.45
Chloride	.10	.22	.34	.42	.55	.80	1.29	1.64	2.44
	.11	.24	.35	.42	.55	.81	1.32	1.61	2.45
(Ster-Bac)									
Acyl									
Colamino	.10	.20	.29	.39	.50		1.02		1.98
Formyl	.10	.22	.30	.40	.51		1.07		1.96
Methyl	.11	.21	.29	.41	.48		1.08		1.97
Pyridinium	.10	.22	.31	.40	.50		1.06		1.96
Chloride									
(Emulsept)									
Alkyl	.09	.20	.29	.39	.51		1.04		2.00
Dimethyl	.10	.21	.29	.39	.50		1.06		1.95
3,4—	.11	.20	.29	.40	.50		1.05		1.96
Dichlorobenzyl	.10	.20	.30	.40	.52		1.07		1.96
Ammonium									
Chloride									
(Tetrosan)									

REFERENCE

1. Harper, W. J., Elliker, R. R., and Moseley, W. K., Quaternary Test, Soap and Sanitary Chemicals, February, 1948.

TAMOL FIELD TEST FOR QUATERNARY AMMONIUM CONTENT IN SANITIZING SOLUTIONS

This is a single reagent test which measures only active quaternary. It is claimed that water hardness, pH, and organic contaminants do not affect the test.

Apparatus

1. Test tubes, graduated at 5 and 10 ml volumes.
2. Indicator bottle with glass dropper calibrated to deliver 20 drop of water per ml.

Reagents

1. Indicator A, 1 per cent Tamol N solution containing 0.10 per cent F. D. & C. brilliant blue dye #1.
2. Indicator B, 0.2 per cent Tamol N solution containing 0.02 per cent F. D. & C. brilliant blue dye #1.

NOTE: Tamol N is the sodium salt of a condensed aryl sulfonic acid and it is obtainable in dry form from Rohm & Haas Company, Washington Square, Philadelphia 5, Pa.

Procedure No. 1

1. For testing Group A

p-Tert. octylphenoxyethyldimethylbenzylammonium chloride (Hyamine 1622 and Hyamine 10X) Benzalkonium chloride (Zylium, Roccal, Rodalon, and B. T. C.) Alkyldimethyl 3, 4 dichlorobenzylammonium chloride (Tetrosan).

Fill test tube to the 5 ml mark with the sanitizing solution of unknown strength, add 2 drops of Tamol indicator A, invert the tube and observe for immediate appearance of tubidity in the solution.

2. Group B—Lauryl pyridinium chloride (Fuldicide)

Fill test tube to the 5 ml mark with the sanitizing solution of unknown strength, add 3 drops of Tamol indicator A and proceed as in #1.

3. Group C—n (Acylcolaminoformylmethyl)-pyridinium chloride (Emulsept)

Fill test tube to the 10 ml mark with the sanitizing solution of unknown strength, add 5 drops to Tamol indicator A and proceed as in #1.

Evaluation of Results: Immediate appearance of turbidity indicates that at least 1:4000 available quaternary is present. At 1:5000 strength there may or may not be a slight turbidity. At 1:6000 or weaker concentrations the solution is perfectly clear.

NOTE: This method cannot be used to test the concentration of 9-Octadecenyldimethylethylammonium bromide (Onyxide) or of Cetylpyridinium chloride (Ceepryn).

Procedure No. 2

For testing concentration of all of the compounds in Groups A, B, and C

1. Measure quaternary sanitizing solution of unknown strength to the 5 ml mark in the test tube.

2. Use Tamol indicator B and add it one drop at a time, inverting the tube once after the addition of each drop. Watch for end-point which is at the development of maximum turbidity.

Evaluate the quaternary content by the number of drops of indicator required to reach maximum turbidity according to Table 19.

TABLE 19

Group	Quaternary Ammonium-Type Compound	No. Drops Tamol Indicator B for Maximum Turbidity — Concentration in Sanitizing Solution			
		100 p.p.m.	150 p.p.m.	200 p.p.m.	250 p.p.m.
A	Hyamine 1622, Hyamine 10-X, Tetrosan	3	5	7	9
A	Roccal, B. T. C., Zylium, Rodalon	3	6	8	10
B	Fuldicide	4	6	8	10
C	Emulsept 2 X	2	3	5	7

REFERENCE

1. Flanagan, T. L., Jr., Drennen, T. J., and Goetchius, G. R. Determining Quaternary Content in Sanitizing Solutions. Soap and Sanitary Chemicals, April, 1948.

PORTLEY PAPER TEST FOR THE RAPID TESTING OF QUATERNARY AMMONIUM SANITIZING SOLUTION

(Suitable as a field test)

Apparatus

1. Supply of Portley Test Paper.
2. Test tubes calibrated and marked at 10 ml volume.
3. Indicator dropper bottle, 1/2 fl. oz.; dropper to deliver approximately 20 drops of water per ml.

Reagents

1. Standardized Portley titrating solution, one drop of which is equivalent to 20 p.p.m. of active quaternary.

NOTE: Portley titrating solution and test paper is obtainable from Fairfield Laboratories, 417 Cleveland Avenue, Plainfield, N. J.

Procedure

A. Indicator Test Paper Method

1. Immerse a strip of test paper in sample sanitizing solution for 15 seconds.

2. Remove paper and compare with color test standards on chart which accompanies paper. If the color of the paper is dark pink, it denotes a quaternary concentration of 200 p.p.m. or more. A peach or light pink color indicates 100 p.p.m. or less.

B. Titration Method

1. Immerse strip of test paper (about 1 inch long) in solution being tested for approximately 15 seconds.

2. Remove paper with clean, dry fingers, and remove excess quaternary solution by shaking. If the color is light pink, peach or dark pink, proceed with next step.

3. Place wet paper in test tube containing 10 ml tap water. Then add titrating solution dropwise and occasionally agitate by rotating tube between fingers. When the color of the paper changes to a very light pink, proceed with caution because this indicates close approach to end-point. Add one more drop of titrating solution, agitate, and wait five seconds for the end-point which is a lemon yellow. Repeat this until paper turns yellow. Usually two drops are sufficient.

4. Determine the p.p.m. of active quaternary by multiplying the number of drops of titrating solution used by twenty (20).

NOTE: Concentrations greater than 1000 p.p.m. may be determined by diluting the solution with distilled water and multiplying the dilution factor by the number of p.p.m. determined as outlined above.

STONE-MARSHALL PAPER TEST FOR APPROXIMATE STRENGTH OF QUATERNARY COMPOUNDS[1]

(Suitable as a field test)

Apparatus

1. Supply of Stone-Marshall test paper A.
2. Medicine dropper, delivering 16 drops of water per ml.
3. Prescription bottle, 2 oz. size, marked at the 0.5, 1.0, 1.5, and 2.0 oz. fill points.

Procedure

1. Place 2 drops of 10 per cent concentration of the quaternary to be tested, into the prescription bottle.

NOTE: Always use the quaternary to be tested at 10 per cent concentration. If labeled 10.5 per cent consider it 10 per cent. If labeled 12.5 per cent, add 2 drops of water to 8 drops of the product in the test bottle. If labeled 15 per cent add 4 drops of water to 8 drops of the product in the test bottle.

2. Add water to the 0.5 oz. fill point in the bottle, test with S-M paper and observe the color. This color represents 400 p.p.m. concentration of the quaternary.

NOTE. Use daylight or its equivalent to observe the color.

3. Add sufficient water to bring the test bottle contents to the 1 oz. level, test with S-M paper and observe the color. This color represents 200 p.p.m.

4. Adjust the level of the test bottle contents to the 1.5 oz. mark with water, test with S-M paper and observe the color. This color represents approximately 150 p.p.m. (actually 140 p.p.m.)

5. Finally adjust the test bottle contents to the 2 oz. mark with water, test with S-M paper and observe the color. This color represents approximately 100 p.p.m. (actually 104 p.p.m.)

6. Test the quaternary sanitizing solution as it is being applied and estimate the concentration by comparing the color shown by the S-M paper on the unknown with the color represented by the known solutions in 2, 3, 4, and 5.

A color chart A is available for use with S-M test paper and this chart is most adaptable to use with compounds such as Roccal, Zylium, B.T.C., Rodalon, NOPCO-QCL or any other quaternaries made up from alkyl dimethyl benzyl ammonium chloride. This color chart ranges from somewhat light lilac-like brown for zero concentration to a light gray for 100 p.p.m., to a light greenish gray for 200 p.p.m., and a light bluish green for 300 p.p.m.

Stone-Marshall paper is obtainable from Cargille Scientific Co., 118 Liberty Street, New York 6, New York.

REFERENCES

1. Stone, R. V. A Paper Test for Quaternary Compounds, The Sanitarian, Nat. Assn. of Sanitarians, March-April, 1947.

Testing Brine for Purity, Strength, and Corrosion Inhibitor

Solutions of calcium or sodium chloride are utilized as refrigerating brines and calcium chloride brine is by far the most widely used for refrigerating purposes in the dairy industry. If not carefully managed and controlled brine solutions may become corrosive to iron pipes, zinc coated cans, and other metal parts in refrigerating systems. The rate of corrosion of brine[1] is influenced by such factors as impurities (acidity, ammonia), specific gravity, concentration of dissolved oxygen, and temperature.

Ammonia which may gain entrance to chloride brines through leaks in the system forms ammonium chloride which is corrosive to iron and zinc. Brines are usually alkaline when freshly made but due to gradual absorbtion of carbon dioxide upon exposure to air these solutions tend to become acid in reaction. In general the more concentrated the brine solution the lower the oxygen solubility and the less the relative corrosive effect. The rate of corrosion of brine is accelerated as the temperature of the solution rises.

Corrosion of metal by refrigerating brines may be retarded by holding oxygen concentration at a minimum, by maintaining an alkaline reaction, and by treatment with corrosion inhibitor.

DETERMINATION OF CHEMICAL REACTION OF BRINE

Filter the brine solution if it is turbid or let stand for the sediment to settle out. Transfer a small amount of the clear solution to a test tube and add a few drops of phenolphthalein indicator. If the resulting solution is colorless, the reaction is acid; if, on the other hand, the color is deep red, the reaction is too strongly alkaline. In either case the solution must be corrected by adding an alkali in the first instance and an acidic substance in the latter. A faint pink color indicates that for all practical purposes the brine reaction is correct.

DETERMINATION OF pH OF BRINE (COLORIMETRIC METHOD)

Apparatus

1. Test tubes.
2. Graduated cylinder, 10 ml.

Reagents

· Universal indicator.

Procedure

Obtain a sample of clear brine either by filtration or sedimentation. Place 10 ml in a test tube and add one drop of universal indicator. Mix thoroughly and observe the color in strong light against a white background. Estimate the pH from table 20.

TABLE 20

Reaction	Color	pH
Acid	Orange red	5.0
	Orange	5.5
	Orange yellow	6.0
	Yellow	6.5
Neutral	Greenish yellow	7.0-7.5
Alkaline	Green	8.0
	Bluish green	8.5
	Greenish blue	9.0
	Blue	9.5
	Violet	10.0

DETERMINATION AND ADJUSTMENT OF pH OF BRINE
(Potentiometric Method)

Apparatus

1. Beckman pH meter or equivalent.
2. Calomel electrode.
3. Glass electrode.
4. Supply of beakers, 50 ml and 150 ml.
5. Burette, 10 ml graduated in 0.1 ml.
6. Small rheostadt controlled electric stirrer with glass paddle.

Reagents

1. Buffer solution for standardizing pH meter.
2. Hydrochloric acid (HCl) 0.1 N
3. Sodium hydroxide (NaOH) 0.1 N

Procedure

1. Standarize the pH meter against buffer solution following exactly, the instructions given for the model in use.

2. Set the instrument to the temperature of the sample to be tested and adjust the pH meter according to the printed directions.

3. Spray the electrodes with distilled water, wipe with absorbent tissue, and immerse the electrodes to the proper level in a small quantity of filtered brine in a 50 ml beaker.

4. With the brine in constant agitation by the glass stirrer, hold down the push button and rotate the slide-wire to bring the needle to zero, stop the rotation, release the push button, and record the pH reading on the dial of the instrument.

NOTE: A pH level in the range 7.5 to 8.5 may be regarded as satisfactory, values above or below this range indicate a need for adjustment.

ADJUSTMENT OF pH OF BRINE

1. Standardize the electrodes and the instrument for temperature, if necessary.

2. Immerse the electrodes to proper level in 100 ml of filtered brine in a 150 ml beaker and determine the pH as directed above.

3. With the brine solution in constant agitation adjust the pH to 8.0 by slowly adding from the burette a measured quantity of 0.1 N NaOH or 0.1 N HCl depending upon whether the initial pH was below or above 8.0.

4. Record the quantity of decinormal NaOH or HCl required to adjust the pH.

CALCULATION: (Assume the brine was slightly acid and it required 1 ml of 0.1 N NaOH to adjust the 100 ml of brine to pH 8.0)

1 ml of 0.1 N NaOH = 0.004 gm NaOH

then 0.004 gm NaOH is required for 100 ml of brine

0.004 gm : 100 ml. :: x : 1000 gal (3,784,000ml)

$$100\ x = 15{,}136$$

$$x = 151.36 \text{ gm NaOH}$$

151.36 gm NaOH=0.333 lb NaOH required to adjust 1000 gal of brine to pH 8.0.

Similarly it may be calculated that when 1 ml of 0.1 N HCl is required to adjust 100 ml of brine to pH 8.0 it takes 1.234 lbs of $Na_2Cr_2O_7{\cdot}2H_2O$ to adjust 1000 gal of brine to pH 8.0.

If the pH value of the brine is to be increased it may be accomplished by the addition of sodium hydroxide. Should it be desirable to reduce the pH of the brine it is often proposed that it be done by the addition of carbonic acid gas from a cylinder of compressed carbon dioxide. This is effective but slow since the gas must be dissolved in the brine. Another method makes use of the acidic property of sodium bichromate which is commonly used in conjunction with sodium hydroxide for the purpose of inhibiting corrosion. Hydrochloric acid should not be used to reduce the pH of brine.

Table 21 indicates the pounds of NaOH or $Na_2Cr_2O_72H_2O$ required per 1000 gal of brine when the number of ml of 0.1 N solution to adjust 100 ml of brine to pH 8.0 is known:

DETERMINATION OF PER CENT OF CORROSION INHIBITOR IN BRINE

Chemical corrosion inhibitor in the form of inorganic chromates and organic chrom-glucocates have been used for refrigerating brine. The American Society of Refrigerating Engineers recommends the treatment of brine with neutral sodium chromate by following the practice of dissolving 100 pounds of commercial sodium bichromate and 27 pounds of caustic soda in 20 gallons of water and adding this solution to 1000 cubic feet (7500 gallons) of brine.[2]

The following test determines the per cent of inhibitor in terms of sodium bichromate, $Na_2Cr_2O_7{\cdot}2H_2O$.

Apparatus

1. Erlenmeyer Flask, 250 ml capacity.
2. Burette, 50 ml calibrated in 0.1 ml.
3. Torsion Balance.
4. Graduated pipette, 10 ml in 0.1 ml.

Reagents

1. Sodium thiosulphate ($Na_2S_2O_3{\cdot}5H_2O$) 0.1 N

TABLE 21—POUNDS OF SODIUM HYDROXIDE OR SODIUM BICHROMATE REQUIRED TO ADJUST 1000 GAL. OF BRINE TO pH 8.

Center Column Shows ml of 0.1N Solution Required to Adjust 100 ml of Brine to pH 8.0

POUNDS OF SODIUM HYDROXIDE PER 1000 GALLONS OF BRINE (FOR ADJUSTMENT OF ACID BRINES TO pH 8.0)	ml	POUNDS OF SODIUM BICHROMATE PER 1000 GALLONS OF BRINE (FOR ADJUSTMENT OF ALKALINE BRINES TO pH 8.0)
0.033	0.1	0.123
0.066	0.2	0.247
0.100	0.3	0.370
0.133	0.4	0.493
0.166	0.5	0.617
0.200	0.6	0.740
0.233	0.7	0.864
0.267	0.8	0.987
0.300	0.9	1.111
0.333	1.0	1.234
0.367	1.1	1.358
0.400	1.2	1.481
0.433	1.3	1.605
0.467	1.4	1.728
0.500	1.5	1.852
0.534	1.6	1.975
0.567	1.7	2.098
0.600	1.8	2.222
0.634	1.9	2.345
0.667	2.0	2.469
0.700	2.1	2.592
0.734	2.2	2.716
0.767	2.3	2.839
0.801	2.4	2.963
0.834	2.5	3.086
0.867	2.6	3.210
0.901	2.7	3.333
0.934	2.8	3.457
0.967	2.9	3.580
1.001	3.0	3.704
1.034	3.1	3.827
1.068	3.2	3.950
1.101	3.3	4.073
1.154	3.4	4.196
1.167	3.5	4.319
1.200	3.6	4.444
1.233	3.7	4.567
1.268	3.8	4.690
1.301	3.9	4.813
1.334	4.0	4.938
1.367	4.1	5.061
1.400	4.2	5.184
1.433	4.3	5.307
1.468	4.4	5.432
1.501	4.5	5.555
1.534	4.6	5.678
1.567	4.7	5.801
1.602	4.8	5.926
1.635	4.9	6.049
1.668	5.0	6.172

Adjust acid brines with amounts of Na OH indicated in column on left.

Adjust alkaline brines with amounts of $Na_2\ Cr_2\ O_7\ 2H_2O$ indicated in column on right.

2. Starch solution (1 gram in 200 ml. distilled water 0.2 gram salicylic acid).

NOTE: The addition of salicylic acid prevents deterioration of the starch solution.

3. Potassium Iodide (KI) 15% solution.
4. Concentrated hydrochloric acid (HCl) C.P.

Procedure

1. Weigh 50 grams of filtered brine into a 250 ml. Erlenmeyer flask.
2. Add 5 ml of 15% solution of KI and then 3.5 to 5 ml of concentrated HCl and mix with a rotary motion.
3. Add 0.1 N $Na_2S_2O_3$ from the burette until the brown color nearly disappears.
4. Add 0.5 to 1.0 ml of the starch solution, mix, and carefully complete the titration with 0.1 N $Na_2S_2O_3$ until the color changes from dark blue to bright green. End point of titration occurs rapidly and bright green color should persist for at least 30 sec.
5. Record number of ml of 0.1 N $Na_2S_2O_3$ required to reach a positive end point.

1 ml of 0.1 N $Na_2S_2O_3$ = 0.0049675 gm of $Na_2Cr_2O_7{\cdot}2H_2O$ and *when a 50 gm sample* is used the per cent inhibitor in terms of sodium bichromate equals the number of ml of 0.1 N $Na_2S_2O_3{\cdot}5H_2O$ required divided by 100.

Example: 18.3 ml 0.1 N $Na_2S_2O_3$ required

$$\text{then } \frac{18.3 \text{ ml } 0.1 \text{ N } Na_2S_2O_3}{100} = 0.183\% \; Na_2Cr_2O_7{\cdot}2H_2O$$

If other than a 50 gm sample of brine is used the per cent of sodium dichromate may be calculated as follows:

$$\text{Per cent } Na_2Cr_2O_7{\cdot}2H_2O = \frac{\text{ml } 0.1 \text{ N } Na_2S_2O_3{\cdot}5H_2O \times \overset{\text{milliequivalent wt of } Na_2Cr_2O_7{\cdot}2H_2O}{(0.0049675)}}{\text{weight of sample}} \times 100$$

It is recommended that the concentration of corrosion inhibitor (as sodium bichromate) be maintained at 0.14 to 0.18 per cent.

DETECTION OF FREE AMMONIA IN BRINE

Apparatus

1. Strips of red litmus paper.
2. Erlenmeyer flask, 250 ml.

Reagents

1. Sodium hydroxide solution (NaOH) conc.
2. Sodium carbonate solution (Na_2CO_3) saturated.

3. Nessler solution: obtainable from chemical supply houses or prepared as follows: Dissolve 50 gm of KI in a minimum quantity of cold, *ammonia free* water (about 35 ml). Add a saturated solution of $HgCl_2$ until a slight permanent precipitate forms. Add 400 ml of 50 per cent KOH or 35 per cent NaOH solution made by dissolving the caustic material and allowing the solution to settle clear before using. Dilute to one liter, let settle until clear and decant. The Nessler solution is then ready for use.

NOTE: Store in a dark glass bottle and protect from light. Efficacy of the solution increases with age.

Procedure (detection by odor or litmus paper)

1. Usually presence of free ammonia may be detected by odor hence, the first step is to smell the brine sample.

2. Place 50 to 100 ml of brine in an Erlenmeyer flask; add a small amount of NaOH solution to help free any NH_3 present, suspend a strip of moist red litmus paper in the mouth of the flask and heat the brine sample.

3. Note any color change in the red litmus paper which will turn blue if even a small amount of NH_3 is present.

NOTE: Do not dip red litmus paper in the brine because it may turn blue due to the alkalinity of the brine.

Procedure (using Nessler solution)

(Use of Nessler solution for detection of NH_3 in calcium chloride is a more delicate test—it will detect as little as 0.01 part of NH_3 per million parts of brine).

1. Add enough sodium carbonate to about 30 ml of brine to precipitate all the calcium and then filter.

2. Add several drops of Nessler solution to about 20 to 25 ml of the filtrate.

CAUTION: The fingers must not contact the brine sample or the reagent. The reagent is very delicate and any NH_3 in the air invalidates the test.

3. Observe color change, if ammonia is present a yellow to dark brown color will appear depending upon the concentration of the ammonia present.

Ammonia in calcium chloride brine causes no precipitation but it may become "fixed" due to absorption of carbon dioxide gas which forms ammonium carbonate. The most practical and positive method of removing ammonia from brine is to heat the brine to high temperatures which drives off "free" ammonia—"fixed" ammonia may be decomposed to "free" ammonia by the addition of a little caustic soda. Lacking the facilities for heating the brine to drive off the ammonia, the desired effect may be obtained by neutralizing free ammonia with carbon dioxide which forms a weak carbonic acid in solution, and tends to "fix" the ammonia as ammonium carbonate. It will not be practical to neutralize a heavy concentration of ammonia by the application of carbon dioxide.

DETERMINATION OF THE SPECIFIC GRAVITY AND FREEZING POINT OF BRINES

Apparatus

1. Hydrometer, specific gravity scale 1.000 to 2.000.
2. Salometer, 0 to 120 degrees.
3. Cylinder (to float hydrometer or salometer in brine.)

Procedure

1. Filter brine, adjust temperature to 60 F (15.6 C) and pour brine into cylinder to float the hydrometer or salometer.

2. After spindle comes to rest read the graduated scale at the point where it coincides with the surface of the brine. (not at the top of the meniscus)

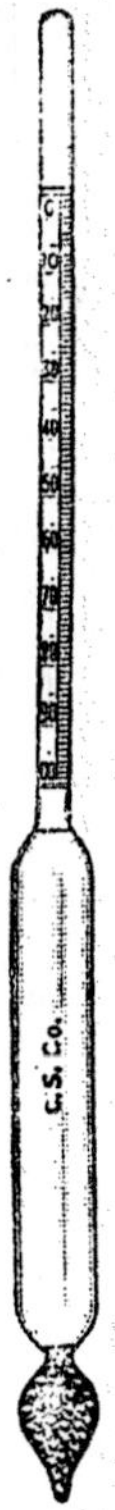

Figure 131.
Sodium chloride salometer.

Figure 132.
Calcium chloride salometer.

TABLE 22—TEMPERATURE CORRECTIONS FOR HYDROMETER READINGS IN CALCIUM CHLORIDE BRINE

Hydrometer Reading	Corrections to be applied to hydrometer readings of calcium chloride solutions at various temperatures, using a hydrometer calibrated at 60°/60°F.										
	0°F −17.8°C	10°F −12.2°C	20°F −6.7°C	30°F −1.1°C	40°F +4.4°C	50°F +10°C	60°F +15.6°C	70°F +21.1°C	80°F +26.7°C	90°F +32.2°C	100°F +37.8°C
1.000					.002	.001	No Corrections	.001	.002	.004	.005
1.010					.002	.001		.001	.002	.004	.005
1.020					.002	.001		.001	.002	.004	.006
1.030				.003	.002	.001		.001	.002	.004	.006
1.040				.003	.002	.001		.001	.002	.004	.006
1.050				.004	.003	.001		.001	.003	.004	.006
1.060				.004	.003	.001		.001	.003	.005	.007
1.070				.004	.003	.002		.001	.003	.005	.007
1.080				.004	.003	.002		.001	.003	.005	.007
1.090				.005	.003	.002		.001	.003	.005	.007
1.100				.005	.003	.002		.001	.003	.005	.007
1.110			.006	.005	.004	.002		.002	.003	.005	.008
1.120			.006	.005	.004	.002		.002	.003	.006	.008
1.130			.006	.005	.004	.002		.002	.004	.006	.008
1.140			.007	.006	.004	.002		.002	.004	.006	.008
1.150			.007	.006	.004	.002		.002	.004	.006	.008
1.160		.009	.007	.006	.004	.002		.002	.004	.006	.009
1.170		.009	.008	.006	.004	.002		.002	.004	.006	.009
1.180		.010	.008	.006	.004	.002		.002	.004	.007	.009
1.190		.010	.008	.007	.005	.003		.002	.004	.007	.009
1.200	.011	.010	.009	.007	.005	.003		.002	.004	.007	.009
1.210	.012	.011	.009	.007	.005	.003		.002	.005	.007	.010
1.220	.012	.011	.009	.007	.005	.003		.002	.005	.007	.010
1.230	.013	.011	.009	.007	.005	.003		.002	.005	.007	.010
1.240	.013	.011	.009	.007	.005	.003		.003	.005	.008	.010
1.250	.013	.012	.010	.008	.005	.003		.003	.005	.008	.010
1.260	.014	.012	.010	.008	.005	.003		.003	.005	.008	.011
1.270	.014	.012	.010	.008	.005	.003		.003	.005	.008	.011
1.280	.015	.013	.010	.008	.006	.003		.003	.005	.008	.011
1.290	.015	.013	.011	.008	.006	.003		.003	.006	.008	.011
1.300	.015	.013	.011	.008	.006	.003		.003	.006	.009	.011
1.310	.016	.014	.011	.009	.006	.003		.003	.006	.009	.012
1.320	.016	.014	.011	.009	.006	.003		.003	.006	.009	.012
1.330	.017	.014	.012	.009	.006	.003		.003	.006	.009	.012
1.340	.017	.015	.012	.009	.006	.003		.003	.006	.009	.012
1.350	.018	.015	.012	.009	.006	.003		.003	.006	.009	.012
1.360			.013	.009	.006	.004		.003	.006	.010	.013
1.370				.010	.007	.004		.003	.006	.010	.013
1.380					.007	.004		.003	.007	.010	.013
1.390						.004		.004	.007	.010	.013
1.400						.004		.004	.007	.010	.013
1.410								.004	.007		

Temperatures below 60° F.
Observed Hydrometer Reading *minus* Temperature Correction = Specific Gravity at 60°/60° F.

Temperatures above 60° F.
Observed Hydrometer Reading *plus* Temperature Correction = Specific Gravity at 60°/60° F.

3. Refer to Table 23 or 24 for concentrations and freezing points of sodium and calcium chloride brines respectively.

NOTE: Correct calcium chloride brine readings made at other than 60 F (15.6 C) according to Table 22.

Strength of Brines: Brines should always be carried at sufficient strength to prevent freezing in the refrigeration system. As strength of brine increases the specific heat decreases and less refrigeration is carried per pound of brine pumped. A strong brine is less corrosive than a weak one because less air will dissolve in strong brines. Brine at a specific gravity of 1.200 is recommended to keep dissolved oxygen at a relatively low concentration and to hold corrosion in check.

[1]Speller, F. N., Corrosion, Causes and Prevention, 2nd Ed., pp. 374 to 383, McGraw-Hill Book Company, Inc., New York.

[2]Solvay Technical Engineering Service: Calcium Chloride in Refrigeration, Bul. No. 4, Fifth Edition, 1941.

TABLE 23. BRINE TABLE—SODIUM CHLORIDE BRINE

Specific gravity at 39° F.	Degrees baumé at 60° F.	Degrees salometer at 60° F.	Pounds of salt per gallon of solution	Pounds of salt per cu. ft.	Percentage of salt by weight	Freezing-point, ° F.	Specific heat	Weight per gallon at 39° F.
1.007	1	4	0.084	0.628	1	31.8	0.992	8.40
1.015	2	8	0.169	1.264	2	29.3	0.984	8.46
1.023	3	12	0.256	1.914	3	27.8	0.976	8.53
1.030	4	16	0.344	2.573	4	26.6	0.968	8.59
1.037	5	20	0.433	3.238	5	25.2	0.960	8.65
1.045	6	24	0.523	3.912	6	23.9	0.946	8.72
1.053	7	28	0.617	4.615	7	22.5	0.932	8.78
1.061	8	32	0.708	5.295	8	21.2	0.919	8.85
1.068	9	36	0.802	5.998	9	19.9	0.905	8.91
1.076	10	40	0.897	6.709	10	18.7	0.892	8.97
1.091	12	48	1.092	8.168	12	16.0	0.874	9.10
1.115	15	60	1.389	10.389	15	12.2	0.855	9.26
1.155	20	80	1.928	14,421	20	6.1	0.829	9.64
1.187	24	96	2.376	17.772	24	1.2	0.795	9.90
1.196	25	100	2.488	18,610	25	0.5	0.783	9.97
1.204	26	104	2.610	19.522	26	1.1	0.771	10.04

TABLE 24. DENSITIES, FREEZING POINTS AND CONCENTRATIONS OF CALCIUM CHLORIDE BRINES

*Specific gravity 60/60° F.	°Baumé at 60° F. (am. std.)	Salometer reading at 60° F.	Weight of brine at 60° F.		% $CaCl_2$ actual	Freezing point ° F.	Lbs. 73-75% Solid $CaCl_2$		Lbs. 77-80% Flake $CaCl_2$		**Lbs. per gallon of water at 60° F.		***Final volume in gals. at 60° F.
			Lbs. per gallon	Lbs. per cu. ft.			Lbs. per gallon of brine at 60° F.	Lbs. per cu. ft. of brine at 60° F.	Lbs. per gallon of brine at 60° F.	Lbs. per cu. ft. of brine at 60° F.	Solid 73-75 $CaCl_2$	Flake 77-80 $CaCl_2$	
1.000	1.0	4.0	8.34	62.37	0	+32.0	—	—	—	—	—	—	1.000
1.009	1.3	5.2	8.41	62.93	1.0	+31.1	0.11	0.86	0.10	0.81	0.11	0.10	1.004
1.010	1.4	5.6	8.42	62.99	1.2	+31.0	0.14	1.03	0.13	1.97	0.14	0.13	1.006
1.018	2.5	10.0	8.49	63.49	2.0	+30.4	0.23	1.73	0.22	1.63	0.23	0.22	1.008
1.020	2.8	11.2	8.50	63.61	2.3	+30.2	0.27	1.99	0.25	1.88	0.27	0.25	1.011
1.026	3.7	14.8	8.55	63.99	3.0	+29.5	0.35	2.61	0.33	2.46	0.36	0.33	1.014
1.030	4.2	16.8	8.59	64.24	3.5	+29.1	0.41	3.06	0.39	2.88	0.42	0.40	1.017
1.034	4.8	19.2	8.62	64.49	4.0	+28.6	0.47	3.51	0.44	3.31	0.48	0.45	1.020
1.040	5.6	22.4	8.67	64.86	4.6	+28.0	0.54	4.06	0.51	3.83	0.55	0.52	1.022
1.043	6.0	24.0	8.70	65.05	5.0	+27.7	0.59	4.43	0.56	4.17	0.61	0.57	1.024
1.050	6.9	27.6	8.75	65.48	5.9	+27.0	0.70	5.26	0.66	4.96	0.72	0.68	1.031
1.052	7.1	28.4	8.77	65.61	6.0	+26.8	0.72	5.36	0.68	5.05	0.75	0.70	1.031
1.060	8.2	32.8	8.84	66.11	7.0	+25.9	0.84	6.30	0.79	5.94	0.88	0.82	1.036
1.069	9.4	37.6	8.91	66.67	8.0	+24.6	0.97	7.26	0.91	6.84	1.02	0.95	1.043
1.070	9.5	38.0	8.92	66.73	8.1	+24.6	0.98	7.35	0.92	6.93	1.03	0.96	1.043
1.078	10.5	42.0	8.99	67.23	9.0	+23.5	1.10	8.23	1.04	7.76	1.16	1.09	1.049
1.080	10.7	42.8	9.00	67.36	9.2	+23.2	1.13	8.43	1.06	7.94	1.20	1.11	1.050
1.087	11.6	46.4	9.06	67.79	10.0	+22.3	1.23	9.22	1.16	8.69	1.31	1.22	1.055
1.090	12.0	48.0	9.09	67.98	10.3	+21.9	1.27	9.53	1.20	8.89	1.36	1.28	1.058
1.096	12.7	50.8	9.14	68.35	11.0	+20.8	1.37	10.23	1.29	9.64	1.47	1.37	1.062
1.100	13.2	52.8	9.17	68.60	11.4	+20.3	1.42	10.64	1.34	10.03	1.53	1.43	1.065
1.105	13.8	55.2	9.21	68.92	12.0	+19.3	1.50	11.25	1.41	10.60	1.62	1.51	1.069
1.110	14.3	57.2	9.25	69.23	12.4	+18.7	1.56	11.68	1.47	11.01	1.69	1.57	1.071
1.115	15.0	60.0	9.30	69.54	13.0	+17.6	1.64	12.30	1.55	11.59	1.79	1.67	1.076
1.120	15.5	62.0	9.34	69.85	13.5	+16.5	1.72	12.83	1.62	12.09	1.88	1.75	1.080
1.125	16.1	64.4	9.38	70.16	14.0	+15.5	1.79	13.36	1.69	12.59	1.97	1.83	1.084
1.130	16.7	66.8	9.42	70.47	14.5	+14.5	1.86	13.90	1.75	13.10	2.05	1.90	1.087
1.135	17.2	68.8	9.46	70.79	15.0	+13.5	1.93	14.45	1.82	13.62	2.14	1.99	1.092
1.140	17.8	71.2	9.50	71.10	15.6	+12.2	2.02	15.09	1.90	14.22	2.25	2.08	1.096
1.145	18.4	73.6	9.55	71.41	16.0	+11.2	2.08	15.54	1.96	14.64	2.32	2.16	1.099

1.150	18.9	75.6	9.59	71.72	16.6	+ 9.7	2.17	16.20	2.04	15.27	2.44	2.25	1.104
1.155	19.4	77.6	9.63	72.03	17.0	+ 8.6	2.23	16.66	2.10	15.70	2.51	2.33	1.107
1.160	20.0	80.0	9.67	72.35	17.6	+ 7.0	2.32	17.32	2.19	16.32	2,63	2.44	1.115
1.165	20.5	82.0	9.71	72.66	18.0	+ 5.9	2.38	17.79	2.24	16.76	2.71	2.50	1,116
1.170	21.1	84.4	9.75	72.97	18.6	+ 4.1	2.47	18.47	2.33	17.40	2.83	2.62	1.124
1.175	21.6	86.4	9.80	73.28	19.0	+ 2.8	2.53	18.94	2.38	17.85	2.90	2.68	1.124
1.180	22.1	88.4	9.84	73.59	19.5	+ 1.2	2.61	19.52	2.46	18.39	3.01	2.78	1.130
1.186	22.7	90.8	9.89	73.97	20.0	− 0.4	2.69	20.13	2.53	18.97	3.12	2.87	1.133
1.190	23.2	92.8	9.92	74.22	20.6	− 2.0	2.78	20.80	2.62	19.60	3.25	2.99	1.142
1.195	23.7	94.8	9.96	74.53	21.0	− 3.9	2.85	21.29	2.69	20.06	3.34	3.08	1.147
1.200	24.2	96.8	10.00	74.84	21.5	− 5.8	2.93	21.89	2.76	20.63	3.45	3.18	1.152
1.206	24.8	99.2	10.05	75.21	22.0	− 7.8	3.01	22.51	2.84	21.21	3.56	3.28	1.156
1.210	25.2	100.8	10.09	75.46	22.4	− 9.4	3.08	23.00	2.90	21.67	3.67	3.36	1.160
1.218	25.9	103.6	10.15	75.96	23.0	−11.9	3.18	23.77	3.00	22.40	3.80	3.50	1.167
1.220	26.2	104.8	10.17	76.09	23.3	−13.2	3.22	24.12	3.03	22.73	3.86	3.54	1.168
1.228	26.9	107.6	10.24	76.59	24.0	−16.2	3.34	25.01	3.15	23.57	4.04	3.71	1.176
1.230	27.1	108.4	10.25	76.71	24.2	−17.1	3.37	25.26	3.18	23.80	4.08	3.75	1.180
1.239	28.0	112.0	10.33	77.27	25.0	−21.0	3.51	26.28	3.31	24.76	4.29	3.93	1.188
1.240	28.1	112.4	10.34	77.33	25.1	−21.5	3.53	26.41	3.33	24.89	4.32	3.96	1.190
1.250	29.0	116.0	10.42	77.96	26.0	−25.8	3.69	27.58	3.48	25.99	4.57	4.18	1.202
1.260	29.9	119.6	10.50	78.58	27.0	−31.2	3.86	28.87	3.64	27.20	4.84	4.42	1.215
1.270	30.8	—	10.59	79.21	27.8	−37.1	4.01	29.96	3.79	28.23	5.08	4.65	1.227
1.272	31.0	—	10.60	79.33	28.0	−37.8	4.04	30.22	3.81	28.48	5.13	4.68	1.228
1.280	31.7	—	10.67	79.83	28.7	−44.3	4.17	31.17	3.93	29.37	5.35	4.86	1.237
1.283	32.0	—	10.70	80.02	29.0	−49.4	4.22	31.57	3.98	29.75	5.43	4.94	1.241
1.290	32.6	—	10.75	80.45	29.6	−59.8	4.33	32.40	4.08	30.53	5.62	5.10	1.250
1.295	33.0	—	10.80	80.76	30.0	−50.8	4.41	32.96	4.16	31.06	5.76	5.23	1.256
1.300	33.5	—	10.84	81.08	30.5	−41.8	4.50	33.65	4.24	31.71	5.92	5.36	1.264

* Hydrometers giving specific gravities and freezing points of calcium chloride brines can be obtained from any Solvay Sales Coropration office.

** Pounds of solid or flake calcium chloride per cubic foot of water at 60° F. may be obtained by multiplying the quantity in pounds per gallon by 7.481.

*** Final volume in gallons at 60° F. when quantity of flake calcium chloride shown in previous column is added to one gallon of water.

Courtesy The Solvay Process Co.

CHAPTER 32

PHYSICAL TESTS APPLIED TO GLASS MILK BOTTLES

Much benefit can accrue to companies using large volumes of milk bottles by the application of physical tests to new bottles as received. Physical testing routine should include a weight and capacity measurement, a polariscopic examination, an internal hydrostatic pressure test, a thermal shock test, and an external impact test.

These tests should be conducted on an adequately representative number of milk bottles picked at random from each shipment received by the milk company. Before the bottles are subjected to any physical tests they should be closely examined for existing defects. Defects that may commonly exist and the usual reason for their occurrence are listed in table 25.

TABLE 25. BOTTLE DEFECTS

TYPE	USUAL REASON
Improper glass distribution	Due to overheating of glass.
Split rings	Glass adhering to neck ring mould causing fissure or break upon opening and releasing bottle from the mould.
Light sides	Due to overheating of glass which renders it unworkable.
Checks or splits at bottom of bottle	Due to adhesion to bottom of blow mould or improper heat in lehr while in process of annealing.
Splits on bottom	Due to cold bottom plates in mould.
Iron stains (stoned)	Due to exposed parts of the iron frame of furnace tank melting into the glass.
Overpressed cap seat	Glass overweight causing an excess of glass to form in the mould.

There are numerous other defects to be found such as crooked finish, air bubbles, smeared lettering, discolored glass, sharp edges around mouth, rocker bottoms, crooked bumper rolls, etc. Bottles that are found with defects such as splits, checks, iron stains, overpressed cap seats, crooked finish, or numerous air bubbles should not be subjected to any physical tests because they are almost certain to fail.

Figure 133.
Setup of U. S. Bureau of Standards type burettes for measuring capacity of different size containers.

Capacity Measurement

All bottles from each lot should be measured for capacity. Such measuring is greatly facilitated by the use of a "Bureau of Standards" type burette. (Figure 133.) A burette for each size bottle should be used. The last 2 ounces of these burettes should be calibrated in one sixteenth ounces. Capacities of each test lot of bottles should be recorded and identified according to the manufacturers' plant mark and mould number.

The following is taken from the National Bureau of Standards Handbook H 29 (1942).

The tolerance to be allowed in excess and in deficiency on milk bottles shall be as follows:

[1]Speller, F. N., Corrosion, Causes and Prevention, 2nd Ed. pp. 374 to 383, McGraw-Hill Book Company, Inc., New York.

[2]Solvay Technical Engineering Service: Calcium Chloride in Refrigeration, Bul. No. 4, Fifth Edition, 1941.

1. When the test comprises less than 25 bottles of the same capacity and ownership, the tolerance shall be those given in table A below.

2. When a test comprises 25 or more bottles of the same capacity and ownership, the tolerance shall be applied not only to the individual bottle but also to the average capacity of at least 25 such bottles, these to be taken at random when the whole supply available is not tested. The error on any individual bottle among those tested shall not exceed the values shown in table A below. The average error on the bottles tested shall not exceed the values shown in table B below. Provided however that in the case of bottles already in use, if the average error is greater than that above specified, then if desired, all of the bottles of the particular size and ownership in question may be treated as individual measures, in which case all of these bottles shall be separately tested and the tolerance shown in table 26 shall be applied.

TABLE 26. BOTTLE TOLERANCES

Capacity	TABLE A Tolerance on Individual Bottles		TABLE B Tolerance on Average Capacity		TABLE C Tolerance on Individual Bottles Already in Use			
					In Excess		In Deficiency	
	Liquid drams	Fluid ozs.	Liquid drams	Fluid ozs.	Liquid drams	Fluid ozs.	Liquid drams	Fluid ozs.
2 Quarts	6	¾	1.5	3/16	6	¾	3	⅜
1 Quart	4	½	1.0	⅛	4	½	2	¼
1 Pint	3	⅜	0.75	3/32	3	⅜	1.5	3/16
½ Pint	2	¼	0.5	1/16	2	¼	1	⅛

Annealing Test

The grading of bottles by means of the polariscope is not only an objective test but also a subjective test in that it depends upon the observer's color vision. On the objective side, it depends on the color of the light used and also on its intensity. The Glass Container Association grading discs have an accurately standardized retardation and reproduce the appearance of a bottle bottom. It is usual to grade bottles by the appearance of the base, as seen by peeping axially down through the neck. (See figure 134.)

Temper #1 (G.C.A. Standard Grading Discs) bottles are sensibly of one color throughout.

Temper #2 and #3 bottles are in three colors as recognized by an average observer. The central parts are field tint, usually a purple; two quadrants verge toward blue at the margin and the other two toward orange.

Temper #4 bottles, in each quadrant, have two recognizable colors as judged by an average observer and the zone of field color is begining to take on the form of a recognizable cross.

Temper #5 and #6 bottles have the cross conspicuous and the quadrants brilliantly illuminated with a rainbow of colors.

Generally speaking temper #5 and #6 are considered sub-standard.

Figure 134.
Observation of the degree of annealing by the polariscopic appearance of a milk bottle.

Hydrostatic Internal Pressure Test

Bottles that are required to withstand internal pressure in use should be tested for resistance to internal pressure. The strength of a bottle under internal pressure depends upon the duration of the pressure as well as on its intensity. A bottle will not hold for 1 minute as much pressure as it will hold for 1 second, nor will it hold for an hour as much as it will hold for a minute, or as much for a day as it will hold for an hour. It is best to use a constant pressure for a definite length of time to see whether or not the bottle breaks.

Figure 135.

Single head sustained pressure testing machine for hydrostatic internal pressure test on milk bottles. Preston Laboratories, Butler, Pa.

The type of apparatus recommended for internal hydrostatic pressure testing is a Single Head Sustained Pressure Testing Machine manufactured by the Preston Laboratories. (Figure 135.) This automatic single-head pressure testing machine is intended to test bottles or other similar glassware by subjecting them to a definite internal hydrostatic pressure for a specified period of time. Pressure of 25 lb. per sq. in. or, of any internal multiple of 12½ lb/sq. in. above 25, up to and including 287½ lb/sq. in. may be used. The duration of the pressure may be automatically controlled for any desired period of time from ten seconds to one minute.

The following average standards for the Hydrostatic Internal Pressure Test have been found to be reasonable.

For 1 quart square bottles – 43 lbs per sq. in. sustained pressure for 15 seconds.

For 2 quart oblong bottles – 32 lbs. per sq. in. sustained pressure for 15 seconds.

For all round bottles – 150 lbs. per sq. in. sustained pressure for 15 seconds.

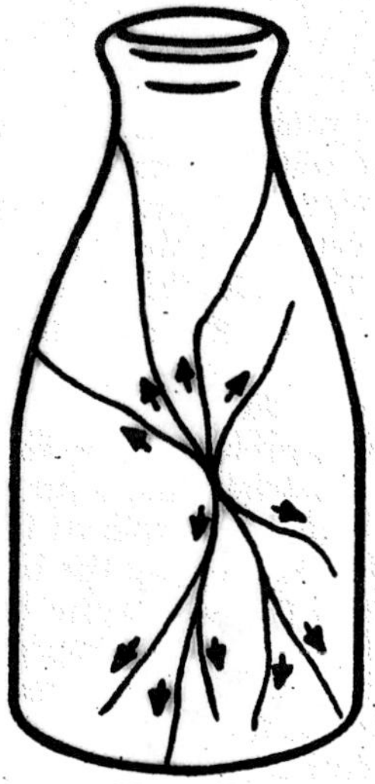

Figure 136.
Example of weak bottle.

A Typical Internal Hydrostatic Pressure Break

Figure 136 is an example of a weak bottle that broke at a moderate pressure. In consequence, the amount of forking of the cracks is moderate also. When a bottle breaks under high pressure the number of fissures is vastly greater. The origin of this break is in normal position on the barrel of the bottle and normally oriented, i.e., the initial fissure runs straight up and down. The nucleus of the origin is some irregularity or bruise on the outer surface of the bottle wall and the origin in its entirety is square to the face of the glass.

Figure 137.
Arrangement of water baths for thermal shock test of milk bottles.

Thermal Shock Test

After the bottles have passed the internal pressure test (at the specified pressure for that type of bottle), the water is emptied out, and the bottles loaded into the thermal shock testing apparatus. (See figure 137.) The most important factor in determining the outcome of the thermal shock test is the temperature differential between the two baths, but other factors should be watched. Thus water is always the quenching medium but the properties of water vary with temperature level; therefore the absolute temperature of the cold bath (not merely its differential from the hot bath) is significant. The time elapsing in transfer between the two baths is important and must be standardized. The degree of agitation of the baths, the depth of immersion, and so on, make a difference. Obviously, the time the bottle stays in the hot bath is important. It is good practice to use a 5 minute soak in the hot water bath at 160 F and to adjust the temperature of the cold at 80 F for an 80 degree differential. The time interval between the two baths should be 15 seconds and the bottles should soak in the cool water for at least 30 seconds before they are examined.

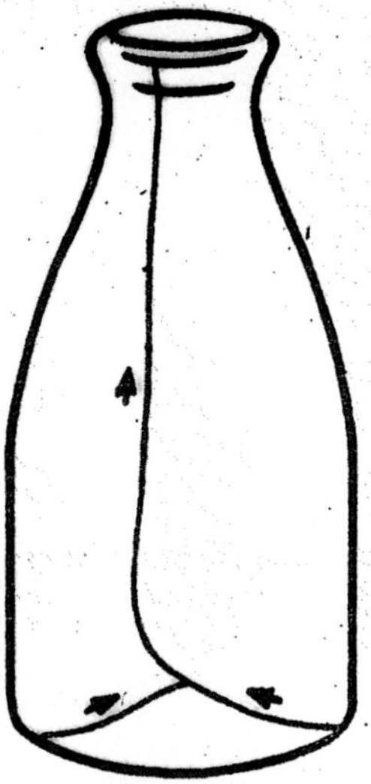

Figure 138.
Typical thermal shock crack.

A typical thermal shock break

Figure 138 illustrates a fairly typical thermal shock crack made by quenching in water with an 80 degree differential. The arrows show the direction of propagation of the crack. The origin is not shown in this figure. Typically the origin is in the base; often in the baffle mark. In this instance the base did not separate completely from the rest of the bottle and the exact position of the origin was not determined. The crack tends to be very smooth and brightly polished throughout most of its extent, without forks, striated only where it turns at a flat angle to the face.

Impact Test

An impact testing machine which is being perfected by the Preston Laboratories is recommended for the impact test on milk bottles.

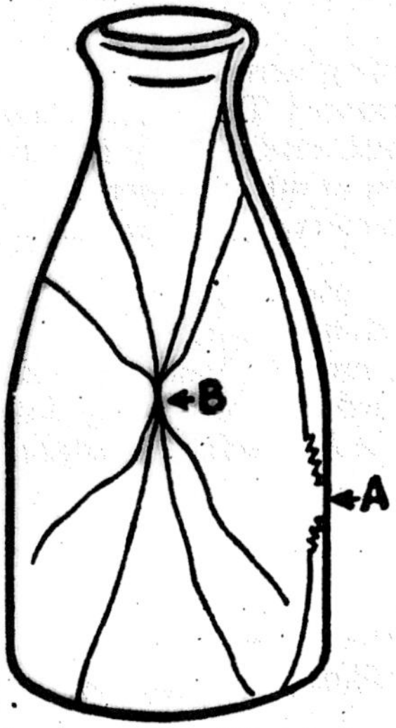

Figure 139.
Example of impact crack.

An impact break

A blow delivered on the outer face of the bottle may knock a conical hole through the glass. The cone may come out in one piece with a broad internal base and a small apex on the outer face such as A in figure 139. On the other hand the blow by deflecting the bottle well inward may produce tension on the inner face of the bottle of such an intensity that an origin is started there which forks into a number of radiating or star cracks such as B in figure 139.

REFERENCE

Preston, F. W.: Bottle Breakage—Causes and Types of Fractures, Bul. American Ceramic Society, Vol. 18, No. 2, 1939.

CHAPTER 33

SUGAR SYRUP TESTS
(Including Maple Syrup)

Cane Sugar Syrup

Specific Gravity

Apparatus

1. Brix hydrometer with a scale reading from 60 to 70 in 1/10 degree graduations at 68 F (20 C) (see figure 140).
2. Cylinder, 300 mm by 40 mm.

Procedure

A representative sample of sugar syrup to be tested is poured carefully into the cylinder. If air bubbles are present, the syrup must stand until the bubbles have risen to the surface. This may be hastened by placing the cylinder in warm water. The hydrometer must be clean and dry. Lower carefully into the syrup, taking care not to incorporate air bubbles. Immerse the hydrometer 2 or 3 mm deeper than the point at which it will come to rest.

The point to be read is obtained by sighting across the level of the surface of the syrup, disregarding the meniscus. The hydrometers are calibrated for reading at 69 F (20 C) and read directly in percentage of sugar solids. If readings are taken at other temperatures, tables of temperature corrections of saccharometers must be consulted.

If it is desired to make complete analysis of sugar syrup by determining the moisture, sucrose, invert sugar, pH, acid, ash, and alkalinity of ash, consult any good sugar handbook such as *Sugar Analysis* by C. A. Brown and F. W. Zerban, published by John Wiley and Sons, Inc., New York, 3rd edition, 1941, or the *Methods of Analysis, A.O.A.C.*, 6th edition (1945). p. 556.

Reference

Methods of Analysis, A.O.A.C., 6th edition (1945), p. 821.

To determine the weight of sugar per gallon of syrup, use the table on page 460.

Figure 140.
Brix hydrometer.

TABLE 27. POUNDS OF SUGAR PER GALLON OF SYRUP AT VARIOUS TEMPERATURES

Brix Reading at 20 C (68 F)	Pounds of Sugar per Gallon					
	30 C (86 F)	25 C (77 F)	22.5 C (72.5 F)	20 C (68 F)	17.5 C (63 F)	15 C (69 F)
67.0	11.037	11.058	11.073	11.079	11.09	11.099
67.2	11.047	11.069	11.082	11.089	11.10	11.110
67.4	11.059	11.079	11.092	11.099	11.11	11.120
67.6	11.069	11.900	11.101	11.110	11.12	11.131
67.8	11.080	11.100	11.111	11.120	11.13	11.141
68.0	11.089	11.110	11.120	11.130	11.14	11.152
68.2	11.100	11.120	11.130	11.140	11.15	11.162
68.4	11.109	11.130	11.141	11.151	11.16	11.173
68.6	11.120	11.140	11.151	11.161	11.18	11.183
68.8	11.130	11.151	11.162	11.172	11.18	11.194
69.0	11.140	11.161	11.172	11.182	11.19	11.204

Brix Reading at 20 C (68 F)	Pounds of Sugar per Gallon					
	12.5 C (54.5 F)	10 C (50 F)	7.5 C (44 F)	5 C (41 F)	2.5 C (36.5 F)	0 C (32 F)
67.0	11.109	11.118	11.127	11.137	11.146	11.155
67.2	11.120	11.128	11.138	11.147	11.156	11.165
67.4	11.130	11.139	11.148	11.158	11.167	11.176
67.6	11.141	11.149	11.159	11.168	11.177	11.186
67.8	11.151	11.160	11.169	11.179	11.188	11.197
68.0	11.162	11.170	11.180	11.189	11.198	11.207
68.2	11.173	11.180	11.191	11.200	11.208	11.217
68.4	11.183	11.191	11.201	11.210	11.219	11.227
68.6	11.194	11.201	11.212	11.221	11.229	11.238
68.8	11.204	11.212	11.222	11.231	11.240	11.248
69.0	11.215	11.222	11.232	11.242	11.250	11.259

NOTE: This test is designed to be used for finding the amount of sugar solids in cane sugar syrup. This usually contains from 61 to 65 per cent sucrose and from 1 to 4 per cent invert sugar.

Accurate determination of sugar solids may be obtained in any concentration by using the proper hydrometer. The hydrometers are made ranging from 0 to 70 degrees. Usually each will have a range of 10 degrees. This test does not distinguish between sucrose and invert sugar but gives only the percentage of sugar solids.

Maple Syrup

SPECIFIC GRAVITY

Apparatus

Baumé hydrometer, reading to 36.

Procedure

The same as under cane sugar syrups. If crystals are present, warm the syrup to dissolve them. Maple syrup should read 36 degrees Baumé at 60 F (15.6 C). Make the following corrections for temperature:

40 F subtract ½ degree from hydrometer reading.
80 F add ½ degree to hydrometer reading.
100 F add 1 degree to hydrometer reading.

If maple syrup has a greater concentration than 36 degrees, there will be crystallization; if less than 36 degrees, fermentation will take place.

PURITY

It is difficult to detect slight adulterations or artificial flavors in maple syrup. Additions of from 5 to 15 per cent of sucrose cannot be detected.

The best policy is to purchase maple syrup which is packed under supervision of a responsible marketing agency, such as the products sold under the New England Quality Farm Products Label. Six New England states have set up official grades, and producers and handlers of unadulterated maple syrup and sugar are registered with the respective state departments of agriculture. A new test is made each year and the use of the label is a form of certification by the state. For further information write to the State Departments of Agriculture of Vermont, Connecticut, Rhode Island, Massachusetts, Maine, or New Hampshire.

The tests usually employed to determine the probable purity of the sample are the total ash, soluble and insoluble ash with respective alkalinities, and estimation of the malic acid value. The minimum for total ash should be 0.5 per cent and for insoluble ash, 0.15 per cent. The malic acid figure should be not less than 0.6 per cent for syrup weighing 11 pounds per gallon.

REFERENCE

Methods of Analysis, A.O.A.C., 6th edition (1945), p. 585.
Vermont maple sugar and syrup, Il. map; *Vt. Ag. Dept. Bull. 38* (1930), 1-45.

CHAPTER 34

GELATIN EXAMINATION

Water Absorption Property

Apparatus

1. Beakers.
2. Stirring rods.
3. Laboratory balance or small scales.
4. Cover glass.
5. Graduate, 25 ml.

Procedure

1. Weigh 20 gm of gelatin and pour without stirring into 180 ml of cold water in beaker.

2. Insert and hold lightly a glass rod vertically in the center of the beaker. Note the time required to form a sufficiently stiff gel to hold the rod erect.

A high quality gelatin will take up water very rapidly in this concentration (10 per cent) and will form a stiff gel capable of supporting a rod in a few minutes. The gelatin however is not in solution at this point.

NOTE: The water absorption test, while helpful in many ways in determining quality of gelatin, is not always dependable and is likely to give erratic results.

Alternative Method

For a more accurate measure of water absorption power, place 5 gm of the gelatin in a 200 ml graduated cylinder and add enough water at 69.8 F (21 C) to bring to the 100 ml mark. Stir for one minute to prevent lumping and then allow the jar to stand for ten minutes, after which the cubic milliliters of swell can be directly observed and recorded.

Rate of Solution

Place a beaker containing 20 gm of gelatin and 180 ml of water in a water bath at 158 F (70 C) and stir until dissolved. Note the time required for the complete solution of the gelatin. Gelatins showing a slow solubility rate may cause local gumminess in a mix, resulting in uneven stabilization of the ice cream.

REFERENCE

Combs, W. B.: *The Ice Cream Review*, 10:76 (1926).

Organoleptic Examination

A. ODOR OF GELATIN

When the solution prepared in the previous test is complete, cover the beaker with a watch glass and allow to stand in the water bath two minutes. Then remove the sample from the bath, take off the cover glass and immediately note the odor. A good gelatin should have a mild, fresh broth odor and should have no indication of a glue-like odor or of a chemical odor caused by the use of preservatives.

B. COLOR

Note the color of the gelatin solution previously prepared. In this concentration, a high quality gelatin will have a slight amber color blending to an orange brown for the lower grades.

C. FOREIGN MATTER

Foreign matter will remain undissolved and can be readily seen. It may consist of two distinct kinds of insoluble matter—fibrous matter which floats, such as wood and hemp, and solids which settle, such as pieces of coal and iron. These may enter the gelatin during the manufacturing process from the air and from the materials employed in handling. To make the test, allow the gelatin solution to stand for an hour and then examine it for fibrous matter by holding it to the light, avoiding direct sunlight.

Note the appearance of the bottom for sediment. A high quality gelatin will have only a faint trace of fibrous matter and sediment.

Moisture Determination

Apparatus

1. Flat-bottom metal dish about 55 mm in diameter provided with tightly fitting friction cover.
2. Drying oven, thermostatically controlled.
3. Analytical balance.
4. Desiccator.

Procedure

1. Heat the dish and its cover to constant weight at 100 C (212 F), cool and weigh.
2. Add approximately 2 gm of the sample, cover loosely and reweigh accurately.
3. Dry the uncovered dish at 100 C (212 F) for 6 hours, cover the dish firmly, cool in a desiccator, and weigh.
4. Report the percentage loss in weight as moisture.

Gelatin normally has a moisture content of about 10 or 12 per cent but this figure will vary over a wide range (from 7 to 15 per cent), depending on the porosity of the gelatin and its previous history, that is, the extent of drying, time of storage, humidity of room during storage, and other conditions.

Ash Determination

Apparatus

1. Porcelain ignition dish.
2. Ring stand, triangle, and asbestos square.

Procedure

1. Weight approximately 1 gm of a well mixed gelatin sample directly into a porcelain ignition dish which has been dried to constant weight and cooled to room temperature in a desiccator.
2. Heat directly over a Bunsen flame to a dull red until all carbon is oxidized.
3. Transfer the crucible to an asbestos mat supported by a triangle and continue the heating.
4. When ashing is complete, cool in a desiccator to room temperature, weigh, and record the results. Reheat, cool, and weigh until a constant weight is obtained. The weight of the ash divided by the weight of the sample multiplied by 100 equals the percentage of ash.

pH Value Determination

Apparatus

Potentiometer or colorimeter.

Procedure

Dissolve 1 gm of gelatin in 99 ml of distilled water by heating to 70 C (158 F). The pH value of this 1 per cent solution may be determined either colorimetrically or electrometrically (see page 258 and 260).

Acidity Determination

Reagents

1. Sodium hydroxide, 0.1 N.
2. Phenolphthalein.

Procedure

Dissolve 5 gm of gelatin in 45 ml of distilled water. Titrate with 0.1 N NaOH, using phenolphthalein as indicator. Acidity and pH of aqueous solutions of gelatin are not of much importance to the ice cream manufacturer because the gelatin is used in such small quantities that it assumes the pH value of the mix.

Gel Strength Determination

BLOOM GELOMETER METHOD

Apparatus

1. Bloom gelometer. This instrument may be purchased from the Glue and Gelatin Makers Association, New York. Complete instructions are furnished with the machine.

2. Water bath, properly controlled for dissolving the gelatin.

3. Refrigerator equipped with a very sensitive thermostat and containing a low temperature water bath.

4. Wide mouth bottle, 59 mm inside diameter, 66 mm outside diameter, 85 mm high, to take No. 9 stopper, 150 ml capacity.

Procedure

1. Weigh 7.15 gm of the gelatin and transfer to a 150 ml wide-mouth bottle. Add 100 ml of distilled water at 20 C (68 F) and stir thoroughly to prevent clumping of the gelatin. Allow the gelatin to soak until the water is absorbed. (For ground gelatin not more than ten minutes is necessary, but for gelatin in the form of flakes or sheets considerably longer periods, often approaching ten or more hours, are required.)

2. Transfer to a hot water bath and dissolve at 70 C (158 F).

3. When solution is complete the foam is removed from the surface of the gelatin with a spoon and the bottle is tightly closed with a rubber stopper.

4. Place the bottle in a refrigerator containing a water bath regulated at 10 C (50 F) $\pm$ 0.1 C (0.2 F) and allow to stand for 16 hours, at the end of which time the sample is ready for testing. To guard against error due to minor variations in procedure, temperature, or in the setting of the instrument, a set of standardized samples should be run together with the unknown sample, and adjustment made in accordance with the deviation of the known samples from their known values.

NOTE: The water bath in the refrigerator is to offset fluctuation in temperature of the air due to the opening of the ice box as each sample is removed for testing.

REFERENCE

Alexander, Jerome: *Glue and Gelatin,* Chemical Catalog Co., New York, 1923.

DAHLBERG METHOD

This method depends on the premise that a sufficient amount of gelatin is present in ice cream when a weak gel is formed. The gel is broken up in the freezer and reforms in the hardening room. The test determines the minimum amount of gelatin necessary to give the full effect.

Test conditions should approach the condition prevailing in ice cream. This means a low temperature, 0 C (32 F), and skim milk as a solvent to correct the acidity and the influence of milk solids on gelatin strength.

Apparatus

1. Ice water bath.
2. Graduated cylinders, 3 of 100 ml capacity.
3. Scale, sensitive to 0.01 gm. A cream test scale or butter moisture scale is satisfactory.
4. Test tubes, one-half inch (50 or more).
5. Water bath.

Reagent

Fresh skim milk.

Procedure

1. Carefully mix and weigh exactly 10 gm sample of gelatin.

2. Weigh 990 gm of fresh sweet milk or measure 956 ml. The skim milk may be prepared from dry skim milk, provided it is completely soluble, by dissolving 100 gm in 900 gm of water. Insolubility increases the apparent strength of the gelatin.

3. Add the gelatin to the cold skim milk, stir well, and allow to stand for ½ hour, making certain that all parts of the gelatin become thoroughly wet. Place the mixture in a water bath at 71-77 C (160-170 F) and stir while the gelatin solution is being heated to 60 C (140 F).

4. Prepare a series of dilutions ranging from 0.30 per cent to 0.60 per cent, graduated in 0.05 per cent increments by mixing 1 per cent stock gelatin solution and skim milk in the proportions shown below.

Ml of 1 per cent gelatin solution	Ml of skim Milk	Final strength of gelatin, %
30	70	0.30
35	65	0.35
40	60	0.40
45	55	0.45
50	50	0.50
55	45	0.55
60	40	0.60

5. The test tubes containing the solutions of varying concentrations should be placed in a suitable container packed with sufficient ice to fill the container above the level of the surface of the gelatin solutions and cold water then poured over the ice. The container with its content should be placed in the refrigerator at 40 F (4.4 C). The gelatin solutions will be held at a temperature of 32 F (0 C) if sufficient ice is present.

6. Make readings after from 18 to 20 hours. Beginning with the tube with the lowest percentage of gelatin, the tubes should be inverted in the order of increasing gelatin content. If the dilutions are properly made the first tubes will show no evidence of jelling. Finally a tube of gelatin will be found which will not flow on inversion. Its concentration represents the minimum amount of gelatin needed to form a gel of sufficient strength to maintain itself in an inverted position for at least 10 seconds.

7. The concentration of gelatin in the tube that withstands inversion without flowing is the minimum quantity to use in ice cream, but a correction must be made for the difference in the water content of ice cream and skim milk. Ice cream contains 65 per cent of water and skim milk 90 per cent. Therefore, 65/90 or 0.7 is the correction factor by which the result should be multiplied. For example, if the observed optimum concentration of gelatin in skim milk is 0.5, then 0.7 times 0.5 equals 0.35 which is the concentration to use in ice cream. The figure represents a minimum quantity to be used in a mix containing 12 per cent of fat. If a manufacturer desires a maximum amount of gelatin in the ice cream the quantity should be increased about 0.05 per cent. If the ice cream contains from 8 to 10 per cent of fat the quantity of gelatin should be increased from 0.05 to 0.07 per cent depending on its strength.

REFERENCE

Dahlberg, A. C.: New Facts on the Use of Gelatin in Ice Cream Manufacturing, *Ice Cream Trade J., 24:* No. 5 (1928); also, *1927 Ann. Conv. Rept. Internat. Assoc. Ice Cream Mfgrs.; Abstract of Literature on the Manufacture and Distribution of Ice Cream,* International Association of Ice Cream Manufacturers *2:* 95 (1928).

Viscosity Determination

The viscosity of gelatin is determined by melting the solution made up for the Bloom test through a standard viscosity pipette at 140 F (60 C) and timing the flow between two marks by means of a stop watch. Any pipette or tube may be used for this purpose but it must be calibrated by the running of standards so that the viscosity of the gelatin can be expressed directly in millipoises. For this purpose a chart is drawn up showing the relationship between viscosity in seconds and millipoises for the particular pipette being employed.

CHAPTER 35

VANILLA FLAVOR TESTS

Specific Gravity

Apparatus

1. Pycnometer.
2. Analytic balance and weights.

Reagents

1. Saturated solution of chromic acid in concentrated sulfuric acid. (See page 540.)
2. Alcohol, 95 per cent.
3. Ether.

Procedure

Determine the specific gravity at 20/4 C (in vacuo) with a pycnometer as follows:

1. Carefully clean the pycnometer by filling with a saturated solution of chromic acid in concentrated sulfuric acid, allowing to stand for several hours, emptying, and rinsing thoroughly with water.

2. Fill the pycnometer with recently boiled distilled water previously cooled to 16-18 C (61-64 F), place in a water bath cooled to the same temperature, and allow the bath to warm slowly to 20 C (68 F).

3. Adjust the level of the water to the proper point on the pycnometer, replace the perforated cap or stopper, remove from the bath, wipe dry with a clean cloth and, after allowing to stand for from 15 to 20 minutes, weigh.

4. Empty, rinse several times with alcohol and then with ether, remove the ether fumes, allow the instrument to become perfectly dry, and reweigh.

5. Ascertain the weight of the contained water at 20 C in air (W of the formula) by subtracting the weight of the empty pycnometer from its weight when full.

6. Cool the sample to 16-18 C, adjust the level of the liquid to the proper point on the pycnometer, replace the perforated cap or stopper, wipe dry, and weigh as before. Ascertain the weight of the contained sample at 20 C in air (S of the formula) by subtracting the weight of the empty pycnometer from its weight when filled with the sample.

Calculate the specific gravity in vacuo by the following formula:

$$G = \frac{S + 0.00105W}{1.00282W}$$ in which

G = corrected specific gravity of sample at 20/4 C in vacuo;
W = weight of contained water at 20 C (68 F) in air; and
S = weight of contained sample at 20 C in air.

REFERENCE

Methods of Analysis, A.O.A.C., 6th edition (1945), p. 365.

Alcohol Content

Apparatus

1. Pycnometer or a small accurately graduated hydrometer.
2. Distillation flask.
3. Condenser.
4. Analytical balance and weights.

Procedure

1. Alcohol by Weight—Official

Weigh from 20 to 25 gm of the sample into a distillation flask, dilute with 100 ml of water, and distil nearly 100 ml. Weigh the distillate or make to volume at 20 C (68 F) and, in either case, determine the specific gravity as has just been directed. Obtain the corresponding percentage of alcohol by weight obtained from an appropriate table; multiply this figure by the weight of the distillate; and divide by the weight of the sample taken to obtain the percentage of alcohol by weight.

2. Alcohol by Volume:

Method 1—Official

From the specific gravity of the distillate, obtained as explained, ascertain the corresponding percentage of alcohol by volume obtained from appropriate tables. Multiply this figure by the volume of distillate and divide by the volume of the sample (calculated from the specific gravity) to obtain the percentage of alcohol by volume in the original sample.

Method 2—Official

Measure 25 ml of the sample at 20 C (68 F) into a distillation flask, dilute with 100 ml of water, distil nearly 100 ml, make to volume at 20 C (68 F) and determine the specific gravity. Obtain from appropriate tables the corresponding percentage of alcohol by volume in the distillate and multiply by 4 to obtain the percentage of alcohol by volume in the original substance.

NOTE: The alcohol content of the distillate may be checked by determining the Zeiss immersion refractometer reading and obtaining from appropriate tables the percentage of alcohol present. (See table 20, page 708 of following reference.)

REFERENCE

Method of Analysis, A.O.A.C., 6th edition (1945), pp. 365-368.

Gravimetric Test for Vanillin and Coumarin

This test is not applicable to concentrated vanillin and coumarin preparations. For such products, use a smaller quantity of sample and dilute to 50 ml.

Apparatus

1. Analytical balance.
2. Separatory funnels.
3. Beakers, 250 ml, marked to show the volumes of 80 and 50 ml.
4. Water bath.

Reagents

1. Lead acetate solution: Dissolve 80 gm of neutral lead acetate in distilled water that has been recently boiled, dilute to 1 liter, and filter the solution if it is not clear.
2. Diethyl ether which has been washed twice with an equal volume of distilled water.
3. Ammonium hydroxide (NH_4OH), dilute, 1 : 11.
4. Hydrochloric acid (HCl):
   (a) Concentrated
   (b) Diluted (1 to 2).
5. Petroleum ether [boiling point 104 F (40C) or below].
6. Resorcinol crystals.
7. Iodine solution, approximately 0.1 N. (See page 529.)

Preparation of Solution

1. Measure 50 ml of the extract at 20 C (68 F) into a 250 ml beaker marked to volumes of 80 ml and 50 ml; dilute to 80 ml and evaporate to 50 ml in a water bath kept at 70 C (158 F) or below.
2. Dilute again with water to 80 ml and evaporate to 50 ml.
3. Transfer to a 100 ml flask, rinsing the beaker with hot water; add 25 ml of 8 per cent neutral lead acetate solution; make up to the mark with water, shake, and allow to stand 18 hours (over night) at 37 — 40 C (98.6 — 104 F).
4. Decant into a small dry filter, reserving the filtrate for the determination of vanillin and coumarin, the lead number, and the residual color.

Procedure for Vanillin

1. Transfer a 50 ml aliquot of the filtrate to a separatory funnel and extract with four successive 15 ml portions of ether (previously washed twice with an equal volume of water to remove alcohol).

2. Wash the combined ether solutions four or five times with dilute NH_4OH (1 : 11), using 10 ml the first time and 15 ml thereafter. Reserve the ether solution for the determination of coumarin.

3. Slightly acidify the combined ammoniacal solutions with dilute HCl (1 : 2), cool, and extract in a separatory funnel with four portions of washed ether, using about 40 ml altogether.

4. Evaporate the ethereal solutions at room temperature, dry over sulfuric acid, and weigh.

NOTE: The vanillin residue often appears first in the form of oil-like droplets, which on standing crystallize into light colored masses. If after standing in a desiccator the residue is considerably discolored or gummy, extract the vanillin from it by treating with at least fifteen successive portions of boiling petroleum ether [boiling point 40 C (104 F) or below]; combine the petroleum ether extracts, evaporate to dryness, and weigh.

The residue, if pure vanillin, should be white crystals melting at approximately 80 C (176 F). Dissolve a small quantity of the residue in 2 drops of strong hydrochloric acid and add a crystal of resorcin. Vanillin gives a pink coloration.

Procedure for Coumarin

1. Evaporate at room temperature the original ether extract under (2) above from which the vanillin has been removed by means of dilute NH_4OH, dry over sulfuric acid, and weigh. The residue, if pure coumarin, melts at approximately 67 C (152.6 F).

2. Dissolve a small quantity of the residue in not more than 0.5 ml of hot water and add a few drops of 0.1 N iodine solution. Coumarin yields a brown precipitate which finally gathers in green flecks leaving a clear brown colored solution. The reaction is especially marked if the reagent is applied with a glass rod to a few drops of the solution on a white plate or tile.

NOTE: If peperonal is present it will appear in the ether extract and subsequently in the residue which contains the coumarin. Peperonal melts at 37 C (98.6 F). It can also be distinguished from coumarin by its odor.

REFERENCE

Methods of Analysis, A.O.A.C., 6th edition (1945), p. 365.

Colorimetric Method for Vanillin

Apparatus

1. Colorimeter with accessories.
2. Satisfactory source of light.
3. Volumetric flasks, 100 ml to 50 ml.

Reagents

1. Vanillin, standard solution: Dissolve 0.1 gm of pure vanillin in water and dilute to 1 liter.

2. Phosphotungstic-phosphomolybdic acid: To 100 gm of pure sodium tungstate and 20 gm of phosphomolybdic acid (free from nitrates and ammonium salts), add 100 gm of syrupy phosphoric acid (containing 85 per cent H_3PO_4) and 700 ml of water. Boil over a free flame for 1½ - 2 hours and then cool. Filter if necessary, and make up with water to 1 liter (an equivalent amount of pure molybdic acid may be substituted for the phosphomolybdic acid).

3. Sodium carbonate, saturated solution.

4. Lead acetate solution: Dissolve 50 gm each of basic and neutral lead acetate in water and make up to 1 liter.

Procedure

1. Transfer to a 100 ml volumetric flask a quantity of the sample that contains from 8 to 12 mg of vanillin (usually 5 ml).

2. Add 75 ml of tap water, at room temperature, and 4 ml of the lead solution. Dilute to 100 ml with water and mix.

3. Filter through a dry filter paper and pipette 5 ml of the clear filtrate into a 50 ml volumetric flask.

4. Into another 50 ml volumetric flask pipette 5 ml of the standard vanillin solution. To each of these flasks add with a pipette 5 ml of the phosphotungstic-phosphomolybdic acid reagent, allowing the reagent to flow down the neck of the flask in such a way as to wash down the vanillin solution that may be on the sides of the flask.

5. Mix the contents of the flask by rotating and after five minutes dilute the contents to 50 ml with the sodium carbonate solution. Mix thoroughly by inverting the flasks several times and allow to stand for at least ten minutes so that the precipitate that forms may separate completely.

6. Filter the solutions through dry filter papers and compare the blue colors of the clear solutions in a colorimeter. Report the result as grams of vanillin per 100 milliliters of extract.

NOTE: Fresh standard must be used for each reading as the color slowly darkens after the addition of alkali. Coumarin and acetanilid give no color with the reagent and the presence of sugar, carmel, or glycerin have no effect.

REFERENCES

Folin, O. and Denis, W.: A New Colorimetric Method for the Determination of Vanillin in Vanilla Extract, *Ind. Eng. Chem. 4:* 670 (1912).

May, P.: Half a Century of Progress, *Perfumery and Essential Oil Record 15:* 351-8 (1924).

Sale, J. W. (Referee); Report on Flavors and Non-Alcoholic Beverages; Vanillin—Method II, *J. Assoc. Offic. Agr. Chemists* 8: 688-9 (1925); Flavoring Extracts, *Ibid. 10:* 43 (1927).

Mojonnier Method for Vanillin

Apparatus

1. Mojonnier tester and accessories. (See page 209.)
2. Cream test bottles, 18 gm.
3. Petri dishes.
4. Desiccator.
5. Babcock centrifuge.

Reagents

1. Lead acetate solution. Dissolve 80 gm. of lead acetate in 1 liter of water.
2. Acetic acid, glacial.
3. Diethyl ether which has been washed twice with distilled water.

Procedure

1. Pipette 25 ml of the extract into a Petri dish, cover and add 10 ml of distilled water. Evaporate to about 25 ml under the suction fan on the hot plate of the Mojonnier tester at a temperature not in excess of 140 F (60 C).

2. Add 10 ml of distilled water and again evaporate to approximately 25 ml.

3. By means of a funnel, transfer the de-alcoholized product to an 18 gm cream test bottle which has previously been calibrated to 50 ml at 68 F (20 C). Rinse the residue from the dish and funnel into the test bottle with warm distilled water.

4. Add 4 drops of glacial acetic acid and shake thoroughly, followed by the addition of 10 ml of lead acetate solution.

5. Make up to the 50 ml mark with water and shake again.

6. Place a stopper in the cream test bottle and centrifuge in a Babcock centrifuge for five minutes at the same speed as that used in the determination of fat (see page 183).

7. Draw off with pipette 25 ml of the clear supernatant liquid into a Mojonnier fat extraction flask and reserve the remainder for the determination of the volume of lead acetate precipitate.

8. Extract the vanillin solution with three 20 ml portions of ether (previously washed twice with distilled water). Use the Mojonnier centrifuge to aid in the separation of the ether layer after each extraction and pour the ether extracts into a weighed solids dish.

9. Evaporate the ether solution of vanillin on the hot plate at a temperature not above 131 F (55 C). Dry over sulfuric acid and weigh. Calculate the vanillin as grams per 100 milliliters, or as percentage by weight if the specific gravity is known.

NOTE: It may be necessary to lower the dividing line in the fat extraction flasks, after the last ether extraction of vanillin, by pipetting off part of the lower layer of liquid.

This method is not applicable if either coumarin or peperonal is present in the flavors, as they would appear in the ether extract and subsequently in the residue with the vanillin.

Lead Number

This determination is intended to be made in conjunction with the official vanillin and coumarin determination.

Apparatus

1. Gooch crucibles.
2. Analytic balance and weights.

Reagents

1. Lead acetate, 8 per cent solution, prepared as indicated under the test for vanillin.
2. Sulfuric acid (H_2SO_4) dilute, 1 : 9.
3. Ethyl alcohol, 95 per cent.
4. Acetic acid, glacial.

Procedure

1. To 10 ml aliquot of the filtrate from the lead acetate precipitate, as obtained under the gravimetric test for vanillin, add 15 ml of distilled water, 10 ml of H_2SO_4 (1 : 9), and 100 ml of 95 per cent alcohol. Allow to stand over night.
2. Filter through a weighed Gooch crucible, wash and precipitate with 95 per cent alcohol, dry at a moderate heat, ignite at low redness for three minutes, taking care to avoid the reducing flame, and weigh.
3. Conduct a blank determination, using distilled water containing 4 or 5 drops of glacial acetic acid instead of the sample. Calculate the lead number by the following formula:

$$P = \frac{100 \times 0.6832}{5} (S\text{-}W) = 13.66\ (S\text{-}W),$$

in which

P = Lead number (gm of metallic lead in the precipitate obtained from 100 ml of the sample);

S = Gm of lead sulfate corresponding to 2.5 ml of the lead acetate solution as determined in a blank analysis; and

W = Gm of lead sulfate obtained in 10 ml of the filtrate from the lead acetate precipitate, as obtained under the gravimetric test for vanillin.

REFERENCES

Methods of Analysis, A.O.A.C., 6th edition (1945), p. 366.

Towt, L. V.: The Use of the Mojonnier Milk Tester for the Routine Determination of Vanillin with Preliminary Discussion of the Towt Lead Number. *J. Dairy Sci. 12:* 469 (1929).

Total Solids

Apparatus

1. Flat-bottom dish, approximately 60 mm in diameter.
2. Desiccator.
3. Analytical balance and weights.

Reagents

1. Hydrochloric acid (HCl), concentrated.
2. Pure quartz sand, 40 to 60 mesh, digested with strong HCl, washed free from acid, and ignited. Preserve in stoppered bottle.

Procedure

1. Place from 25 to 30 gm of the prepared sand and a short stirring rod in the flat-bottom dish, dry thoroughly, cool in a desiccator, and weigh.
2. Add 10 ml of the sample to be tested and mix thoroughly with the sand.
3. Heat on a steam bath for 15 to 20 minutes, stirring at intervals of 2 or 3 minutes, or until the mass becomes too stiff to manipulate readily.
4. Dry at a pressure not to exceed 100 mm of mercury, making trial weighings at 2 hour intervals toward the end of the drying period until the change in weight does not exceed 2 mg. Report the percentage of the sample retained as total solids.

Quality of Vanilla Flavor

Add 5 ml of the vanilla extract to be judged to 100 ml of distilled water at 68 F - 70 (20 - 21.1C). The water dilutes the alcohol so as not to interfere with the true taste and odor of the extract. Judge the strength and quality of the extract by the aroma and flavor. Do not swallow the solution in tasting. The color is unimportant. Rinse the mouth with water after tasting each sample.

Adulteration with vanillin, coumarin, etc., can be detected. Moldy, musty, woody, prune, and other off-flavors are readily detected. The results should be verified on the selected samples by testing in ice cream. A standard pure vanilla extract should require about 4 ounces of flavor to a ten gallon batch of sweet cream to make an ice cream with a mild flavor. If the resultant flavor in the ice cream under these conditions is pronounced, there is good reason to suspect re-enforcement with synthetic materials.

REFERENCE

Methods of Analysis, A.O.A.C., 6th edition (1945), pp. 367 and 557.

CHAPTER 36

CHOCOLATE AND COCOA TESTING

Apparatus

1. Electric muffle.
2. Porcelain crucible, 30 ml.
3. Electric oven, 650 F.
4. Ashless filter paper.
5. Soxhlet fat extraction apparatus.
6. Microscope.
7. Glass slide, 6 inches square.
8. Sieve, 100 mesh.
9. Chemical balance.
10. Flasks, 150 ml.
11. Pipette, 1 ml.
12. Petri dishes.
13. Fermentation tubes.
14. Flask for fat extractions, 60 ml.

Preparation of sample

1. Cocoa powder requires no preliminary treatment.

2. Chocolate in cake form should be carefully shaved with a knife or a nutmeg grater. Carefully procure a representative sample, keeping hand contamination to a minimum.

Moisture Test

Weigh 2 gm of the sample on a watch glass to an accuracy of 0.1 milligram. Place in an electric oven at 221 F (105 C) and dry to constant weight, requiring approximately five hours. (Solids content may be checked by Mojonnier method.)

Express loss of weight in percentage of the original as the percentage of moisture.

Total Ash

Weigh a 2 gm sample in a porcelain crucible and heat gently until thoroughly charred in a standard oven, followed by heating to low redness until a white ash is obtained. Cool in a desiccator or Mojonnier cooling oven and weigh. It is essential to bear in mind that the ash should not be heated too highly. Temperatures from 600 to 650 F give a good ash determination. It requires several hours to produce a good ash of uniformly white or gray color, free from fused lumps or particles or unburned carbon. A red or green ash generally denotes chemically treated or alkaline cocoa.

Express the weight of the ash in terms of the percentage of the original sample.

Soluble and Insoluble Ash

ASH INSOLUBLE IN WATER

To determine the ash insoluble in water, weigh the total ash and add about 25 ml of water to the crucible. Cover it with a watch glass to avoid loss by spattering, and heat nearly to boiling. Filter through an ashless filter and wash with an equal volume of hot water. Place the filter paper and residue in the crucible, again ignite, cool, and weigh.

From the loss of weight calculate the water-soluble ash, and by difference, the water-insoluble ash.

Express the results in terms of the percentage of the original sample.

ASH INSOLUBLE IN ACID

Reagent

Hydrochloric acid (HCl), 10 per cent.

Procedure

The weighed residue in the determination of ash insoluble in the water may be used, or a new sample may be weighed, ignited, and used directly. To the ash in either case add 25 ml of 10 per cent HCl. Cover with a watch glass and boil gently over a low flame or on a hot plate for 5 minutes, and filter through an ashless filter. Wash with hot distilled water, and return the filter and residue to the dish. Ignite, cool, and weigh.

Express the results in terms of the percentage of the original sample.

NOTE: This determination is important and of value to show added mineral matter, such as sand, dirt, and talc. The limit allowed is 0.4 per cent.

Alkalinity of Total Ash

ALKALINITY OF WATER SOLUBLE ASH

Reagents

1. Hydrochloric acid (HCl), 0.1N.
2. Methyl orange indicator.

Procedure

Allow the filtrate from the insoluble ash determination to cool and titrate with 0.1N HCl, using methyl orange as indicator. Express the alkalinity in terms of the number of milliliters 0.1N HCl required to neutralize the water-soluble ash per gram of chocolate. (Since the original sample contained 2 gm, ½ the titration value would be used).

ALKALINITY OF INSOLUBLE ASH

Reagents

1. Hydrochloric acid (HCl), 0.1N.
2. Sodium hydroxide (NaOH), 0.1N.
3. Methyl orange indicator.

Procedure

Add 15 ml of 0.1N HCl to the weighed sample of the insoluble ash, cover with a watch glass and heat cautiously, nearly to boiling, over a low flame. Allow to cool. Titrate the excess acid with 0.1N NaOH, using methyl orange as indicator. The number of milliliters of 0.1N NaOH subtracted from the 15 ml of 0.1N HCl gives the number of milliliters of HCl required to neutralize the alkalinity of the ash.

The alkalinity of the ash is important in the analysis of chocolate, as indicating adulteration. A decreased quantity of alkali salts in the ash would indicate adulteration. An increase in alkali may indicate chemical treatment.

Detection of Alkali

Alkali treated cocoa will usually be indicated by the increased portion of soluble ash and a high alkalinity of the ash, especially the soluble portion.

TABLE 28. ASH CONTENT OF UNTREATED AND ALKALI TREATED COCOA

	Untreated cocoa		Alkali-treated cocoa	
	Max. %	Ave. %	Max. %	Ave. %
Total ash	6.08	5.25	9.25	7.49
Water soluble ash	3.15	2.32	7.62	5.79
Alkalinity of ash, ml 0.1N acid for 1 gm	6.85	6.04	12.29	9.36

Note that the alkalinity of the ash refers to the alkalinity of the total ash, not of the soluble portion. This would be distinctly lower, and with untreated cocoa will seldom exceed 3.75 ml of 0.1N acid for 1 gm of sample. In an untreated cocoa or chocolate the soluble ash is considerably less than the insoluble ash, usually about one third of the total ash.

Percentage of Cocoa Butter

Reagent

Petroleum ether.

Procedure

1. Weigh 2 gm sample of chocolate on a watch glass and transfer carefully to filter paper (Whatman No. 50).

2. Weigh a small Soxhlet flask and add from 15 to 20 ml of petroleum ether. Record the number and weight of the flask and of the corresponding sample.

3. Heat until ether boils slowly and a steady stream of drops from the condenser falls on the sample. Continue the extraction for 14 to 16 hours, after which all the cocoa butter from the sample will be in the ether extract.

4. Remove the flask from the extraction apparatus, distil off the ether on a water bath and place the flask in an electric oven at 105 C (221 F) until all the ether is evaporated (can be determined by the odor).

5. Weigh. The difference between the two weighings represents the weight of the cocoa butter in the sample. Record the weight of cocoa butter in terms of percentage of the original sample.

ALTERNATE METHOD

Mojonnier

Although the Mojonnier method (see page 209) may be used, chocolate manufacturers have found that with poorly milled chocolates the fat is difficult to extract by this procedure and recommend the continuous extraction method for accurate work.

TABLE 29. FATS WHICH MAY BE PRESENT IN COCOA AND CHOCOLATE

	Specific Gravity	Melting point	Saponification value	Reichert Meisel number	Iodine value
Cocoa butter	0.964-0.974 at 15 C (59 F)	30-34 C	192-195	0.2-0.9	32-42
Cotton seed stearin	0.867 at 100 C (212 F)	27-45 C	194-195	0.8-1.0	89-93
Cocoanut oil	0.9259 at 15 C (59 F)	20-28 C	246-254	6.6-8.4	8.0-9.0
Palm oil	0.8731 at 15.5 C (60 F)	23-30 C	242-255	5.0-6.8	10.5-17.5
Peanut oil	0.911-0.926 at 15.5 C (60 F)	....	186-194	0.5	83-101.0

Test for Adulteration of Cocoa with Shells, Fiber, Carbon, Foreign Starches, and Dyes

1. Place a small amount of melted sample at 38 C (100 F) between two glass slides. Place the slides before a strong light, such as an electric bulb, and examine for fiber, carbon, and coarse matter.

2. To compare the color of samples, place the melted samples side by side on one glass slide and allow to cool; reverse the glass and note the color comparison.

3. Extract 5 gm of sample with petroleum ether. Centrifuge or decant the solvent. Prepare the residue for microscopic examination

with a saturated solution of chloral hydrate. Examine for foreign starch, fat, crystals, shell, and carbon.

Test for Fineness

Weigh 30 gm of sample and place in a 100-mesh sieve. Place the sieve in a bath of hot water and allow it to stand until the chocolate is thoroughly softened, about 20 minutes. Wash through the sieve, using a steady, strong stream of hot water until the residue does not seem to diminish. Transfer the residue with a spatula to the palm of the hand and rub. Very carefully remove the sludge and fiber from the hand and wash again through the sieve. Repeat this operation until there is only fiber remaining. Remove the fiber residue from the sieve with a spatula to a watch glass of known weight. Dry the residue in the oven at 105 C (221 F) to constant weight, about two hours. Cool and weigh the residue, calculate the percentage on a fat-free and moisture-free basis.

TABLE 30. ANALYSIS OF CHOCOLATE LIQUORS

	Maximum	Minimum
Total ash	3.7	2.7
Soluble ash	1.6	1.1
Ash insoluble in acid	0.4	0.1
Ash insoluble in water	2.6	1.6
Alkalinity of ash, 0.1 N acid for 1 gm	4.4	0.55
Fat	54.3	49.5
Moisture	1.5	0.8

Analysis of Cocoa Powder

	Maximum	Minimum
Total ash	5.10	4.4
Ash insoluble in water	4.5	2.0
Alkalinity of ash, 0.1 N acid, ml	3.5	1.5
Ash acid insoluble	0.4	0.1
Crude fiber	7.0	3.5
Fat	26.0	18.0
Moisture	5.0	3.1

Bacteriological Analysis of Chocolate Products

Procedure

1. Prepare a sample from the chocolate by scraping the surface; another sample by cutting through the mass, employing aseptic precautions. Keep the samples cold and mix thoroughly with a sterile knife or spatula.
2. Weigh 1 gm of the sample on sterile filter paper.
3. Transfer to a sterile flask containing 99 gm of water.
4. Warm the flask to 37.5 C (100 F) to melt chocolate and shake vigorously.
5. Proceed to make dilutions and agar plates as described on page 42.

CHAPTER 37

FRUIT TESTS

Canned Fruit Grades

Apparatus

1. Screen and pan for determining drained weight; ⅛ in. mesh, 12 in. in diameter, set in a frame with a vertical side higher than the level of the product on the screen.
2. Thermometer.
3. Balling hydrometer, 0 to 70 degrees.
4. Cylinder, 500 ml.

Procedure

In order to determine the grade and quality of the fruit, the drained weight, count per can, and concentration of syrup in the can should be determined. The appearance of the fruit should also be noted and compared to the standards required for the use intended.

Determination of Drained Weight

To determine drained weight, the contents of the can should be emptied on the circular screen and should be distributed so as to form a layer of uniform depth, this being accomplished so far as possible by the manner of emptying from the can. When necessary, the fruit in halves should be turned over with the pit side down to permit draining of any liquid present in the pit cavity, but this is to be done in such a way as to express no additional amount of liquor from the material. The period of draining is 2 minutes.

The drained weights should be equal to or greater than the claim stated on the labels of the cans.

TABLE 31. CONCENTRATION OF SYRUP AND NUMBER OF PIECES PER NO. 10 CAN OF FRUITS

Grade	Apricots			Cling Peaches			Pears		
	Pieces per can	Syrup when canned	Syrup when opened Baumé	Pieces per can	Syrup when canned Baumé	Syrup when opened Baumé	Pieces per can	Syrup when canned Baumé	Syrup when opened Baumé
Fancy	66-86	55	27	30-43	55	26	30-43	40	22
Choice	83-108	40	21	36-54	40	20	36-54	30	18
Standard	122-151	25	17	54-76	25	15	54-76	30	15
Seconds	....	10	12	....	10	12	....	10	12

Grade	Plums			Royal Anne Cherries			Prunes		
	Pieces per can	Syrup when canned Baumé	Syrup when opened Baumé	Pieces per can	Syrup when canned Baumé	Syrup when opened Baumé	Pieces per can	Syrup when canned Baumé	Syrup when opened
Fancy	55	27	248-306	40	23	Varies	20	....
Choice	40	21	306-378	30	20	with	to	....
Standard	25	17	378-522	20	15	size of	30	....
Seconds	10	12	....	10	12	fruit used		

	Pie, Ounces	Solid Pack Pie, Ounces
Apricots	70	92
Bartlett pears	70	92
Yellow cling peaches	76	92
Sliced yellow cling peaches	76	92
Yellow free peaches	70	92
Muscat grapes	68	84
Royal Anne cherries	68	84
Black and white cherries	68	84
Plums	68	88

REFERENCES

Cruess, W. V.: *Commercial Fruit and Vegetable Products,* McGraw-Hill Book Company, New York, 1938.

Cruess, W. V., and Christie, A. W.: *A Laboratory Manual for Fruit and Vegetable Products,* McGraw-Hill Book Company, New York, 1922.

The Specifications of the Canners' League of California.

NOTE: Material prepared by E. M. Mrak of the University of California.

Determination of Syrup Concentration

The syrup concentration is determined with the Balling hydrometer, which indicates the percentage of sugar, provided the syrup is tested at 15.5 C or 60 F. It is therefore practically always necessary to make a temperature correction by using the accompanying table 32.

To use the table, the apparent sugar percentage (Balling degree) of the syrup is determined by the hydrometer, and the temperature is observed. Opposite the indicated temperature and in the column below the indicated sugar percentage will be found the amount to be added or substracted from the observed reading.

Example:

Observed sugar ....60 per cent
Observed temperature ....46.1 C (115 F)

Opposite 46.1 C (115 F) and below 60 per cent is found 2.32. This is then added to 60 per cent, making 62.32, the correct Balling degree.

TABLE 32
Temperature Corrections for Balling Hydrometer

Temp., Fahr.	Temp., Cent.	Degrees Balling 10°	20°	30°	40°	60°	75°
	Corrections to be Subtracted from Degree Balling						
32	0	.41	.62	.82	.98	1.22	1.29
41	5	.37	.52	.65	.75	.88	.94
50	10	.29	.36	.42	.49	.54	.61
54	10.2	.22	.26	.31	.34	.40	.46
57	13.9	.16	.18	.21	.22	.26	.32
61	16.1	.08	10	.11	.12	.14	.18
62	16.7	.03	.03	.04	.04	.05	.06
	Corrections to be Added to Degree Balling						
64	17.78	.03	.03	.03	.03	.03	.02
68	20.00	.08	.09	.10	.10	.10	.10
72	22.33	.29	.31	.32	.33	.32	.25
75	23.89	.41	.44	.46	.47	.46	.40
79	26.11	.54	.58	.61	.62	.62	.55
82	27.78	.68	.72	.76	.78	.78	.70
86	30.00	.82	.92	.94	.98	.88	.86
90	32.23	.98	1.03	1.08	1.10	1.10	.98
93	33.89	1.14	1.21	1.24	1.28	1.26	1.17
97	36.11	1.32	1.38	1.41	1.46	1.42	1.33
100	37.78	1.49	1.55	1.59	1.64	1.60	1.49
104	40.00	1.67	1.73	1.79	1.82	1.78	1.65
108	42.23	1.89	1.93	1.99	2.00	1.96	1.81
110	43.34	1.96	2.03	2.09	2.10	2.06	1.89
112	44.45	2.06	2.13	2.19	2.20	2.14	1.97
115	46.11	2.27	2.34	2.39	2.40	2.32	2.15
117	47.23	2.38	2.45	2.49	2.50	2.41	2.24
119	48.34	2.49	2.56	2.59	2.60	2.50	2.33
121	49.45	2.60	2.67	2.69	2.70	2.60	2.42
122	50.00	2.71	2.78	2.80	2.80	2.70	2.51
124	51.11	2.81	2.89	2.90	2.91	2.80	2.60
126	52.23	2.92	3.00	3.01	3.02	2.90	2.69
128	53.34	3.03	3.11	3.12	3.13	3.00	2.78
130	54.45	3.14	3.22	3.23	3.24	3.10	2.87
131	55.00	3.26	3.33	3.33	3.29	3.20	2.93
133	56.11	3.39	3.44	3.44	3.46	3.30	3.05
135	57.23	3.52	3.55	3.55	3.57	3.40	3.14
137	59.34	3.64	3.66	3.66	3.68	3.50	3.23
139	59.45	3.76	3.77	3.77	3.79	3.60	3.32
140	60.00	3.82	3.88	3.88	3.90	3.70	3.41
149	65.00	4.53	4.51	4.49	4.48	4.21	3.88
158	70.00	5.18	5.14	5.10	5.06	4.32	4.35
167	75.00	6.00	5.84	5.74	5.66	5.27	4.84
176	80.00	6.62	6.54	6.38	6.26	5.82	5.33
185	85.00	7.44	7.30	7.10	6.92	6.39	5.85
194	90.00	8.26	8.06	7.85	7.68	6.58	6.37
203	95.00	9.4	8.89	8.61	8.35	7.59	6.90
212	100.00	10.10	9.72	9.39	9.03	8.22	7.42

Detection of Chemical Preservatives

The chemical preservatives most likely to be found in fruits and flavors are sulfurous acid, sulfites, and sodium benzoate.

SULFUROUS ACID AND SULFITES

The sulfurous acid and sulfites are commonly employed and frequently sulfur dioxide is used as such in the processing of fruits.

Apparatus

1. Condenser and accessories.
2. Gooch crucible.
3. Separatory funnel.

Reagents

1. Phosphoric acid (H_3PO_4), 20 per cent.
2. Bromine water, saturated.
3. Sodium bicarbonate ($NaHCO_3$).
4. Hydrochloric acid (HCl), dilute 1 : 3.
5. Barium chloride ($BaCl_2$), 10 per cent.

Procedure

1. Weigh 20 to 100 gm sample of fruit into a 500 ml flask, adding recently distilled water if necessary, and 1 gm of $NaHCO_3$.

2. Connect with the distillation apparatus immediately.

3. Through the separatory funnel add 10 ml of the 20 per cent H_3PO_4.

4. Collect 150 ml of the distillate in 100 ml of the saturated bromine water, allowing the tip of the adapter to dip below the surface.

5. When the distillation is finished, boil off the excess bromine, dilute the solution to about 250 ml, add 5 ml of the dilute HCl, heat to boiling, and precipitate the H_2SO_4 with a slight excess of the $BaCl_2$ solution.

6. Boil a few minutes longer, allow to stand over night in a warm place, filter through a weighed Gooch crucible or through an ashless filter paper, wash with hot water, ignite at a dull red heat, and weigh as barium sulfate.

7. Conduct a blank determination on the apparatus and reagents and correct the result.

Calculation

$$\frac{\text{Wt. of BaSO}_4\text{ ppt.} \times 0.27443}{\text{Wt. of Sample}} \times 100 = \%\ SO_2$$

However, the SO_2 content is often expressed in terms of the milligrams per kilogram or parts per million (p. p. m.):

$$\frac{\text{Wt. of BaSO}_4\text{ ppt.} \times 0.27443}{\text{Wt. of Sample}} \times 1{,}000{,}000 = \text{p. p. m. of } SO_2$$

NOTE: The calculation of results can be simplified by using that weight of the sample which is a multiple of the gravimetric factor, 0.27443, such as 27.44 gm or 54.88 gm. Both the free and combined, or the total, sulfurous acid are determined in this method. The free acid only may be determined by the same procedure if the phosphoric acid is not added.

SODIUM BENZOATE AND BENZOIC ACID

NOTE: Benzoic acid as its sodium salt is very frequently used in food products and is permitted under the Federal Food and Drug law, provided that the presence and amount be indicated on the label.

Qualitative

Apparatus

1. Separatory funnels.
2. Calibrated burette.
3. Volumetric flask, 500 ml.
4. Porcelain evaporating dish.
5. Desiccator containing sulfuric acid.

Reagents

1. Hydrochloric acid (HCl), dilute 1 : 3.
2. Ammonium hydroxide (NH_4OH), 1 per cent.
3. Chloroform ($CHCl_3$).
4. Ether.
5. Petroleum ether.
6. Alcohol, 95 per cent, neutral to phenolphthalein.
7. Ferric chloride ($FeCl_3$), 0.5 per cent.
8. Sodium chloride (NaCl), saturated.
9. Sodium hydroxide (NaOH), 10 per cent.
10. Sodium hydroxide, 0.05 N.
11. Phenolphthalein indicator.

Procedure

If the material is a liquid, acidify 25 or 50 ml with 5 ml of dilute HCl and extract with 50 to 100 ml of ether in a separatory funnel. Be careful not to form an emulsion.

If the sample is solid or semisolid, macerate or grind it in a mortar with distilled water, transfer to a 500 ml volumetric flask, and fill to the mark with distilled water. Shake thoroughly, acidify 25 or 50 ml of this solution, and extract with ether as explained above. If the mixture emulsifies, add 10 or 15 ml of petroleum ether and shake not too vigorously. If this treatment fails to break the emulsion, centrifuge or allow it to stand until a considerable portion of the aqueous layer has separated. Decant, shake, and again allow to separate. Wash the ether layer with two 5 ml portions of water, evaporate the greater portion of the ether in a porcelain dish on a steam bath, allow the remainder to evaporate spontaneously, and add a drop of 0.5 per cent $FeCl_3$. A brownish precipitate of ferric benzoate indicates the presence of benzoic acid. (A violet color indicates salicylic acid.)

Quantitative Determination

Procedure

For the quantitative determination, the use of chloroform for extraction rather than ether is preferable because only traces of mineral acids, tannin, salts, and other interfering substances are obtained.

1. Mix the sample thoroughly, grinding if solid or semisolid.

2. Transfer 150 ml or 150 gm to a 500 ml volumetric flask, saturate the solution with NaCl, make alkaline to litmus paper with 10 per cent NaOH, and dilute to the mark with a saturated NaCl. Shake thoroughly, allow to stand for at least two hours, with frequent shaking, and filter.

3. When the sample contains large quantities of fat, add a few milliliters of the NaOH to the filtrate and extract with ether. Retain the aqueous portion.

4. Measure 100 or 200 ml of the filtrate or aqueous portion into a separatory funnel. Neutralize to litmus paper with dilute HCl, and add 5 ml in excess. Extract carefully with $CHCl_3$, using successively portions of 70, 50, 40, and 30 ml. To avoid the formation of an emulsion, shake cautiously each time. The $CHCl_3$ usually separates readily after standing a few minutes. If an emulsion forms, break it by stirring the $CHCl_3$ layer with a glass rod, by drawing it off into a second funnel and giving one or two sharp shakes from one end of the funnel to the other, or by centrifuging for a few minutes. As this is a progressive extraction, draw off carefully as much of the clear $CHCl_3$ as possible after each extraction, but do not draw off any of the emulsion with the $CHCl_3$ layer. If this precaution is taken, the $CHCl_3$ extract need not be washed.

5. Transfer the combined $CHCl_3$ extracts to a porcelain evaporating dish, rinse the container several times with a few milliliters of $CHCl_3$, and evaporate to dryness at room temperature in a current of dry air.

6. Dry the residue over night in a desiccator containing H_2SO_4.

7. Dissolve the residue of benzoic acid in 30 to 50 ml of neutral 95 per cent alcohol; add about one fourth of this volume of water

and 1 or 2 drops of phenolphthalein indicator, and titrate with 0.05N NaOH. One ml of 0.05N NaOH is equivalent to 0.0072 gm of anhydrous sodium benzoate.

REFERENCE

Methods of Analysis, A.O.A.C., 6th edition (1945), p. 382.

Determination of Total Solids

Procedure

1. Weigh exactly 10 gm of a finely ground sample of fruit into a flat, tared dish, and dry for 12 hours in a vacuum oven at 70 C (158 F), and not over 4 inches of mercury pressure. Reweigh and express results as percentage of total solids.

For syrups and juices, an estimation of the sugar or solids content may be secured by using a Balling hydrometer or a specific gravity spindle, using table 32, page 483.

Microscopic Examination for Bacteria, Yeast, and Molds

The estimation of the contamination of fruits by bacteria, yeasts, and molds has not been as thoroughly studied as for other foods, such as dairy and tomato products. A rough guide to the amount of such contamination may be secured by adapting the Howard method to the fruit or fruit product. Reference should be made to Microscopic Studies on Tomato Products, by B. J. Howard, *U. S. Department of Agriculture, Bulletin* 581, (1917).

Apparatus

1. Microscope with 10X and 12.5X eye pieces, and 16 mm, 8 mm, 4 mm, objectives.
2. Hemocytometer.
3. Howard counting chamber.
4. Stage micrometer.
5. Pipettes.

Reagents

1. Stains.

(a) Loeffler's methylene blue (see page 526).

(b) Carbolfuchsin. (Dissolve 1 gm basic fuchsin and 5 gm carbolic acid in 90 ml distilled water and add 10 ml absolute alcohol.)

2. Formaldehyde (HCHO, 40 per cent).

ESTIMATION OF MOLDS

Procedure

1. Equip the microscope with a 10X eye piece and a 16 mm objective. Place a stage micrometer under the objective and adjust the draw tube until the field of vision is 1.4 mm in diameter. This equals a field area of 1.5 sq mm. If a stage micrometer is not available, a hemocytometer may be used to measure the diameter of the field. Each small square of the hemocytometer is 0.05 mm in diameter. The magnification should be from 90 to 100 diameters.

2. Spread a drop of the well mixed sample on the disk of the Howard counting chamber. Cover with the heavy cover glass and press it evenly against the slide. This forms a layer of material 1 mm thick.

3. Examine carefully 50 separate fields for mold filaments. This will usually require two mountings. All fields showing mold filaments aggregating one sixth or more of the diameter of the field are considered positive. Multiply the number of positive fields by two to obtain "percentage of fields showing mold." With some samples it may be necessary to dilute the sample in the proportion of 2 ml. of water to each gram of sample. The number of positive fields is then multiplied by six.

YEAST CELLS AND MOLD SPORES

1. Fit the microscope with a 10X eye piece and an 8 mm (1/3 in.) objective. This gives a magnification of about 200 diameters. Set the draw tube at 160 mm. Mix thoroughly 1 gm of sample with 3 ml of water and allow to settle a few minutes. Mount a drop of the liquid on the disk of a hemocytometer. Cover with a large cover glass and press evenly against the slide.

2. Count the number of mold spores and yeast cells in 200 of the small squares of the hemocytometer, i. e. eight blocks of 25 squares each.

3. Each of the smallest squares is 0.05 mm square, or 0.0025 sq mm in area. Since the liquid is 0.1 in depth the volume of liquid above each square is 0.00025 cu mm. Therefore, the volume above 200 squares is $0.00025 \times 200 = 0.05$ mm, or 1/20 cu mm. Since the sample is diluted to one-third its original concentration, the above estimation represents the number of yeast cells and mold spores per 1/60 gm.

BACTERIA (HOWARD METHOD)

1. Fit the microscope with a 12.5X eye piece and a 4 mm (1/6 in.) objective. This combination gives a magnification of about 570 at draw tube setting of 160 mm.

2. Use the same mounting as for yeasts and spores. Count the number of distinct rod-shaped bacteria in 25 small squares; i. e., five groups of five linear adjacent small squares. Do not count spherical forms. Mount two more samples and repeat the examination for each and determine the average number of bacteria per small square.

3. Since one small square represents 0.00025 (or 1/4,000) cu mm of liquid, one bacterium to a square represents 4,000 per cubic millimeter, or 4,000,000 per milliliter. Since the sample was diluted to one third of its original concentration, one bacterium per square represents 12,000,000 bacteria per gram of original sample.

BACTERIA (MILLER MODIFICATION)

1. Transfer 10 gm of the sample to a 100 ml beaker.
2. Add 2 ml of Loeffler's methylene blue stain.
3. Boil three minutes.
4. Add 2 ml of Ziehl-Nielson carbolfuchsin stain.
5. Boil three minutes.
6. Allow to cool slightly.
7. Add 3 or 4 drops of 40 per cent solution of formaldehyde.
8. Make up to 60 ml with water, mix, and allow to settle.
9. Examine as in (3), the Howard method. The staining renders the bacteria more clearly visible.

REFERENCES

Hiss, P. H., Sr., and Zinsser, Hans: *Textbook of Bacteriology,* 4th edition, D. Appleton & Co., New York, 1918.

Cruess, W. V.: *Commercial Fruit and Vegetable Products,* McGraw-Hill Book Co., New York, 2nd edition, (1938).

Cruess, W. V., and Christie, A. W.: *A Laboratory Manual for Fruit and Vegetable Products,* McGraw-Hill Book Co., New York, 1922.

CHAPTER 38

TIN DETERMINATIONS

Determination of Tin Thickness on Tin Plated Cans

This method is suitable for the determination of the weight of tin coating on tin plate used in evaporated milk cans.

Apparatus

1. Stamping die for stamping out exactly a two inch square of tin plate (4 sq. inches of surface on *each* side of plate.)
2. Vacuum oven.
3. Erlenmeyer flasks, 300 ml.
4. Watch glasses.
5. Rubber stoppers.
6. Ice water bath, see figure 141 on page 491.
7. Analytical balance.
8. Glass tubing bent twice at right angles and fitted to a rubber stopper. (See figure 141).
9. Hot plate, electric.

Reagents

1. HCl C. P. sp. gr. 1.18.
2. Starch indicator, dissolve 2 gm of soluble starch in 200 ml of water.
3. Iodine solution—0.1N (accurate to fourth decimal place).

Dissolve 12.692 gm of c. p. iodine in a small quantity of distilled water, in which is dissolved 20 gm of KI. Make up to 1 liter and standardize against chemically pure tin, using a sample weighing approximately 0.2 gm and following the procedure which will be used in making the actual test.

4. Sodium bicarbonate (saturated solution).
5. Carbon tetrachloride.
6. Ethyl Alcohol.
7. Air-free water made by dissolving 12 gm of $NaHCO_3$ in one liter of water, adding 20 ml of conc. HCl and allowing the resulting gas to escape. Keep in a tightly stoppered bottle.

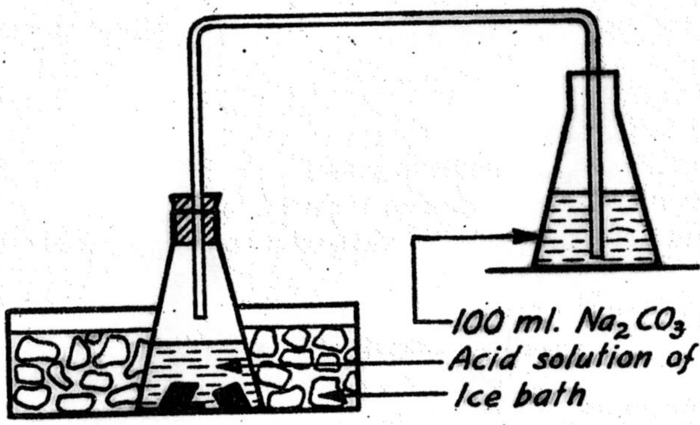

Figure 141.
Apparatus for tin determination.

Procedure

1. Stamp out sample of tin plate exactly 2 square inches in size. Bend to permit subsequent insertion in 300 ml Erlenmeyer flask, clean for two minutes in carbon tetrachloride and then for two minutes in ethyl alcohol.

2. Dry for 20 minutes in vacuum oven, cool, weigh and place in the Erlenmeyer flask.

3. Add two or three gm of sodium bicarbonate to the flask and then 75 ml of conc. HCl and 25 ml of water.

4. Place the flask on an electric hot plate under the hood and with gentle agitation of the flask dissolve the tin at boiling temperature. Cover flask with a watch glass to prevent escape of HCl fumes.

5. As soon as solution is complete connect the flask by means of a rubber stopper and a glass tube, bent twice at right angles, to a flask containing sodium bicarbonate (see figure 141).

6. Place the Erlenmeyer flask in an ice and water bath (see figure 141), so that as the sample cools, sodium bicarbonate is drawn into the flask. Upon coming in contact with the HCl, CO_2 is liberated, thereby preventing any possibility of oxidation of the stannous chloride.

7. After cooling completely, remove flask from bath and wash down inside of flask with 50 ml of air-free water.

8. Add 5 ml of starch indicator and titrate with the O.1N iodine solution until the end-point appears, which is a purple starch-iodine color.

9. Record the number of milliliters of 0.1N iodine solution used.

10. Calculations:

a. Required data for per cent of Sn:

1. Weight of 2″ square sample in grams.

2. Milliliters of 0.1 iodine solution required for titration.

3. Equivalent of one ml of 0.1N iodine solution, which is 0.005935 gm of tin.

Then:

$$\text{Per cent of Sn} = \frac{\text{ml } 0.1\text{N iodine} \times 0.005935 \text{ gm}}{\text{wt. of sample in gm}} \times 100$$

b. Thickness of tin plate (tin coating)

where:

T = thickness
W = weight of coating
A = area of coating
7.2 = density of Sn (gm per cu. cm)
118 = density of Sn (gm per cu. in.)

$$T \text{ (inches)} = \frac{W \text{ (gm)}}{A \text{ (sq. in)} \times 118} \quad \text{or} \quad T \text{ (cm)} = \frac{W \text{ (gm)}}{A \text{ (sq. cm)} \times 7.2}$$

example:

W = 10 ml 0.1N iodine solution × 0.005935 = 0.05935 gm Tin.
A (sq. in.) = 4 sq. in. each side or 8 sq. in. total area.

$$\text{then: } T = \frac{0.05935}{8 \times 118} = 0.000063 \text{ in., thickness of tin coating.}$$

c. $$\text{Pounds of Tin per base box} = \frac{\text{ml } 0.1\text{N iodine} \times 0.005935 \times 70 \times 112}{453.59}$$

NOTE: This equation is correct *only* when the sample analyzed is exactly two inches square and the sheets of tin plate are 14x20 inches, with 112 sheets to the base box.

When the "Pounds of tin per base box" is known the thickness of tin coating may be estimated by multiplying this value by the factor 0.00006.

The following is quoted from Burns and Schuh[1]:

"The weight of the coating on tin plate is expressed as the weight of tin per "base box." The base box is essentially a unit of area, and is the equivalent of 112 sheets, 14″ x 20″, that is, 31,360 square inches of sheet or 62,720 square inches of surface to be coated" (page 140).

"The tin-plate from which most cans are made is of the coke quality, that is, the thickness of the tin is from 0.00006 to 0.00015 inch. This corresponds approximately to 1 to 2.5 lbs. of tin per base box or from 1.0 to 2.5 milligrams per square centimeter of surface" (page 145).

REFERENCE

1. Burns, R. M., and Schuh, A. E., Protective Coatings for Metals. 1939, Reinhold Pub. Corp., N. Y. City.

The Determination of the Porosity of Tin Coatings on Steel

FIRST METHOD

This method consists of exposing samples of tin plate to hot distilled water for some hours. Adherent rust spots form wherever there are pin holes in the tin coating. These are clearly visible and can be counted.

Reagents

1. Carbon tetrachloride (CCl_4).
2. Sodium silicate solution, 0.1%, should have a ratio $Na_2O:SiO_3 = 1:2$. Sodium silicate "C" marketed by the Philadelphia Quartz Co. of Philadelphia, Pa., corresponds to this ratio.

Procedure

1. Degrease the tin plate with CCl_4 and immerse for 1 minute in a boiling 0.1 per cent solution of sodium silicate.
2. Rinse thoroughly the degreased plate with tap water.
3. Immerse in distilled water at 200 degrees F and maintain at this temperature for 6 hours. The period in hot water may be reduced to 3 hours if the article is left to cool to room temperature in the water over night.

Reddish brown spots with black nuclei indicate pores in the tin plate.

SECOND METHOD

This method detects the presence of pores in the tin plate by the development over them of blue spots of ferrous ferricyanide on a sheet of paper saturated with potassium ferricyanide.

Reagents

1. Carbon tetrachloride (CCl_4).
2. Sodium silicate solution, same strength as described in First Method.
3. Potassium ferricyanide solution, 1.0 per cent, in distilled water containing sodium chloride 0.5 per cent.

Procedure

1. Thoroughly degrease the tin plate as described in (1) and (2) of the First Method.
2. Spread a sheet of smooth typewriter paper or paper of similar quality on a glass plate and wet it uniformly by brushing it with the

freshly prepared solution of potassium ferricyanide. The paper should be completely wet but not to such an extent that any liquid will drip from it when held vertically.

3. Place the moist paper on the degreased surface and bring into close contact by means of a brush dampened with the reagent. After ten to fifteen minutes, depending on the temperature and humidity of the air in the room, the paper may become somewhat dry. It should be slightly dampened by stroking with a brush which has been dipped in distilled water. This treatment may be repeated at intervals during the test.

4. After one hour remove the paper, wash well to remove soluble salts, and suspend in the air to dry.

The test paper provides a permanent record of the porosity. To facilitate counting the pores which are indicated by blue spots, the paper may be ruled into one inch squares. A low power lens or reading glass should be employed in counting in order to detect the smallest blue spots.

REFERENCE

1. Macnaughtan, D. J., Clarke, S. G., and Prytherch, J. C.: The Determination of the Porosity of Tin Coatings on Steel, Technical Publications of the International Tin Research and Development Council, London Series A No. 7, 1935.

CHAPTER 39

BIOCHEMICAL OXYGEN DEMAND DETERMINATION

The biochemical oxygen demand of sewage, commonly called the "B.O.D." value, as the name suggests, refers to the amount of dissolved oxygen used by the microorganisms in the destruction of the organic matter present in the sewage.

Usually the test is run over a five-day period in which 68 per cent of the total B.O.D. is satisfied. The test consists briefly of making determinations of the dissolved oxygen in various concentrations of sewage before and after a five-day period of incubation. The amount of dissolved oxygen present after five days (called the F.D.O.—final dissolved oxygen) is subtracted from the amount of dissolved oxygen at the beginning of the test (called the I.D.O.—initial dissolved oxygen). The amount of oxygen required (difference between I.D.O. and F.D.O.) is multiplied by the suitable concentration factor—(dilution) to obtain the B. O. D. in parts per million for each concentration. From these B.O.D. values obtained for each dilution, the final value expressing the B.O.D. for the original sewage is calculated in the following procedure.

It is important to avoid the presence of interfering substances which would retard or restrict bacterial growth such as acids, alkalies, antiseptics, etc. or substances such as nitrites which interfere with the determination of dissolved oxygen. Extremely acid or alkaline sewage samples must be neutralized before beginning the test. Ordinarily the high dilutions used tend to reduce the effect of acids or alkalies present in the minimal amounts. The pH should be about 7.2.

Nitrates if present are destroyed by the use of Sodium azide which has previously been dissolved in one of the dissolved oxygen reagents subsequently described.

Apparatus

1. Incubator adjusted to 20 C (68 F).
2. Bottles of following capacities 5 gal, 1 gal, 500 ml (glass stoppered), B.O.D. bottles, 135 ml (numbered 1-8).
3. Graduate cylinders 500 ml, 250 ml, and 100 ml.
4. Erlenmeyer flasks, 250 ml wide mouth, 4.
5. Burettes, 50 ml preferred, 2.
6. Wash bottle, 1 l, for distilled water.
7. Siphon for delivery of diluting water.

8. Dispenser consisting of a two-hole rubber stopper which fits the 500 ml bottle, with 2 six-inch glass tubes, 1 of which is the delivery tube which is inserted into the stopper from the top, the other the vent tube which is inserted into the stopper from the bottom.

Reagents

1. Manganous sulfate reagent—dissolve 400 gm $MnSO_4$ ctr. 2 H_20 in about 900 ml of distilled water and dilute to 1 l.
2. Alkaline—iodide—azide reagent (see below).
3. Sulfuric acid, c.p. concentrated H_2SO_4.
4. Starch solution, 0.5 per cent.
5. Thiosulfate solution· N/10.
6. Thiosulfate solution, N/80.
7. Potassium bi-odate solution, N/10.
8. Potassium bi-odate solution, N/80.
9. Theresult-Nicols (T.N.) diluting water (see below).

Preparation of alkaline—iodide—azide reagent. Dissolve 500 gm of c.p. sodium hydroxide (NOH) and 150 gm c.p. potassium iodide (KI) in distilled water and dilute to 900 ml and cool. Dissolve 8 gm sodium azide (NaN_2) in 50 ml distilled water. Filter if necessary and add to the cool alkaline-iodide solution. Mix thoroughly and dilute to 1 liter. Transfer to pyrex liter bottle with rubber stopper and label.

Preparation Therbault-Nicols (T.N.) diluting water.

This diluting water is made by adding certain minerals and a buffer solution to a well stabilized distilled water just prior to making the various concentrations for the B.O.D. test. A good stabilized distilled water is one that has a dissolved oxygen content between 8.0 and 9.0 p.p.m. after having been aerated 10 to 20 minutes daily for about 20 days and has a five-day depletion of about 0.2 to 0.4 part for million of dissolved oxygen.

The mineral solution is made by adding 0.25 ml of a 1 per cent ferric chloride solution ($FeCl_3$) to about 200 ml of distilled water, followed by the addition of 10 ml each of a 7.5 per cent solution of Calcium Chloride ($CaCl_2$) and a 7.5 per cent solution of magnesium sulfate ($MgSO_47H_2O$) and diluting to 250 ml with distilled water. This mineral solution is referred to as Stock A solution.

The buffer solution is made by dissolving 8.5 gm of potassium acid phosphate (KH_2PO_4) in 125 ml distilled water and adding about 44 ml of N NaOH to give a pH of 7.2. Add 0.375 gms of ammonium sulfate (NH_4) ($_2$SO4) and dilute to 250 ml with distilled water. This buffer solution is referred to as Stock B solution. The amounts of these stock solutions to be added to each liter of well-stabilized distilled water are 10 ml. of stock A, and 2 ml. of Stock B. Mix.

Procedure for Diluting Samples

1. Arrange the numbered bottles consecutively on the work bench.

2. Remove stoppers and add to each bottle for the purpose of seeding on drop of stale sewage previously strained through cheese cloth.

3. Fill carefully numbers 1 and 2 with T.N. diluting water by means of the siphon. Stopper and set aside No. 1, for immediate dissolved oxygen (I.D.O.) and No. 2 for incubation at 20 C.

4. Make a 5 per cent concentration of the sample of sewage to be analyzed by diluting 25 ml sewage to 500 ml in a glass-stoppered bottle with T.N. dilution water. Insert stopper, shake, remove stopper and let stand for the entrained air to escape.

5. Insert the "dispenser." Invert the bottle holding the index fingers over the tube openings. Then, fill carefully B.O.D. bottles No. 3 and No. 4. Stopper, mark bottles "5 per cent conc." and set aside placing the odd numbers together and the even numbers together.

6. Pour out the remainder of the 5 per cent concentration into a marked 250 ml beaker. Wash the bottle with distilled water, discarding the rinse.

7. Make a 2.5 per cent concentration of sewage by diluting 150 ml of the 5 per cent concentration to 300 ml in the 500 ml graduate. Transfer to the 500 ml bottle. Stopper, shake and allow the entrained air to escape.

8. Remove stopper and insert the "dispenser". Invert bottle as above and fill B.O.D. bottles No. 5 and No. 6. Stopper and mark bottle "2.5 per cent conc." Set aside as above. Discard the remainder of the 2.5 per cent concentration.

9. Make a 1 per cent concentration of sewage by diluting 60 ml of the 5 per cent concentration to 300 ml in the 500 ml graduate. Transfer to the 500 ml bottle. Stopper, shake and allow entrained air to escape.

10. Remove stopper and insert the "dispenser". Invert bottle as above and fill B.O.D. bottles No. 7 and No. 8. Stopper, mark bottle "1.0 per cent conc." and set aside as above.

11. Place all of the even numbered bottles, consisting of one of diluting water and one from each set of the 5, 2.5 and 1 per cent concentrations, into the incubator and incubate at 20 C. for 5 days.

12. Determine the initial dissolved oxygen on the odd numbered bottles by the following procedure.

Procedure for D.O. Determination of B.O.D. Samples

1. Remove stoppers from bottles Nos. 1, 3, 5 and 7. Add 0.5 ml $MnSO_4$ solution to each bottle under the surface of the liquid by means of a pipette.

2. Add 0.5 ml of the alkaline-iodide-azide reagent in like manner. Insert stoppers· mix thoroughly by quickly inverting the bottles several times. Allow the floc to settle.

3. Remove stoppers and add 0.5 ml concentrated H_2SO_4. Insert stoppers and shake by inverting bottles several times as above to dissolve the precipitate and to liberate iodine uniformly throughout the solution.

4. Remove stoppers, and by means of the 100 ml graduate, transfer the 100 ml portions of iodine samples to the 250 ml Erlenmeyer flasks of corresponding numbers.

5. Fill a 50 ml burette with N/80 thiosulfate, draining by means of the stopcock so that the bottom of the meniscus rests on the zero mark. Titrate the liberated iodine of No. 1 flask over white paper or porcelain until a pale yellow color remains in the sample.

6. Add approximately 1 ml of the starch solution and continue to titrate to the disappearance of the blue color. Note the final burette reading. Disregard any return of blue color.

7. The number of milliliters used in the titration, representing the difference between the initial and final burette readings, equals the parts per million of initial dissolved oxygen (I.D.O.).

8. Titrate the liberated iodine of the remaining samples Nos. 3, 5 and 7, similarly as above, always recording the initial and final burette reading in order to arrive at the number of milliliters N/80 thiosulfate used in each titration. Record the results in the corresponding places on a work card in the column marked I.D.O.

9. At the end of the 5 day period, determine the final dissolved oxygen on the even numbered bottles, following the above procedure for D.O. Record results as parts per million on the work card in the column marked F.D.O.

Procedure for determining B.O.D.

1. Find the depletion of D.O. in the diluting water by subtracting the final D.O. from the initial D.O. Record in column marked depletion on the line opposite diluting water. This value obtained is known as the blank. Likewise, find and record the depletions, similarly, for the various concentrations.

2. Subtract the blank from the depletion of each concentration and record in the column marked depletion less blank.

3. Calculate the per cent oxygen depleted by dividing the depletion of the concentration by its initial dissolved oxygen and multiply by 100. Record these results in the column marked per cent oxygen depletion.

4. Find the concentration factor for each concentration used by dividing the particular per cent concentration into 100 per cent. Re-

cord these values in their respective places in the column marked conc. factor.

5. Calculate the B.O.D. value for each concentration by multiplying the value of depletion less blank by its conc. factor. Record results in their respective places in column marked B.O.D. conc.

Rules for selecting the B.O.D. of the sample from the values obtained from the various concentrations.

Rule I. Consider only the B.O.D. values from the concentrations whose depletions of dissolved oxygen are between 25 per cent and 75 per cent of their initial dissolved oxygen.

Rule II. Select any B.O.D. value where the oxygen depletion is between 40 per cent and 60 per cent. If there are more than one, take the average. It is believed that most reliable results are obtained when there is about a 50 per cent depletion.

Rule III. Select the average of B.O.D. values where one depletion is above 60 per cent and one below 40 per cent.

Rule IV. If two or more depletions are either above 60 per cent or below 40 per cent in which Rules II and III cannot be applied, select the one nearest to either the 60 per cent or the 40 per cent depletion.

Example

5-DAY B.O.D. CALCULATION
(On back of Laboratory Work Card)

Concentration of Sample	I.D.O. Bot. No.	F.D.O. Bot. No.	Depletion	Depletion less Blank	% Oxygen Depletion	Conc. Factor	B.O.D. of Conc. 5-day	B.O.D. of Sample 5-day
T.N. Dil Water	1 8.2	2 7.9	.3					
5.0%	3 7.6	4 0.0	7.6	7.3	100	20	146+	
2.5%	5 7.9	6 1.9	6.0	5.7	76	40	228	
1.0%	7 8.2	8 5.4	2.8	2.5	34	100	250	
Applying Rule III.								239

In making use of the rules in selecting the B.O.D. of the sample of sewage in the above example, it can be seen easily that:

Rule I does not apply because the oxygen depletion is 100 per cent;

Rule II cannot be applied because there are no depletions between 40 per cent and 60 per cent;

Rule III can be applied because there is one depletion above 60 per cent and one below 40 per cent;

Rule IV cannot be applied because there are neither two above 60 per cent nor two below 40 per cent depletion;

In applying Rule III, it was assumed that the 76 per cent depletion of the 2.5 per cent concentration was within the limit since it was very close to the upper limit.

NOTES

The advantages of this procedure are (1) it eliminates the dissolved oxygen determination on the sample being tested (2) it avoids the cumbersome mathematical computations or equations in the intensive calculation of results (3) it automatically eliminates the determination and calculation for the initial oxygen demand, since 30 minutes usually elapse from the time of preparation of the concentrations to the actual determination of the initial dissolved oxygen, and (4) a set of 4 simple rules assists the analyst to select uniformly the B.O.D. of the sample from the B.O.D. results of the various concentrations.

The Therbault-Nicols (T.N.) diluting water has been found to give satisfactory results on sewage, sewage effluents, and on samples which are deficient in nitrogen. Since this diluting water has been recommended for standard use, by the committee on Standard Methods of the Federation of Sewage Works Association, it has been adopted for this work.

The use of the small bottles and the proportional quantities of reagents is in accord with Standard Methods.

Sewages having 5 day B.O.D. values ranging from about 50 p.p.m. to above 650 p.p.m. may be handled by this procedure. For sewage effluents and sewage-polluted streams having low 5 day B.O.D. values modify this procedure to include as many of the higher concentrations, such as 50, 25 and 10 per cent, as may be necessary.

REFERENCE

Rogers, D. Paul, Laboratory Guide for Sewage Works Operators, Pennsylvania Sewage Works Association, 1946. 92 p. The outline of this procedure for B.O.D. determinations has been reprinted from the book published by the Pennsylvania Sewage Works Association with full permission. The reader is referred to this book for information on procedures of related subjects dealing with sewage problems.

PART VII

PREPARATION OF MEDIA AND REAGENTS

The preparation of culture media requires great care, especially as to reaction, or pH value. A slight variation in pH value causes marked differences in results. Every batch should be tested for pH value.

Unless considerable quantities are required, laboratories may find it advisable to secure the prepared media in dehydrated form from reliable sources.

Care must be taken to prevent evaporation from the culture flasks and tubes. Sterilizing temperatures must be carefully controlled. Too low temperatures may permit contaminated media, whereas too high temperatures may cause the breaking down of the ingredients and may produce a change in pH value. The length of time of maintaining the sterilizing temperature also is important. Each laboratory should use a constant, even temperature for sterilizing media.

CHAPTER 40

CULTURE MEDIA

Hydrogen Ion Determination

The hydrogen ion concentration of a culture medium is the true index of the actual acidity or alkalinity. Accordingly it gives information not obtainable by titration of a sample with acid or alkali. It is expressed for convenience in terms of pH units just as we use degrees Fahrenheit or Centigrade for expressing temperature.

The pH is arbitrarily defined as the logarithm of the reciprocal of the hydrogen ion concentration, and may be determined either potentiometrically or colorimetrically. The potentiometric method measures the pH in terms of millivolts and is the more dependable and accurate of the two methods. In recent years much progress has been made in simplifying the instruments used in making the electrical method more accurate, and in reducing the cost of the equipment, especially those units using the glass electrode. The various firms supplying potentiometers give detailed instructions for the operation of their specialized apparatus.

Apparatus

1. Comparator block.
2. Set of color standards.
3. Test tubes.

Reagents

1. Bromthymol blue indicator solution, supplied by the manufacturer of color standards.
2. Bromphenol blue indicator solution.
3. Sodium hydroxide, 1 N.
4. Sodium hydroxide, 0.1 N.
5. Tartaric acid solution, 10 percent aqueous solution.

Procedure

1. Add 5 ml of culture medium to 5 ml of distilled water in a test tube of appropriate size for the comparator block. If agar is being tested it should be melted and cooled to 45 - 50 C (113 - 122 F); likewise the water should be 50 C (122 F) and the tube should be kept at approximately this temperature during the determination.

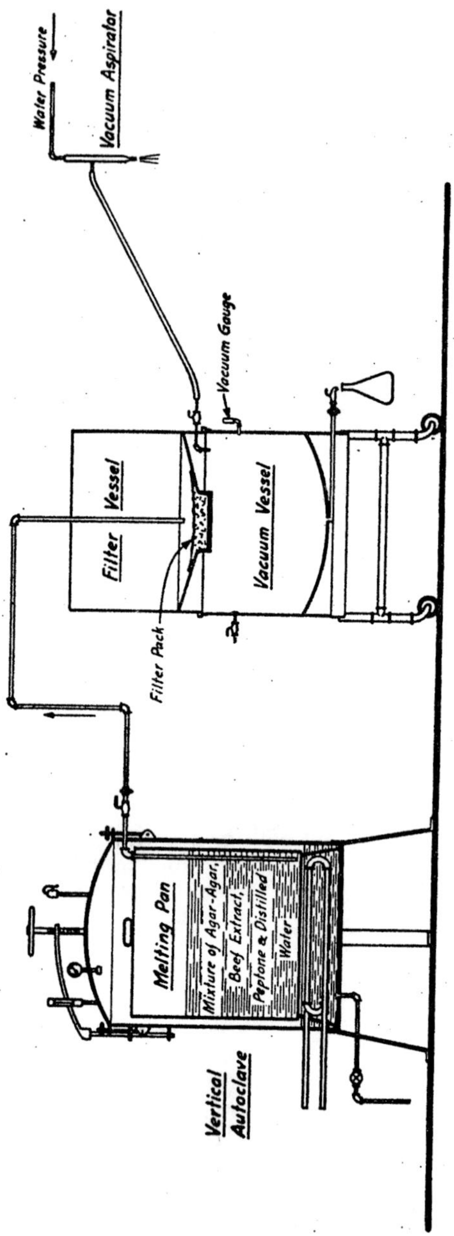

Figure 142.
Apparatus for preparation of nutrient agar on large scale.

2. Add 10 drops of the indicator, or an amount which will give the same concentration of indicator in the unknown as is present in the color standard, and then mix thoroughly.

3. The following diagram shows the arrangement of the test tubes in the comparator block.

SOURCE OF LIGHT
(Through Ground Glass Plate)

U	U & I	U
S	W	S

U—Unknown solution (diluted with an equal volume of water).
I—Ten drops of indicator solution.
W—Distilled water.
S—Color standard (buffer solution plus 10 drops of indicator solution.)

NOTE: Sealed color standards, such as are supplied by the LaMotte, Taylor, an Hellige companies, are satisfactory for this work, provided such color standards are compared every few months with standards which are known to be correct.

Standard Nutrient Agar

Borden Method

The ingredients are used in the proportions recommended in "Standard Methods" and about 60 liters are made at one time in the apparatus illustrated in figure 142. A 60-liter volume of distilled water is placed in the melting pan and warmed up for about 5 minutes in flowing stream, the beef extract, the tryptone, and the agar are then mixed with it in the order named. The original 60 liters finally becomes approximately 58.5 liters owing to evaporation during the process. The batch in next autoclaved at 15 pounds for 35 minutes. The pressure is then allowed to drop and the necessary standardization is accomplished at this point with normal NaOH. Before standardization the batch is generally about pH 6.0 and approximately 25 ml of normal NaOH is required to effect a 0.10 increase in pH. The batch is brought to pH 6.60 and the cover of the autoclave is again fastened down. The air which is thus entrapped expands because of the heat and the pressure of the expansion is sufficient to force all of the solution up through the side pipe and over the filter vessel. The filter pack which has been carefully prepared by whipping Whatman No. 4 paper in warm water is already in place, and when sufficient nutrient agar has entered the filter vessel, the vacuum aspirator is started, and vacuum filtration begins. It commonly takes from an hour to an hour and one-half to filter a batch of medium with this method. The medium may be drawn off and sterilized as best suited to local conditions.

Bowman Method

This method of preparation differs from others in that it makes use of a centrifugal clarifier to remove sediment from the medium.

The ingredients are brought into solution in a steam jacket aluminum kettle which is so situated that the medium will flow directly from it to the bottom of a Sharpless clarifier. The medium is caught in any suitable receptacle transferred to Erlenmeyer flasks and sterilized in an autoclave.

TRYPTONE-GLUCOSE-EXTRACT-MILK AGAR

In view of the widespread use of this new medium, the following directions are taken from *Standard Methods for the Examination of Dairy Products,* 9th edition, 1948, with the permission of the American Public Health Association.

Formula

1. Agar, best quality, not oven dried............1.5%
2. Beef extract............0.3%
3. Tryptone............0.5%
4. Glucose............0.1%
5. Skim milk............1.0%
6. Reaction............pH 6.6 – 7.0
   Preferred reaction............pH 7.0

One per cent of skim milk is to be added just before final sterilization in all cases in which dilutions greater than 1 : 10 are to be made.

Preparation

1. Either granulated or shredded agar may be used. After weighing, soak, if desired, the shredded agar in distilled water to remove the salts and dirt.
2. Drain.
3. Add the 3 gm of beef extract, the 5 gm of tryptone, the 1 gm of glucose, and the 15 gm of agar (undried market product as purchased) to 1,000 ml of distilled water.
4. Dissolve by boiling over a free flame with stirring (to prevent burning on the bottom of the container), or by heating in a flask (or other suitable container) with flowing steam in a steam chest or autoclave.
5. Make up the lost weight with distilled water.
6. When the dilutions that are to be made are greater than 1 : 10, add 10 ml of good quality skim milk just before final sterilization. (The milk may be kept in stock by storing in a sterile condition in test tubes, bottles, or flasks.)

An equivalent amount of spray process skim milk powder may be substituted for the skim milk. Dissolve 10 gm of milk powder in 100 ml of water. Use 10 ml of this reconstituted milk per liter of agar. Care should be taken when powder is used to avoid troublesome precipitates.

NOTE: Some workers have found it advantageous to omit the skim milk and add sterilized skim to the flasks of melted medium just prior to pouring the plates. This avoids troublesome precipitate which makes counting of colonies difficult.

Bacto-tryptone-glucose-extract agar, dehydrated, or other dehydrated agars of the composition specified above may be used. Add the skim milk, if required, during the preparation of the agar.

Adjustment of Reaction

State the reaction of the culture medium in terms of the hydrogen ion concentration as expressed in pH values. Determine the hydrogen ion concentration either electrometrically or colorimetrically. Colorimetric methods are usually satisfactory for this purpose.

Test the reaction of the medium by the method described below, and adjust the same if it falls outside the pH range of 6.6 - 7.0. The preferred reaction is 7.0.

To test the reaction:

1. Add 5 ml of distilled water at 50 C (122 F) to each of two clean test tubes, similar in size, shape, and color to the tubes used for color standards; 6 x ½ inch tubes are recommended.

2. Withdraw 10 ml of medium to be adjusted, and add 5 ml to each of these tubes.

3. To one of these tubes, add 5 drops of a solution of an indicator which will adequately cover the desired range, preferably bromthymol blue. The amount of indicator in the test solution should be the same as in the standard.

4. Using a comparator block, superimpose the tube containing the diluted medium plus the indicator over the tube of distilled water, and superimpose the tube of diluted medium without the indicator over the color standard of the desired pH.

5. Titrate the tube of diluted medium plus the indicator with an accurate 1 : 10 dilution (1 part NaOH solution to 9 parts of H_2O) of an approximately normal NaOH solution until the color viewed through the distilled water tube matches the color of the pH standard as observed through the diluted medium without the indicator.

6. Calculate the amount of normal NaOH solution which must be added to the medium to reach the desired pH.

7. Add the amount so calculated to the agar, mix thoroughly, and again test the reaction.

8. If the reaction is incorrect, make the proper addition of acid or alkali to correct the error. If the errors are excessive, it may be advisable to discard this particular batch of agar.

Sterilization and Storage

Before final sterilization, bring the medium to a boiling temperature, stir frequently, restore the lost weight with hot distilled water, and clarify if deemed advisable. The necessity for clarification depends upon the intended use, the inherent cloudiness, and the preference of the individual worker. In clarifying use any method which yields a medium suitable for the detection of all bacterial colonies and which at the same time will not remove or add nutritive ingredients. The usual methods are centrifugalization, sedimentation, or filtration of the melted agar through paper, cotton, cheesecloth, or towels. Do not clarify with egg albumin.

After clarification distribute the medium in suitable containers and sterilize in the autoclave. Flasks and bottles are convenient receptacles for the sterilized agar although some prefer to pour the melted agar from test tubes containing the measured 10 – 12 ml quantities. Limit the amount of agar in large containers so that no part of it will be more than 2.5 cm from the glass, except that containers more than 5 cm in diameter may be filled to a depth not exceeding 5 cm. If flasks or bottles containing unmeasured quantities are stored, prevent excessive evaporation and contamination during storage by securely fitting pliable metal foil, parchment paper, or viscose caps over the cotton plugs and necks of the containers. If crown cork and seal type caps are used, such hazards are reduced in some degree. To prevent recontamination, sterilize the coverings chemically when placing them over the caps before sterilization.

Remove a portion of each batch of sterilized agar aseptically immediately before use and determine its final reaction. Record this as the official reaction of the medium. The hydrogen ion concentration is likely to increase during sterilization in a pH decrease of 0.2 to 0.4. After solution of the ingredients, the minimum exposure to heat consistent with freedom from precipitates and with absolute sterility is essential. Check the sterility of each lot of agar, whenever a portion is used, by pouring at least one control plate at the end of the plating process. The sterility of the dilution water should be tested by pouring a control plate with each series of samples examined and for each lot of dilution blanks used with these samples.

For large scale preparation of media, two methods are briefly described on page 505.

NUTRIENT AGAR

(For water analysis)

Formula

1. Beef extract .... 3 gm.
2. Peptone .... 5 gm.
3. Agar of high quality .... 15 gm.
4. Distilled water to find volume of .... 1,000 ml.

Reaction (after sterilization), pH 6.6 (± 0.2)

Preparation

1. Weigh the beef extract, peptone, and agar, and dissolve in distilled water.

2. Heat in an autoclave at 15 lb steam pressure for a sufficient time to dissolve completely the agar (20 min.).

3. Restore the loss of water and adjust the pH if necessary to 6.6 (± 0.2).

4. Filter through cotton or clarify, with a suitable centrifugal machine.

5. Dispense in suitable portions in tubes, flasks or screw-capped bottles (bakelite caps, white rubber linings), and sterilize at 121 C (250 F) (15 lb steam pressure) for 20 minutes. The medium should be clear, free of particles, and should have a final reaction of pH 6.6 (± 0.2).

6. Instead of preparing nutrient agar in the manner described, dehydrated nutrient agar prepared according to the directions on the bottle may be used.

Media for Hemolytic Streptococci

Many of the hemolytic streptococci will grow on the following media.

VEAL (OR HEART) INFUSION BROTH

Formula

1.	Finely ground veal (or heart muscle)	500 gm
2.	Sodium chloride	5 gm
3.	Tryptone	5 gm
4.	Distilled water	1,000 ml

Preparation

1. Infuse the meat over night in the ice box in 1 liter of distilled water.
2. Boil until the infusion is clear and the coagulum is brown.
3. Strain through cheesecloth and filter through paper.
4. Add now the 5 gm each of sodium chloride and the tryptone.
5. Adjust the reaction to pH 7.2.
6. Filter the infusion.
7. Sterilize in autoclave at 121 C (250 F) for 20 min at 15 lb pressure.

VEAL (OR HEART) INFUSION AGAR

Formula

1. Agar, shredded .... 20 gm
2. Veal infusion broth (prepared as above) ....1,000 ml
3. Defibrinated horse (see page 511) or rabbit blood...... 54 gm

Preparation

1. Dissolve 20 gm agar in 1,000 ml of the veal infusion broth, by boiling and stirring.
2. Restore to original volume.
3. Adjust reaction to pH 7.2.
4. Place in flasks or tubes.
5. Sterilize in autoclave at 121 C (250 F) for 20 min at 15 lb pressure.
6. Under aseptic conditions, add 5 per cent (54 gm) of defibrinated horse or rabbit blood to the agar which has previously been been cooled to 45 C (115 F).

BLOOD AGAR

Formula

1. Defibrinated horse blood
2. Veal infusion agar

Preparation

1. Place 0.6 ml of sterile defibrinated horse blood into a sterile Petri dish.
2. Immediately pour about 10 ml of sterile veal infusion agar which has been melted and cooled to 45 C (113 F).
3. Rotate the plate to mix the blood and agar thoroughly.

After the agar has solidified, the plate may be used for a streak culture. If it is desired to make a poured plate culture, the inoculum should be added to the Petri dish before the agar is poured.

BLOOD AGAR SLANTS

Formula

1. Defibrinated horse blood
2. Veal infusion agar, sterile

Preparation

1. Place 0.3 ml of defibrinated horse blood in a sterile test tube.
2. Add 5 ml of sterile veal infusion agar which has been melted and cooled to 45 C (113 F).
3. Mix well, and place tube in slanting position in refrigerator.

TRYPTONE BLOOD AGAR

Formula

1.	Agar, best quality, not oven dried	20 gm
2.	Tryptone	5 gm
3.	Sodium chloride	5 gm
4.	Defibrinated horse or rabbit blood	54 gm
5.	Distilled water	1,000 ml

Preparation

1. Dissolve the agar, the tryptone, and the sodium chloride in 1,000 ml of distilled water by boiling and stirring.
2. Restore to original volume.
3. Adjust the reaction to pH of 7.2.
4. Place in flasks or tubes.
5. Sterilize in autoclave at 121 C (250 F) for 20 min at 15 lbs pressure.
6. Under aseptic conditions add 5 per cent (54 gm) of defibrinated horse or rabbit blood to the agar which has been cooled to 45 C (113 F) prior to pouring the plates.
7. The agar without the blood may be stored in small flasks for plating, or in test tubes in 10 ml amounts.
8. If Burri agar slants are to be used, the blood should be added to the tubes of melted agar (45 C) under aseptic conditions, the tubes slanted, and allowed to dry in a horizontal position until no water of condensation is present.

Collecting and Defibrinating Blood

Collecting horse blood. Prepare a bottle for the collection of the blood by placing some dry glass beads and pieces of glass rod about 2 or 3 inches long in the bottom of a dry 1-liter, flat-bottom flask. Insert a rubber tube about 3/8 inch in diameter and 16 inches long about 8 inches into the flask. Wind sufficient nonabsorbent cotton around that portion of the tube which is in the neck of the flask to hold it securely in place. Make a hole the size of the rubber tubing in a square piece of cheesecloth. Put the tube through this hole and the cloth then covers the cotton, which extends above the neck of the flask. At the other end of the tube attach a Leur slip hypodermic needle (13 gage, 3½ inches long). Cover the needle with a test tube which is held in place by cotton placed between the rubber tubing and the test tube. Put a screw pinchcock on the tubing. Sterilize the whole apparatus in the autoclave at 15 pounds pressure for 30 minutes.

Prepare the horse for bleeding by shaving an area about 5 cm in diameter over the jugular vein about midway between the head and shoulders. Swab the shaved area with alcohol or iodine solution. Place a twist on the horse's nose, which is to be held by an assistant. Place a rope about the horse's neck near the shoulders, and under this insert a smooth stick to be used for twisting the rope. This serves as a tourniquet and causes the bulging of the vein. When the jugular vein becomes quite prominent, plunge the needle quickly into it. Then remove the tourniquet and allow the blood to flow into the sterile bottle until it is about half full. Withdraw the needle and shake the bottle vigorously until the blood is completely defibrinated. The whole process must be carried out as aseptically as possible to avoid contamination of the blood.

At the laboratory, transfer the defibrinated blood to sterile test tubes by means of a sterile 25 or 50 ml pipette. Shake the bottle frequently during the transfer, so that the test tubes will have the normal amount of corpuscles (40 per cent) and serum (60 per cent). From four of the tubes, plate 0.6 ml of the blood, using veal infusion agar (10 ml). Incubate all the tubes of blood over night at 37 C (98.6 F). The next morning plate 0.6 ml of blood from the same four tubes, and if found sterile on incubation, the blood is satisfactory for use. The blood should be kept under refrigeration at all times.

NOTE: Where a relatively small number of blood agar plates are poured, a rabbit serves as a satisfactory source of blood. The rabbit may be bled directly from the heart by means of a sterile 20 ml syringe equipped with a long 22 gage needle.

CARBOHYDRATE MEDIA FOR FERMENTATION REACTIONS

Sugar-Free Base Medium

Formula

1. Veal, lean, ground .... 1¼ lb.
2. Culture of *Escherichia coli,* 10-hour culture .... 15 ml
3. Distilled water ....1,000 ml
4. Peptone (Difco) .... 10 gm
5. Horse serum, sterile
6. Various fermentable sugars in 10 per cent aqueous solution, sterilized at 10 lb pressure for 30 min.

Preparation

1. Mix the ground lean veal in 1,000 ml of distilled water.
2. Add 15 ml of a 10 hour culture of *Esch. coli.*
3. Incubate for 18 hours at 37 C (98.6 F), to destroy muscle sugars.
4. Steam mixture for 1 hr to destroy the *Esch. coli.*
5. Strain through cheesecloth under considerable pressure.
6. Make up to 1,000 ml with distilled water.
7. Add the 10 gm of Difco peptone.

8. Steam for 45 min to coagulate the albumin and dissolve the peptone.
9. Filter through paper.
10. Adjust the reaction of pH 7.8.
11. Steam again.
12. Filter through paper.
13. Tube in 5 ml amounts.
14. Sterilize at 15 lb pressure for 30 min.
15. Under aseptic conditions, add 2 drops of sterile horse serum to each tube.
16. To each tube add aseptically 0.05 ml of each sugar solution, which is to be used for fermentation studies.
17. Incubate the tubes over night to check sterility.

Serum Broth

Formula and Preparation

Prepared as described above for the base carbohydrate medium including the addition of the horse serum but omitting the addition of the fermentable sugar.

Sodium Hippurate Medium

Formula

1. Pork, lean and ground .... 1 lb
2. Peptone .... 10 gm
3. Sodium chloride .... 5 gm
4. Sodium hippurate .... 10 gm
5. Horse serum, sterile........2 drops per tube (about 20 ml)
6. Distilled water ....1,000 ml

Preparation

1. Mix the pork with 1,000 ml of distilled water, and refrigerate over night.
2. Press through cheesecloth.
3. Steam for 1 hour.
4. Cool to below 60 C (140 F).
5. Add the peptone and the sodium chloride, and dissolve.
6. Adjust the reaction to pH 7.9.
7. Boil for 10 minutes, and filter through paper.
8. A portion is sterilized in flasks and stored in refrigerator.
9. To the remainder add 1 per cent of sodium hippurate.
10. Tube in 5 ml quantities.
11. Sterilize at 15 lb pressure for 30 min.
12. Incubate over night and check for sterility.
13. Add aseptically 2 drops of sterile horse serum.
14. Incubate again over night, repeating the check for sterility.

CERTIFIED MILK MEDIUM

(Recommended by the American Association of Medical Milk Commissions)

Formula

1.	Finely ground lean pork, veal, or beef	500 gm
2.	Peptone	10 gm
3.	Sodium chloride	5 gm
4.	Granulated agar, U. S. P.	15 gm
5.	Distilled water	1,000 ml

Preparation

1. Mix the meat with the water, infuse for 1 hour, and then boil for 5 minutes.
2. Strain through wire or cotton gauze and filter through paper.
3. Restore the volume to 1,000 ml.
4. Add the peptone and the NaCl.
5. Adjust the reaction to pH 7.8.
6. Boil for 15 – 20 minutes.
7. Filter through paper and restore the volume.
8. Add the agar, and heat to boiling with constant stirring.
9. Add 5 ml of 1 N NaOH.
10. Autoclave for 15 minutes.
11. Restore the volume with boiling water and stir well.
12. Place a sterile lid over the vessel and allow the medium to stand at room temperature over night.
13. Next morning remove the solidified agar from the vessel and cut off and discard the sediment.
14. Melt the agar and distribute into tubes or flasks and sterilize.
15. The final reaction should be pH 7.2–7.5.

Media for the Determination of Coliform Types

BRILLIANT GREEN LACTOSE PEPTONE BILE (2 PER CENT)

This medium contains brilliant green in a concentration of 1 : 75,000 and 2 per cent of dehydrated bile.

Formula

1.	Peptone	10 gm
2.	Lactose	10 gm
3.	Ox bile, fresh	200 ml
	dehydrated	20 gm
4.	Brilliant Green (certified dye) solution of 0.1 per cent	13.3 ml
5.	Water, distilled	1,000 ml

Preparation

1. Dissolve the 10 gm of peptone and the 10 gm of lactose in not more than 500 ml of distilled water.

2. Add 200 ml of *fresh* ox bile or 20 gm of dehydrated ox bile dissolved in 200 ml of distilled water. (No dehydrated ox bile should be used which has a pH below 7.0).

3. Make up to at least 975 ml by adding distilled water.

4. Adjust reaction to a pH of 7.4.

5. Add 13.3 ml of a 0.1 per cent solution of brilliant green dye in water.

6. Make up to a total of 1,000 ml.

7. Filter through cotton.

8. Distribute the medium in tubes provided with inverted vials, in 10 ml quantities for tests of 1 ml or less of sample. If samples of 10 ml or more are to be tested, the medium should contain 1½ times the proportions of the various ingredients indicated above, and should be placed in tubes in 20 ml amounts.

9. Sterilize tubes of medium in autoclave at 15 lb pressure, 121 C (250 F) for 15 min.

The reaction after sterilization must not be less than pH 7.1 and should not be more than pH 7.4. This reaction should ordinarily be determined by potentiometric methods using the hydrogen or glass electrode, although it is reported that colorimetric methods may be used with satisfaction before the addition of brilliant green if the medium is diluted 1 : 4 with distilled water and phenol red is selected as the indicator.

The medium described above may be used in the dehydrated form in the proportion of 40 gm of the dehydrated material to 1,000 ml of distilled water when 1 ml or less of the milk or cream sample is tested; and in the proportion of 60 gm to 1,000 ml of distilled water and poured into tubes in 20 ml quantities, when 10 ml samples are tested.

DESOXYCHOLATE AGAR*

Formula

1.	Peptone Nutri	10 gm
2.	Agar (granular, U. S. P.)	16 gm
3.	Sodium hydroxide (N)	2 ml
4.	Sodium desoxycholate	1 gm
5.	Sodium chloride	5 gm
6.	Dipotassium phosphate	2 gm
7.	Lactose	10 gm
8.	Ferric ammonium citrate, ferric sodium citrate, or ferric potassium citrate	2 gm
9.	Neutral red, certified, 1 per cent aqueous solution	3.3 ml
10.	Distilled water	1,000 ml

*NOTE: This formula has been adopted by the American Association of Medical Milk Commissions. The medium, in dehydrated form, with slight variations, is available from Baltimore Biological Laboratory, Baltimore, Md., and Difco Laboratories, Detroit, Mich. The dehydrated media will be found more convenient and uniform as to results.

Preparation

1. Dissolve 10 gm peptone in 1,000 ml of distilled water by boiling a short time, and filter through paper. Neutralize if necessary.

2. Add 17 gm of agar and 2 ml of 1 N NaOH.

3. Heat in Arnold steamer until the agar is dissolved. Restore the original volume with distilled water and add the following in the order given: Sodium desoxycholate 1 gm; sodium chloride, 5 gm; dipotassium phosphate, 2 gm; lactose, 10 gm; and 2 gm of ferric ammonium citrate, or ferric sodium citrate, or ferric potassium citrate.

4. Adjust the reaction to pH 7.5, and add 3.3 ml neutral red.

5. Place 10 ml portions in tubes and sterilize for 15 min. in the Arnold steamer; a longer time is required if distributed in flasks. Autoclaving is not necessary because Gram-positive, spore-forming bacteria do not grow in the medium. The medium should be heated as little as possible and only sufficiently to kill vegetative cells. Store in the dark because neutral red is decolorized by light.

ENDO MEDIUM

Formula for the Stock Medium for Endo's Plate Method

1.	Meat extract	10 gm
2.	Agar, shredded	25 gm
3.	Sodium sulfite, 2.5 per cent solution	25 ml
4.	Lactose, 10 per cent solution	100 ml
5.	Fuchsin, alcoholic solution	5 ml
6.	Distilled water	1,000 ml

NOTE: Dissolve ingredients by heating in autoclave and adjust the reaction to pH 7.0 by adding 1 N NaOH and using phenolphthalein as the indicator. Filter the medium and place 100 ml portions in Erlenmeyer flasks or in 4 to 6 ounce bottles.

Preparation

1. To each 100 ml of melted agar just before pouring plates, add 10 ml of 2.5 per cent solution of sodium sulfite, 10 ml of 10 per cent solution of lactose in 15 ml of water, and 0.5 ml of alcoholic fuchsin solution.

2. Mix and pour 10 ml portions of the reddish medium into sterile Petri dishes.

3. The plates are ready to use as soon as the Endo's medium has solidified.

EOSIN METHYLENE BLUE AGAR

Formula

1.	Peptone	10 gm
2.	Dibasic Potassium phosphate (K_2HPO_4)	2 gm
3.	Lactose, C. P.	10 gm

4. Agar .... 15 gm
5. Eosin .... 0.4 gm
6. Methylene blue .... 0.1 gm
7. Distilled water .... 1,000 ml
pH after sterilization 7.0

Preparation

1. Mix the first four ingredients and dissolve in 1,000 ml water by boiling until the agar is dissolved.

2. Filter through paper pulp by means of a Buchner funnel and suction.

3. Add the last two ingredients and dissolve completely by boiling.

4. Dispense in flasks or tubes and sterilize at 15 lb for 30 min.

NOTE: This medium should be made up for immediate use only.

FORMATE RICINOLEATE LACTOSE PEPTONE BROTH

Formula

1. Peptone .... 5 gm
2. Lactose .... 5 gm
3. Sodium formate .... 5 gm
4. Sodium ricinoleate .... 1 gm
5. Distilled water .... 1,000 ml

Preparation

1. Add the 5 gm of peptone, 5 gm of lactose, 5 gm of sodium formate, and 1 gm of sodium ricinoleate to 1,000 ml of distilled water.

2. Heat slowly on water bath with constant stirring until ingredients are dissolved.

3. Add distilled water to make up the volume to 1,000 ml.

4. Adjust the reaction so that the pH after sterilization will be 7.3 – 7.5.

5. Distribute the medium in tubes provided with inverted vials in 10 ml quantities for tests of 1 ml or less of sample. If samples of 10 ml or more are to be tested, the medium must contain 1½ times the proportions of the various ingredients indicated above, and should be tubed in 20 ml amounts.

6. Sterilize the tubes of medium in autoclave at 10 – 15 lb for not longer than 15 min.

The medium described above may be used in the dehydrated form in the proportion of 16 gm to 1,000 ml of distilled water when 1 ml or less of the milk or cream samples is tested, and in the proportion of 24 gm to 1,000 ml of distilled water poured into tubes in 20 ml quantities, when 10 ml samples are tested.

GENTIAN VIOLET LACTOSE PEPTONE BILE MEDIUM

Formula

1.	Peptone	10 gm
2.	Lactose	10 gm
3.	Bacto ox gall	10 gm
4.	Gentian violet (Coleman & Bell)	0.04 gm
5.	Distilled water	1,000 ml

Preparation

1. Mix the ingredients thoroughly and dissolve in 1,000 ml of distilled water.
2. Adjust the reaction to pH 7.6.
3. Place in fermentation tubes for sterilization.
4. The final reaction should be pH 7.6.

VIOLET RED BILE AGAR*

This medium may be prepared from the various ingredients in published formulas, but in the interest of uniformity it is recommended that it be used in the dehydrated form in the proportion of 41 gm of dehydrated violet red bile agar to 1,000 ml of distilled water. It should be sterilized in the autoclave for 15 min. at a pressure of 15 lb.

*NOTE: May be procured from Difco Laboratories, Detroit, Mich.

Lactose Broth

Formula

1.	Beef extract	3 gm
2.	Peptone	5 gm
3.	Lactose	5 gm
4.	Distilled water to make	1,000 ml
	Reaction, pH 6.8 – 7.0.	

Preparation

1. Dissolve the ingredients by heating in flowing steam until the solution is complete.
2. Adjust the pH if necessary to 6.8 – 7.0.
3. Distribute 10 ml quantities into tubes, containing Durham fermentation vials.
4. Sterilize at 15 lb pressure for 20 min.

Potato Dextrose Agar

Formula

1.	Infusion from 200 gm potato	1,000 ml
2.	Dextrose, commercial	20 gm
3.	Agar shredded	15 gm.
4.	Tartaric Acid, U. S. P., 10 per cent solution, sterilized.	
	Reaction at time of pouring plates, pH 3.5	

Preparation

1. Place 200 gm of potato, previously peeled or scraped and sliced, in 1,000 ml of distilled water.
2. Boil for 1 hour.
3. Strain through double thickness of clean cheesecloth.
4. Restore to original volume.
5. Add 20 gm commercial dextrose.
6. Add 15 gm agar.
7. Dissolve in autoclave by heating at 15 lb pressure for 20 min.
8. Filter through cotton.
9. Dispense in flasks, 100 ml per flask.
10. Sterilize at 121 C (250 F) (15 lb pressure) for 20 min.
11. To the hot sterile agar (just prior to pouring) add sufficient sterile tartrate acid solution to give a reaction of pH 3.5 ± 0.1. The amount of tartaric acid solution to be added must be determined for each batch of medium, as it is unnecessary to test the pH on each bottle of adjusted agar. Test the reaction with the indicator bromphenol blue. Usually, the amount runs about 2.0 – 3.0 ml of 10 per cent U. S. P. tartaric acid per 100 ml of medium.

NOTE: Adjust the reaction for only the amount of agar needed for the samples which are to be plated the same day. Since remelting the acidified agar destroys its solidifying properties.

Do not heat the agar after the addition of the acid, as heating at pH 3.5 may destroy the jelling properties of the agar.

Tomato Juice Agar

(For counting viable *Lactobacillus acidophilus*)

Formula

1.	Tomato Juice	400 ml
2.	Peptonized milk	10 gm
3.	Peptone	10 gm
4.	Agar	15 gm
5.	Distilled water, to make	1,000 ml

Reaction, pH 6.0 to 6.5

Preparation

1. Filter the juice from canned tomatoes through a filter paper (No. 12 Whatman is satisfactory).
2. Neutralize the juice to pH 7.0 with sodium hydroxide.

NOTE: Kulp and White suggest adjusting the juice to pH values ranging from 6.0 - 6.2, since tomato juice is more stable to heat on the acid side and the final reaction will not be appreciably changed.

3. Mix all ingredients and heat sufficiently to dissolve the agar (from 30 to 40 minutes at boiling temperature).
4. Dispense in test tubes, 18 ml per tube (this relatively large amount is advisable for acidophilus work).
5. Sterilize at 15 lb for 20 min.

NOTE: Agar plates for quantitative estimation of the number of viable *L. acidophilus* should be incubated under an atmosphere containing approximately 10 per cent carbon dioxide. (See figures 28 and 29, pages 69 and 71.)

REFERENCE

Kulp, W. L., and White, Vinton: *Science 76*: 17-18 (1946).

Tribútyrin Agar

(Alternate medium for detection of lipolytic organisms)

Formula

1. Nutrient beef extract agar (See page 508) .... 1,000 ml
2. Sodium taurocholate .... 1 gm
3. Tributyrin (Eastman) .... 10 ml

Final reaction, pH 6.8 - 7.0.

Preparation

1. To 1,000 ml of melted nutrient agar, which has been adjusted to pH 6.8 to 7.0 and filtered, add 1 gm of sodium taurocholate and dissolve thoroughly in the agar.

2. Dispense in 100 ml quantities in screw-capped bottles or in flasks with cotton plugs.

3. To each 100 ml quantity of the hot melted agar add 1 ml of tributyrin and thoroughly emulsify by shaking.

NOTE: The tributyrin should be washed free of any butyric acid and glycerin which may be present before it is added to the nutrient agar. This may be done most easily by shaking the tributyrin with distilled water, allowing the tributyrin to settle to the bottom of the bottle or tube, and drawing off the supernatant aqueous layer. This process may be repeated two or three times, using 10 parts of water to 1 part of the tributyrin.

4. Sterilize by autoclaving at 121 C (250 F) (15 lb pressure) for 20 min.

NOTE: Before pouring the plates, disperse the tributyrin uniformly through the agar which has been melted and cooled to 43 C (109.4 F).

Trypsin Digest Agar (Modified)

(Alternate medium for plating acidophilus milk)

Although the trypsin digest agar has proved satisfactory for routine quality control of acidophilus milk, a superior medium, described below, may be prepared by fortifying the dehydrated trypsin digest agar with tomato juice.

Formula

1. Bacto trypsin digest agar (dehydrated) .... 59 gm
2. Filtered tomato juice, unseasoned .... 300 ml
3. Distilled water .... 700 ml

Final reaction, pH 6.1 ($\pm$ 0.2)

Preparation

1. Filter unseasoned canned tomato juice through glass wool in a large funnel until all the red color is removed. Do not use suction.

2. To 700 ml of distilled water add 59 gm of trypsin digest agar, dehydrated, and dissolve by placing in the autoclave at 15 lb pressure for 20 min. or by heating in a double boiler.

3. Restore loss of water if necessary.

4. Add 300 ml of the hot filtered tomato juice, to the 700 ml of trypsin digest agar.

5. Adjust the reaction of the mixture to pH 6.2 to 6.4 by adding about 10 ml (more or less as required) of 1 N sodium hydroxide per liter of the medium.

6. Sterilize in tubes or flasks at 121 C (250 F) (15 lb steam pressure) for 20 min.

7. The final pH should be 6.1 (±0.2).

Whey Agar

(For Making Bacteria Counts on Lactic Starters)

Formula

1. Whey .... 1,000 ml
2. Peptone .... 5 gm
3. Agar .... 15 gm

Adjust reaction after sterilization to pH 6.5

Preparation

1. Skim milk is heated to from 80 C – 90 C (176 F – 194 F) and enough hydrochloric acid (10 per cent solution) is added completely to coagulate the proteins. (Two or three preliminary tests should be made on 5 ml portions of the skim milk to determine the amount of acid required to bring about complete coagulation.)

2. The acid is thoroughly mixed with the skim milk and the coagulated protein is allowed to settle.

3. The whey is filtered first through absorbent cotton and then through paper pulp, by means of a Büchner funnel and suction.

4. The whey is neutralized with sodium hydroxide to pH 6.8 – 7.0.

5. Add 5 gm peptone and 15 gm agar.

6. Heat at 100 C (212 F) until the agar is dissolved, then heat in the autoclave at 15 lb for 15 min.

7. Filter through paper pulp, using a Büchner funnel and suction.

8. Test the reaction, and if necessary adjust to pH 6.8 – 7.0.

9. Dispense 100 ml quantities into flasks and sterilize at 15 lb pressure for 20 min.

Yeast Dextrose Agar

Formula

1.	Beef extract	3 gm
2.	Yeast extract	1 gm
3.	Peptone	5 gm
4.	Tryptone	2.5 gm
5.	Dextrose	1 gm
6.	Agar	15 gm
7.	Distilled water	1,000 ml

Final reaction pH, 7.0 (±0.1).

Preparation

1. Dissolve all the ingredients listed in 1,000 ml of distilled water by heating in an autoclave, or by boiling.
2. Restore loss of water if necssary.
3. Dispense in tubes or flasks.
4. Sterilize in the autoclave at 121 C (250 F) (15 lb pressure) for 20 min.

Bacto Nutritive Caseinate Agar

(Casein Agar)

Formula

Bacto peptonized milk	7 gm
Bacto isoelectric casein	3 gm
Special agar	12 gm
	22 gm

Preparation

1. Dissolve 22 gm of above in 1,000 ml of distilled water.
2. Heat and boil for one or two minutes.
3. Place in tubes or flasks.
4. Autoclave for 20 min. at 15 lbs. pressure.
5. Adjust to pH 6.5.

NOTE: Many laboratories prefer skim milk agar (see page below) for detection of proteolytic bacteria of milk. Above procedure is much simpler than the casein agar originally advocated by W. C. Frazier & P. Rupp. (Studies on the Proteolytic Bacteria of Milk), Journal of Bacteriology, Vol. XVI, pp. 57-78, July 1928.

Skim Milk Nutrient Agar

Formula

Tryptone-glucose extract agar	100 ml
Skim milk (in tubes, 5 ml per tube)	5 ml

Preparation

1. Prepare and sterilize tryptone glucose extract agar in the usual manner in 100 ml quantities in Erlenmeyer flasks or screw-capped bottles.
2. Select fresh skim milk of good quality, having a low acidity.
3. Sterilize the skim milk in test tubes, 5 ml per tube, at 121 C (15 lbs. pressure) for 20 min.
4. After the agar has been cooled to 45 C (113 F), and just before pouring the plates, add aseptically the 5 ml of sterile skim milk to 100 ml of tryptone-glucose extract agar and mix thoroughly.

Burri Medium

This medium is standard nutrient agar, page 506, except 20 gm (or 2 per cent) of agar are used instead of 1.5 per cent. Since this medium is usually used as agar slants in test tubes, the added agar gives a firmer slant.

Buffered Phosphate Stock Solution

(For Bacterial distilled water blanks or dilutions)

Dissolve 34 grams potassium dihydrogen phosphate (KH_2PO_4) in 500 ml of distilled water, add about 175 ml of normal NaOH and dilute to one liter with distilled water. Adjust to pH 7.2. Use at rate of 1 ml of this stock solution to 800 ml of boiled and cooled distilled water.

Litmus Milk

Formula

1. Skim milk .... 1000 ml
2. Concentrated litmus solution (10 per cent) .... 80 ml

NOTE: Since litmus powder varies in strength the above quantity of litmus solution to be added to skim milk may have to be varied to give the desired color.

Preparation

1. Prepare a concentrated solution of litmus by adding 10 gm of litmus powder to 100 ml of distilled water and heating 30 min. in a steamer, the flask being shaken at intervals.
2. Allow the mixture to stand over night.
3. Filter through Whatman No. 12 folded filter paper or other suitable paper.
4. Dispense in vials (screw-capped) and sterilize at 121 C (250 F) (15 lbs pressure) for 20 min. These tubes may be stored for future use.
5. Add the filtered concentrated litmus solution to good quality fresh skim milk at the rate of 80 ml per liter.
6. Mix thoroughly and dispense in tubes, 10 ml per tube.
7. Sterilize at 121 C (250 F) (15 lbs steam pressure) for 20 min.

CHAPTER 41

STAINS

Acid Stain for Breed Smears

(Mallman and Churchill)

Formula

1. Methylene blue (certified for bacteria use) .... 1 gm
2. 95 per cent ethyl alcohol .... 500 ml
3. Hydrochloric acid .... 0.5 to 5.0 ml

(The larger amount of acid employed, the clearer the background of the smear.)

Staining Procedure

1. Allow the slide to remain in the staining solution 3-5 min. Remove and dip in tap water only long enough to wash away the excess stain.

2. Decolorize in 95% alcohol very carefully to avoid decolorization of the background.

Reference

Mallman, W. L., and Churchill, Elbert. A Rapid Test for Determining Bacteria in Liquid Eggs. *U. S. Egg and Poultry Mag. 48*: 406 (1942).

Differential Color Stain

Formula

1. Ethyl alcohol, 95 per cent.. 54 ml
2. Tetrachlorethane, technical .... 40 ml
3. Acetic acid, glacial.. 5 ml
4. Neutral red, powder (certified*) .... 1 gm
5. Brilliant green powder (certified*)............................... 0.5 gm

Procedure

1. Dissolve dyes in tetrachlorethane in stoppered flask or bottle, held in water bath at 50 C (122 F).

NOTE: Unless powdered stains are thoroughly dissolved, reddish blurs may appear on the dried film.

2. Add alcohol, and in a few minutes glacial acetic acid.

3. Allow to stand over night and filter.

*Certified by the Commission on Standardization of Biological Stains.

Gram Stain
(Burke Modification)
Formula

Solution 1

Crystal violet (certified*)	1 gm
Distilled water	100 ml

Solution 2

Sodium bicarbonate c.p.	5 gm
Distilled water	100 ml

Solution 3

Iodine c.p., crystals	1 gm
Potassium iodide, c.p., crystals	2 gm
Distilled water	100 ml

Solution 4

Acetone, c.p.	50 ml
Ethyl ether, c.p.	50 ml

Solution 5

Safranin 0 (certified*)	2 gm
Distilled water	100 ml

Solution 6

Xylol, c.p.

Preparation

Prepare solutions 1, 2, 3, and 5 as indicated. In the case of solution 4, the two solvents are merely measured and mixed.

NOTE: Solutions 1, 3, and 5 should be kept in glass-stopper brown bottles. Clear glass bottles may be used for solutions 2 and 4.

Procedure for Using Gram Stain

1. Air-dry rapidly and fix the film or smear by passing the slide through a flame three times.

2. Flood the smear with solution 1 (crystal violet) and add to the dye on the slide 3 to 8 drops of solution 2 (sodium bicarbonate). Allow the mixture to stand 2 to 3 min.

3. Flush off the excess stain with solution 3 (iodine) and cover the smear with fresh iodine solution. Allow the iodine to remain on the slide 1 min. or longer.

4. Wash the slide in water.

5. Blot all free water from around the film or smear until the surface is nearly free of water, but do not allow the film to become dry.

*Certified by the Commission on Standardization of Biological Stains.

6. Decolorize by adding solution 4 (acetone-ether mixture) to the slide a few drops at a time until the decolorizing solution flows from the slide practically uncolored. This usually requires less than 10 seconds but may take longer.

7. Air dry.

8. Counter stain 5 to 15 seconds (may take longer) with solution 5 (Safranin).

9. Wash off excess stain by a short exposure to water.

10. Blot excess water from around the film and air dry.

11. Immerse the slide in xylol 2-5 min. remove, and allow to dry.

Gram-positive cells are stained purple whereas Gram-negative cells are stained red

Loeffler's Modified Methylene Blue Stain

(for Breed direct mircroscopic count)

Formula

1. Saturated alcoholic solution of methylene blue............ 30 ml
2. Potassium hydroxide 0.01 per cent solution............... 100 ml

Modified Newman-Lampert Stain

Formula

1. Methylene blue, powder (certified*)............................ 1 gm
2. Ethyl alcohol, 95 per cent.. 54 ml
3. Tetrachlorethane, technical (Eastman)........................ 40 ml
4. Acetic acid, glacial.. 6 ml

Preparation

1. Add the alcohol to the tetrachlorethane and by means of a water bath warm the mixture for 10 min. at a temperature not to exceed 70 C (158 F). (Keep away from the flame.)

2. Add the powdered methylene blue, and shake vigorously until the dye is completely dissolved.

3. Add the glacial acetic acid drop by drop.

4. Filter the entire volume through a hard filter paper.

5. Keep in stoppered glass bottles.

NOTE: Satisfactory results may be obtained with denatured alcohol.
*Certified by Biological Stains Commission.

CHAPTER 42

STANDARD SOLUTIONS

Preparation of Standard Solutions

The preparation of standard solutions requires the utmost care and accuracy, and should be undertaken only by experienced chemists having available the proper laboratory facilities. If only small quantities are required or if a trained personnel and suitable facilities are not available, it is advisable to purchase standard solutions from reliable sources. Standard solutions should be adjusted and recorded to the second decimal place.

A standard solution may be defined as any solution of known concentration.

A normal solution is a special case of a standard solution in which the concentration bears a definite relationship to the molecular weight of the substance. A normal solution may be defined as one in which there is dissolved one hydrogen equivalent in grams per liter of solution. The following will illustrate the principle.

1. HCl which has a hydrogen equivalent of one would contain one molecular weight (36.46 gm) per liter.

2. H_2SO_4 which has a hydrogen equivalent of two would contain one-half molecular weight (49.04 gm) per liter.

3. H_3PO_4 which had a hydrogen equivalent of three would contain one-third molecular weight (32.68 gm) per liter.

A 0.1 N solution contains one-tenth the amount of solute as a normal solution, a 0.5 N solution contains one-half the amount, etc.

In general the procedure is to make up a solution of the unknown which is slightly more concentrated than is required, then titrate the unknown against a known standard solution. From the values obtained in the titration the exact amount of water required to dilute the unknown is calculated. As a matter of practice, it is customary to calculate the exact amount of dilution, then add slightly less than the amount required and the titration is repeated, thereby approaching the final dilution in stages. It is necessary of course to keep an accurate record of the amounts of the solution removed and the amounts of water added in each titration.

To calculate the exact amount of water to be added apply the following general formula.

$$W = \frac{C \times (A - B)}{B}$$

in which:

W = the ml of water to be added to the unknown.
A = ml of known standard solution used in the titration.
B = ml of the solution to be standardized which were used in the titration.
C = total volume in ml of the unknown solution.

After making the required dilutions (in stages) it is then desirable to calculate the actual normality of the newly standardized solution.

To calculate the normality apply the following general formula.

$$\text{Normality of the unknown} = \frac{\text{Normality of known solution} \times \text{ml of the known used}}{\text{ml of the unknown used.}}$$

Sample Calculation:

Assume that 10 liters of an approximate 0.1 N NaOH solution have been prepared and a total of 100 ml have previously been removed from the container for titration, leaving a total of 9900 ml of the NaOH. Also assume that 24 ml of this 0.1 N NaOH were required to neutralize 25 ml of a known 0.1 N H_2SO_4. To calculate the amount of water to add to the NaOH:

$$W = \frac{9900\ (25 - 24)}{24} = \frac{9900}{24} = 412.5 \text{ ml of water to add}$$

As a matter of practice it would be advisable not to add the full amount (412.5 ml) but add for example, 400 ml.

After the addition of 400 ml of water to the 9900 ml of NaOH, there would be a new total of 10,300 ml. Assume that 100 ml of this solution is now removed for a second titration with the known H_2SO_4, and it is found that 24.9 ml of the NaOH are required to neutralize 25 ml of the known H_2SO_4. To calculate the exact normality of the NaOH:

$$\text{Normality of the NaOH} = \frac{0.1 \times 25}{24.9} = 0.1004$$

Standard solutions unless properly cared for and used within a reasonable time do not remain standard. It is therefore essential that all standard solutions be frequently tested for accuracy.

In general the following suggestions are offered in caring for standard solutions:

1. Keep all solutions tightly stoppered to prevent evaporation.

2. Avoid unnecessary exposure to light. Certain types of solution, such as silver nitrate and sodium thiosulfate, are easily decomposed by light and therefore must be kept in brown bottles or in the dark.

3. Preferably keep solutions at a reasonably uniform temperature (60-70 F, 15.5-21.1 C).

4. Check or restandardize all standard solutions at frequent intervals (at least once a month).

Hydrochloric Acid Solutions

Apparatus

1. Same equipment as listed under Standard Sodium Hydroxide.

Reagents

1. Standard sodium hydroxide (NaOH).
2. Phenolphthalein indicator.
3. Hydrochloric acid, sp. gr. 1.18 (HCl).

Procedure

1. Prepare one of the solutions of the approximate normality desired:

460 ml of HCl to 10 liters distilled water = 0.5 N (approx.).
230 ml of HCl to 10 liters distilled water = 0.25 N (approx.).
95 ml of HCl to 10 liters distilled water = 0.10 N (approx.).

A preliminary titration should give slightly higher normalities than indicated here.

2. Standardizing the solutions:

Use the previously standardized NaOH solution in standardizing the acid. Proceed as described under Standard Sodium Hydroxide (Procedure 4 to 10, pages 535 to 536, making the necessary dilutions, and using the formula:

$$\frac{C \times (A-B)}{B} = \text{ml of distilled } H_2O \text{ to be added to make standard solution.}$$

In Which:
A = ml of standard NaOH.
B = ml of HCl used in titration.
C = total number of ml of HCl solution.

Calculate the normality of the standard solution by the formula:

$$\text{Normality of HCl} = \frac{\text{Normality of NaOH used}}{\text{ml of HCl used}}$$

Iodine Solution—Tenth Normal

Apparatus

1. Same general equipment as listed under Standard Sodium Hydroxide, (page 535).

Reagents

1. Sublimed iodine.
2. Potassium iodide, (KI) c. p.
3. Starch indicator solution (page 539).
4. Sodium thiosulfate solution, 0.1 N $Na_2S_2O_3$ (page 537).

Procedure

1. Weigh into a small weighing bottle 13 gm of iodine.

NOTE: Iodine vapors will corrode the balance and should therefore never be weighed on watch glasses or in open containers.

2. Weigh 20 to 25 gm of c. p. K1 and dissolve in as little water as possible (25 to 30 ml).

3. Add the iodine to the KI solution and, after it has dissolved, transfer to a liter volumetric flask.

4. Make up to volume with distilled water.

NOTE: Do not add the KI solution to the water as precipitation of iodine will take place.

5. Titrate 25 ml of the iodine solution with 0.1 N $Na_2S_2O_3$.

6. When the yellow color of iodine has almost disappeared, add 1 ml of starch solution, and continue the titration until the blue color disappears. Slightly more than 25 ml of $Na_2S_2O_3$ solution should be required.

7. Proceed as outlined under Standard Sodium Hydroxide on page 536, procedure 5 and 8 to 10 inclusive, adding a 2.5 per cent solution of KI in distilled water instead of water in making the necessary adjustments to dilute to 0.1 N.

To calculate the amount of 2.5 per cent of KI solution to be added, apply the formula:

$$W = \frac{C \times (A - B)}{A}$$

in which:

W = ml of 2.5 per cent KI solution to add
A = ml of standard $Na_2S_2O_3$
B = ml of iodine in titration
C = total number of ml of iodine solution

To calculate the normality of the iodine solution apply the formula:

$$\text{Normality of iodine solution} = \frac{\text{Normality of } Na_2S_2O_3 \times \text{ml of } Na_2S_2O_3 \text{ used}}{\text{ml of iodine used}}$$

NOTE: Keep the iodine solution in an amber-colored glass-stopper bottle, the stopper of which has been lightly greased with petroleum jelly. Contact with rubber or rubber compounds must be avoided to prevent rapid deterioration of the solution.

Molybdate Solution (for Phosphorus Determination)

Apparatus

1. Pyrex beakers, 600 ml and 2,000 ml.
2. Graduated cylinders, 250 ml and 500 ml.
3. Reagent bottle, glass-stopper, 2 1.

Reagents

1. Molybdic acid, c. p., 85 per cent.
2. Ammonia water, (NH_4OH) concentrated (sp. gr. 0.9).

3. Nitric acid, (HNO_3), concentrated (sp. gr. 1.42).
4. Ammonium phosphate.
5. Ammonium nitrate washing solution.

Add 250 ml of distilled water and 355 ml of concentrated NHO_3 to a solution of 300 ml of concentrated NH_4OH in 250 ml. of distilled water.

Procedure

1. Place in the 600 ml beaker 100 gm of 85 per cent molybdic acid and 240 ml of distilled water. Stir well and add 140 ml of concentrated NH_4OH.
2. Filter and add 60 ml of concentrated HNO_3.
3. In the 2 liter beaker mix 400 ml of concentrated HNO_3 and 960 ml of distilled water.
4. When the solutions prepared in steps 2 and 3 have cooled, pour slowly the molybdate solution into the acid (never the reverse), stirring constantly. When mixed add 0.01 gm of ammonium phosphate dissolved in 10 ml of water and allow the solution to stand for at least 24 hours.

Potassium Acid Phthalate Solution—Tenth Normal

Apparatus

1. 500 ml glass-stopper volumetric flask, calibrated at 20 C (68 F).
2. Weighing bottle, 2 in. x 1 in., glass-stopper.
3. Funnel, 3 in.
4. Watch glass, 4 in.

Reagents

1. Potassium acid phthlate, c. p. crystals.
2. Distilled water, recently boiled and cooled.

Procedure

1. Spread approximately 12 gm of the pure potassium acid phthlate on a watch glass and dry at 100 C (212 F) for 30 min.
2. Cool in a desiccator for 30 min. and then weigh exactly 10.2095 gm into the small weighing bottle.
3. Carefully transfer the weighed potassium acid phthlate, using the funnel, to the 500 ml volumetric flask, which should contain approximately 200 ml of previously boiled and cooled [24 C (75 F)] distilled water, and throughly rinse the bottle and funnel with not more than 100 ml of the boiled distilled water.
4. Insert stopper and shake the flask until the potassium acid phthlate is completely in solution. Add the previously boiled and cooled distilled water until it just reaches the mark and allow time for the solution to adjust itself to room temperature [21 C (70 F)].

5. If necessary, adjust the volume to 500 ml and transfer the solution to a clean, dry, glass-stopper bottle. Grease the stopper with a little petroleum jelly, and label the bottle.

NOTE: When standardizing one solution against the other the temperature of the two solutions should be within ± 2 degrees F of each other.

Potassium Dichromate Solution—Tenth Normal

Apparatus

1. Same general equipment as listed under Standard Sodium Hydroxide, (page 535).

Reagent

1. Potassium dichromate crystals, ($K_2Cr_2O_7$) c. p.

Procedure

1. Weigh accurately 4.9035 gm of the $K_2Cr_2O_7$ crystals (previously dried for ½ hour at 100 C (212 F).
2. Dissolve in a few hundred ml of distilled water, transfer to a liter volumetric flask, and make up to the mark.

This solution will be exactly 0.1 N and if kept in a glass-stopper bottle at normal room temperature will keep indefinitely.

Potassium Permanganate Solution—Tenth Normal

Apparatus

1. Same equipment as listed under Standard Sodium Hydroxide (page 535).
2. Büchner funnel, 2 in.
3. Small filter flask.

Reagents

1. Potassium permanganate crystals, U. S. P.
2. Sodium oxalate, 0.1 N.
3. Sulfuric acid, concentrated (H_2SO_4), c. p.
4. Asbestos fiber.

Procedure

1. Stock solution: Dissolve 33 gm of $KMnO_4$ crystals in a liter of distilled water by heating to 70–80 C (158–176 F). Allow to stand in a stoppered bottle for several days and then filter through an asbestos mat on the small Büchner funnel. Keep in a glass-stopper bottle.
2. Place 110 ml of the stock solution in a liter volumetric flask, and make up to the mark with recently boiled and cooled distilled water.

3. Pipette 25 ml of 0.1 N sodium oxalate solution into a 300 ml Erlenmeyer flask, add 25 ml of distilled water and 15 ml of H_2SO_4.

4. Heat to boiling, and titrate with the permanganate solution to the first appearance of a faint but permanent pink color.

5. Make the necessary dilution as calculated under Standard Sodium Hydroxide, procedure (5), adding slightly less water (15 ml) than the calculated amount, using the formula:

$$\frac{C \times (A - B)}{B} = \text{ml of distilled water to be added to make standard solution}$$

in which:

A = ml of standard sodium oxalate
B = ml of $KMnO_4$ used in titration
C = total number of ml of $KMnO_4$ solution.

6. Make final titrations and dilution and let this solution stand in an amber-color bottle for a week or more; otherwise calculate the factor, after making duplicate titrations, using a burette instead of the pipette for measuring the 0.1 N sodium oxalate. Use the formula:

$$\text{Normality of } KMnO_4 = \frac{\text{Normality of Oxalate} \times \text{ml of oxalate used}}{\text{Ml of } KMnO_4 \text{ used}}$$

Calculation

Twenty-five ml 0.1 N sodium oxalate divided by the number of milliliters of $KMnO_4$ solution required for the titration equals the normality factor. After the first few weeks of standing, the permanganate solution should be filtered and again standardized. This solution will remain nearly constant for a period of 30 days if kept in an amber-colored glass-stopper bottle. For longer periods of time it must be re-standardized and the factor determined.

Silver Nitrate Solution—Tenth Normal

Apparatus

1. Same equipment as listed under Standard Sodium Hydroxide, (page 535).
2. Casserole, 375 ml.

Reagents

1. Silver nitrate crystals ($AgNO_3$), c. p.
2. Sodium chloride (NaCl), 0.1 N.
3. Potassium chromate (K_2CrO_4) indicator (see page 539).

Procedure

1. Dissolve 17.50 gm of $AgNO_3$ crystals in 200 ml of distilled water. Place in a liter volumetric flask and make up to the mark.

2. Pipette 25 ml of the 0.1 N NaCl into the casserole, add 25 ml of distilled water and 0.50 ml of K_2CrO_4 indicator.

3. Titrate with the $AgNO_3$ solution to a light brick red color.

NOTE: Two drops of the silver nitrate solution should give a decided brick red color after the first faint reddish tinge is observed.

The $AgNO_3$ solution should be slightly stronger than the 0.1 N NaCl.

4. Proceed as described under Standard Sodium Hydroxide, (page 536, 5 and 8 to 10 inclusive, adding slightly less water (10 ml) than the calculated amount, using the formula:

$$\frac{C \times (A - B)}{B} = \text{ml of distilled water to be added to make standard solution}$$

in which:

A = ml of standard NaCl.
B = ml of $AgNO_3$ used in titration.
C = total number of ml of $AgNO_3$ solution.

5. Make final titration in duplicate, using a burette for measuring the 0.1 N NaCl, or a pipette which has been calibrated to deliver exactly 25 ml, and adjust the solution to exactly 0.1 N.

6. Calculate the normality by using the formula:

$$\text{Normality of } AgNO_3 = \frac{\text{Normality of NaCl} \times \text{ml of NaCl used}}{\text{ml of } AgNO_3 \text{ used}}$$

7. Keep in an amber-color glass-stopper bottle.

Silver Nitrate Solution

(For Determination of Salt in Butter)

The solution is always standardized to contain 29.06 gm of silver nitrate per liter. One ml of the $AgNO_3$ solution is therefore equivalent to 1.710 ml of 0.1 N NaCl.

NOTE: Solutions containing chlorides must be neutral and free from phosphates before titration with $AgNO_3$ solution, otherwise the end-point is indefinite and the error may be considerable. Paranitrophenol indicator is the most satisfactory for adjusting the solution to neutrality.

Sodium Chloride Solution—Tenth Normal

Apparatus

1. Volumetric flask, 500 ml.
2. Weighing bottle, 2 in. x 1 in. glass-stopper.
3. Watch glass or dish, 4 in.
4. Funnel, 3 in.
5. Beaker, 250 ml.

Reagents

1. Sodium chloride, c. p. crystals (NaCl).

Procedure

1. Weigh exactly 2.923 gm of the freshly dried NaCl.
2. Dissolve in 100–200 ml of distilled water.
3. Transfer to the 500 ml volumetric flask, washing beaker and funnel with the distilled water, and make up to volume at 20–25 C (68 - 77 F).

Sodium Hydroxide Solution

Apparatus

1. Analytical balance.
2. Counter balance.
3. 2 Burettes, 50 ml, Bureau of Standards calibrated.
4. Volumetric flasks, 1,000 ml and 500 ml.
5. Pipettes, 100, 50, 25, and 10 ml.
6. Erlenmeyer flask, 300 ml.
7. Graduated cylinder, 250 ml.
8. Beakers, 1,000 ml and 400 ml.
9. Weighing bottle, 1.25 x 2.00 in. with ground-in glass stopper.
10. Balanced watch glasses.
11. Spatula, 6 in. stainless steel blade.
12. Dropping bottle, with Barnes' pipette stopper.
13. Bottle, 3 or 5 gallon.

Reagents

1. Sodium hydroxide, reagent quality containing less than 5 per cent sodium carbonate.
2. Standard potassium acid phthalate solution (see Page 531).
3. Standard sulfuric acid, solution (see Page 538).
4. Phenolphthalein indicator solution.

Procedure

1. Prepare a concentrated stock solution by dissolving 400 gm of NaOH in 400 ml of distilled water. Temper by placing the beaker in warm water (40 C, 104 F), and transfer to a small-mouth bottle as soon as possible. Stopper the bottle with a rubber stopper and allow to stand for several days, or until perfectly clear.

2. Syphon the clear liquid into another clean bottle. The standard NaOH is to be prepared from this bottle of solution.

3. The quantities of concentrated alkali solution to use in preparing 10 liters of standard solution of the corresponding approximate normality are given as follows:

350 ml made to 10 liters = 0.50 N (approx.).
175 ml made to 10 liters = 0.25 N (approx.).
70 ml made to 10 liters = 0.10 N (approx.).

It is advantageous to rinse the graduate with a little of the distilled water.

4. Standardization of the solution:

Pipette 25 ml of the NaOH solution and titrate with a standard acid solution (either potassium acid phthalate or sulfuric acid), of the same normality using phenolphthalein as the indicator. Repeat the titrations until duplicates agree with ±0.05 ml. The NaOH solution should test stronger than the standard acid.

5. Calculate the amount of distilled water to be added to the standard NaOH solution, as described on Page 528, using formula:

$$\frac{C \times (A - B)}{B} = \text{ml of distilled water to be added to make standard solution}$$

in which:

A = ml of standard acid used in titration.
B = ml of NaOH used in titration.
C = total number of ml of NaOH solution.

6. Add about 100 ml less than the calculated amount of water and shake thoroughly for several minutes.

7. Draw into the Erlenmeyer flask from the acid burette 26 ml of standard acid and titrate the acid with the NaOH solution, using the alkali burette and phenolphthalein indicator. Since the titration has been reversed by titrating the standard acid with the NaOH solution, the amount of alkali solution used should be slightly less than 25 ml (the amount of standard acid used). Duplicate titrations are made and should agree exactly.

8. Record accurately the amount of solution used in the titrations, as well as the water added, and calculate the total amount of NaOH solution remaining.

9. To calculate the final amount of water to be added use the formula in 5.

10. Add the additional water as calculated in 5 and shake for several minutes to mix thoroughly.

11. As a final check on the accuracy of the standardization of the NaOH weigh in duplicate one of the following amounts of potassium acid phthalate, depending on the normality of the sodium hydroxide solution just prepared:

2.5518 gm for a 0.50 N NaOH.
1.2759 gm for a 0.25 N NaOH.
0.5104 gm for a 0.10 N NaOH.

Dissolve each of the two weighed quantities in 25 ml of recently boiled and cooled distilled water and titrate with the standard NaOH solution from the burette, using phenolphthalein indicator. This titration should require exactly 25 ml of the alkaline solution. If any further adjustments are required use this titration as the basis for determining the amount.

Calculate the normality of the standard solution by the formula:

$$N = \frac{\text{gm potassium acid phthalate}}{\text{ml NaOH} \times 204.136/1000}$$

Sodium Oxalate Solution—Tenth Normal

Apparatus

1. Same general equipment as listed under Standard Sodium Hydroxide (page 535).

Reagents

1. Sodium oxalate, special powder for standardizing.

Procedure

1. Weigh accurately 6.700 gm of the freshly dried special sodium oxalate.
2. Dissolve in 200 to 300 ml of hot distilled water.
3. Transfer to an accurately calibrated liter volumetric flask. This solution will be exactly 0.1 N.

Sodium Thiosulfate—Tenth Normal

Apparatus

1. Equipment listed under Standard Sodium Hydroxide, (page 535).
2. Glass-stopper bottle, 350 ml.
3. Cylinder of graduated pipette, 10 ml.
4. Filter paper or absorbent cotton.

Reagents

1. Sodium thiosulfate ($Na_2S_2O_3.5H_2O$), c. p.
2. Potassium dichromate ($K_2Cr_2O_7$) 0.1 N.
3. Potassium iodide, (KI) c. p.
4. Borax (20 Mule Team).
5. Hydrochloric acid (HCl) c. p., concentrated.
6. Starch indicator solution (See page 539).
7. Distilled water.

Procedure

1. Weigh 130 gm of the $Na_2S_2O_3.5H_2O$ and dissolve in 1 liter of distilled water.
2. Weigh 19 gm of Borax and dissolve in 500 ml of distilled water.
3. Mix the two solutions, make up to 5 liters, and let stand in the dark for 24 hours. Filter before standardizing.
4. Make solution of KI by dissolving 15 gm in distilled water and make up to 150 ml.
5. Place in a 350 ml glass-stopper bottle 50 ml of distilled water and 15 ml of KI solution.
6. Add from a burette 25 ml of 0.1 N $K_2Cr_2O_7$ and immediately follow with 4 ml of concentrated HCl. Shake and allow to stand in the dark for 10 min.
7. Rinse the stopper carefully and titrate the contents of the bottle with the thiosulfate solution. When the yellow color of iodine has almost disappeared, add about 1 ml of starch solution and continue the titration until the deep blue color of the solution changes to sea-green. One additional drop of this thiosulfate solution is sufficient

to effect this change. Twenty-five ml of the thiosulfate solution is required for the titration when it is exactly 0.1 N. The solution, however, should be stronger than 0.1 N, therefore the proper dilution is calculated at the top of page 528, the ml of thiosulfate used being subtracted from the 25 ml of 0.1 N, $K_2Cr_2O_7$, using the formula:

$$\frac{C \times (A - B)}{B} = \text{ml of distilled } H_2O \text{ to be added to make standard solution}$$

in which:

A = ml of standard $K_2Cr_2O_7$.
B = ml of thiosulfate used in titration.
C = total number of ml of thiosulfate solution.

Calculate the normality of the standard solution by the formula:

$$\text{Normality of } Na_2S_2O_3 . 5\,H_2O = \frac{\text{Normality of } K_2Cr_2O_7 \times \text{ml of } K_2Cr_2O_7 \text{ used}}{\text{ml } Na_2S_2O_3 \text{ used.}}$$

NOTE: Keep solution in tightly stoppered bottle in dark, cool place. If sediment appears, prepare a fresh solution.

Sulfuric Acid Solutions

Apparatus

1. Same equipment as listed under Standard Sodium Hydroxide.

Reagents

1. Standard sodium hydroxide prepared according to method given on page 533.
2. Phenolphthalein indicator solution.
3. Sulfuric acid sp. gr. 1.84.

Procedure

1. Prepare one of the solutions of the approximate normality desired:

145 ml of H_2SO_4 to 10 liters distilled water = 0.50 N (approx.).
72 ml of H_2SO_4 to 10 liters distilled water = 0.25 N (approx.).
29 ml of H_2SO_4 to 10 liters distilled water = 0.10 N (approx.).

A preliminary titration should give higher normalities than indicated here.

2. Standardization of the solution:

Use the previously standardized NaOH solution for standardizing the acid. Proceed as described under Standard Sodium Hydroxide (Procedure 4 - 10, pages 535 to 536) making the necessary dilutions, using the formula:

$$\frac{C \times (A - B)}{B} = \text{ml of distilled } H_2O \text{ to be added to make standard solution.}$$

in which

A = ml standard NaOH.
B = ml H_2SO_4 used in titration.
C = total number of ml of H_2SO_4 solution.

Calculate the normality of the standard solution by the formula:

$$\text{Normality of } H_2SO_4 = \frac{\text{Normality of NaOH} \times \text{ml of NaOH used}}{\text{ml of } H_2SO_4 \text{ used.}}$$

CHAPTER 43

INDICATORS AND REAGENTS

1. Indicators

Phenolphthalein: Dissolve 5 gm of phenolphthalein in 375 ml of 95 per cent alcohol and dilute with distilled water to 500 ml. Neutralize by adding 0.1 N alkali until a faint pink color appears. This is a 1 per cent solution.

Methyl Orange: Dissolve 1 gm of methyl orange in distilled water and dilute to 1 liter. Filter if necessary.

Methyl Red: Dissolve 1 gm of methyl red in 100 ml of 95 per cent alcohol.

Bromphenol Blue: Dissolve 0.20 gm of bromphenol blue in 200 ml of distilled water.

Paranitrophenol: Dissolve 1 gm paranitrophenol in 100 ml of distilled water and filter.

Alizarin Red (Alizarin monosodium sulfonate): Dissolve 0.5 gm of alizarin red in 100 ml of distilled water and filter.

Starch Solution: Grind with mortar and pestle 1 gm of arrow root starch with a little cold distilled water and then add, with constant stirring, 100 ml of boiling distilled water. Cool and decant the clear solution. Add 2 ml of chloroform as a preservative.

Potassium or Sodium Chromate: Dissolve 25 gm of pure dry potassium chromate (K_2CrO_4) or 21 gm of pure dry sodium chromate (Na_2CrO_4) in a small amount of distilled water; make up to 250 ml with distilled water and filter.

Glymol or Cream Test Reader.

A. Reagents:

1. Clear mineral oil, sp. gr. not to exceed 0.85 at 20 C (68 F)
2. Oil soluble, red dye
3. Oil soluble blue dye

} When ordering, specify oil soluble technical aniline dyes in powdered form.

B. Preparation:

1. A red dye and a blue dye are listed. Red has long been a popular color, but blue has been finding favor because a more distinct line is noted between blue and yellow than between red and yellow. For best results, use 6 gm of the red dye to 1 gallon of oil, or 2 gm of the blue dye to 1 gallon of oil.

2. Dissolve the required amount of dye in about 100 ml of the oil, warming the oil, and allowing to stand with frequent stirring. When dissolved, dilute to 1 gallon with oil of the main batch and stir well.

2. Reagents

Acetic Acid, Glacial: Contains 99.5 per cent CH_3COOH. The sp. gr. is 1.058.

Ammonium Hydroxide, Conc.: The solution contains 28 to 29 per cent of NH_3 by weight. Sp. gr. is 0.90.

Ammonium Hydroxide, Dilute: Mix 200 ml of concentrated NH_4OH with 350 ml of distilled water. The solution contains about 10 per cent of NH_3 by weight.

Ammonium Oxalate (4%): Dissolve 10 gm of $(NH_4)_2C_2O_4 \cdot H_2O$ in distilled water and make up to 250 ml.

Barium Chloride (10%): Dissolve 25 gm of $BaCl_2 \cdot 2H_2O$ in distilled water and dilute to 250 ml.

Chromic Acid Cleaning Solution: Concentrated technical sulfuric acid is saturated with technical potassium dichromate. The solution is kept in a covered glass container.

Cochineal Glycerin Solution: Boil 0.40 gm of cochineal in 90 ml of distilled water for 30 minutes. Filter hot and when cool make up to a volume of 90 ml. One ml of this solution is added to 10 ml of glycerin.

Prepared Asbestos: Washed and ignited asbestos fiber is soaked for several days in concentrated hydrochloric acid, then washed on a Büchner funnel first with clean water and finally with distilled water until free from HCl or chlorides. The asbestos mat on the funnel is removed after washing, cut with scissors into ½ inch lengths or pieces, and placed in a jar. The small pieces as required are placed in a 500 ml bottle and shaken with distilled water, so that the asbestos suspended in water may be easily poured. A Dumore type of mixer is recommended for mixing the asbestos and water.

Fehling's Alkaline Tartrate Solution: Dissolve 346 gm of Rochelle salts ($NaKC_4H_4O_6 \cdot 4H_2O$) and 100 gm of NaOH in distilled water, and dilute to 1 liter. Let stand for a few days and filter through prepared asbestos.

Fehling's Copper Solution: Dissolve 69.28 gm of $CuSO_4 \cdot 5H_2O$ in distilled water, dilute to 1 liter, and filter.

Ferric Chloride (10%): Dissolve 41.5 gm of $FeCl_3 \cdot 6H_2O$ in 210 ml of distilled water.

Hydrochloric Acid, Conc.: Contains 35.5 – 37.5 per cent of HCl by weight. sp. gr. is 1.18–1.19.

Nitric Acid, Conc.: Contains 69–70 per cent of HNO_3.

Nitric Acid, Red Fuming: Concentrated HNO_3 saturated with nitrogen peroxide.

Sulfuric Acid, Conc.: Contains 94–95 per cent H_2SO_4. The sp. gr. is about 1.84.

Sulfuric Acid, Dilute: Into 400 ml of water pour cautiously 100 ml of concentrated H_2SO_4 with constant stirring. This solution contains about 20 per cent H_2SO_4.

Washed Sand: Several pounds of clean, hard, dry sand are screened through a 40 mesh sieve. The fine sand is then removed by screening through a 60 mesh sieve. The sand retained on the 60 mesh is allowed to stand over night in a 5 - 10 per cent solution of hydrochloric acid, then washed free from acid, and dried. A good quality white sea sand screened through a piece of ordinary window screen is satisfactory, after treating with dilute HCl, throughly washing, and drying.

APPENDIX

CONTENTS

See Table of Contents, pages xix to xx for complete list of Tables.

CONVERSION TABLES

Compiled from Cardullo's Practical Thermodynamics, by Alan Leighton, Chemist, U. S. Dept. of Agriculture

LENGTH

	Feet	Inches	Meters	Centi-meters
Foot............	1	12	0.30480	30.48000
Inch............	0.08333	1	0.02540	2.54001
Meter...........	3.28083	39.37996	1	100
Centimeter.......	0.032808	0.3937	0.01	1
Millimeter........	0.003281	0.03937	0.001	0.1

AREA

Square	Feet	Inches	Meters	Centi-meters
Foot............	1	144	0.092903	929.03
Inch............	0.0006944	1	0.00064516	6.4516
Meter...........	10.7639	1550.0	1	10000
Centimeters......	0.00107639	0.15500	0.0001	1

MASS

	Pounds	Kilograms	Grams
Pound...............	1	0.453592	453.592
Kilogram............	2.04622	1	1000
Gram...............	0.00204622	0.001	1

VOLUME (FLUID MEASURES)

Units	Cu. In.	Cu. Ft.	Cu. Yd.	M.L. C.C.*	Cu. D.M. Liter	Cu. M.	Pt.	Qt.	Gal.	Liq. Oz.
Cubic Inches....	1	.0005787	.000031433	16.387	.016387	.000016387	.034632	.0173160	.004329	.554112
Cubic Foot.....	1728	1	.037037	28317	28.3170	.028317	59.88	29.92	7.481	958.08
Cubic Yards....	46656	27	1	764559	764.559	.76456	1615.76	807.88	201.97	25868
Milliliter* Cubic Centimeter	.061023	.000035314	.000001308	1	.001	.000001	.0021134	.0010567	.000264178	.033814
Cubic Decimeter* Liter...........	61.023	.035314	.0013079	1000	1	.001	2.11342	1.0567	.264178	33.814
Cubic Meter....	61023	35.3144	1.3079	1,000,000	1000	1	2113.4	1056.7	264.178	33814
Pint...........	28.875	.01670	.0006187	473.167	.473167	.000473167	1	.5	.125	16
Quart..........	57.75	.03340	.0012374	946.334	.946334	.000946334	2	1	.25	32
Gallon.........	231	.13360	.0049496	3785.33	3.78533	.00378533	8	4	1	128
Liquid Ounce...	1.8047	.0010437	.00003868	29.573	.029572	.000029572	.0625	.03125	.0078125	1

*The cubic decimeter and the liter, and the milliliter and the cubic centimeter are considered as being equal in this table. One cubic decimeter equals .001 Cu.M., the liter, .001 Cu.M.

1 Bbl. (fluid)=4.2109 cu. ft.; 31.5 gal.
1 dram=3.697 ml.; 60 minims, 0.125 oz.
1 gal.=0.83268 gal. (British).
1 l=270.5 dr.; 35.196 oz. (British).

1 minim=0.06161 ml.; 0.0167 dr., 0.0021 oz.
1 oz.=1.0408 oz. (British fluid).
1 pt.=128 dr.
1 gal. water (62 F, 16.7 C) weighs 8.3378 lb.

VOLUME AND CAPACITY (DRY MEASURES)

	Cu. In.	Pt.	L.	Qt.	Pk.	Cu. Ft.	Bu.	Bbl.
Cubic Inch	1	0.02976	0.016387	0.014881	0.00186	0.000579	0.00046	0.00014
Pint	33.6	1	0.5506	0.5	0.0625	0.0194	0.01563	0.00476
Liter	61.025	1.816	1	0.9081	0.1135	0.035	0.02838	0.0086
Quart	67.2	2	1.101	1	0.125	0.0389	0.03125	0.0095
Peck	537.6	16	8.81	8	1	0.3111	0.25	0.0762
Cubic Foot	1728	51.428	28.316	25.714	3.21428	1	0.80357	0.245
Bushel	2150.4	64	35.24	32	4	1.2444	1	0.3048
Barrel	7055.9	210	115.62	105	13.12	4.0833	3.281	1

1 Bu=0.96895 Bu (British)
1 l =0.027497 Bu (British)

PRESSURE

	Pounds per sq. in.	Pounds per sq. ft.	Inches of Hg	Atmospheres	Kilos per sq. cm.	Millimeters of Hg	Dynes per sq. cm.
1 pound per sq. in.	1	144	2.0360	0.068044	0.070307	51.713	68,947
1 pound per sq. ft.	0.0069441	1	0.014140	0.00047255	0.00048825	0.35920	578.80
1 inch of Hg.	0.49117	70.721	1	0.033420	0.034532	25.400	33,863
1 atmosphere	14.696	2116.3	29.921	1	1.0333	760	1,130,300
1 kilo per sq. cm.	14.223	2048.2	28.958	0.9678	1	735.54	980,665
1 mm. of Hg.	0.019337	2.7840	0.03937	0.0013158	0.0013595	1	1,333.3
1 dyne per sq. cm.	0.000014505	0.002106	0.000026532	0.0000009868	0.0000010198	0.00075000	1

ENERGY

	Foot-pounds	B.T.U.	Kilogram-meters	Greater calories*	Joules**	Kilowatt hours	Horse-power hours
Foot-pound	1	0.0012861	0.13826	0.0003241	1.3558	0.00000037662	0.00000050505
B.T.U.	777.5	1	107.50	0.25200	1054.2	0.0002928	0.0003927
Kilogram-meter	7.2330	0.009302	1	0.0023442	9.80665	0.0000027244	0.0000036532
Greater calorie	3086.0	3.9683	426.6	1	4183.4	0.0011621	0.0015583
Joule	0.73756	0.0009486	0.10197	0.0002390	1	0.00000027778	0.00000037251
Kilowatt hour	2,655.200	3415	367,080	860.5	3,600,000	1	1.3410
Horse-power hour	1,980,000	2547	273,740	641.7	2,684,500	0.7457	1

*1 greater calorie 1000 lesser calories. A greater calorie is the heat necessary to raise 1 kg. of water 1° C.
**1 Joule 10,000,000 Ergs.

AVOIRDUPOIS WEIGHTS

	Gr.	Gm.	Dr.	Oz.	Lb.
Grain	1	0.0648	0.03657	0.002286	0.000143
Gram	15.432	1	0.5644	0.03527	0.002205
Dram	27.34	1.772	1	0.0625	0.0039
Ounce	437.5	28.35	16	1	0.0625
Pound	7000	453.6	256	16	1

1 oz. (Troy or Apoth.) = 31.103 gm.; 480.0 grains
1 lb. (Troy) = 373.2 gm.

FORCE

	Pounds	Kilograms	Dynes
Pound	1	0.453592	444,800
Kilogram	2.04622	1	980,665
Dyne	0.0000020866	0.00000101975	1

METRIC WEIGHTS AND MEASURES

Length

10 mm	=	1 centimeter
10 cm	=	1 decimeter
10 decim	=	1 meter
10 m	=	1 decameter
10 dm	=	1 hectometer
10 hm	=	1 kilometer

Capacity

10 ml	=	1 centileter
10 cl	=	1 deciliter
10 dl	=	1 liter
10 l	=	1 decaliter
10 decl	=	1 hectoliter
10 hl	=	1 kiloliter

Area

1 centiare	=	1 sq. m.
1 are	=	100 sq. m.
1 hectare	=	10,000 sq. m.
(1 hectare	=	2.471 acres)

Weights

10 mg	=	1 centigram
10 c	=	1 decigram
10 dg	=	1 gram
10 gm	=	1 decagram
10 decg	=	1 hectogram
10 hg	=	1 kilogram

TROY WEIGHTS

(Gold, silver, & precious metals)

24 grains (gr.)	=	1 pennyweight
20 pwt	=	1 ounce
12 oz	=	1 pound

APOTHECARIES' WEIGHTS

(for compounding medicines)

20 grains (gr.)	=	1 scruple
3 scr	=	1 dram
8 dr	=	1 ounce
12 oz	=	1 pound

(grain, ounce and pound have same value as Troy)

AVOIRDUPOIS WEIGHT

27 11/32 grains	=	1 dram
16 dr	=	1 ounce
16 oz	=	1 pound
2000 lbs	=	1 ton

TABLE FOR COMPUTING POUNDS OF MILK FROM CASES AND CANS

No. of Cases or Cans	Lbs. Milk in 12 qt. Cases	Lbs. Milk in 8-gal. Cans	Lbs. Milk in 10-gal. Cans
1	25.8	68	86
2	52	136	172
3	77	204	258
4	103	272	344
5	129	340	430
6	155	408	516
7	181	476	602
8	206	544	688
9	232	612	774
10	258	680	860
20	516	1,360	1,720
30	774	2,040	2,580
40	1,032	2,720	3,440
50	1,290	3,400	4,300
60	1,548	4,080	5,160
70	1,806	4,760	6,020
80	2,064	5,440	6,880
90	2,322	6,120	7,740
100	2,580	6,800	8,600
200	5,160	13,600	17,200
300	7,740	20,400	25,800
400	10,320	27,200	34,400
500	12,900	34,000	43,000
600	15,480	40,800	51,600
700	18,060	47,600	60,200
800	20,640	54,400	68,800
900	23,220	61,200	77,400
1000	25,800	68,000	86,000
2000	51,600	136,000	172,000

B. A. E. EQUIVALENTS

The following conversion factors are used by the Bureau of Agricultural Economics, U. S. D. A., in some current releases. The quantity of milk actually used to produce one pound of each product depends chiefly on the butterfat test of the milk, and this varies in different sections of the country, and to some extent with the season.

One lb. Equals	lbs. Whole Milk	
Butter	21.0	
Whole Milk Cheese	10.0	
Condensed Milk	2.2	
Evaporated Milk	2.2	
Malted Milk	2.6	
Whole Milk Powder	8.0	
Cream Powder	19.0	
Ice Cream (1 gal.)	12.0	(15.0 including butter & conc. Milks)
Condensed Skim Milk	3.0	lbs. skim milk
Condensed Buttermilk	3.0	lbs. buttermilk
Skim Milk Cheese (Cottage, Pot and Bakers)	6.25	lbs. skim milk
Skim Milk Powder	11.0	lbs. skim milk
Whey Powder	16.0	lbs. whey
Buttermilk Powder	11.0	lbs. buttermilk
Casein	35.7	lbs. skim milk
Lactose	40	lbs. whey

COMPARISON OF THERMOMETER SCALES

Cent.	Reau.	Fahr.	Cent.	Reau.	Fahr.	Cent.	Reau.	Fahr.
−40	−32.0	−40.0	21	16.8	69.8	62	49.6	143.6
−38	−30.4	−36.4	22	17.6	71.6	63	50.4	145.4
−36	−28.8	−32.8	23	18.4	73.4	64	51.2	147.2
−34	−27.2	−29.2	24	19.2	75.2	65	52.0	149.0
−32	−25.6	−25.6	25	20.0	77.0	66	52.8	150.8
−30	−24.0	−22.0	26	20.8	78.8	67	53.6	152.6
−28	−22.4	−18.4	27	21.6	80.6	68	54.4	154.4
−26	−20.8	−14.8	28	22.4	82.4	69	55.2	156.2
−24	−19.2	−11.2	29	23.2	84.2	70	56.0	158.0
−22	−17.6	−7.6	30	24.0	86.0	71	56.8	159.8
−20	−16.0	−4.0	31	24.8	87.8	72	57.6	161.6
−18	−14.4	−0.4	32	25.6	89.6	73	58.4	163.4
−16	−12.8	+3.2	33	26.4	91.4	74	59.2	165.2
−14	−11.2	6.8	34	27.2	93.2	75	60.0	167.0
−12	−9.6	10.4	35	28.0	95.0	76	60.8	168.8
−10	−8.0	14.0	36	28.8	96.8	77	61.6	170.6
−8	−6.4	17.6	37	29.6	98.6	78	62.4	172.4
−6	−4.8	21.2	38	30.4	100.4	79	63.2	174.2
−4	−3.2	24.8	39	31.2	102.2	80	64.0	176.0
−2	−1.6	28.4	40	32.0	104.0	81	64.8	177.8
0	0.0	32.0	41	32.8	105.8	82	65.6	179.6
+1	+0.8	33.8	42	33.6	107.6	83	66.4	181.4
2	1.6	35.6	43	34.4	109.4	84	67.2	183.2
3	2.4	37.4	44	35.2	111.2	85	68.0	185.0
4	3.2	39.2	45	36.0	113.0	86	68.8	186.8
5	4.0	41.0	46	36.8	114.8	87	69.6	188.6
6	4.8	42.8	47	37.6	116.6	88	70.4	190.4
7	5.6	44.6	48	38.4	118.4	89	71.2	192.2
8	6.4	46.4	49	39.2	120.2	90	72.0	194.0
9	7.2	48.2	50	40.0	122.0	91	72.8	195.8
10	8.0	50.0	51	40.8	123.8	92	73.6	197.6
11	8.8	51.8	52	41.6	125.6	93	74.4	199.4
12	9.6	53.6	53	42.4	127.4	94	75.2	201.2
13	10.4	55.5	54	43.2	129.2	95	76.0	203.0
14	11.2	57.2	55	44.0	131.0	96	76.8	204.8
15	12.0	59.0	56	44.8	132.8	97	77.6	206.6
16	12.8	60.8	57	45.6	134.6	98	78.4	208.4
17	13.6	62.6	58	46.4	136.4	99	79.2	210.2
18	14.4	64.4	59	47.2	138.2	100	80.0	212.0
19	15.2	66.2	60	48.0	140.0			
20	16.0	68.0	61	48.8	141.8			

$$\text{Fahr.} = 32 + \frac{9}{5}\text{Cent.} = 32 + \frac{9}{4}\text{Reau.}$$

Freezing point on Fahrenheit scale is +32 degrees; boiling point, 212 degrees.

Freezing point on Centigrade scale is 0 degrees; boiling point, 100 degrees.

Freezing point on Reaumur scale is 0 degrees; boiling point, 80 degrees.

Of water at sea level at normal barometer pressure (29.9 inch).

The "absolute zero" of temperature denotes that condition of matter at which heat ceases to exist. At this point a body would be wholly deprived of heat and a gas would exert no pressure.

The absolute zero on the Fahrenheit scale is about 461 degrees below zero.

The absolute zero on the Centigrade scale is about 274 degrees below zero.

The absolute zero on the Reaumur scale is about 219 degrees below zero.

An English unit of heat (B.T.U.) is the quantity of heat required to raise one pound of water one degree Fahrenheit. A metric unit of heat or metric caloric (M.C.) is the quantity of heat required to raise one liter of water one degree Centigrade.

—York Mfg. Co., York, Pa.

BAUME CONVERSION TABLES

Specific Gravity at $\frac{60° \text{ F.}}{60° \text{ F.}}$ corresponding to degrees Baume for liquids heavier than water.

$$\text{Sp. Gr. } 60°\text{F.} = \frac{145}{145 - \text{Deg. Baume}} \qquad \text{Baume degrees} = 145 - \frac{145}{\text{Sp. Gr. } \frac{60°\text{F.}}{60°\text{ F.}}}$$

SPECIFIC GRAVITY 60° F

Degrees Baume	TENTHS OF DEGREES									
	0	1	2	3	4	5	6	7	8	9
0	1.0000	1.0007	1.0014	1.0021	1.0028	1.0035	1.0042	1.0049	1.0055	1.0062
1	1.0069	1.0076	1.0083	1.0090	1.0097	1.0105	1.0112	1.0119	1.0126	1.0133
2	1.0140	1.0147	1.0154	1.0161	1.0168	1.0175	1.0183	1.0190	1.0197	1.0204
3	1.0211	1.0218	1.0226	1.0233	1.0240	1.0247	1.0255	1.0262	1.0269	1.0276
4	1.0284	1.0291	1.0298	1.0306	1.0313	1.0320	1.0328	1.0335	1.0342	1.0350
5	1.0357	1.0365	1.0372	1.0379	1.0387	1.0394	1 0402	1.0409	1.0417	1.0424
6	1.0432	1.0439	1.0447	1.0454	1.0462	1.0469	1.0477	1.0484	1.0492	1.0500
7	1.0507	1.0515	1.0522	1.0530	1.0538	1.0545	1.0553	1.0561	1.0569	1.0576
8	1.0584	1.0592	1.0599	1.0607	1.0615	1.0623	1.0630	1.0638	1.0646	1.0654
9	1.0662	1.0670	1.0677	1.0685	1 0693	1.0701	1.0709	1.0717	1.0725	1.0733
10	1.0741	1.0749	1.0757	1.0765	1.0773	1.0781	1.0789	1.0797	1.0805	1.0813
11	1.0821	1.0829	1.0837	1.0845	1.0853	1.0861	1.0870	1.0878	1.0886	1.0894
12	1.0902	1.0910	1.0919	1.0927	1.0935	1.0943	1.0952	1.0960	1.0968	1.0977
13	1.0985	1.0993	1.1002	1.1010	1.1018	1.1027	1 1035	1 1043	1.1052	1.1060
14	1.1069	1.1077	1.1086	1.1094	1.1103	1.1111	1.1120	1.1128	1.1137	1.1145
15	1.1154	1.1162	1.1171	1.1180	1.1188	1.1197	1.1206	1.1214	1.1223	1.1232
16	1.1240	1.1249	1.1258	1 1267	1.1275	1.1284	1 1293	1.1302	1.1310	1.1319
17	1.1328	1.1337	1.1346	1.1355	1.1364	1 1373	1.1381	1.1390	1.1399	1.1408
18	1.1417	1.1426	1.1435	1.1444	1.1453	1.1462	1.1472	1.1481	1.1490	1.1499
19	1.1508	1.1517	1.1526	1.1535	1.1545	1.1554	1.1563	1.1572	1.1581	1.1591
20	1.1600	1.1609	1.1619	1.1628	1.1637	1.1647	1.1656	1.1665	1.1675	1.1684
21	1.1694	1.1703	1.1712	1.1722	1.1731	1.1741	1.1750	1.1760	1.1769	1.1779
22	1.1789	1.1798	1.1808	1.1817	1.1827	1.1837	1.1846	1.1856	1.1866	1.1876
23	1.1885	1.1895	1.1905	1.1915	1.1924	1.1934	1.1944	1.1954	1.1964	1.1974
24	1.1983	1.1993	1.2003	1.2013	1.2023	1.2033	1.2043	1.2053	1.2063	1.2073
25	1.2083	1.2093	1.2104	1.2114	1.2124	1.2134	1.2144	1.2154	1.2164	1.2175
26	1.2185	1.2195	1.2205	1.2216	1.2226	1.2236	1.2247	1.2257	1.2267	1.2278
27	1.2288	1.2299	1.2309	1.2319	1.2330	1.2340	1.2351	1.2361	1.2372	1.2383
28	1.2393	1.2404	1.2414	1.2425	1.2436	1.2446	1.2457	1.2468	1.2478	1.2489
29	1.2500	1.2511	1.2522	1.2532	1.2543	1.2554	1.2565	1.2576	1.2587	1.2598
30	1.2609	1.2619	1.2630	1.2641	1.2652	1.2663	1.2674	1.2685	1.2697	1.2798
31	1.2719	1.2730	1.2741	1.2752	1.2763	1.2775	1.2786	1.2797	1.2808	1.2820
32	1.2831	1.2842	1.2854	1.2866	1.2877	1.2888	1.2900	1.2912	1.2923	1.2934
33	1.2946	1.2957	1.2968	1.2979	1.2991	1.3004	1.3016	1.3028	1.3040	1 3052
34	1.3063	1.3075	1.3087	1.3098	1.3110	1.3122	1.3134	1 3146	1.3.58	1.3170
35	1.3182	1.3194	1.3206	1.3218	1.3230	1.3242	1.3254	1.3266	1.3278	1.3290
36	1.3302	1.3314	1.3326	1.3339	1.3352	1.3364	1.3376	1.3389	1.3401	1 3414

SPECIFIC GRAVITY 60° F.

Degrees Baume	TENTHS OF DEGREES									
	0	1	2	3	4	5	6	7	8	9
37	1.3426	1.3438	1.3451	1.3464	1.3476	1.3488	1.3500	1.3512	1.3525	1.3528
38	1.3551	1.3564	1.3577	1.3589	1.3602	1.3615	1.3627	1.3640	1.3653	1.3666
39	1.3679	1.3692	1.3705	1.3718	1.3731	1.3744	1.3757	1.3770	1.3783	1.3796
40	1.3809	1.3822	1.3836	1.3849	1.3862	1.3875	1.3888	1.3902	1.3915	1.3928
41	1.3942	1.3955	1.3969	1.3982	1.3996	1.4009	1.4023	1.4036	1.4050	1.4064
42	1.4078	1.4091	1.4105	1.4118	1.4132	1.4146	1.4160	1.4174	1.4188	1.4202
43	1.4216	1.4230	1.4244	1.4258	1.4272	1.4286	1.4300	1.4314	1.4328	1.4342
44	1.4356	1.4370	1.4385	1.4399	1.4413	1.4428	1.4442	1.4456	1.4471	1.4485
45	1.4500	1.4514	1.4529	1.4543	1.4558	1.4573	1.4588	1.4602	1.4617	1.4632
46	1.4646	1.4661	1.4676	1.4691	1.4706	1.4721	1.4736	1.4751	1.4766	1.4781
47	1.4796	1.4811	1.4826	1.4841	1.4856	1.4871	1.4887	1.4902	1.4917	1.4933
48	1.4948	1.4963	1.4979	1.4994	1.5010	1.5026	1.5041	1.5057	1.5073	1.5088
49	1.5104	1.5120	1.5136	1.5142	1.5167	1.5182	1.5199	1.5215	1.5231	1.5247
50	1.5263	1.5279	1.5295	1.5311	1.5327	1.5343	1.5360	1.5376	1.5392	1.5409
51	1.5425	1.5442	1.5458	1.5474	1.5491	1.5508	1.5525	1.5541	1.5558	1.5574
52	1.5591	1.5608	1.5625	1.5642	1.5659	1.5676	1.5693	1.5710	1.5727	1.5744
53	1.5761	1.5778	1.5795	1.5812	1.5829	1.5847	1.5864	1.5882	1.5899	1.5916
54	1.5934	1.5951	1.5969	1.5986	1.6004	1.6022	1.6040	1.6057	1.6075	1.6093
55	1.6111	1.6129	1.6147	1.6165	1.6183	1.6201	1.6219	1.6237	1.6255	1.6274
56	1.6292	1.6310	1.6328	1.6347	1.6365	1.6384	1.6402	1.6421	1.6439	1.6458
57	1.6477	1.6496	1.6515	1.6534	1.6552	1.6571	1.6590	1.6609	1.6628	1 6647
58	1.6666	1.6686	1.6705	1.6724	1.6743	1.6763	1.6782	1.6802	1.6821	1.6840
59	1.6860	1.6879	1.6899	1.6919	1.6939	1.6959	1.6979	1.6999	1.7019	1.7039
60	1.7059	1.7079	1.7099	1.7119	1.7139	1.7159	1.7180	1.7200	1.7221	1.7241
61	1.7262	1.7282	1.7303	1.7324	1.7344	1.7365	1.7386	1.7407	1.7428	1.7449
62	1.7470	1.7491	1.7512	1.7533	1.7554	1.7576	1.7597	1.7619	1.7630	1.7661
63	1.7683	1.7704	1.7726	1.7748	1.7769	1.7791	1.7813	1.7835	1.7857	1.7879
64	1.7901	1.7923	1.7945	1.7967	1.7990	1.8012	1.8034	1.8057	1.8080	1.8102
65	1.8125	1.8148	1.8170	1.8193	1.8216	1.8239	1.8262	1.8285	1.8308	1.8331
66	1.8354	1.8377	1.8401	1.8424	1.8447	1.8471	1.8494	1.8518	1.8542	1.8566
67	1.8590	1.8614	1.8638	1.8662	1.8686	1.8710	1.8734	1.8758	1.8782	1.8806
68	1.8831	1.8855	1.8880	1.8905	1.8929	1.8954	1.8979	1.9904	1.9029	1.9054
69	1.9079	1 9104	1.9129	1.9154	1.9180	1.9205	1.9231	1.9256	1.9281	1.9307
70	1.9333	1.9359	1.9385	1.9411	1.9437	1.9463	1.9489	1.9515	1.9542	1.9568

DEFINITION OF CHEMICAL TERMS

An Acid is any substance which yields hydrogen ions.

The Active Mass of a substance is the number of gram-molecular-weights per liter in solution or in gaseous form.

Adsorption—The ability of a solid to condense gases, liquids, or dissolved substances on their surfaces is called adsorption. It is a manifestation of the force of adhesion.

An Atom is the smallest unit quantity of an element that is capable of entering into chemical combination.

A Base is any substance which yields hydroxyl ions.

A Balanced or Reversible Action is one which can be caused to proceed in either direction by suitable variation in the conditions of temperature, volume, pressure or of the quantities of reacting substances.

A Catalytic Agent is a substance which by its mere presence alters the velocity of a reaction, and may be recovered unaltered in nature or amount at the end of the reaction.

A Colligative Property is a property numerically the same for a group of substances, independent of their chemical nature.

A Constitutive Property is a property which depends on the constitution or structure of the molecule.

A Cryohydrate is the solid which separates when a saturated solution freezes. It contains the solvent and the solute in the same proportions as they were in the saturated solution.

The Combining Weight of an element or radicle is its atomic weight divided by its valence.

A Eutectic is that alloy of two or more metals which has the lowest melting point.

The Hydrogen Equivalent of a substance is the number of replaceable hydrogen atoms in one molecule or the number of atoms of hydrogen with which one molecule could react.

The Heat of Combustion of a substance is the amount of heat evolved by the combustion of one gram of the substance.

An Ion is a charged atom or group of atoms in solution. Solutions always contain equivalent numbers of positive and negative ions.

A Molecule is the smallest unit quantity of matter which can exist by itself and retain all the properties of the original substance.

A Molar Solution contains one gram molecular weight of dissolved substance per liter of solution.

A Normal Solution contains one gram molecular weight of dissolved substance divided by the hydrogen equivalent of the substance per liter of solution.

Oxidation is any process which increases the proportion of oxygen or acid-forming element or radicle in a compound.

Reduction is any process which increases the proportion of hydrogen or base-forming elements or radicle in a compound.

A Salt is any substance which yields ions, other than hydrogen or hydroxyl ions.

The Solubility Product or precipitation value is the product of the concentrations of the ions of a substance in a saturated solution of the substance.

—**Handbook of Chemistry and Physics.**

INTERNATIONAL ATOMIC WEIGHTS 1941

Name	Symbol	Atomic Weight	Valence
Aluminum	Al	26.97	3
Antimony, stibium	Sb	121.77	3, 5
Argon	A	39.944	0
Arsenic	As	74.91	3, 5
Barium	Ba	137.36	2
Beryllium, glucinum	Be	9.02	2
Bismuth	Bi	209.00	3, 5
Boron	B	10.82	3
Bromine	Br	79.916	1
Cadmium	Cd	112.41	2
Caesium	Cs	132.91	1
Calcium	Ca	40.08	2
Carbon	C	12.01	2, 4
Cerium	Ce	140.13	4, 3
Chlorine	Cl	35.457	1
Chromium	Cr	52.01	2, 3, 6
Cobalt	Co	58.94	2, 3
Columbium, niobium	Cb	92.91	3, 5
Copper	Cu	63.57	1, 2
Dysprosium	Dy	162.46	3
Erbium	Er	167.2	3
Europium	Eu	152.0	3
Fluorine	F	19.0	1
Gadolinium	Gd	156.9	3
Gallium	Ga	69.72	3
Germanium	Ge	72.60	4
Gold	Au	197.2	1, 3
Hafnium, celtium	Hf	178.6	4
Helium	He	4.003	0
Holmium	Ho	164.94	3
Hydrogen	H	1.008	1
Indium	In	114.76	3
Iodine	I	126.92	1
Iridium	Ir	193.1	3, 4
Iron, ferrum	Fe	55.85	2, 3
Krypton	Kr	83.7	0
Lanthanum	La	138.92	3
Lead, plumbum	Pb	207.21	2, 4
Lithium	Li	6.94	1
Lutecium	Lu	174.99	3
Magnesium	Mg	24.32	2
Manganese	Mn	54.93	2, 4, 6, 7
Mercury, hydrargyrum	Hg	200.61	1, 2

INTERNATIONAL ATOMIC WEIGHTS 1941 (Cont'd)

Name	Symbol	Atomic Weight	Valence
Molybdenum	Mo	95.95	3, 4, 6
Neodymium	Nd	144.27	3
Neon	Ne	20.183	0
Nickel	Ni	58.69	2, 3
Nitrogen	N	14.008	3, 5
Osmium	Os	190.2	2, 3, 4, 8
Oxygen	O	16.0	2
Palladium	Pd	106.7	2, 4
Phosphorus	P	30.98	3, 5
Platinum	Pt	195.23	2, 4
Potassium, kalium	K	39.096	1
Praseodymium	Pr	140.92	3
Protactinium	Pa	231.0	5
Radium	Ra	226.05	2
Radon, niton	Rn	222.0	0
Rhenium	Re	186.31	4, 6, 7
Rhodium	Rh	102.91	3
Rubidium	Rb	85.48	1
Ruthenium	Ru	101.7	3, 4, 6, 8
Samarium	Sa	150.43	3
Scandium	Sc	45.10	3
Selenium	Se	78.96	2, 4, 6
Silicon	Si	28.06	4
Silver, argentum	Ag	107.88	1
Sodium, natrium	Na	22.997	1
Strontium	Sr	87.63	2
Sulfur	S	32.06	2, 4, 6
Tantalum	Ta	180.88	5
Tellurium	Te	127.61	2, 4, 6
Terbium	Tb	159.2	3
Thallium	Tl	204.39	1, 3
Thorium	Th	232.12	4
Thulium	Tm	169.4	3
Tin, stannum	Sn	118.70	2, 4
Titanium	Ti	47.90	3, 4
Tungsten, wolframium	W	183.92	6
Uranium	U	238.07	4, 6
Vanadium	V	50.95	3, 5
Xenon	Xe	131.3	0
Ytterbium	Yb	173.04	3
Yttrium	Y	88.92	3
Zinc	Zn	65.38	2
Zirconium	Zr	91.22	4

BOILING POINT OF SOME LIQUIDS AT THE PRESSURE OF THE ATMOSPHERE

Liquids	Deg. F.	Liquids	Deg. F.
Ether	95	Linseed oil	597
Alcohol (pure)	173	Naphthalene	428
Ammonia	−28	Paraffin	570
Benzine	195–212	Kerosene	302
Chloride of lime solution (saturated)	356	Phosphorus	552
Chloroform	142	Mercury	675
Fats	570–615	Nitric acid	266
Glycerin	554	Sulphur	839
Chloride of sodium solution (saturated)	226	Sulphuric acid	615
Carbonic acid	−108	Carbon bisulphide	115
		Oil of turpentine	318
		Water	212

—International Ass'n of Ice Cream Mfg.

PEARSON SQUARE METHOD FOR STANDARDIZING MILK AND CREAM

In the center of the square, as illustrated, place the desired percentage of fat after standardization. The figure at the upper left-hand corner is the actual percentage of fat to be standardized. At the lower left-hand corner is the percentage of fat in the product to be used for standardization. For example, in this illustration 35 represents the fat test of the final cream; 40 represents the fat test of the cream before standardization, and 4 represents the fat test of milk that will be mixed with the 40% cream.

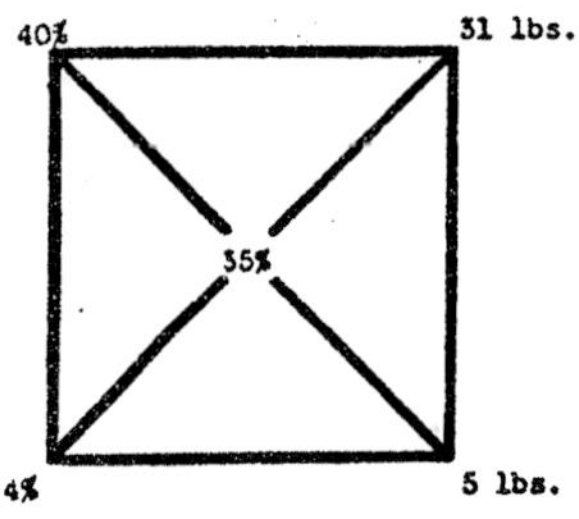

By subtracting cornerwise the ratio in which the two products on hand must be mixed to produce the desired fat content as obtained. The figure in the upper right-hand corner, which is arrived at by subtracting 4 from 35, represents the proportional amount of 40% cream to use. The figure in the lower right-hand corner, which is obtained by subtracting 35 from 40, represents the proportional amount of 4% milk to use. In this instance 5 lbs. of milk will be added to every 31 lbs. of cream.

TABLE FOR CORRECTING THE QUEVENNE LACTOMETER READING ACCORDING TO TEMPERATURE

Degrees of Lactometer	DEGREES OF THERMOMETER (FAHRENHEIT)														
	45	46	47	48	49	50	51	52	53	54	55	56	57	58	59	60
20	19.0	19.0	19.1	19.1	19.2	19.2	19.3	19.4	19.4	19.5	19.6	19.7	19.8	19.9	19.9	......
21	19.9	20.0	20.0	20.1	20.2	20.3	20.3	20.3	20.4	20.5	20.6	20.7	20.8	20.9	20.9	......
22	20.9	21.0	21.0	21.1	21.2	21.2	21.3	21.3	21.4	21.5	21.6	21.7	21.8	21.9	21.9	......
23	21.9	22.0	22.0	22.1	22.2	22.2	22.3	22.3	22.4	22.5	22.6	22.7	22.8	22.8	22.9	......
24	22.9	22.9	23.0	23.1	23.2	23.2	23.3	23.3	23.4	23.5	23.6	23.6	23.7	23.8	23.9	......
25	23.8	23.9	24.0	24.0	24.1	24.1	24.2	24.3	24.4	24.5	24.6	24.6	24.7	24.8	24.9	......
26	24.8	24.9	24.9	25.0	25.1	25.1	25.2	25.2	25.3	25.4	25.5	25.6	25.7	25.8	25.9	......
27	25.8	25.9	25.9	26.0	26.1	26.1	26.2	26.2	26.3	26.4	26.5	26.6	26.7	26.8	26.9	......
28	26.7	26.8	26.8	26.9	27.0	27.0	27.1	27.2	27.3	27.4	27.5	27.6	27.7	27.8	27.9	......
29	27.7	27.8	27.8	27.9	28.0	28.0	28.1	28.2	28.3	28.4	28.5	28.6	28.7	28.8	28.9	......
30	28.6	28.7	28.7	28.8	28.9	29.0	29.1	29.1	29.2	29.3	29.4	29.6	29.7	29.8	29.9	......
31	29.5	29.6	29.6	29.7	29.8	29.9	30.0	30.1	30.2	30.3	30.4	30.5	30.6	30.8	30.9	......
32	30.4	30.5	30.5	30.6	30.7	30.9	31.0	31.1	31.2	31.3	31.4	31.5	31.6	31.7	31.9	......
33	31.3	31.4	31.4	31.5	31.6	31.8	31.9	32.0	32.1	32.3	32.4	32.5	32.6	32.7	32.9	......
34	32.2	32.3	32.3	32.4	32.5	32.7	32.9	33.0	33.1	33.2	33.3	33.5	33.6	33.7	33.9	......
35	33.0	33.1	33.2	33.4	33.5	33.6	33.8	33.9	34.0	34.2	34.3	34.5	34.6	34.7	34.9	......

Degrees of Lactometer	DEGREES OF THERMOMETER														
	61	62	63	64	65	66	67	68	69	70	71	72	73	74	75
20	20.1	20.2	20.2	20.3	20.4	20.5	20.6	20.7	20.9	21.0	21.1	21.2	21.3	21.5	21.6
21	21.1	21.2	21.3	21.4	21.5	21.6	21.7	21.8	22.0	22.1	22.2	22.3	22.4	22.5	22.6
22	22.1	22.2	22.3	22.4	22.5	22.6	22.7	22.8	23.0	23.1	23.2	23.3	23.4	23.5	23.7
23	23.1	23.2	23.3	23.4	23.5	23.6	23.7	23.8	24.0	24.1	24.2	24.3	24.4	24.6	24.7
24	24.1	24.2	24.3	24.4	24.5	24.6	24.7	24.9	25.0	25.1	25.2	25.3	25.5	25.6	25.7
25	25.1	25.2	25.3	25.4	25.5	25.6	25.7	25.9	26.0	26.1	26.2	26.4	26.5	26.6	26.8
26	26.1	26.2	26.3	26.5	26.6	26.7	26.8	27.0	27.1	27.2	27.3	27.4	27.5	27.7	27.8
27	27.1	27.3	27.4	27.5	27.6	27.7	27.8	28.0	28.1	28.2	28.3	28.4	28.6	28.7	28.9
28	28.1	28.3	28.4	28.5	28.6	28.7	28.8	29.0	29.1	29.2	29.4	29.5	29.7	29.8	29.9
29	29.1	29.3	29.4	29.5	29.6	29.8	29.9	30.1	30.2	30.3	30.4	30.5	30.7	30.9	31.0
30	30.1	30.3	30.4	30.5	30.7	30.8	30.9	31.1	31.2	31.3	31.5	31.6	31.8	31.9	32.1
31	31.2	31.3	31.4	31.5	31.7	31.7	31.8	32.0	32.2	32.4	32.5	32.6	32.8	33.0	33.1
32	32.2	32.3	32.5	32.6	32.7	32.9	33.0	33.2	33.3	33.4	33.6	33.7	33.9	34.0	34.2
33	33.2	33.3	33.5	33.6	33.8	33.9	34.0	34.2	34.3	34.5	34.6	34.7	34.9	35.1	35.2
34	34.2	34.3	34.5	34.6	34.8	34.9	35.0	35.2	35.3	35.5	35.6	35.8	36.0	36.1	36.3
35	35.2	35.3	35.5	35.6	35.8	35.9	36.1	36.2	36.4	36.5	36.7	36.8	37.0	37.2	37.3

TABLE FOR DETERMINING TOTAL SOLIDS IN MILK FROM ANY GIVEN SPECIFIC GRAVITY AND PERCENTAGE OF FAT

(Per Cent Total Solids)

Percentage of fat	LACTOMETER READING AT 60° F. (Quevenne degrees)										
	26	27	28	29	30	31	32	33	34	35	36
2.00	8.90	9.15	9.40	9.65	9.90	10.15	10.40	10.66	10.91	11.16	11.41
2.05	8.96	9.21	9.46	9.71	9.96	10.21	10.46	10.72	10.97	11.22	11.47
2.10	9.02	9.27	9.52	9.77	10.02	10.27	10.52	10.78	11.03	11.28	11.53
2.15	9.08	9.33	9.58	9.83	10.08	10.33	10.58	10.84	11.09	11.34	11.59
2.20	9.14	9.39	9.64	9.89	10.14	10.39	10.64	10.90	11.15	11.40	11.65
2.25	9.20	9.45	9.70	9.95	10.20	10.45	10.70	10.96	11.21	11.46	11.71
2.30	9.26	9.51	9.76	10.01	10.26	10.51	10.76	11.02	11.27	11.52	11.77
2.35	9.32	9.57	9.82	10.07	10.32	10.57	10.82	11.08	11.33	11.58	11.83
2.40	9.38	9.63	9.88	10.13	10.38	10.63	10.88	11.14	11.39	11.64	11.89
2.45	9.44	9.69	9.94	10.19	10.44	10.69	10.94	11.20	11.45	11.70	11.95
2.50	9.50	9.75	10.00	10.25	10.50	10.75	11.00	11.26	11.51	11.76	12.01
2.55	9.56	9.81	10.06	10.31	10.56	10.81	11.06	11.32	11.57	11.82	12.07
2.60	9.62	9.87	10.12	10.37	10.62	10.87	11.12	11'38	11.63	11.88	12.13
2.65	9.68	9.93	10.18	10.43	10.68	10.93	11.18	11.44	11.69	11.94	12.19
2.70	9.74	9.99	10.24	10.49	10.74	10 99	11.24	11.50	11.75	12.00	12.25
2.75	9.80	10.05	10.30	10.55	10.80	11.05	11.31	11.56	11.81	12.06	12.31
2.80	9.86	10.11	10.36	10.61	10.86	11.11	11.37	11.62	11.87	12.12	12.37
2.85	9.92	10.17	10.42	10.67	10.92	11.17	11.43	11.68	11.93	12.18	12.43
2.90	9.98	10.23	10.48	10.73	10.98	11.23	11.49	11.74	11.99	12.24	12.49
2.95	10.04	10.29	10.54	10.79	11.04	11.30	11.55	11.80	12.05	12.30	12.55
3.00	10.10	10.35	10.60	10.85	11.10	11.36	11.61	11.86	12.11	12.36	12.61
3.05	10.16	10.41	10.66	10.91	11.17	11.42	11.67	11.92	12.17	12.42	12.68
3.10	10.22	10.47	10.72	10.97	11.23	11.48	11.73	11.98	12.23	12.48	12.74
3.15	10.28	10.53	10.78	11.03	11.29	11.54	11.79	12.04	12.29	12.55	12.80
3.20	10.34	10.59	10.84	11.09	11.35	11.60	11.85	12.10	12.35	12.61	12.86
3.25	10.40	10.65	10.90	11.16	11.41	11.66	11.91	12.16	12.42	12.67	12.92
3.30	10.46	10.71	10.96	11.22	11.47	11.72	11.97	12.22	12.48	12.73	12.98
3.35	10.52	10.77	11.03	11.28	11.53	11.78	12.03	12.28	12.54	12.79	13.04
3.40	10.58	10.83	11.09	11.34	11.59	11.84	12.09	12.34	12.60	12.85	13.10
3.45	10.64	10.89	11.15	11.40	11.65	11.90	12.15	12.40	12.66	12.91	13.16
3.50	10.70	10.95	11.21	11.46	11.71	11.96	12.21	12.46	12 72	12.97	13.22
3.55	10.76	11.02	11.27	11.52	11.77	12.02	12.27	12.52	12 78	13.03	13.28
3.60	10.82	11.08	11.33	11.58	11.83	12.08	12.33	12.58	12.84	13.09	13.34
3.65	10.88	11.14	11.39	11.64	11.89	12.14	12.39	12.64	12.90	13.15	13.40
3.70	10.94	11.20	11.45	11.70	11.95	12.20	12.45	12.70	12.96	13.21	13.46
3.75	11.00	11.26	11.51	11.76	12.01	12.26	12.51	12.76	13.02	13.27	13.52
3.80	11.06	11.32	11.57	11.82	12.07	12.32	12.57	12.82	13.08	13.33	13.58
3.85	11.12	11.38	11.63	11.88	12.13	12.38	12.63	12.88	13.14	13.39	13 64
3.90	11.18	11.44	11.69	11.94	12.19	12.44	12.69	12.94	13.20	13.45	13.70
3.95	11.24	11.50	11.75	12.00	12 25	12.50	12.75	13.00	13.26	13.51	13.77

(Per Cent Total Solids)

Per-centage of fat	LACTOMETER READING AT 60° F. (Quevenne degrees)										
	26	27	28	29	30	31	32	33	34	35	36
4.00	11.30	11.56	11.81	12.06	12.31	12.56	12.81	13.06	13.32	13.57	13.83
4.05	11.36	11.62	11.87	12.12	12.37	12.62	12.87	13.12	13.38	13.63	13.89
4.10	11.42	11.68	11.93	12.18	12.43	12.68	12.93	13.18	13.44	13.69	13.95
4.15	11.48	11.74	11.99	12.24	12.49	12.74	12.99	13.25	13.50	13.76	14.01
4.20	11.54	11.80	12.05	12.30	12.55	12.80	13.05	13.31	13.56	13.82	14.07
4.25	11.60	11.86	12.11	12.36	12.61	12.86	13.12	13.37	13.62	13.88	14.13
4.30	11.66	11.92	12.17	12.42	12.67	12.92	13.18	13.43	13.68	13.94	14.19
4.35	11.72	11.98	12.23	12.48	12.73	12.98	13.24	13.49	13.74	14.00	14.25
4.40	11.78	12.04	12.29	12.54	12.79	13.04	13.30	13.55	13.80	14.06	14.31
4.45	11.84	12.10	12.35	12.60	12.85	13.10	13.36	13.61	13.86	14.12	14.37
4.50	11.90	12.16	12.41	12.66	12.91	13.16	13.42	13.67	13.92	14.18	14.43
4.55	11.97	12.22	12.47	12.72	12.97	13.22	13.48	13.73	13.98	14.24	14.49
4.60	12.03	12.28	12.53	12.78	13.03	13.28	13.54	13.79	14.04	14.30	14.55
4.65	12.09	12.34	12.59	12.84	13.09	13.34	13.60	13.85	14.10	14.36	14.61
4.70	12.15	12.40	12.65	12.90	13.15	13.40	13.66	13.91	14.16	14.42	14.67
4.75	12.21	12.46	12.71	12.96	13.21	13.46	13.72	13.97	14.22	14.48	14.73
4.80	12.27	12.52	12.77	13.02	13.27	13.52	13.78	14.03	14.28	14.54	14.79
4.85	12.33	12.58	12.83	13.08	13.33	13.58	13.84	14.09	14.34	14.60	14.85
4.90	12.39	12.64	12.89	13.14	13.39	13.64	13.90	14.15	14.40	14.66	14.91
4.95	12.45	12.70	12.95	13.20	13.45	13.70	13.96	14.21	14.46	14.72	14.97
5.00	12.51	12.76	13.01	13.26	13.51	13.76	14.02	14.27	14.52	14.78	15.03
5.05	12.57	12.82	13.07	13.32	13.57	13.83	14.08	14.33	14.58	14.84	15.09
5.10	12.63	12.88	13.13	13.38	13.63	13.89	14.14	14.39	14.64	14.90	15.15
5.15	12.69	12.94	13.19	13.44	13.69	13.95	14.20	14.45	14.70	14.96	15.21
5.20	12.75	13.00	13.25	13.50	13.75	14.01	14.26	14.51	14.76	15.02	15.27
5.25	12.81	13.06	13.31	13.56	13.81	14.07	14.32	14.57	14.82	15.08	15.33
5.30	12.87	13.12	13.37	13.62	13.87	14.13	14.38	14.63	14.88	15.14	15.39
5.35	12.93	13.18	13.43	13.68	13.93	14.19	14.44	14.70	14.95	15.20	15.45
5.40	12.99	13.24	13.49	13.74	14.00	14.25	14.50	14.76	15.01	15.26	15.51
5.45	13.05	13.30	13.55	13.80	14.06	14.31	14.56	14.82	15.07	15.32	15.57
5.50	13.11	13.36	13.61	13.86	14.12	14.37	14.62	14.88	15.13	15.38	15.63
5.55	13.17	13.42	13.67	13.93	14.18	14.43	14.69	14.94	15.19	15.44	15.69
5.60	13.23	13.48	13.73	13.99	14.24	14.49	14.75	15.00	15.25	15.50	15.75
5.65	13.29	13.54	13.79	14.05	14.30	14.55	14.81	15.06	15.31	15.56	15.81
5.70	13.35	13.60	13.85	14.11	14.36	14.61	14.87	15.12	15.37	15.62	15.87
5.75	13.41	13.66	13.91	14.17	14.42	14.68	14.93	15.18	15.43	15.68	15.93
5.80	13.47	13.72	13.97	14.23	14.48	14.74	14.99	15.24	15.49	15.74	15.99
5.85	13.53	13.78	14.04	14.29	14.54	14.80	15.05	15.30	15.55	15.80	16.06
5.90	13.59	13.84	14.10	14.35	14.60	14.86	15.11	15.36	15.61	15.86	16.12
5.95	13.65	13.90	14.16	14.41	14.66	14.92	15.17	15.42	15.67	15.92	16.18

(Per Cent Total Solids)

Percentage of fat	LACTOMETER READING AT 60° F. (Quevenne degrees)										
	26	27	28	29	30	31	32	33	34	35	36
6.00	13.71	13.96	14.22	14.47	14.72	14.98	15.23	15.48	15.73	15.98	16.24
6.05	13.77	14.02	14.28	14.53	14.78	15.04	15.29	15.54	15.79	16.04	16.30
6.10	13.83	14.08	14.34	14.59	14.84	15.10	15.35	15.60	15.85	16.10	16.35
6.15	13.89	14.14	14.40	14.65	14.90	15.16	15.41	15.66	15.91	16.16	16.42
6.20	13.95	14.20	14.46	14.71	14.96	15.22	15.47	15.72	15.97	16.22	16.48
6.25	14.01	14.26	14.52	14.77	15.02	15.28	15.53	15.78	16.03	16.28	16.54
6.30	14.07	14.32	14.58	14.83	15.08	15.34	15.59	15.84	16.09	16.34	16.60
6.35	14.13	14.38	14.64	14.90	15.14	15.40	15.65	15.90	16.15	16.40	16.66
6.40	14.19	14.44	14.70	14.96	15.20	15.46	15.71	15.96	16.21	16.46	16.72
6.45	14.25	14.50	14.76	15.02	15.26	15.52	15.77	16.02	16.27	16.52	16.78
6.50	14.31	14.56	14.82	15.08	15.32	15.58	15.83	16.08	16.33	16.58	16.84
6.55	14.37	14.62	14.88	15.14	15.38	15.64	15.89	16.14	16.39	16.64	16.90
6.60	14.43	14.68	14.94	15.20	15.44	15.70	15.95	16.20	16.45	16.70	16.96
6.65	14.49	14.74	15.00	15.26	15.50	15.76	16.01	16.26	16.51	16.76	17.02
6.70	14.55	14.80	15.06	15.32	15.56	15.82	16.07	16.32	16.57	16.82	17.08
6.75	14.61	14.86	15.12	15.38	15.62	15.88	16.13	16.38	16.63	16.88	17.14
6.80	14.67	14.92	15.18	15.44	15.68	15.94	16.19	16.44	16.69	16.94	17.20
6.85	14.73	14.98	15.24	15.50	15.74	16.00	16.25	16.50	16.75	17.00	17.26
6.90	14.79	15.04	15.30	15.56	15.80	16.06	16.31	16.56	16.81	17.06	17.32
6.95	14.85	15.10	15.36	15.62	15.86	16.12	16.37	16.62	16.87	17.12	17.38

Proportional Parts

Lactometer Fraction	Fraction to Be Added to Total Solids	Lactometer Fraction	Fraction to Be Added to Total Solids	Lactometer Fraction	Fraction to Be Added to Total Solids
0.1	0.03	0.4	0.10	0.7	0.18
0.2	0.05	0.5	0.13	0.8	0.20
0.3	0.08	0.6	0.15	0.9	0.23

Directions for Using the Table—If the specific gravity as expressed in Quevenne degrees is a whole number, the percentage of total solids is found at the intersection of the vertical column headed by this number with the horizontal column corresponding to the percentage of fat.

If the specific gravity as expressed in Quevenne degrees is a whole number and a decimal, the percentage of total solids corresponding to the whole number is first found, and to this is added the fraction found opposite the tenth under "Proportional Parts." Two examples may suffice for illustration:

(1) Fat, 3.8 per cent; specific gravity, 32. Under column headed 32, 12.57 per cent is found corresponding to 3.8 per cent fat.

(2) Fat, 3.8 per cent; specific gravity, 32.5. The percentage of total solids corresponding to this percentage of fat and a specific gravity of 32 is 12.57. Under "Proportional Parts" the fraction 0.13 appears opposite 0.5. This added to 12.57 makes 12.70, which is the desired percentage.

An inspection of the table shows that the percentage of total solids increases practically at the rate of 0.25 for each lactometer degree and 1.2 for each unit of percentage of fat. This gives rise to Babcock's simple formula: Total solids = $\frac{1}{4}$ L + 1.2 F. (L = lactometer reading in Quevenne degrees and F = percentage of fat.)

To illustrate the use of the formula the following example is given: Fat, 4 per cent; specific gravity, 32. In this case one-quarter of 32 is 8; 1.2 multiplied by 4 is 4.8; 8 plus 4.8 equals 12.8, which represents the percentage of total solids.

PERCENTAGE OF TOTAL SOLIDS IN MILK

NOTE: Lactometer readings according to New York Board of Health Lactometer. All tests to be made at a temperature of 60 degrees Fahrenheit.

Per Cent of Fat	Lactometer Readings 95	96	97	98	99	100	101	102	103	104	105	106	107
2.8	10.25	10.32	10.39	10.46	10.54	10.61	10.68	10.76	10.83	10.90	10.97	11.05	11.12
2.9	10.37	10.44	10.51	10.58	10.66	10.73	10.80	10.88	10.95	11.02	11.09	11.17	11.24
3.0	10.49	10.56	10.63	10.70	10.78	10.85	10.92	11.00	11.07	11.14	11.21	11.29	11.36
3.1	10.61	10.68	10.75	10.82	10.90	10.97	11.04	11.12	11.19	11.26	11.33	11.41	11.48
3.2	10.73	10.80	10.87	10.94	11.02	11.09	11.16	11.24	11.31	11.38	11.45	11.53	11.60
3.3	10.85	10.92	10.99	11.06	11.14	11.21	11.28	11.36	11.43	11.50	11.57	11.65	11.72
3.4	10.97	11.04	11.11	11.18	11.26	11.33	11.40	11.48	11.55	11.62	11.69	11.77	11.84
3.5	11.09	11.16	11.23	11.30	11.38	11.45	11.52	11.60	11.67	11.74	11.81	11.89	11.96
3.6	11.21	11.28	11.35	11.42	11.50	11.57	11.64	11.72	11.79	11.86	11.93	12.01	12.08
3.7	11.33	11.40	11.47	11.54	11.62	11.69	11.76	11.84	11.91	11.98	12.05	12.13	12.20
3.8	11.45	11.52	11.59	11.66	11.74	11.81	11.88	11.96	12.03	12.10	12.17	12.25	12.32
3.9	11.57	11.64	11.71	11.78	11.86	11.93	12.00	12.08	12.15	12.22	12.29	12.37	12.44
4.0	11.69	11.76	11.83	11.90	11.98	12.05	12.12	12.20	12.27	12.34	12.41	12.49	12.56
4.1	11.81	11.88	11.95	12.02	12.10	12.17	12.24	12.32	12.39	12.46	12.53	12.61	12.68
4.2	11.93	12.00	12.07	12.14	12.22	12.29	12.36	12.44	12.51	12.58	12.65	12.73	12.80
4.3	12.05	12.12	12.19	12.26	12.34	12.41	12.48	12.55	12.63	12.70	12.77	12.85	12.92
4.4	12.17	12.24	12.31	12.38	12.46	12.53	12.60	12.68	12.75	12.82	12.89	12.97	13.04
4.5	12.29	12.36	12.43	12.50	12.58	12.65	12.72	12.80	12.87	12.94	13.01	13.09	13.16
4.6	12.41	12.48	12.55	12.62	12.70	12.77	12.84	12.92	12.99	13.06	13.13	13.21	13.28
4.7	12.53	12.60	12.67	12.74	12.82	12.89	12.96	13.04	13.11	13.18	13.25	13.33	13.40
4.8	12.65	12.72	12.79	12.86	12.94	13.01	13.08	13.16	13.23	13.30	13.37	13.45	13.52
4.9	12.77	12.84	12.91	12.98	13.06	13.13	13.20	13.28	13.35	13.42	13.49	13.57	13.64
5.0	12.89	12.96	13.03	13.10	13.18	13.25	13.32	13.40	13.47	13.54	13.61	13.69	13.76
5.1	13.01	13.08	13.15	13.22	13.30	13.37	13.44	13.52	13.59	13.66	13.73	13.81	13.88
5.2	13.13	13.20	13.27	13.34	13.42	13.49	13.56	13.64	13.71	13.78	13.85	13.93	14.00
5.3	13.25	13.32	13.39	13.46	13.54	13.61	13.68	13.76	13.83	13.90	13.97	14.05	14.12
5.4	13.37	13.44	13.51	13.58	13.66	13.73	13.80	13.88	13.95	14.02	14.09	14.17	14.24
5.5	13.49	13.56	13.63	13.70	13.78	13.85	13.92	14.00	14.07	14.14	14.21	14.29	14.36
5.6	13.61	13.68	13.75	13.82	13.90	13.97	14.04	14.12	14.19	14.26	14.33	14.41	14.48
5.7	13.73	13.80	13.87	13.94	14.02	14.09	14.16	14.24	14.31	14.38	14.45	14.53	14.60
5.8	13.85	13.92	13.99	14.06	14.14	14.21	14.28	14.36	14.43	14.50	14.57	14.65	14.72
5.9	13.97	14.04	14.11	14.18	14.26	14.33	14.40	14.48	14.55	14.62	14.69	14.77	14.84
6.0	14.09	14.16	14.28	14.30	14.38	14.45	14.52	14.60	14.67	14.74	14.81	14.89	14.96

PERCENTAGE OF TOTAL SOLIDS IN MILK
(Continued)

NOTE: Lactometer readings according to New York Board of Health Lactometer. All tests to be made at a temperature of 60 degrees Fahrenheit.

Per Cent of Fat	Lactometer Readings 108	109	110	111	112	113	114	115	116	117	118	119	120
2.8	11.19	11.26	11.34	11.41	11.48	11.55	11.63	11.70	11.77	11.84	11.92	11.99	12.06
2.9	11.31	11.38	11.46	11.53	11.60	11.67	11.75	11.82	11.89	11.96	12.04	12.11	12.18
3.0	11.43	11.50	11.58	11.65	11.72	11.79	11.87	11.94	12.01	12.08	12.16	12.23	12.30
3.1	11.55	11.62	11.70	11.77	11.84	11.91	11.99	12.06	12.13	12.20	12.28	12.35	12.42
3.2	11.67	11.74	11.82	11.89	11.96	12.03	12.11	12.18	12.25	12.32	12.40	12.47	12.54
3.3	11.79	11.86	11.94	12.01	12.08	12.15	12.23	12.30	12.37	12.44	12.52	12.59	12.66
3.4	11.91	11.98	12.06	12.13	12.20	12.27	12.35	12.42	12.49	12.56	12.64	12.71	12.78
3.5	12.03	12.10	12.18	12.25	12.32	12.39	12.47	12.54	12.61	12.68	12.76	12.83	12.90
3.6	12.15	12.22	12.30	12.37	12.44	12.51	12.59	12.66	12.73	12.80	12.88	12.95	13.02
3.7	12.27	12.34	12.42	12.49	12.56	12.63	12.71	12.78	12.85	12.92	13.00	13.07	13.14
3.8	12.39	12.46	12.54	12.61	12.68	12.75	12.83	12.90	12.97	13.04	13.12	13.19	13.26
3.9	12.51	12.58	12.66	12.73	12.80	12.87	12.95	13.02	13.09	13.16	13.24	13.31	13.38
4.0	12.63	12.69	12.78	12.85	12.92	12.99	13.07	13.14	13.21	13.28	13.36	13.43	13.50
4.1	12.75	12.82	12.90	12.97	13.04	13.11	13.19	13.26	13.33	13.40	13.48	13.55	13.62
4.2	12.87	12.94	13.02	13.09	13.16	13.23	13.31	13.38	13.45	13.52	13.60	13.67	13.74
4.3	12.99	13.06	13.14	13.21	13.28	13.35	13.43	13.50	13.57	13.64	13.72	13.79	13.86
4.4	13.11	13.18	13.26	13.33	13.40	13.47	13.55	13.62	13.69	13.76	13.84	13.91	13.98
4.5	13.23	13.30	13.38	13.45	13.52	13.59	13.67	13.74	13.81	13.88	13.96	14.03	14.10
4.6	13.35	13.42	13.50	13.57	13.64	13.71	13.79	13.86	13.93	14.00	14.08	14.15	14.22
4.7	13.47	13.54	13.62	13.69	13.76	13.83	13.91	13.98	14.05	14.12	14.20	14.27	14.34
4.8	13.59	13.66	13.74	13.81	13.88	13.95	14.03	14.10	14.17	14.24	14.32	14.39	14.46
4.9	13.74	13.78	13.86	13.93	14.00	14.07	14.15	14.22	14.29	14.36	14.44	14.51	14.58
5.0	13.83	13.90	13.98	14.05	14.12	14.19	14.27	14.34	14.41	14.48	14.56	14.63	14.70
5.1	13.95	14.02	14.10	14.17	14.24	14.31	14.39	14.46	14.53	14.60	14.68	14.75	14.82
5.2	14.07	14.14	14.22	14.29	14.36	14.43	14.51	14.58	14.65	14.72	14.80	14.87	14.94
5.3	14.19	14.26	14.34	14.41	14.48	14.55	14.63	14.70	14.77	14.84	14.92	14.99	15.06
5.4	14.31	14.38	14.46	14.53	14.60	14.67	14.75	14.82	14.89	14.96	15.04	15.11	15.18
5.5	14.43	14.50	14.58	14.65	14.72	14.79	14.87	14.94	15.01	15.08	15.16	15.23	15.30
5.6	14.55	14.62	14.70	14.77	14.84	14.91	14.99	15.06	15.13	15.20	15.28	15.35	15.42
5.7	14.67	14.74	14.82	14.89	14.96	15.03	15.11	15.18	15.25	15.32	15.40	15.47	15.54
5.8	14.79	14.86	14.94	15.01	15.08	15.15	15.23	15.30	15.37	15.44	15.52	15.59	15.66
5.9	14.91	14.98	15.06	15.13	15.20	15.27	15.35	15.42	15.49	15.56	15.64	15.71	15.78
6.0	15.03	15.10	15.18	15.25	15.32	15.39	15.47	15.54	15.61	15.68	15.76	15.83	15.90

VOLUME OF AMMONIA GAS (CUBIC FEET) THAT MUST BE PUMPED PER MINUTE TO PRODUCE 1 TON OF REFRIGERATION IN 24 HOURS

Suction Pressure (Gauge) and Corresponding Temperature of Gas		Condenser Pressures (Lb. per Sq. In. Gauge) and Temperatures (F.)								
Lbs. per Sq. In.	°F.	103 (65°)	115 (70°)	127 (75°)	139 (80°)	153 (85°)	168 (90°)	184 (95°)	200 (100°)	218 (105°)
4	—20	5.84	5.90	5.96	6.03	6.09	6.16	6.23	6.30	6.43
6	—15	5.35	5.40	5.46	5.52	5.58	5.64	5.70	5.77	5.83
9	—10	4.66	4.73	4.76	4.81	4.86	4.91	4.97	5.05	5.08
13	— 5	4.09	4.12	4.17	4.21	4.25	4.30	4.35	4.40	4.44
16	0	3.59	3.63	3.66	3.70	3.74	3.78	3.83	3.87	3.91
20	5	3.20	3.24	3.27	3.30	3.34	3.38	3.41	3.45	3.49
24	10	2.87	2.90	2.93	2.96	2.99	3.02	3.06	3.09	3.12
28	15	2.59	2.61	2.65	2.68	2.71	2.73	2.76	2.80	2.82
33	20	2.31	2.34	2.36	2.38	2.41	2.44	2.46	2.49	2.51
39	25	2.06	2.08	2.10	2.12	2.15	2.17	2.20	2.22	2.24
45	30	1.85	1.87	1.89	1.91	1.93	1.95	1.97	2.00	2.01
51	35	1.70	1.72	1.74	1.76	1.77	1.79	1.81	1.83	1.85

(Add 33 per cent for probable actual volume.)

—From Dairy Engineering (Farrell)

WEIGHT OF AMMONIA NEEDED IN A SYSTEM

The quantity of anhydrous ammonia needed in a compression plant may be estimated by assuming the whole space of the condenser to be filled with ammonia vapor at 165 lb. pressure, the ammonia expansion pipes to be filled with ammonia vapor at 20 lb. pressure, and allowing also 25 lb. of liquid ammonia for every cubic foot of liquid receiver capacity. The following table gives the weight of anhydrous ammonia which experience shows is required for the compression side of refrigerating plants:

Tons of refrigeration	5	10	20	40	100	200	300	500
Ammonia, pounds	110	150	230	300	440	620	840	1,215

The anhydrous ammonia required per 100 running feet of pipe on the expansion side is approximately as follows:

Size of pipe, inches	1	1¼	1½	2
For refrigerating plants, direct expansion and brine cooling coils, pounds	14	18	20	25
For ice plants, expansion coils for can and plate use, pounds	8	11	12	15

For flooded-type evaporators allow sufficient refrigerant to fill the evaporator entirely to the float level, in addition to that necessary for the compression side of the refrigeration machine. The ammonia level in the receiver should be carried between ⅓ and ¾ full at all times if possible.

—From Dairy Engineering (Farrell)

TEMPERATURES OF SATURATED STEAM AT VARYING PRESSURES

Temperature °C.	Pressure of Sat. Steam (less atmospheric pressure) (lbs. per sq. in)	Temperatur °F.
101	0.02	212
101	0.65	213.8
102	1.20	215.6
103	1.76	217.4
104	2.34	219.2
105	2.94	221
106	3.56	222.8
107	4.19	224.6
108	4.84	226.4
109	5.51	228.2
110	6.20	230
111	6.91	231.8
112	7.63	233.6
113	8.38	235.4
114	9.15	237.2
115	9.94	239
116	10.75	240.8
117	11.58	242.6
118	12.44	244.4
119	13.31	246.2
120	14.21	248
121	15.14	249.8
122	16.08	251.6
123	17.06	253.4
124	18.06	255.2
125	19.08	257
126	20.13	258.8
127	21.20	260.6
128	22.30	262.4
129	23.43	264.2
130	24.59	266

LOGARITHMIC TABLE

Instructions: Report bacteria counts as follows:

1-100	—actual
100-1,000	—to the nearest 10
1,000-100,000	—to the nearest 1,000
100,000-1,000,000	—to the nearest 10,000
1,000,000-100,000,000	—to the nearest 100,000

The logarithms of all bacterial counts to be reported from 1 to 100,000,000 are given directly in the accompanying table. The logarithms should be listed opposite the counts on the work sheet. To find the logarithmic average, find the arithmetic average of the logarithms. Then find the bacterial count in the table which is opposite the arithmetic average of the logarithms. Following is an example:

Counts	Logarithms
35,000	4.54
11,000	4.04
9,000	3.95
95,000	4.98
	17.51

17.51÷4 counts=4.38 average logarithm. 4.38 in the table is opposite 24,000, which is therefore, the average bacterial count.

If it is found that the average logarithm occurs opposite more than one bacterial count in the table, take the lowest bacterial count as the average.

Cts.	Logarithms	Cts.	Logarithms	Cts.	Logarithms	Cts.	Logarithms	Cts.	Logarithms	Cts.	Logarithms
1	0.00	32	1.51	63	1.80	94	1.97	350	2.54	660	2.82
2	0.30	33	1.52	64	1.81	95	1.98	360	2.56	670	2.83
3	0.48	34	1.53	65	1.81	96	1.98	370	2.57	680	2.83
4	0.60	35	1.54	66	1.82	97	1.99	380	2.58	690	2.84
5	0.70	36	1.56	67	1.83	98	1.99	390	2.59	700	2.85
6	0.78	37	1.57	68	1.83	99	1.99	400	2.60	710	2.85
7	0.85	38	1.58	69	1.84	100	2.00	410	2.61	720	2.86
8	0.90	39	1.59	70	1.85	110	2.04	420	2.62	730	2.86
9	0.95	40	1.60	71	1.85	120	2.08	430	2.63	740	2.87
10	1.00	41	1.61	72	1.86	130	2.11	440	2.64	750	2.88
11	1.04	42	1.62	73	1.86	140	2.15	450	2.65	760	2.88
12	1.08	43	1.63	74	1.87	150	2.18	460	2.66	770	2.89
13	1.11	44	1.64	75	1.88	160	2.20	470	2.67	780	2.89
14	1.15	45	1.65	76	1.88	170	2.23	480	2.68	790	2.90
15	1.18	46	1.66	77	1.89	180	2.26	490	2.69	800	2.90
16	1.20	47	1.67	78	1.89	190	2.28	500	2.70	810	2.91
17	1.23	48	1.68	79	1.90	200	2.30	510	2.71	820	2.91
18	1.26	49	1.69	80	1.90	210	2.32	520	2.72	830	2.92
19	1.28	50	1.70	81	1.91	220	2.34	530	2.72	840	2.92
20	1.30	51	1.71	82	1.91	230	2.36	540	2.73	850	2.93
21	1.32	52	1.72	83	1.92	240	2.38	550	2.74	860	2.93
22	1.34	53	1.72	84	1.92	250	2.40	560	2.75	870	2.94
23	1.36	54	1.73	85	1.93	260	2.42	570	2.76	880	2.94
24	1.38	55	1.74	86	1.93	270	2.43	580	2.76	890	2.95
25	1.40	56	1.75	87	1.94	280	2.45	590	2.77	900	2.95
26	1.42	57	1.76	88	1.94	290	2.46	600	2.78	910	2.96
27	1.43	58	1.76	89	1.95	300	2.48	610	2.79	920	2.96
28	1.45	59	1.77	90	1.95	310	2.49	620	2.79	930	2.97
29	1.46	60	1.78	91	1.96	320	2.51	630	2.80	940	2.97
30	1.48	61	1.79	92	1.96	330	2.52	640	2.81	950	2.98
31	1.49	62	1.79	93	1.97	340	2.53	650	2.81	960	2.98

LOGARITHMIC TABLE (Cont'd)

Cts.	Logarithms	Cts.	Logarithms	Cts.	Logarithms	Cts.	Logarithms	Cts.	Logarithms
970	2.99	58,000	4.76	280,000	5.45	880,000	5.94	5,800,000	6.76
980	2.99	59,000	4.77	290,000	5.46	890,000	5.95	5,900,000	6.77
990	2.99	60,000	4.78	300,000	5.48	900,000	5.95	6,000,000	6.78
1,000	3.00	61,000	4.79	310,000	5.49	910,000	5.96	6,100,000	6.79
2,000	3.30	62,000	4.79	320,000	5.51	920,000	5.96	6,200,000	6.79
3,000	3.48	63,000	4.80	330,000	5.52	930,000	5.97	6,300,000	6.80
4,000	3.60	64,000	4.81	340,000	5.53	940,000	5.97	6,400,000	6.81
5,000	3.70	65,000	4.81	350,000	5.54	950,000	5.98	6,500,000	6.81
6,000	3.78	66,000	4.82	360,000	5.56	960,000	5.98	6,600,000	6.82
7,000	3.85	67,000	4.83	370,000	5.57	970,000	5.99	6,700,000	6.83
8,000	3.90	68,000	4.83	380,000	5.58	980,000	5.99	6,800,000	6.83
9,000	3.95	69,000	4.84	390,000	5.59	990,000	5.99	6,900,000	6.84
10,000	4.00	70,000	4.85	400,000	5.60	1,000,000	6.00	7,000,000	6.85
11,000	4.04	71,000	4.85	410,000	5.61	1,100,000	6.04	7,100,000	6.85
12,000	4.08	72,000	4.86	420,000	5.62	1,200,000	6.08	7,200,000	6.86
13,000	4.11	73,000	4.86	430,000	5.63	1,300,000	6.11	7,300,000	6.86
14,000	4.15	74,000	4.87	440,000	5.64	1,400,000	6.15	7,400,000	6.87
15,000	4.18	75,000	4.88	450,000	5.65	1,500,000	6.18	7,500,000	6.88
16,000	4.20	76,000	4.88	460,000	5.66	1,600,000	6.20	7,600,000	6.88
17,000	4.23	77,000	4.89	470,000	5.67	1,700,000	6.23	7,700,000	6.89
18,000	4.26	78,000	4.89	480,000	5.68	1,800,000	6.26	7,800,000	6.89
19,000	4.28	79,000	4.90	490,000	5.69	1,900,000	6.28	7,900,000	6.90
20,000	4.30	80,000	4.90	500,000	5.70	2,000,000	6.30	8,000,000	6.90
21,000	4.32	81,000	4.91	510,000	5.71	2,100,000	6.32	8,100,000	6.91
22,000	4.34	82,000	4.91	520,000	5.72	2,200,000	6.34	8,200,000	6.91
23,000	4.36	83,000	4.92	530,000	5.72	2,300,000	6.36	8,300,000	6.92
24,000	4.38	84,000	4.92	540,000	5.73	2,400,000	6.38	8,400,000	6.92
25,000	4.40	85,000	4.93	550,000	5.74	2,500,000	6.40	8,500,000	6.93
26,000	4.42	86,000	4.93	560,000	5.75	2,600,000	6.42	8,600,000	6.93
27,000	4.43	87,000	4.94	570,000	5.76	2,700,000	6.43	8,700,000	6.94
28,000	4.45	88,000	4.94	580,000	5.76	2,800,000	6.45	8,800,000	6.94
29,000	4.46	89,000	4.95	590,000	5.77	2,900,000	6.46	8,900,000	6.95
30,000	4.48	90,000	4.95	600,000	5.78	3,000,000	6.48	9,000,000	6.95
31,000	4.49	91,000	4.96	610,000	5.79	3,100,000	6.49	9,100,000	6.96
32,000	4.51	92,000	4.96	620,000	5.79	3,200,000	6.51	9,200,000	6.96
33,000	4.52	93,000	4.97	650,000	5.81	3,300,000	6.52	9,300,000	6.97
34,000	4.53	94,000	4.97	660,000	5.82	3,400,000	6.53	9,400,000	6.97
35,000	4.54	95,000	4.98	670,000	5.83	3,500,000	6.54	9,500,000	6.98
36,000	4.56	96,000	4.98	630,000	5.80	3,600,000	6.56	9,600,000	6.98
37,000	4.57	97,000	4.99	690,000	5.84	3,700,000	6.57	9,700,000	6.99
38,000	4.58	98,000	4.99	680,000	5.83	3,800,000	6.58	9,800,000	6.99
39,000	4.59	99,000	4.99	700,000	5.85	3,900,000	6.59	9,900,0000	6.99
42,000	4.62	100,000	5.00	740,000	5.87	4,000,000	6.60	10,000,000	7.00
40,000	4.60	110,000	5.04	730,000	5.86	4,100,000	6.61	11,000,000	7.04
41,000	4.61	130,000	5.11	750,000	5.88	4,200,000	6.62	12,000,000	7.08
43,000	4.63	140,000	5.15	760,000	5.88	4,300,000	6.63	13,000,000	7.11
45,000	4.65	150,000	5.18	770,000	5.89	4,400,000	6.64	14,000,000	7.15
46,000	4.66	160,000	5.20	780,000	5.89	4,500,000	6.65	15,000,000	7.18
47,000	4.67	170,000	5.23	790,000	5.90	4,600,000	6.66	16,000,000	7.20
44,000	4.64	180,000	5.26	800,000	5.90	4,700,000	6.67	17,000,000	7.23
48,000	4.68	640,000	5.81	210,000	5.32	4,800,000	6.68	18,000,000	7.26
49,000	4.69	120,000	5.08	810,000	5.91	4,900,000	6.69	19,000,000	7.28
50,000	4.70	190,000	5.28	820,000	5.91	5,000,000	6.70	20,000,000	7.30
51,000	4.71	200,000	5.30	830,000	5.92	5,100,000	6.71	30,000,000	7.48
52,000	4.72	220,000	5.34	840,000	5.92	5,200,000	6.72	40,000,000	7.60
53,000	4.72	230,000	5.36	850,000	5.93	5,300,000	6.72	50,000,000	7.70
54,000	4.73	240,000	5.38	860,000	5.93	5,400,000	6.73	60,000,000	7.78
55,000	4.74	250,000	5.40	870,000	5.94	5,500,000	6.74	70,000,000	7.85
56,000	4.75	260,000	5.42	710,000	5.85	5,600,000	6.75	80,000,000	7.90
57,000	4.76	270,000	5.43	720,000	5.86	5,700,000	6.76	90,000,000	7.95
								100,000,000	8.00

EXAMINATION OF PLANT PRODUCTS

STATION WEATHER TEMP. 10:00 A.M. DATE

	Fat	T.S.	Bact.	Taste	Condition of		Temp.	Icing	Visible Dirt	Crm. Line	Acid	Crm. Whip	Body	Past. Chart	Holding Chart				Cooling Chart	In Coffee
					B	C									1	2	3	4		
Grades Milk																				
CREAM																				
Light																				
Medium																				
Heavy																				
Ex. Heavy																				
Guernsey Sp. Ex. Hvy.																				
Sr. Cream																				
Cond. Skim																				
Skimmilk																				
I. C. Mix Unswt.																				
I. C. Mix Swt.																				
Buttermilk																				
Butter																				

Tank	Alkalinity	Chlor. P.P.M.	Temp. B.W.	Tem. W. B. at Filler
1				
2				
3				
4				

Pounds of glass broken per 1000 bot. fill.			
No. Bots. Fill. Dly.	Before Washing	At. discharge end Washer & Filler	Total

Signature

DAILY PLANT OPERATING RECORD

STATION .... DATE ....

PRODUCTION	RECEIVED		PROCESSED						TOTAL In Quarts or Pounds	Time Started	Time Finished	Time Consumed
	40	20	40	8	Quarts	1/3	Pints	½				
Milk												
Milk												
Pasteurizing												
Separator												
Skimmilk												
Buttermilk												
Cream												
Sour Cream												
Ice Cream Mix												
Condensed Skim												
Butter												
Can Washing												
Platform												
Dump Room												
Bottle Washing												
Total												

SPECIFY IN DETAIL AS TO TIME LOST IF ANY IN OPERATING

....

....

....

....

....

NUMBER OF MEN EMPLOYED

At Work..

Days Off..

....

TOTAL..

Signed..

FIRST AID SUGGESTIONS

ANTIDOTES OF POISONS

Acetic Acid—Emetics, magnesia, chalk, soap, oil.

Arsenic, Rat Poison, Paris Green—Milk, raw egg, sweetoil, lime water, flour and water.

Carbolic Acid—Any soluble nontoxic sulphate, after provoking vomiting with zinc sulphate; uncooked white of egg in abundance, milk of lime, saccharate of calcium, olive or castor oil with magnesia in suspension, ice, washing the stomach with equal parts water and vinegar; give alcohol or whiskey or about four fluid ounces camphorated oil at one dose.

Chloroform, Chloral, Ether—Dash cold water on head and chest, artificial respiration.

Hydrochloric Acid—Magnesia, alkali carbonates, albumin, ice.

Hydrocyanic or Prussic Acid—Hydrogen peroxide internally, and artificial respiration, breathing ammonia or chlorine from chlorinated lime, ferrous sulphate followed by potassium carbonate, emetics, warmth.

Iodine—Emetics, stomach siphon, starchy foods in abundance, sodium thiosulphate.

Lead Acetate—Emetics, stomach siphon, sodium, potassium or magnesium sulphates, milk, albumin.

Mercuric Chloride or Corrosive Sublimate—Zinc sulphate, emetics, stomach siphon, white of egg, milk, chalk, castor oil, table salt, reduced iron.

Nitrate of Silver—Salt and water.

Nitric Acid—Same as for hydrochloric acid.

Opium, Morphine, Laudanum, Paregoric, Etc.—Strong coffee, hot bath. Keep awake and moving at any cost.

Phosphoric Acid—Same as for hydrochloric.

Sodium Hydroxide or Potassium Hydroxide—Vinegar, lemon juice, orange juice, oil, milk.

Sulphuric Acid—Same as for hydrochloric acid with the addition of soap or oil.

Sulphurous Acid or Sulphur Dioxide—Mustard plaster on chest; narcotics, expectorants.

—Handbook of Chemistry and Physics.

THE SERUM POINT METHOD OF PROPORTIONING BATCHES

The serum point method is presented as the simplest and shortest method available for ice cream batch proportioning. It involves only straight arithmetical handling of two formulas with minor intervening steps and a final simple subtraction. Use of the method will be made still easier by figuring batches on a 100-pound basis, though this is not necessary.

As set up to handle a straight three-milk-products mix, the steps are as follows:

1. To find serum solids concentrate (condensed or dry skimmilk) requirement—

$$\frac{\text{batch weight} \times \text{\% ss desired} - \text{batch weight} \times \frac{\text{\% serum in batch}}{100} \times \text{\% solids in plain serum}}{\text{\% ss in ss concentrate} - \text{\% solids in plain serum} \times \frac{\text{\% serum in ss concentrate}}{100}}$$

2. To find fat concentrate (cream or butter) requirement—

$$\frac{\text{batch weight} \times \text{remaining \% bf desired} - \text{required lbs. milk \& fat concentrate} \times \text{\% fat in milk}}{\text{\% fat in fat concentrate} - \text{\% fat in milk}}$$

3. To find the third product (milk) requirement—
required lbs. milk & fat concentrate − lbs. fat concentrate

Between Steps 1 and 2 it is necessary to find the fat, if any, brought in by the serum solids concentrate and to subtract that from the total fat desired to arrive at the required figure in Step 2.

on "remaining % bf desired." It is also necessary to subtract from total pounds milk products required the pounds milk products in serum solids concentrate coming into the batch, as found in the first formula, to furnish the needed figure on "required lbs. milk and fat concentrate."

The "% serum in batch" figure called for in this method is the percentage of milk products in batch minus percentage butterfat in batch. The "% solids in plain serum," which is safely assumed to be the same for the three milk products, should be the average figure found by the user's tests on his own supplies. Where there is a lack of facilities for making frequent tests, 9 is a fairly sound figure for most sections. A slight discrepancy in this key figure will result in only an insignificant error in the calculation of milk products requirements. The "% serum in ss concentrate" called for in this method is the percentage of milk products in serum solids concentrate minus percentage butterfat in serum solids concentrate.

If sweetened condensed is used and, in the amount required, brings in too much sugar, adjustment is made by throwing out part of the condensed and making up the shortage in serum solids thus brought about by bringing in another serum solids concentrate not carrying sugar. The procedure is as follows:

Find the weight of condensed that will carry just the right amount of sugar for the batch and subtract this from the requirement as first found; multiply the difference by the figure representing the excess serum solids in the sweetened condensed and divide this product (which is the shortage in points due to dropping the condensed) by the figure representing excess serum solids in dry skimmilk or other unsweetened concentrate, for the pounds of the new material to come in. If the unsweetened concentrate coming in carries fat, find the ss excess by the formula—

$$\%\text{ ss in ss concentrate} - 9 \times \frac{\%\text{ serum in ss concentrate}}{100}$$

Then find the remainder of fat to come in and the remaining pounds of milk products required and find the proportions of milk and cream, or other fat concentrate, in the usual way.

—From The Ice Cream Trade Journal, September, 1922.

SERUM POINT METHOD SIMPLIFIED

The serum point method (1) of calculating ice cream mixes may be simplified and shortened by the use of the following table. The table gives the weights of three common serum solids concentrates to use per 100 pounds of mix for mixes of almost any desired composition.

A type problem will be presented to make the application of the table clear. Assume it is desired to prepare 100 pounds of ice cream mix which will test 12 per cent butterfat, 10 per cent serum solids, 15 per cent sugar, and .4 per cent gelatin. The calculation of the pounds of the serum solids concentrate required would ordinarily involve the use of the following formulae:

$$\text{Pounds of Serum Solids Concentrate} = \frac{\text{Pounds of Serum Solids Desired} - \text{Pounds of}^{*}\ \text{Serum in Mix} \times .09}{\text{Per cent of Serum Solids in Concentrate} - \text{Per cent Serum in Concentrate} \times .09}$$

Through the use of the table it is only necessary to solve the numerator of this fraction or:

$$\text{Pounds of Serum Solids Desired} - \text{Pounds of Serum in mix} \times .09$$

Substituting in this formula:

$$10-(72.6 \times .09) \text{ or}$$
$$10-6.534=3.466$$

Referring to table on next page and looking down the column headed numerator until 3.466 is found it will be observed that 19.255 pounds of plain condensed skimmilk testing 27 per cent serum solids, 15.21 pounds of sweetened condensed skimmilk testing 28 per cent serum solids, 42 per cent sugar or 3.928 pounds of skimmilk powder testing 97 per cent serum solids would be required for this particular mix. Similarly the amount of serum solids concentrate to use per 100 pounds of mix for almost any desired composition may be determined.

Obviously the table is only applicable when one of the three serum solids concentrates specified in the table is to be used and when these products conform to the specified composition.

(1) See appendix, page 409, for a complete discussion of the serum point method.

*Pounds of serum in mix is calculated by the following formula:
Batch weight—(lbs. fat.+lbs. of sugar+lbs. gelatin)=lbs. of serum in the mix.

Showing weights of three common serum solids concentrates to use per 100 pounds of ice cream mix.

Numerator	Plain Condensed Skimmilk	Sweetened Condensed Skimmilk	Skimmilk Powder
Composition of Concentrate	27% S. S.	28% S. S. 42% Sugar	97% S. S.
1.466	8.144	6.435	1.6659
1.566	8.70	6.848	1.772
1.666	9.25	7.313	1.893
1.766	9.811	7.75	2.00
1.866	10.366	8.191	2.12
1.966	10.922	8.630	2.234
2.066	11.477	9.069	2.347
2.166	12.033	9.508	2.461
2.266	12.588	9.947	2.575
2.366	13.144	10.386	2.688
2.466	13.70	10.825	2.802
2.566	14.25	11.264	2.9159
2.666	14.811	11.676	3.029
2.766	15.366	12.142	3.1431
2.866	15.922	12.581	3.256
2.966	16.477	13.02	3.370
3.066	17.033	13.459	3.484
3.166	17.588	13.898	3.597
3.266	18.144	14.337	3.711
3.366	18.70	14.776	3.825
3.466	19.255	15.21	3.938
3.566	19.811	15.654	4.052
3.666	20.366	16.093	4.165
3.766	20.922	16.532	4.279
3.866	21.477	16.971	4.393
3.966	22.033	17.410	4.506
4.066	22.588	17.848	4.62
4.166	23.144	18.287	4.734
4.266	23.70	18.726	4.847
4.366	24.255	19.165	4.961
4.466	24.811	19.604	5.075
4.566	25.366	20.043	5.188
4.666	25.922	20.456	5.302
4.766	26.477	20.921	5.415
4.866	27.033	21.360	5.529
4.966	27.588	21.789	5.643
5.066	28.144	22.238	5.756
5.166	28.70	22.677	5.870
5.266	29.256	23.116	5.984
5.366	29.811	23.555	6.097
5.466	30.366	23.994	6.211
5.566	30.922	24.4337	6.325
5.666	31.477	24.872	6.438
5.766	32.033	25.31	6.552
5.866	32.588	25.75	6.665

—Professors W. H. Martin and W. J. Caulfield,
Department of Dairy Husbandry, Kansas State Agr. College.

THE BALANCE METHOD OF PROPORTIONING ICE CREAM MIXES

One satisfactory method for proportioning the ingredients in making ice cream mixes is "the balance method," a term originating in connection with the work reported in U. S. D. A. Bul. 1123. It is a method that can be easily understood, is applicable to all combinations of ingredients, and reduces to a minimum the chances of error in the calculations. Furthermore, it furnishes a record of the ingredients used. The proportions obtained by this method are based on five conditions:

1. The amount (pounds) of mix that will be necessary to produce the number of gallons of ice cream desired.

2. The composition (standard) of ice cream desired.

3. The amount of solid constituents necessary for the mix.

4. The quantity and physical condition of the ingredients on hand.

5. The composition of the ingredients to be used.

An example of this method may be illustrated and explained as follows: Give the proportions for 500 gallons of ice cream testing approximately 12 per cent fat, 15 per cent sugar, 10 per cent milk solids not fat, and 0.3 per cent gelatin, and weighing 5 pounds per gallon. Stock on hand: Sugar, gelatin, 342 pounds 30.5 per cent cream; 608 pounds of 38 per cent cream; 720 pounds of sweetened condensed milk containing 8.2 per cent fat, 42 per cent sugar, and 21 per cent milk solids not fat; and skimmilk powder

Method of Calculating: With a tabulated form similar to that used above first indicate in the respective columns the pounds of mix and the percentage of each constituent desired. From this the total pounds of each constituent required in the mix is calculated by multiplication, i.e. 300, 375, 250 and 7.5 pounds respectively, and the ingredients are listed as shown in the illustration. Next the amount of fat and milk solids not fat furnished by the 720 pounds of condensed milk is calculated by multiplication, i.e. 59.0, 302.0 and 151.0 pounds respectively, and the amount of fat and M. S. N. F. furnished by the 342 pounds of 30.5 per cent cream is also calculated by multiplication, i.e. 199.0 and 30.0 pounds respectively. The difference between the total

Total lbs. desired 5×500= 2500 lbs.	Ingredients and Composition	Constituents Furnished			
		Fat 12% 300 lbs.	Sugar 15% 375 lbs.	M. S. N. F. 10% 250 lbs.	Gelátin 0.3% 7.5 lbs.
73.0 lbs.	Sugar..............	....	73.0	....	....
7.5 lbs.	Gelatin..............	....	....	....	7.5
342.0 lbs.	Cream, 30.5% (0.093% serum solids)......	199.0	....	30.0	.....
110.5 lbs.	Cream, 38.0%........	42.0	....	6.0	.....
720.0 lbs.	Condensed milk, 8.2% fat, 42% sugar, 21% milk solids not fat..	59.0	302.0	151.0	....
65.0 lbs.	Skimmilk pwdr., 97%	....	....	63.0	....
1182.0 lbs.	Water.............	....	....	....	....
2500.0		300.0	375.0	250.0	7.5

pounds of fat already furnished subtracted from the total pounds required, divided by 38.0/100 gives 110.5, the number of pounds of 38 per cent cream necessary to balance the fat requirement, i.e. 300−(199+59) divided by 38/100=110.5. The amount of milk solids not fat in 110.5 pounds of this cream is then determined and the difference between the sum of all that has been added by the cream and condensed milk, and the total amount required determines the amount of skimmilk powder to use, i.e. 250−(30.0+6.0+151.0)=63, 63 divided by 97×100=65.0 pounds. By difference the amount of granulated sugar is also determined. The quantity of gelatin is the same as previously indicated. The amount of water necessary is found by subtracting the sum of the ingredients already calculated from the total amount of mix desired, i.e. 2500−(73+7.5+342+110.5+720+65)=1182 pounds of water.

This method becomes very simple and reliable with a little practice.

—From O. E. Williams, Dairy Mfg. Specialists, Bureau of Dairying, U. S. Dept. of Agriculture

CHECK-AND-BALANCE METHOD* OF MIX PROPORTIONING

For Figuring Batch Requirements in Any Order of Calculation

To find, in any desired order of calculation, the requirement of each milk product in a three-milk-products mix, this method first arrives at a divisor figure, which remains constant for all findings, and then at a dividend figure, which varies for each product sought.

To arrive at the fixed divisor figure, set down the three milk products, numbering them 1, 2 and 3 in any desired order, except that sweetened condensed, if used, must not be handled as No. 1. Then make the following calculation:

Diff. between bf %'s of Nos. 1 & 2 × Diff. between ss %'s of Nos. 1 & 3 subtracted or added* with

Diff. between ss %'s of Nos. 1 & 2 × Diff. between bf %'s of Nos. 1 & 3

*Add only if No. 1 ranks second among the three milk products in order of size of bf %, or second in order of size of ss %, not of both; in all other cases, subtract the smaller result from the larger.

To arrive at the variable dividend figure for each milk product, select the one to be sought and set down the others, numbering them 1 and 2 in any desired order, except that sweetened condensed milk, if used, must not be handled as No. 1.
Then make the following calculation:

Diff. between bf % desired × batch wght. and bf % of No. 1 × lbs. milk prod. desired multiplied by Diff. between ss %'s of Nos. 1 & 2 subtracted or added* with

Diff. between ss % desired × batch wght. and ss % of No. 1 × lbs. milk prod. desired multiplied by Diff. between bf %'s of Nos. 1 & 2

*Add only if No. 1 has bf and ss %'s both greater or both smaller than the respective percentages in No. 2 and, used alone to make up the full milk products portion of the batch, would bring in more bf or more ss, not more of both, than desired; or if No. 1 has either, not both, a bf % or a ss % greater than the respective percentage in No. 2 and, used alone to make up the full milk products portion of the batch, would bring in less of both bf and ss than desired; in all other cases, subtract the smaller result from the larger.

To find the requirement of any product, divide the dividend figure for that product by the fixed divisor figure.

In using this method, it will be quickly seen that, after finding the requirement of any one product, the shortest order in which to find those of the others is to find that of the original No 2 product second, handling the same No. 1 product again as No. 1; then to find the requirement of No. 1 by subtracting the total of the first two (exclusive of sugar in condensed) from lbs. milk products.

When sweetened condensed is used as one of the three milk products, a bf or ss figure on No. 1 product, wherever subtracted with the respective figure on sweetened condensed, drops a percentage of its amount equal to the percentage of sugar in the condensed, before being employed in the subtraction.

This method can also be used in handling a mix made up of two milk products and water by placing zeros where bf or ss %'s of the third product are called for. Water should not, of course, be handled as No. 1 product. The method is also applicable to a mix made up of four milk products in which the excess serum solids requirement is split between condensed and dry skimmilk.

—From The Ice Cream Trade Journal, August, 1927.

SIMPLIFYING THE PEARSON SQUARE METHOD

This method follows but decidedly simplifies the familiar Pearson square method. It can best be grasped from a concrete illustration of its application. For purposes of illustration there has been selected a mix made up of cream carrying 40% bf and 6% ss; milk carrying 4% bf and 8% ss; condensed carrying 8% bf and 20% ss; sugar, gelatin and flavor. Solids requirements set are 12% bf; 10% ss, 15% sugar, 0.5% gelatin and 0.5% flavor (84% mp). Round figures and a 100-lb. batch have been handled only to shorten the illustration.

First find bf % and ss % required in milk products of batch—14.286% bf; 11.905% ss.

Step A—Proportion, to meet the bf requirement of the mp batch, a trial combination of the fat concentrate and one fat-deficient product:

$\frac{40-4}{14.286-4}=3.5$ parts, in which 1 of cream takes care of 2.5 of milk.

Step B—Proportion, in the same way, a trial combination of the fat concentrate and the other fat-deficient product:

$\frac{40-8}{14.286-8}=5.091$ parts, in which 1 of cream takes care of 4.091 of condensed.

Step C—Find % of serum solids in Combination A:

$\frac{(1\times6)+(2.5\times8)}{3.5}=7.429$, % ss in Combination A.

Step D—Find % of ss in Combination B:

$\frac{(1\times6)+(4.091\times20)}{5.09}=17.25$, % ss in Combination B.

Step E—Proportion, to meet the ss % requirement of the mp batch a mix made up of Combinations A and B:

$\frac{17.250-7.429}{11.905-7.429}=2.194$ parts, in which 1 of B takes care of 1.194 of A.

Then, to find the mix requirement of either fat-deficient product, multiply together number of parts of that product appearing in the trial combination (A or B) in which it is used, the number of parts of that trial combination appearing in the mix (E) and the pounds of mp going into the desired batch, and divide by the total number of parts in the trial combination (A or B) in which the product is used, multiplied by the total number of parts in the mix (E).

To find milk requirement:

$$\frac{2.5\times 1.194\times 84}{3.5\times 2.194}=32.653 \text{ lbs.}$$

To find condensed requirement:

$$\frac{4.091\times 1\times 84}{5.091\times 2.194}=30.765 \text{ lbs.}$$

To find mix requirement of fat concentrate, subtract weight of two fat-deficient products from lbs. mp:

$$84-(32.653+30.765)=20.582 \text{ lbs. cream.}$$

Where sweetened condensed is employed in the batch the bf and ss %'s in the condensed called for will be the %'s in the mp portion of the condensed and the weight of this product as first found will be lbs. mp in required condensed, from which lbs. condensed can be calculated. Note, however, that the condensed figure to be used in the final calculation to find the fat concentrate requirement of the batch is that expressing lbs. mp in required condensed.

—From The Ice Cream Trade Journal, October, 1927

AMOUNT OF WATER AND ICE AT VARIOUS TEMPERATURES IN ICE CREAM CONTAINING 12% FAT, 10% SERUM SOLIDS, AND 14% SUGAR*

Condition of the Ice Cream	Temp. Deg. C.	Temp. Deg. F.	Pounds of Water and Ice in 100 Lbs. of Ice Cream	
			Water	Ice
Liquid — initial freezing point	−2.20	27.48	64.0	0.0
Drawn from freezer	−2.98	26.62	47.0	17.0
Partially hardened	−7.99	17.62	20.0	44.0
Trifle soft for serving	−10.65	12.83	15.0	49.0
About right for serving	−14.34	6.19	12.5	51.5
Hard during storage	−17.93	−0.27	10.0	54.0

*These calculations were made by A. C. Dahlberg, using the method presented by Alan Leighton, Jour. of Dy. Science, Vol. 10, No. 4, 1927.

CALCULATION OF THE FREEZING POINT OF ICE CREAM MIXES

For practical purpose the freezing point of an ice cream mix may be considered as a summation of the normal freezing point lowering caused by the milk salts and the abnormal lowering caused by the sugars. This means that empirically the freezing point can be calculated as follows:

$$\frac{\text{Parts Milk Solids Not Fat} \times 2.37}{\text{Parts Water}} \quad \text{Salt lowering in degrees Centigrade}$$

$$\frac{100\ (.545\ \text{MSNF} - \text{Parts Sucrose})}{\text{Parts Water}} \quad \text{Parts total sugar to 100 parts of water}$$

By reference to the table for the freezing point lowering, that due to sugar can be found. The freezing point in degrees Centigrade equals the salt lowering plus the sugar lowering. This method will probably not hold exactly if powdered milk is used.

FREEZING POINT LOWERING OF CANE SUGAR SOLUTIONS

Parts cane sugar to 100 parts water	Per cent cane sugar	Lowering °C.	Lowering due to 1 part cane sugar
3.59	3.47	0.21	0.05
6.85	6.41	0.40	0.05
10.84	9.78	0.65	0.06
15.83	13.67	0.95	0.06
19.80	16.53	1.23	0.06
22.58	18.42	1.37	0.06
25.64	20.41	1.58	0.06
28.51	22.19	1.77	0.06
32.22	24.37	1.99	0.06
35.14	26.00	2.15	0.06
37.86	27.46	2.33	0.06
43.72	30.42	2.71	0.06
45.62	31.33	2.82	0.07
50.02	33.35	3.13	0.07
54.74	35.37	3.47	0.07
59.46	37.29	3.81	0.07
64.55	39.23	4.22	0.07
69.74	41.09	4.60	0.07
75.91	43.15	5.07	0.07
82.35	45.16	5.65	0.07
88.67	47.00	6.11	0.07
95.94	48.97	6.76	0.07
102.70	50.65	7.38	0.07
111.30	52.67	8.06	0.07
121.00	54.75	9.02	0.07
131.60	56.82	9.93	0.07
143.10	58.86	10.90	0.07
153.80	60.60	11.69	0.08
165.60	62.35	12.72	0.08
181.70	64.49	13.80	0.08

—From Alan Leighton, Chemist, Dairy Bureau, U. S. Dept. of Agriculture

OVERRUN TABLE

Wt. per 1-gal. can lbs.	Wt. per 3-gal. can lbs.	Wt. per 5-gal. can lbs.	Wt. per qt. Brick ozs.	Wt. per pt. Brick ozs.	% Overrun Mix 9.0 lbs. per gal.	% Overrun Mix 9.1 lbs. per gal.	% Overrun Mix 9.2 lbs. per gal.	% Overrun Mix 9.3 lbs. per gal.
4.00	12.00	20.00	16.0	8.0	125	127	130	133
4.05	12.15	20.25	16.2	8.1	122	125	127	130
4.10	12.30	20.50	16.4	8.2	120	122	124	127
4.15	12.45	20.75	16.6	8.3	117	119	121	124
4.20	12.60	21.00	16.8	8.4	114	117	119	121
4.25	12.75	21.25	17.0	8.5	112	114	116	119
4.30	12.90	21.50	17.2	8.6	110	112	113	116
4.35	13.05	21.75	17.4	8.7	107	109	111	114
4.40	13.20	22.00	17.6	8.8	105	106	109	111
4.45	13.35	22.25	17.8	8.9	102	104	106	109
4.50	13.50	22.50	18.0	9.0	100	102	104	107
4.55	13.65	22.75	18.2	9.1	97	100	102	104
4.60	13.80	23.00	18.4	9.2	96	98	100	102
4.65	13.95	23.25	18.6	9.3	94	96	97	100
4.70	14.10	23.50	18.8	9.4	91	94	95	98
4.75	14.25	23.75	19.0	9.5	89	92	93	96
4.80	14.40	24.00	19.2	9.6	88	90	91	94
4.85	14.55	24.25	19.4	9.7	86	88	89	92
4.90	14.70	24.50	19.6	9.8	84	86	87	90
4.95	14.85	24.75	19.8	9.9	82	84	85	88
5.00	15.00	25.00	20.0	10.0	80	82	84	86
5.05	15.15	25.25	20.2	10.1	78	80	82	84
5.10	15.30	25.50	20.4	10.2	76	78	80	82
5.15	15.45	25.75	20.6	10.3	75	77	78	80
5.20	15.60	26.00	20.8	10.4	73	75	76	79
5.25	15.75	26.25	21.0	10.5	71	73	75	77
5.30	15.90	26.50	21.2	10.6	70	72	73	75
5.35	16.05	26.75	21.4	10.7	68	70	71	74
5.40	16.20	27.00	21.6	10.8	67	68	70	72
5.45	16.35	27.25	21.8	10.9	65	67	68	70
5.50	16.50	27.50	22.0	11.0	64	65	67	69

—Research and Production Dept., Eastern Dairies, Springfield, Mass.

TABLE OF SUGAR (COMMON SUGAR OR MILK SUGAR) SOLUTIONS

Percentage of Sugar by Wt.	Specific Gravity at 68° F.	Weight in Lbs. per Gallon at 68° F.	Freezing Point Centi-grade	Freezing Point Fahren-heit
0	.9982	8.33	0.0	32.
1	1.0021	8.34	−0.06	31.89
5	1.0178	8.47	−0.31	31.44
10	1.0381	8.64	−0.66	30.81
15	1.0591	8.82	−1.11	30.00
20	1.0809	9.00	−1.54	29.22
25	1.1035	9.19	−2.05	28.31
30	1.1269	9.38	−2.70	27.14
35	1.1512	9.58	−3.43	25.79
40	1.1764	9.79	−4.47	23.95
45	1.2025	10.01	−5.62	21.88
50	1.2295	10.24	−7.25	18.95
55	1.2575	10.47	−9.02	15.76
60	1.2864	10.71	−11.52	11.26
65	1.3163	10.96	−13.84	7.07

Table compiled by J. C. Hening. Specific gravities were taken from Van Nostrand's Chemical Annual by Olsen, 4th issue, 1918, D. Van Nostrand Co., Boston. The weights per gallon were calculated from them. The freezing points were calculated from the table of U. S. Pickering, as given by Alan Leighton, Jour. of Dairy Science, Vol. 10, No. 4, 1927.

At a concentration of about 65 per cent and the corresponding temperature given in the table a sucrose (common sugar) solution becomes supersaturated and sugar crystals may form. At a concentration of slightly less than 15 per cent a lactose (milk sugar) solution becomes supersaturated and lactose crystals may form.

—International Ass'n of Ice Cream Mfg

NEUTRALIZING VALUE OF ALKALIS IN STANDARDIZING ACIDITY OF CREAM OR MIXES

Formula	Name	Theoretical Amount to Neutralize One Pound of Acid	Actually Need About
$Ca(OH)_2$	Calcium Hydrate (slaked lime)	0.411	0.440
Mg(OH)	Magnesium Hydrate	0.324	0.380
MgO	Magnesium Oxide	0.224	0.27
$NaHCO_3$	Sodium Bicarbonate	0.933	1.00

—C. D. Dahle, Assoc. Prof. of Dairy Husbandry, Pennsylvania State College.

SOLID CARBON DIOXIDE REQUIRED IN SINGLE SERVICE ICE CREAM CARTONS WINTER WEATHER

Amount of Ice Cream		Solid CO_2 Required to Carry					
		4 Hrs.	8 Hrs.	12 Hrs.	16 Hrs.	20 Hrs.	24 Hrs.
4 Quarts	Bottom	½ lb.	½ lb.	1 lb.	1 lb.	1½ lb.	2 lb.
Brick	Top	1 lb.	1½ lb.	2 lb.	2½ lb.	2½ lb.	3 lb.
8 Quarts	Bottom	½ lb.	1 lb.	1 lb.	1½ lb.	2 lb.	2½ lb.
Brick	Top	1½ lb.	1½ lb.	2½ lb.	3 lb.	3 lb.	3½ lb.
16 Quarts	Bottom	1 lb.	1½ lb.	1½ lb.	2 lb.	2½ lb.	3 lb.
Brick	Top	2 lb.	2½ lb.	3 lb.	3 lb.	3½ lb.	4 lb.
24 Quarts	Bottom	1 lb.	1 lb.	1½ lb.	2 lb.	2½ lb.	3 lb.
Brick	Center	½ lb.	1 lb.	1 lb.	1½ lb.	2 lb.	2 lb.
	Top	1½ lb.	2 lb.	2½ lb.	3 lb.	3½ lb.	4 lb.
5 Gallons	Bottom	1 lb.	1½ lb.	2 lb.	2½ lb.	3 lb.	3½ lb.
Bulk	Top	2 lb.	2 lb.	2½ lb.	3 lb.	3½ lb.	4 lb.
1 Dozen	Bottom	1 lb.	1 lb.	1 lb.	1½ lb.	2 lb.	2½ lb.
Fancy Forms	Top	1½ lb.	2 lb.	2½ lb.	3 lb.	3 lb.	3½ lb.
2 Dozen	Bottom	1½ lb.	2 lb.	2 lb.	2 lb.	2½ lb.	3 lb.
Fancy Forms	Top	2 lb.	2 lb.	2½ lb.	3 lb.	3½ lb.	4 lb.

SOLID CARBON DIOXIDE REQUIRED IN SINGLE SERVICE ICE CREAM CARTONS
SUMMER WEATHER

Amount of Ice Cream		Solid CO_2 Required to Carry					
		4 Hrs.	8 Hrs.	12 Hrs.	16 Hrs.	20 Hrs.	24 Hrs.
4 Quarts	Bottom	1 lb.	1 lb.	1½ lb.	1½ lb.	2½ lb.	2½ lb.
Brick	Top	1½ lb.	2 lb.	2½ lb.	3 lb.	3 lb.	3½ lb.
8 Quarts	Bottom	1 lb.	1½ lb.	2 lb.	2 lb.	2½ lb.	3 lb.
Brick	Top	2 lb.	2 lb.	2½ lb.	3 lb.	3½ lb.	4 lb.
16 Quarts	Bottom	1½ lb.	2 lb.	2½ lb.	3 lb.	3½ lb.	4 lb.
Brick	Top	2½ lb.	3 lb.	3 lb.	4 lb.	4½ lb.	5 lb.
24 Quarts	Bottom	1 lb.	1½ lb.	2½ lb.	3 lb.	3 lb.	3½ lb.
Brick	Center	1 lb.	1 lb.	1½ lb.	1½ lb.	2 lb.	2 lb.
	Top	2 lb.	2½ lb.	3 lb.	3½ lb.	4 lb.	4½ lb.
5 Gallons	Bottom	2 lb.	2 lb.	3 lb.	3 lb.	4 lb.	5 lb.
Bulk	Top	2 lb.	3 lb.	3 lb.	4 lb.	5 lb.	6 lb.
1 Dozen	Bottom	1½ lb.	2 lb.	2 lb.	2½ lb.	3 lb.	3½ lb.
Fancy Forms	Top	2 lb.	2 lb.	2½ lb.	3 lb.	3½ lb.	4 lb.
2 Dozen	Bottom	1½ lb.	2 lb.	2½ lb.	3 lb.	3½ lb.	4 lb.
Fancy Forms	Top	2½ lb.	2½ lb.	3 lb.	3 lb.	3½ lb.	4½ lb.

1. For all family orders, wrap Solid CO_2 for the top with one thickness of wrapping paper.
2. Seal cartons carefully.
3. This data was obtained by using a 16x16 pt. Kraft paper carton, with one double and one single walled liners.

LEGAL STANDARDS

U. S. P. H. S. Definitions

Milk is hereby defined to be the lacteal secretion obtained by complete milking of one or more healthy cows which contain not less than 8 percent of milk solids not fat and not less than 3¼ percent milk fat. It shall be practically colostrum free. Milk fat or Butter fat—Milk fat or butterfat is the fat of milk.

Skimmed Milk — Skimmed milk is milk from which a sufficient portion of milk fat has been removed to reduce its milk fat percentage to less than 3¼ percent.

Milk or Skimmed milk beverage—A milk beverage or a skimmed milk beverage is a food compound or confection consisting of milk or skimmed milk, as the case may be, to which has been added a syrup or flavor consisting of wholesome ingredients.

Buttermilk—Buttermilk is the product which remains when fat is removed from milk or cream in the process of churning. It contains not less than 8 percent of milk solids not fat.

Cultured Buttermilk, Cultured Milk, Cultured Skim Milk—Cultured buttermilk, cultured milk and cultured skimmilk are products resulting from the souring or treatment by a lactic acid or other culture of milk or milk products. They contain not less than 8 percent of milk solids not fat.

Vitamin D Milk—Vitamin D milk is milk, the Vitamin D content of which has been increased by a method and in amount approved by the Health Officer.

Reconstituted or Recombined milk and cream — Reconstituted or recombined milk is a product resulting from the recombining of milk constituents with water, and which complies with the Standards of Milk Fat and Solids-not-fat of Milk as defined herein. Reconstituted or recombined cream is a product resulting from a combination of dried cream, butter or butter fat with cream, milk, skim milk and water.

Milk Products—Milk products shall be taken to mean and include sour cream, cream, goat milk, Vitamin D milk, buttermilk, cultured buttermilk, skimmed milk, reconstituted or recombined milk and cream, milk beverages and skimmed milk beverages.

Pasteurization — The term "Pasteurization," "Pasteurized" and similar terms shall be taken to refer to the process of heating every particle of milk or milk products to at least 143 F. and holding at such temperature for at least 30 minutes or at least 160 F. and holding at such temperature for at least 15 seconds in approved and properly operated equipment.

Homogenized Milk—Homogenized milk is milk which has been treated in such a manner as to insure a break-up of the fat globules to such an extent that after a 48 hour storage, no visible cream separation occurs on the milk and the fat percentage of the top 100 ml. of milk in a quart bottle or a proportionate volume in containers of other sizes does not differ by more than 10% of itself and the fat percentage of the remaining portion as determined after proper mixing. It shall be in a quiescent state 48 hours before testing.

Milk Ordinance and Code Recommended by
U. S. Public Health Service, 1948

FEDERAL STANDARDS FOR BUTTER

"Butter" shall be understood to mean the food product usually known as butter, and which is made exclusively from milk or cream, or both, with or without common salt, and with or without additional coloring matter, and containing not less than 80 per centum by weight of milk fat, all tolerances having been allowed for.

—Act of Congress, Approved, Mar. 4, 1923.

DEFINITIONS AND STANDARDS OF IDENTITY; FILL OF CONTAINER; U. S. FOOD AND DRUG ADMINISTRATION

§ 18.500 *Cream class of food; identity.* Cream is the class of food which is the sweet, fatty liquid or semi-liquid separated from milk, with or without the addition thereto and intimate admixture therewith of sweet milk or sweet skim milk. It may be pasteurized and if it contains less than 30 percent of milk fat as determined by the method hereinafter referred to, it may be homogenized. It contains not less than 18 percent of milk fat, as determined by the method prescribed in "Official and Tentative Methods of Analysis of the Association of Official Agricultural Chemists", Fourth Edition, 1935, page 277, under "Fat, Roese-Gottlieb Method — Official". The word "milk" as used herein means cows' milk.

§ 18.501 *Light cream, coffee cream, table cream; identity.* Light cream, coffee cream, table cream, conforms to the definition and standard of identity prescribed for the cream class of food by § 18.500, except that it contains less than 30 percent of milk fat, as determined by the method referred to in such section.

§ 18.510 *Whipping cream class of food; identity.* Whipping cream is the class of food which conforms to the definition and standard of identity prescribed for the cream class of food by § 18.500, except that it contains not less than 30 percent of milk fat, as determined by the method referred to in such section.

§ 18.511 *Light whipping cream; identity.* Light whipping cream conforms to the definition and standard of identity prescribed for the whipping cream class of food by § 18.510, except that it contains less than 36 percent of milk fat, as determined by the method referred to in § 18.500.

§ 18.515 *Heavy cream, heavy whipping cream; identity.* Heavy cream, heavy whipping cream, conforms to the definition and standard of identity prescribed for the whipping cream class of food by § 18.510, except that it contains not less than 36 percent of milk fat, as determined by the method referred to in § 18.500.

§ 18.520 *Evaporated milk; identity; label statement of optional ingredients.* (a) Evaporated milk is the liquid food made by evaporating sweet milk to such point that it contains not less than 7.9 percent of milk fat and not less than 25.9 percent of total milk solids. It may contain one or both of the following optional ingredients:

(1) Disodium phosphate or sodium citrate or both, or calcium chloride, added in a total quantity of not more than 0.1 percent by weight of the finished evaporated milk.

(2) Vitamin D in such quantity as increases the total vitamin D content to not less than 7.5 U. S. P. units per avoirdupois ounce of finished evaporated milk.

It may be homogenized. It is sealed in a container and so processed by heat as to prevent spoilage.

(b) When optional ingredient (a) (2) is present, the label shall bear the statement, "With Increased Vitamin D Content" or "Vitamin D Content Increased". Such statement shall immediately and conspicuously precede or follow the name "Evaportaed Milk", without intervening written, printed, or graphic matter, wherever such name appears on the label so conspicuously as to be easily seen under customary conditions of purchase.

(c) For the purpose of this section:

(1) The word "milk" means cows' milk.

(2) Such milk may be adjusted, before or after evaporation, by the addition or abstraction of cream or sweet skim milk, or by the addition of concentrated sweet skim milk.

(3) The quantity of milk fat is determined by the method prescribed under "Fat—Official" on page 280, and the quantity of total milk solids is determined by the method prescribed under "Total Solids—Official" on page 279, of "Official and Tentative Methods of Analysis of the Association of Official Agricultural Chemists", Fourth Edition, 1935.

(4) Vitamin D content may be increased by the application of radiant energy or by the addition of a concentrate of vitamin D (with any accompanying vitamin A when such vitamin D in such concentrate is obtained from natural sources) dissolved in a food oil; but if such oil is not milk fat the quantity thereof added is not more than 0.01 percent of the weight of the finished evaporated milk.

(5) The quantity of vitamin D is determined by the method prescribed in the "The Second Supplement to the Pharmacopoeia of the United States of America Eleventh Decennial Revision", pages 132-134 inclusive, and pages 136-138 inclusive, with such modification of the method of feeding as is necessary for evaporated milk instead of an oil. [Reg., June 28, 1940; effective Mar. 1, 1941; 5 F.R. 2444, 3621]

§ 18.525 *Concentrated milk, plain condensed milk; identity; label statement of optional ingredients.* Concentrated milk, plain condensed milk, conforms to the definition and standard of identity, and is subject to the requirements for label statement of optional ingredients, prescribed for evaporated milk by § 18.520, except that:

(a) It is not processed by heat;

(b) Its container may be unsealed; and

(c) Optional ingredient § 18.520 (a) (1) is not used.

§ 18.530 *Sweetened condensed milk; identity.* (a) Sweetened condensed milk is the liquid or semi-liquid food made by evaporating a mixture of sweet milk and refined sugar (sucrose) or any combination of refined sugar (sucrose) and refined corn sugar (dextrose) to such point that the finished sweetened condensed milk contains not less than 28.0 percent of total milk solids and not less than 8.5 percent of milk fat. The quantity of refined sugar (sucrose) or combination of such sugar and refined corn sugar (dextrose) used is sufficient to prevent spoilage.

(b) For the purpose of this section:

(1) The word "milk" means cows' milk.

(2) Such milk may be adjusted, before or after evaporation, by the addition or abstraction of cream or sweet skim milk, or the addition of concentrated sweet skim milk.

(3) Milk fat is determined by the method prescribed in "Official and Tentative Methods of Analysis of the Association of Official Agricultural Chemists", Fourth Edition, 1935, page 281, under "Fat—Official".

§ 18.535 *Condensed milks which contain corn sirup; identity.* *See* p. 50.

§ 18.540[1] *Dried skim milk, powdered skim milk, skim milk powder; identity.* Dried skim milk, powdered skim milk, skim milk powder, is the food made by drying sweet skim milk. It contains not more than 5 percent of moisture, as determined by the method prescribed in "Official and Tentative Methods of Analysis of the Association of Official Agricultural Chemists", Fourth Edition, 1935, page 282, under the caption "Moisture—Tentative". The term "skim milk" as used herein means cows' milk from which the milk fat has been separated.

§ 19.525 *Cottage cheese; identity.* (a) Cottage cheese is the soft uncured cheese prepared by the procedure set forth in paragraph (b) of this section. The finished cottage cheese contains not more than 80 percent of moisture, as determined by the method prescribed under "Moisture—Official" on page 301 of "Official and Tentative Methods of Analysis of the Association of Official Agricultural Chemists", Fifth Edition, 1940.

(b) (1) Sweet skim milk is pasteurized; calcium chloride may be added in a quantity not more than 0.02 percent (calculated as anhydrous calcium chloride) of the weight of such skim milk; harmless lactic-acid-producing bacteria, with or without rennet, are added and it is held until it becomes coagulated. The coagulated mass may be cut; it may be warmed; it may be stirred; it is then drained. The curd may be washed with water and further drained; it may be pressed, chilled, worked, seasoned with salt.

(2) For the purposes of this section skim milk means the milk of cows from which the milk fat has been separated.

§ 19.530 *Creamed cottage cheese, identity.* (a) Creamed cottage cheese is the soft uncured cheese prepared by mixing cottage cheese with pasteurized cream or a pasteurized mixture of cream with milk or skim milk or both. Such cream or mixture is used in such quantity that the milk fat added thereby is not less than 4 percent by weight of the finished creamed cottage cheese. The finished creamed cottage cheese contains not more than 80 percent of moisture as determined by the method prescribed under "Moisture—Official" on page 301 of "Official and Tentative Methods of Analysis of the Association of Official Agricultural Chemists", Fifth Edition, 1940.

(b) For the purposes of this section milk means sweet milk of cows and skim milk means milk from which the milk fat has been separated.

[1]NOTE: Public Law 244, 78th Congress, approved March 2, 1944, provides a statutory definition for this food under the names: "Non-fat dry milk solids" and "defatted milk solids."

LEGAL STANDARDS FOR MILK PRODUCTS BY STATES

Minimum Percentage Composition

State	Fat	Milk S.N.F.	T.M.S.	Milk T.M.S.	Cream	Butter	Sweetened Cond.	Milk
				Skim	Fat	Fat	Fat	T.M.S.
Federal	......	......	......	......	[a]18	80	8.5	28.0
Alabama	3.25	8.50	11.75	8.50	18	[b]80	8.0	28.0
Alaska	3.25	8.00	11.25	8.00	18	Fed.	Fed.	Fed.
Arizona	3.25	8.50	11.75	8.80	[c]18	[b]80	8.5	28.0
Arkansas	3.25	8.00	11.25	8.00	18	80	8.0	28.0
California	3.50	8.15	......	8.50	[d]20	80	8.5	28.0
Colorado	3.20	......	......	......	18	[b]80	7.7	28.0
Connecticut	3.25	8.50	11.75	......	[e]	80	8.0	28.0
Delaware	3.50	......	12.00	......	18	....	8.0	......
Dist. of Columbia	3.50	......	11.50	9.00	20	80	....	......
Florida	3.25	8.50	......	......	18	[b]80	8.0	28.0
Georgia	3.25	8.50	11.75	9.25	18	80	8.0	28.0
Hawaii	3.00	8.50	......	......	[f]18	80	....	......
Idaho	3.20	8.00	11.00	9.30	18	80	....	......
Illinois	3.00	8.50	11.50	9.25	[f]18	80	8.5	28.0
Indiana	3.25	8.00	......	8.50	[f, g]18	80	8.5	28.0
Iowa	3.00	8.50	11.50	[h]	16	80	8.0	28.0
Kansas	3.25	......	......	......	18	80	8.0	28.0
Kentucky	3.25	8.00	......	......	18	80	8.0	28.0
Louisiana	3.50	8.50	12.00	8.00	18	80	8.0	28.0
Maine	3.25	8.50	11.75	......	18	....	[i]	26.5
Maryland	3.50	......	12.00	......	18	80	8.5	28.0
Massachusetts	3.35	......	12.00	9.30	[j]16	80	8.0	24.0
Michigan	3.00	8.50	......	......	[f]18	80	....	......
Minnesota	3.25	......	......	......	[f, g]20	80	8.5	28.0
Mississippi	3.00	8.50	11.75	8.50	18	80	8.0	28.0
Missouri	3.25	8.00	......	......	[f]18	80	8.0	28.0
Montana	3.25	8.00	11.25	8.00	20	80	8.0	28.0
Nebraska	3.00	......	......	9.25	18	80	....	......
Nevada	3.25	8.50	11.75	......	22	80	8.0	28.0
New Hampshire	3.35	......	11.85	8.50	18	[b]80	8.5	28.0
New Jersey	3.00	......	11.50	......	16	[b]80	8.0	28.0
New Mexico	3.25	8.50	11.75	......	18	[b]80	8.0	28.0
New York	3.00	......	11.50	8.50	[l]18	80	8.0	28.0
North Carolina	3.25	8.50	11.75	......	18	80	8.5	28.0
North Dakota	3.25	......	11.50	8.50	20	80	8.5	28.0
Ohio	3.00	......	12.00	......	[f]18	80	[n]	......
Oklahoma	3.25	8.00	11.25	......	18	[b]80	Fed.	Fed.
Oregon	3.20	8.50	11.70	......	18	80	....	......
Pennsylvania	3.25	......	12.00	......	18	80	8.0	28.0
Rhode Island	3.25	......	12.00	......	[o]18	80	8.0	28.0
South Carolina	3.25	8.00	11.25	8.00	18	....	8.0	28.0
South Dakota	3.25	8.50	11.75	9.25	18	80	8.0	28.0
Tennessee	3.25	8.50	12.00	......	[f]18	80	8.0	28.0
Texas	3.25	8.00	......	......	[f]18	80	8.0	28.0
Utah	3.20	8.30	11.50	8.50	[f]18	80	7.8	25.0
Vermont	3.25	8.50	11.75	8.50	18	[b]80	8.0	28.0
Virginia	3.25	8.50	11.75	8.50	18	80	7.8	25.5
Washington	3.25	8.00	......	......	[q]18	80	7.8	25.5
West Virginia	3.00	8.50	11.50	9.00	[f]18	80	7.8	28.0
Wisconsin	3.00	8.25	......	8.50	[f]18	[r]82.5	8.0	28.0
Wyoming	3.00	8.50	11.50	9.25	[f]18	80	7.9	25.9

LEGAL STANDARDS FOR MILK PRODUCTS BY STATES (Continued)

State	Evaporated Milk		Plain Ice Cream		Fruit or Nut Ice Cream		Frozen Custard	
	Fat	T.M.S.	Fat	T.M.S.	Fat	T.M.S.	Fat	T.M.S.
Federal	7.9	25.9	....	....	....	....	....	....
Alabama	7.8	25.5	10	18	8	18	10	18
Alaska	Fed.	Fed.	10	18	8	14	8	....
Arizona	7.9	25.9	10	....	8	....	10	....
Arkansas	7.9	25.5	10	18	8	14	10	14
California	7.9	25.9	10	....	8	....	10	....
Colorado	7.7	......	12	....	10	....	....	....
Connecticut	7.8	25.5	10	....	8	....	....	....
Delaware	7.8	25.5	12	17	10	....	12	....
Dist. of Columbia	....	......	8	....	8	....	8	18
Florida	7.8	25.5	10	18	8	14	10	18
Georgia	7.8	25.5	10	....	8	....	8	....
Hawaii	....	......	12	18	10	15	12	18
Idaho	7.8	25.5	12	18	10	14	....	....
Illinois	7.9	25.9	12	....	10	....	....	....
Indiana	7.9	25.9	10	18	8	14	10	18
Iowa	7.8	25.5	12	20	10	16	....	....
Kansas	7.8	25.5	10	20	10	20	....	....
Kentucky	7.8	25.5	10	18	10	18	10	18
Louisiana	7.9	25.5	10	20	8	20	8	18
Maine	i	26.5	12	....	10	....	....	....
Maryland	7.9	25.9	12	20	8	15	12	20
Massachusetts	7.8	......	10	18.5	8	16.5	10	18.5
Michigan	....	......	12	k	10	k	12	k
Minnesota	7.9	25.9	12	20	10	16	12	20
Mississippi	7.8	25.5	10	....	8	....	8	16
Missouri	7.9	25.9	8	....	8	....	....	....
Montana	7.8	25.5	10	....	9	....	10	....
Nebraska	....	......	12	20	10	18	12	20
Nevada	7.8	25.5	14	....	12	....	....	....
New Hampshire	7.9	25.9	14	....	12	....	12	....
New Jersey	7.8	25.5	10	....	8	....	10	
New Mexico	7.8	25.5	12	20	10	....	....	
New York	7.8	25.5	10	18	8	14	10	18
North Carolina	7.9	25.7	10	....	8	....	8	16
North Dakota	7.9	25.9	12	m	10	m	....	....
Ohio	....	......	10	18	8	14	10	18
Oklahoma	Fed.	Fed.	10	21	....	....	....	....
Oregon	7.8	25.5	12	20	10	18	....	....
Pennsylvania	7.8	25.5	10	....	8	....	10	....
Rhode Island	7.8	25.5	8	....	8	....	8	....
South Carolina	....	......	10	18	10	18	10	18
South Dakota	7.8	25.5	12	k	10	k	12	k
Tennessee	7.8	p25.5	10	18	8	14	8	14
Texas	7.8	25.5	8	....	6	....	....	....
Utah	7.8	25.5	12	....	12	....	....	....
Vermont	7.8	25.5	14	....	12	....	....	....
Virginia	7.8	25.5	10	....	8	....	8	....
Washington	7.8	25.5	10	20	10	20	....	....
West Virginia	7.8	25.5	8	18	8	18	....	....
Wisconsin	8.0	26.15	13	21	11	....	13	....
Wyoming	7.9	25.9	10	18	10	18	10	18

Sources: Bureau of Dairy Industry, Jan., 1950.
Int'l Ass'n of Ice Cream Mfrs., Dec., 1951.

FOOTNOTES FOR TABLE OF LEGAL STANDARDS

[a] Light, coffee, or table cream, not less than 18% milk fat but less than 30% milk fat; light whipping cream, not less than 30% milk fat but less than 36% milk fat; heavy or whipping cream, not less than 36% milk fat.

[b] Butter, water not more than 16%.

[c] Heavy cream, milk fat not less than 30%; extra heavy cream, milk fat not less than 40%.

[d] Whipping cream, milk fat not less than 35%.

[e] Extra light cream, milk fat not less than 12% but less than 16%; light cream, milk fat not less than 16%, but less than 26%; medium cream, milk fat not less than 26% but less than 36%; heavy cream, milk fat not less than 36%

[f] Whipping cream, milk fat not less than 30%.

[g] Heavy cream, milk fat not less than 36%.

[h] Skim milk, total milk solids less than 11.5%, or milk fat less than 3%.

[i] Milk fat not less than 29.5% of milk solids.

[j] Medium cream, milk fat not less than 25%; heavy cream, milk fat not less than 34%; extra heavy cream, milk fat not less than 38%.

[k] Milk solids, not less than 0.9 lb. per gallon.

[l] Medium cream, milk fat not less than 25%; heavy cream, milk fat not less than 36%.

[m] Milk solids, not less than 0.8 lb. per gallon.

[n] Must be made from whole milk.

[o] Heavy cream, milk fat not less than 38%.

[p] Total milk solids plus milk fat not less than 33.7%.

[q] Whipping cream, milk fat not less than 32%.

[r] 2.5% tolerance.

PROPERTIES OF DAIRY AND RELATED PRODUCTS

ANALYSES OF COW'S MILK BY DIFFERENT ANALYSTS

	Van Slyke	Babcock	Richmond	Oliver	Fleischmann	Willoughby	Van Slyke (cheese factory milk)
Water	87.1	87.3	87.1	87.6	87.75	87.0	87.4
Fat	3.9	3.6	3.9	3.25	3.40	3.8	3.75
Casein	2.5	3.0	3.0	3.40	2.80	3.3	2.45
Albumin	0.7	0.6	0.4	0.45	0.70	0.4	0.7
Total protein	3.2	3.8	3.4	3.85	3.50	3.7	3.15
Lactose	5.1	4.5	4.75	4.55	4.60	4.8	5.0
Ash	0.7	0.7	0.75	0.75	0.75	0.7	0.7
Total solids	12.9	12.7	12.9	12.4	12.25	13.0	12.6

SHOWING RATIO OF SOLIDS NOT FAT TO FAT IN AVERAGE MILK OF DIFFERENT BREEDS

Breed	Total Solids Per Cent	Fat Per Cent	Solids Not Fat Per Cent	Ratio of Solids Not Fat to Fat
Holstein	12.29	3.45	8.84	1 : .390
Brown Swiss	12.46	3.68	8.78	1 : .419
Ayrshire	12.98	3.85	9.13	1 : .420
Shorthorn	12.50	3.80	8.70	1 : .437
Devon	14.50	4.60	9.90	1 : .465
Jersey	14.90	5.14	9.76	1 : .527
Guernsey	14.20	4.98	9.22	1 : .540

—From Condensed Milk and Milk Powder (Hunziker).

SPECIFIC HEATS OF MILK AND CREAM

Temperature F.	Temperature C.	Milk	20% Cream	40% Cream	Temperature F.	Temperature C.	Milk	20% Cream	40% Cream
35.6	2.0	....	0.88	....	95.9	35.5	0.93	0.89	0.86
37.4	3.0	0.92	....	0.83	100.4	38.0	....	....	.80
43.7	6.5	.92	.91	.90	105.8	41.0	.92	.87	....
48.2	9.0	....	.92	....	109.4	43.0	....	....	.78
51.8	11.0	.93	....	.96	114.8	46.0	....	.87	....
55.4	13.0	....	.94	....	118.4	48.0	.92	....	.78
59.0	15.0	.94	.95	1.02	123.8	51.0	....	.86	....
66.2	19.0	.95	1.01	1.07	127.4	53.0	....	....	.77
71.6	22.0	.94	.95	....	131.0	55.0	.93	.86	....
75.2	24.0	....	....	.93	136.4	58.0	....	....	.76
78.8	26.0	.93	....	.88	141.8	61.0	.93	.87	....
82.4	28.0	....	.91	....	145.4	63.0	....	....	.72
86.0	30.0	.92	....	.88	150.8	66.0	.94	....	....
89.6	32.0	....	.91	....					

—U. S. Dept. of Agri., Bul. No. 98
"Application of Refrigeration to the Handling of Milk" (Bowen).

RATIO OF FAT TO SOLIDS—NOT—FAT IN MILK OF VARIOUS FAT PERCENTAGES

Fat	SNF	Total Solids	SNF Fat	Fat	SNF	Total Solids	SNF Fat
3.0	8.15	11.15	2.72	4.6	9.14	13.74	1.99
3.1	8.22	11.32	2.65	4.7	9.18	13.88	1.95
3.2	8.29	11.49	2.59	4.8	9.22	14.02	1.92
3.3	8.36	11.66	2.53	4.9	9.26	14.16	1.89
3.4	8.43	11.83	2.48	5.0	9.30	14.30	1.86
3.5	8.50	12.00	2.43	5.1	9.33	14.43	1.83
3.6	8.57	12.17	2.38	5.2	9.36	14.56	1.80
3.7	8.64	12.34	2.34	5.3	9.39	14.69	1.77
3.8	8.71	12.51	2.29	5.4	9.42	14.82	1.74
3.9	8.78	12.68	2.25	5.5	9.45	14.95	1.72
4.0	8.85	12.85	2.21	5.6	9.47	15.07	1.69
4.1	8.90	13.00	2.17	5.7	9.49	15.19	1.66
4.2	8.95	13.15	2.13	5.8	9.51	15.31	1.64
4.3	9.00	13.30	2.09	5.9	9.53	15.43	1.62
4.4	9.05	13.45	2.06	6.0	9.55	15.55	1.59
4.5	9.10	13.60	2.02				

CHLORIDES IN MILK

Chloride content of normal milk from individual cows. All colostrum milk, milk from diseased udders, and milk within one month of the end of lactation, were eliminated.

Lactation time months.	Number of cows represented	Average 0/0	Chlorides in the Milk Highest 0/0	Lowest 0/0
1	16	.093	.109	.076
2	13	.102	.120	.083
3	8	.099	.116	.080
4	14	.098	.121	.081
5	14	.098	.112	.081
6	17	.103	.122	.083
7	11	.112	.130	.095
8	16	.111	.130	.094
9	17	.117	.143	.088

SPECIFIC HEATS OF MILK AND MILK DERIVATIVES
(Including Heat Required to Melt Fat if This Factor Enters)

	At 0° (32° F.)	At 15° (59° F.)	At 40° (104° F.)	At 60° (140° F.)
Whey	0.978	0.976	0.974	0.972
Skimmilk	0.949	0.943	0.952	0.963
Whole milk	0.920	0.938	0.930	0.918
15% cream	0.750	0.923	0.899	0.900
20% cream	0.723	0.940	0.880	0.886
30% cream	0.673	0.983	0.852	0.860
45% cream	0.606	1.016	0.787	0.793
60% cream	0.560	1.053	0.721	0.737
Butter	(0.512)	(0.527)	0.556	0.580
Butterfat	(0.445)	(0.467)	0.500	0.530

(For butter and butterfat)

Values in parentheses were obtained by extrapolation, under assumption that the specific heat is about the same in the solid and liquid states.

—Iowa Agr. Sta. Res. Bul. No. 14, "The Specific Heat of Milk and Milk Derivatives" (Hammer and Johnson).

ACIDITY OF FRESH CREAM

Assume the acidity of 4% milk is .16%, then the acid in cream would be

$100-4=96\%$ serum carrying .16 lbs. acid

or $\frac{.16}{96}\times 100=.166\%$ in serum

$40\%=100-40=60$ lbs. serum

$\frac{60\times .166}{100}=.09960$ lbs. or .09960% acid in 40% cream

10% cream = .1494% acid
15% cream = .1411% acid
20% cream = .1328% acid
25% cream = .1245% acid
30% cream = .1162% acid
35% cream = .1078% acid
40% cream = .0996% acid
45% cream = .0913% acid
50% cream = .083% acid

—C. D. Dahle, Professor of Dairy Mfrs., Penn. State College.

THE WATER, FAT AND SOLIDS NOT FAT CONTENT OF DIFFERENT DAIRY PRODUCTS DERIVED FROM A CERTAIN WHOLE MILK, IN PERCENTAGES

Constituents	Whole Milk	Skim-milk	Cream	Butter (un-salted)	Butter-milk	Sweet-ened Con-densed Whole Milk	Plain Bulk Con-densed Skim-milk	Plain Bulk Con-densed Whole Milk	Evap-orated Milk	Ice Cream Mix	Whole Milk Powder	Skim-milk Powder
Water	87.75	91.08	54.62	16.00	90.83	26.50 45.50 sugar	74.50	70.00	74.50	66.00 13.50 sugar, gelatin and flavors	5.00	5.00
Fat	3.75	.10	40.00	83.00	.40	8.00	.29	8.00	7.80	8.00	29.09	1.07
Milk solids not fat	8.50	8.82	5.38	1.00	8.77	20.00	25.21	22.00	17.70	12.50	65.91	93.93
Total solids	12.25	8.92	45.38	84.00	9.17	73.50	25.50	30.00	25.50	34.00	95.00	95.00
Pounds of finished product obtained from 1000 lbs. whole milk testing as indicated under whole milk. No allowance for losses in manufacture.	1000	909	91	44	47	425	397	386	481	680	129	85
						3.5 lbs. excess fat to be re-moved in stand-ardiz-ing				17.0 lbs. fat short to be added in stand-ardiz-ing		

—From The Technical Control of Dairy Products (Mojonnier & Troy).

APPROXIMATE WEIGHT PER GALLON OF MILK AND CREAM AT VARIOUS TEMPERATURES

Degrees Fahrenheit	40	45	50	55	60	65	70	75
	Lbs.	Lbs.	Lbs.	Lbs.	Lbs.	Lbs.	Lbs.	Lbs.
Skimmilk	8.66	8.655	8.65	8.645	8.640	8.635	8.63	8.625
4% Milk.......	8.61	8.605	8.60	8.595	8.590	8.585	8.58	8.570
20% Cream.....	8.54	8.525	8.51	8.495	8.480	8.460	8.45	8.435
25% Cream.....	8.50	8.485	8.47	8.450	8.435	8.420	8.40	8.385
30% Cream.....	8.47	8.450	8.43	8.410	8.390	8.370	8.35	8.330
35% Cream.....	8.45	8.425	8.40	8.375	8.350	8.330	8.31	8.290
40% Cream.....	8.42	8.395	8.37	8.345	8.320	8.295	8.27	8.250

Degrees Fahrenheit	80	85	90	95	100	105	110
	Lbs.	Lbs.	Lbs.	Lbs.	Lbs.	Lbs.	Lbs.
Skimmilk	8.620	8.610	8.60	8.590	8.58	8.57	8.555
4% Milk........	8.565	8.560	8.55	8.535	8.52	8.51	8.500
20% Cream.....	8.420	8.405	8.39	8.380	8.37	8.36	8.350
25% Cream.....	8.370	8.355	8.34	8.325	8.31	8.30	8.290
30% Cream.....	8.315	8.295	8.28	8.265	8.25	8.24	8.230
35% Cream.....	8.270	8.250	8.23	8.215	8.20	8.19	8.175
40% Cream.....	8.230	8.210	8.19	8.170	8.15	8.14	8.125

Degrees Fahrenheit	115	120	125	130	135	140	145
	Lbs.	Lbs.	Lbs.	Lbs.	Lbs.	Lbs.	Lbs.
Skimmilk	8.542	8.530	8.517	8.505	8.490	8.480	8.470
4% Milk........	8.485	8.475	8.465	8.450	8.440	8.430	8.420
20% Cream.....	8.335	8.325	8.315	8.300	8.290	8.280	8.270
25% Cream.....	8.280	8.270	8.260	8.250	8.240	8.230	8.220
30% Cream.....	8.215	8.205	8.195	8.185	8.170	8.160	8.150
35% Cream.....	8.165	8.155	8.145	8.130	8.120	8.110	8.100
40% Cream.....	8.110	8.100	8.090	8.075	8.065	8.050	8.045

—From Bureau of Dairy Industry,
U. S. Dept. of Agriculture.

WEIGHT OF MILK PRODUCTS ACCORDING TO U. S DEPARTMENT OF AGRICULTURE, APPROXIMATELY, AT A TEMPERATURE OF 68° F.

	Per Cent Butter-fat	Pounds per Gal.		Per Cent Butter-fat	Pounds per Gal.
Skimmilk	0.025	8.63	Mixed milk and cream	15.00	8.47
Fresh milk	3.00	8.60	Cream testing	20.00	8.43
Fresh milk	3.50	8.59	Cream testing	25.00	8.39
Fresh milk	4.00	8.58	Cream testing	30.00	8.35
Fresh milk	5.00	8.58	Cream testing	35.00	8.31
Fresh milk	6.00	8.57	Cream testing	40.00	8.28
Mixed milk and cream	10.00	8.53			

For all practical purposes milk testing from 3 to 5 per cent butterfat may be figured at 8.6 pounds per gallon.

Average Weight of Milk Products (about 68° F.)

	Pounds to Gallon
1 gal. ice cream mix ready to freeze	9.10
1 gal. of water	8.34
1 gal. of plain condensed skimmilk	9.18
1 gal. of plain condensed whole milk	9.05
1 gal. of evaporated milk	8.89
1 gal. of sweetened condensed skimmilk	11.16
1 gal. of sweetened condensed whole milk	10.90

NOTE: The above table is correct only for milk and cream standardized from skimmilk and cream; it does not apply to normal milk as obtained from the cow.

WEIGHTS PER GALLON OF FRUITS AND SYRUPS

Chocolate syrup (hot)	9.4 lbs.
Chocolate syrup (cold)	9.7 lbs.
Cherries	10.0 lbs.
Nuts in syrup	10.0 lbs.
Pineapple (before cooking)	8.85 lbs.
Frozen pudding	10.3 lbs.
Strawberries (2 to 1 pack)	9.5 lbs.
Scotch Bob	12.0 lbs.
Home-made	10.0 lbs.

Equivalents: To change a given amount of products listed below into quarts liquid, divide amount by factor opposite each product listed:

	Factor	
Lbs. dry nuts	2.5	(For example: 25 lbs. nuts put into mix occupy space of 10 qts. of mix)
Lbs. gelatin	2.5	
Lbs. egg powder or "AA"	2.3	
Lbs. skimmilk powder	2.3	
Lbs. cane sugar or cerelose	3.3	
Lbs. raisins (dry)	2.5	
Doz. bananas (average size)	1.0	(i.e. 1 doz. bananas = 1 qt. pulp)
Lbs. bananas (as purchased)	3.0	
Lbs. shredded cocoanut	2.5	
Lbs. ground macaroons	3.0	
Lbs. peanut candy	3.0	
12-oz. packages of grape nuts	4.0	(i.e. 4 packages = 1 quart)
Lbs. butterscotch	2.5	
Doz. eggs (whole)	1.6	
Doz. eggs (yolks)	4.6	

10 doz. bananas = 10 qts. (or) 30 lbs. bananas = 10 qts.

—From the Research and Production Dept.,
Eastern Dairies, Springfield, Mass.

AVERAGE COMPOSITION AND WEIGHTS PER GALLON OF INGREDIENTS USED IN ICE CREAM MIX

Ingredients	Per Cent Fat	Per Cent Milk S.N.F.	Per Cent Sugar	Per Cent Total Solids	Weight per Gal. Lbs.	Weight per Qt. Lbs.
Water	0.00	0.00	0.00	0.00	8.3	2.075
Skimmilk	0.00	8.5	0.00	8.5	8.6	2.15
Milk	3.5	8.33	0.00	11.33	8.6	2.15
Milk	4.00	8.79	0.00	12.79	8.6	2.15
Milk	5.00	9.10	0.00	14.10	8.6	2.15
Cream	12.00	7.8	0.00	19.80	8.5	2.125
Cream	15.00	7.57	0.00	22.57	8.5	2.125
Cream	18.00	7.31	0.00	25.31	8.5	2.125
Cream	20.00	7.13	0.00	27.13	8.5	2.125
Cream	25.00	6.68	0.00	31.68	8.4	2.1
Cream	30.00	6.24	0.00	36.24	8.4	2.1
Cream	35.00	5.69	0.00	40.79	8.3	2.075
Cream	40.00	5.35	0.00	45.35	8.2	2.05
Evap. milk (canned)	8.00	20.00	0.00	28.00	8.8	2.2
Evap. milk (bulk)	.....	.....	.....	.....	.....	.....
Plain cond.	9.5	23.00	0.00	33.00	9.0	2.25
Sweetened cond.	.....	.....	.....	.....	.....	.....
Whole milk	8.00	26.00	40.00	74.00	11.0	2.75
Skim cond. (unsweet.)	0.00	28.00	0.00	28.00	9.0	2.25
Skim cond. (sweet.)	0.00	28.00	40.00	68.00	11.0	2.75
Skimmilk powder	0.00	95.00	0.00	95.00	4.5	.....
Granulated sugar	0.00	0.00	100.00	100.00	7.5	1.875
Invert sugar syrup	0.00	0.00	71.50	71.50	10.0	.....
Corn sugar	0.00	0.00	70.00	90.00	.....	.....
Corn syrup	0.00	0.00	50.00	64.00	12.0	3.0
Gelatin or ice cream powder	0.00	0.00	0.00	90.00	.....	.....
Butter, unsalted	83.00	1.00	0.00	84.00	.....	.....
Butter, oil	100.00	0.00	0.00	100.00	7.5	.....
Dried egg yolk	45.00	0.00	0.00	95.00	.....	.....
Fluid egg	9.30	0.00	0.00	34.5	1.5 lbs. per doz.	.....
Cocoa	24.00	0.00	0.00	97.00	.....	.....
Chocolate liquor	52.00	0.00	0.00	97.00	.....	.....
Sugared cream	30.00	3.00	25.00	58.00	9.15	2.2875

—From the Research and Production Dept., Eastern Dairies, Springfield, Mass.

AMOUNTS OF NUTRIENTS IN A POUND OF MILK AS COMPARED WITH A POUND OF MEAT, BREAD AND OTHER FOOD PRODUCTS*

Food Materials	Refuse	Edible Portion					Fuel Value
		Nutrients					
	Lb.	Water Lb.	Protein Lb.	Fat Lb.	Carbo-hydrates Lb.	Min. Matter Lb.	Cal.
Milk (1 pint or 1 pound):							
Whole Milk	.....	0.87	0.03	0.04	0.05	0.01	325
Skimmilk (0.3 per cent fat)	.....	0.90	0.04	.....	0.05	0.01	170
Buttermilk	.....	0.91	0.03	0.01	0.05	0.01	165
Other food materials (1 lb. each):							
Cheese	.....	0.34	0.26	0.34	0.02	0.04	1965
Butter	.....	0.11	0.01	0.85	.....	0.03	3605
Beef:							
Round	0.08	0.61	0.18	0.12	.....	0.01	870
Shoulder clod	.....	0.69	0.19	0.11	.....	0.01	835
Sirloin	0.13	0.53	0.16	0.17	.....	0.01	1040
Fore quarters	0.19	0.50	0.14	0.16	.....	0.01	950
Hind quarters	0.16	0.51	0.15	0.17	.....	0.01	1000
Mutton, side	0.19	0.43	0.13	0.24	.....	0.01	1275
Pork:							
Loin	0.16	0.44	0.14	0.25	.....	0.01	1340
Ham	0.14	0.35	0.13	0.34	.....	0.04	1655
Salt, fat	.....	0.07	0.02	0.87	.....	0.04	3715
Chicken	0.35	0.48	0.15	0.01	.....	0.01	325
Codfish:							
Fresh	0.30	0.58	0.11	.....	.....	0.01	205
Salt	0.25	0.40	0.16	.....	.....	0.19	315
Mackerel, salt	0.23	0.38	0.17	0.17	.....	0.10	1050
Oysters, solids	.....	0.88	0.06	0.02	0.03	0.01	235
Wheat flour	.....	0.12	0.11	0.01	0.75	0.01	1645
Corn meal	.....	0.13	0.09	0.02	0.75	0.01	1655
Oatmeal	.....	0.07	0.16	0.07	0.68	0.02	1860
Wheat bread	.....	0.35	0.10	0.01	0.53	0.01	1205
Crackers	.....	0.08	0.11	0.10	0.69	0.02	1895
Dried beans	.....	0.13	0.22	0.02	0.59	0.04	1590
Beets	0.20	0.70	0.01	.....	0.08	0.01	170
Potatoes	0.15	0.67	0.02	.....	0.15	0.01	325
Turnips	0.30	0.62	0.01	.....	0.06	0.01	135
Apples	0.25	0.62	0.01	.....	0.12	.....	255

* U. S. Dept. of Agr., Farmers Bulletin No. 74.

—The Butter Industry (Hunziker)

NUTRIENT MATERIALS IN VARIOUS DAIRY PRODUCTS

Showing Percentage Composition, and Calorie and Vitamin Content Per 100 Grams

Dairy Product	Per Cent by Weight: Butter Fat	Protein	Carbohydrate	Calcium	Phosphorus	Calories per 100 Grams	Vitamin Content per 100 Grams: A I.U.	Thiamin (B_1)-Mg.	D I.U.	Riboflavin (G)-Mg.
Whole Milk	3.5	3.4	5.0	.12	.093	65	141	.054	3	.17
(plain or homogenized[1])	4.0	3.4	4.9	.12	.093	69	161	.054	3	.17
	4.5	3.3	4.9	.12	.093	73	181	.054	4	.17
	5.0	3.3	4.9	.12	.093	78	201	.054	4	.17
	5.5	3.3	4.9	.12	.093	82	221	.054	5	.17
Whole Milk	3.5	3.4	5.0	.12	.093	65	141	.054	44	.17
(Vitamin D 400 U/qt.	4.0	3.4	4.9	.12	.093	69	161	.054	44	.17
plain or homogenized[1])	4.5	3.3	4.9	.12	.093	73	181	.054	45	.17
	5.0	3.3	4.9	.12	.093	78	201	.054	45	.17
	5.5	3.3	4.9	.12	.093	82	221	.054	46	.17
Cream, Sweet	18.0	2.7	4.1	.1	.089	186	1006	.036	13	.2
	20.0	2.7	4.0	.1	.087	207	1120	.036	20	.2
	25.0	2.5	3.8	.095	.083	253	1406	.036	20	.1
	30.0	2.4	3.6	.09	.078	296	1681	.033	27	.1
	35.0	2.3	3.4	.085	.074	340	1966	.033	27	.1
	40.0	2.1	3.2	.08	.07	384	2241	.033	33	.1
	45.0	2.0	3.0	.075	.065	427	2526	.033	40	.1
Cream, Sour	16.0	2.8	4.0	.1	.090	171	900	.036	13	.2
	18.0	2.7	3.9	.1	.089	188	1007	.036	13	.2
	20.0	2.7	3.8	.1	.087	206	1120	.036	20	.2
Chocolate Drink and	1.0	3.3	10.9	.114	.088	66	39	.030	1	.09
Chocolate Milk	1.5	3.3	10.8	.113	.088	70	58	.030	1	.08
(Sugar 6%, Cocoa 1.6%)	2.0	3.2	10.8	.112	.088	74	78	.030	1	.08
	2.5	3.2	10.7	.111	.087	78	97	.030	2	.08
	3.0	3.2	10.7	.110	.086	82	116	.030	2	.08
	3.5	3.2	10.6	.109	.086	86	136	.030	2	.08
	4.0	3.2	10.6	.108	.085	91	155	.030	3	.08
Skimmed Milk	0.1	3.5	5.0	.122	.096	35	nil	.042	0	.14
Buttermilk	0.5	3.5	5.0	.122	.097	38	21	.033	0	.09
	1.0	3.5	5.0	.121	.096	43	42	.033	1	.09
	2.0	3.4	4.9	.120	.095	52	83	.033	2	.09
	3.0	3.4	4.9	.119	.094	60	125	.030	2	.09
	3.5	3.4	4.9	.118	.093	65	146	.030	3	.09
Cottage Cheese	0.5	21.0	4.3	.083	.264	106	23	†	33	†
	1.0	20.9	4.3	.083	.263	110	43	†	66	†
	2.0	20.7	4.3	.082	.260	118	83	†	165	†

	3.0	20.5	4.2	.081	.257	126	125	†	198	†
	4.0	20.3	4.2	.080	.255	134	158	†	297	†
	5.0	20.1	4.1	.080	.252	142	215	†	330	†
	6.0	19.9	4.1	.079	.250	150	248	†	396	†
Ice Cream										
T.S. Sugar										
37% 15.5%	10	4.2	21.6	.15	.12	193	331	.042	11	.12
39% 15.5%	12	4.2	21.6	.15	.12	206	399	.042	14	.12
41% 15.5%	14	4.2	21.6	.15	.12	229	467	.042	17	.12
42% 15.5%	16	3.8	21.0	.147	.117	243	544	.039	19	.12
44% 15.5%	18	3.8	21.0	.144	.115	261	621	.039	21	.11
Sherbet—1.4% MSNF.	1.5	0.9	30.0	.12	.095	138	63	.021	1	.07
Butter	80.5	1.0	0	.02	.02	730	4150	0	60	0
Cheese—Amer. Process 60% T.S.	30.2	22.3	1.7	.87	.61	368	2000	.045	22	.60
Cheese—Amer. Cheddar	32.3	23.9	1.7	.87	.61	393	2000	.045	22	.60
Cream Cheese 45% T.S.	34.0	7.3	1.8	.046	.264	342	2194	.018	60	.11
Cream Cheese Spreads	18.0	8.5	2.0	.053	.306	180	1160	.021	32	.11
	24.0	8.0	1.9	.051	.290	233	1548	.021	42	.11
	30.0	7.6	1.8	.048	.275	286	1936	.018	53	.11
Sweetened Condensed Milk—T.M.S. 29.1%	8	7.3	51.5	.3	.23	307	375	.069	nil	.35
Evaporated Milk T.S. 25.9%	7.9	6.9	9.4	.25	.20	138	325	.057	nil[2]	.31
Dried Whole Milk	26.4	25.7	38.1	.92	.71	494	1778	.339	63	1.62
Dried Skimmed Milk	1.1	35.6	51.9	1.22	.96	360	21	.375	0	1.80
Malted Milk	7.4	12.7	73.4	.36	.34	409	384	.306	7	.49
Dried Whey	4.3	12.9	67.7	.92	.71	360	nil	†	0	†

Vitamin values in the above table were obtained by averaging figures representing the range of values reported in the literature.

Niacin is not included in the above chart as there is not sufficient data available.

Vitamin C is also omitted as this nutrient is not found in any important quantity in dairy products.

Vitamins A and D values of dairy products will tend to be higher in summer produced products and tend to be lower in winter produced products.

Vitamins A and D are expressed in International (U.S.P.) Units, thiamin and riboflavin in milligrams. One milligram thiamin equals 333 International Units, 1 milligram riboflavin equals approximately 333 Bourquin-Sherman Units.

The values for products of varying fat content were calculated from known values for definite fat contents as given in the literature cited.

† Indicates that the vitamin is present but that exact figures are not available.

[1] Specific data concerning the vitamin content of homogenized milk have not been found, but it is usually assumed that homogenization causes no destruction of these vitamins.

[2] In irradiated or fortified evaporated milk the vitamin D content is stated on the can.

—From Sealtest, Inc., New York.

INDEX